Australian Landscapes

Geological Society books refereeing procedures

The Society makes every effort to ensure that the scientific and production quality of its books matches that of its journals. Since 1997, all book proposals have been refereed by specialist reviewers as well as by the Society's Books Editorial Committee. If the referees identify weaknesses in the proposal, these must be addressed before the proposal is accepted.

Once the book is accepted, the Society Book Editors ensure that the volume editors follow strict guidelines on refereeing and quality control. We insist that individual papers can only be accepted after satisfactory review by two independent referees. The questions on the review forms are similar to those for *Journal of the Geological Society*. The referees' forms and comments must be available to the Society's Book Editors on request.

Although many of the books result from meetings, the editors are expected to commission papers that were not presented at the meeting to ensure that the book provides a balanced coverage of the subject. Being accepted for presentation at the meeting does not guarantee inclusion in the book.

More information about submitting a proposal and producing a book for the Society can be found on its web site: www.geolsoc.org.uk.

It is recommended that reference to all or part of this book should be made in one of the following ways:

BISHOP, P. & PILLANS, B. (eds) 2010. *Australian Landscapes.* Geological Society, London, Special Publications, **346**.

QUIGLEY, M. C., CLARK, D. & SANDIFORD, M. 2010. Tectonic geomorphology of Australia. *In*: BISHOP, P. & PILLANS, B. (eds) *Australian Landscapes.* Geological Society, London, Special Publications, **346**, 243–265.

GEOLOGICAL SOCIETY SPECIAL PUBLICATION NO. 346

Australian Landscapes

EDITED BY

P. BISHOP
University of Glasgow, UK

and

B. PILLANS
Australian National University, Australia

2010
Published by
The Geological Society
London

THE GEOLOGICAL SOCIETY

The Geological Society of London (GSL) was founded in 1807. It is the oldest national geological society in the world and the largest in Europe. It was incorporated under Royal Charter in 1825 and is Registered Charity 210161.

The Society is the UK national learned and professional society for geology with a worldwide Fellowship (FGS) of over 10 000. The Society has the power to confer Chartered status on suitably qualified Fellows, and about 2000 of the Fellowship carry the title (CGeol). Chartered Geologists may also obtain the equivalent European title, European Geologist (EurGeol). One fifth of the Society's fellowship resides outside the UK. To find out more about the Society, log on to www.geolsoc.org.uk.

The Geological Society Publishing House (Bath, UK) produces the Society's international journals and books, and acts as European distributor for selected publications of the American Association of Petroleum Geologists (AAPG), the Indonesian Petroleum Association (IPA), the Geological Society of America (GSA), the Society for Sedimentary Geology (SEPM) and the Geologists' Association (GA). Joint marketing agreements ensure that GSL Fellows may purchase these societies' publications at a discount. The Society's online bookshop (accessible from www.geolsoc.org.uk) offers secure book purchasing with your credit or debit card.

To find out about joining the Society and benefiting from substantial discounts on publications of GSL and other societies worldwide, consult www.geolsoc.org.uk, or contact the Fellowship Department at: The Geological Society, Burlington House, Piccadilly, London W1J 0BG: Tel. +44 (0)20 7434 9944; Fax +44 (0)20 7439 8975; E-mail: enquiries@geolsoc.org.uk.

For information about the Society's meetings, consult *Events* on www.geolsoc.org.uk. To find out more about the Society's Corporate Affiliates Scheme, write to enquiries@geolsoc.org.uk.

Published by The Geological Society from:
The Geological Society Publishing House, Unit 7, Brassmill Enterprise Centre, Brassmill Lane, Bath BA1 3JN, UK

(*Orders*:   Tel. +44 (0)1225 445046, Fax +44 (0)1225 442836)
Online bookshop: www.geolsoc.org.uk/bookshop

British Library Cataloguing in Publication Data

A catalogue record for this book is available from the British Library.
ISBN 978-1-86239-314-1

Typeset by Techset Composition Ltd, Salisbury, UK
Printed by CPI Antony Rowe, Chippenham, UK

Distributors

North America
For trade and institutional orders:
The Geological Society, c/o AIDC, 82 Winter Sport Lane, Williston, VT 05495, USA
Orders: Tel. +1 800-972-9892
             Fax +1 802-864-7626
             E-mail: gsl.orders@aidcvt.com

For individual and corporate orders:
AAPG Bookstore, PO Box 979, Tulsa, OK 74101-0979, USA
Orders: Tel. +1 918-584-2555
            Fax +1 918-560-2652
            E-mail: bookstore@aapg.org
            Website: http://bookstore.aapg.org

India
Affiliated East-West Press Private Ltd, Marketing Division, G-1/16 Ansari Road, Darya Ganj, New Delhi 110 002, India
Orders: Tel. +91 11 2327-9113/2326-4180
            Fax +91 11 2326-0538
            E-mail: affiliat@vsnl.com

Contents

Dedication

This volume is dedicated to Martin Williams and John Chappell to mark their enormous contributions to geomorphology in Australia and worldwide over many decades. Jim Bowler, another 'great' of Australian geomorphology, noted the following in a personal thank-you to John and Martin, which he has given permission to be reproduced here.

Twin towers in Australian geomorphology

John Chappell and Martin Williams, I link you jointly in memory of years shared in more youthful contemplations, and am now witnessing, from a slightly distant hill, your joint departures from the scene. You two stand in joint eminence, your departures defining a closing moment in the history of Australian geomorphology. This personal note resonates with memories of those early days at the ANU where the spirit of Joe Jennings cemented a bond of common interest and long-term commitments.

There is a special sense in which your joint retirements define a closure after almost simultaneous arrival on my radar screen some 43 years ago. Your close personal friendship almost seems to parallel the patterns of your separate academic achievements. You constitute a special pair, a special partnership in defining Australia's amazing environmental legacy. Your joint departure invites this joint response.

This note is written with a sense of disappointment. Disappointment not so much with your inevitable winding down as with my own disability to match your regular and exemplary publication records. Although I regret my absence from the volume in your honour, an absence forced on me by various circumstances, I take this opportunity to convey at this personal level a debt I owe you both. It is not as though our lives have not encountered and ridden over the odd bumpy patch. The joy of that journey, however, has been shared both in the travelling and now, in your arrival at a worthy closure, one to be celebrated and honoured in your jointly spectacular achievements. Your legacies are writ large, not only in your numerous and insightful publications but even more importantly in the hearts and minds of the numerous people who have come under your influence.

As one of those, I can but say farewell (valete!), long life, sound health and many thanks for your friendship, inspiration and personal support. I trust you both will continue to ruminate over the wondrous planet whose secrets we have all been so privileged to explore.

May the lights of the Chappell–Williams towers long continue to shine and inspire.

Best wishes and long life to you both.

JIM BOWLER

Introduction: Australian geomorphology into the 21st century

PAUL BISHOP[1]* & BRAD PILLANS[2]

[1]*School of Geographical and Earth Sciences, East Quadrangle, University of Glasgow, Glasgow G12 8QQ, UK*

[2]*Research School of Earth Sciences, Australian National University, Canberra, ACT 0200, Australia*

Corresponding author (e-mail: paul.bishop@ges.gla.ac.uk)

Australian Landscapes is at least the third edited volume to be devoted to Australian geomorphology in the last half-century, two others being *Landform Studies from Australia and New Guinea* (Jennings & Mabbutt 1967) and *Landform Evolution in Australasia* (Davies & Williams 1978). There is a strong thread running between those two volumes and this one: the second volume marked the retirement of Joe Jennings, who was one of the editors of the first volume, and this volume honours one of the editors of the second, Martin Williams, along with John Chappell; both John and Martin retired recently. The legacy of Joe Jennings, who supervised the PhD research of both Martin and John, is summarized briefly in an Appendix to this Introduction. The fact that we, the editors of this current volume, were supervised in our PhD research by John (B.P.) and Martin (P.B.) continues that intertwining of threads.

The apparent directness of the links between these three volumes is matched in some ways by the enduring nature of some themes in the three volumes, themes that have been particularly Australian (or at least Gondwanan) for more than half a century. Thus, at least seven papers in the 1967 volume emphasize the antiquity of the Australian landscape, with statements such as 'The pattern of relief imposed by structure has persisted at least since the Mesozoic' (Twidale 1967, p. 95). That clear grasp of landscape antiquity, along with parallel landmark Gondwanan contributions by researchers such as King (1962), contrasts clearly with the viewpoint of (at least some of) contemporary work in the Northern Hemisphere, such as that of Thornbury (1969), who argued that landscapes are probably no older than the Pleistocene.

The emphasis on landscape antiquity continued in the Davies & Williams (1978) volume, in particular with the paper by Ollier (1978) on the evolution of the eastern Australian highlands, a theme that emerged in the late 1970s and early 1980s, particularly with the work of Bob Young (e.g. Young 1983). A key difference between the Jennings & Mabbutt (1967) volume and that of Davies & Williams (1978) is the increasingly routine application of geochronology in the latter. The Jennings & Mabbutt (1967) volume included three radiocarbon ages in the paper by Davies (one of them apparently reported for the first time in that paper), one previously published radiometric age on granite in Galloway's paper, one previously published radiocarbon age in the paper by Ollier, and a few tens of radiocarbon ages in the papers by Fairbridge and Gill (not unexpectedly in the case of Gill, given his central role in establishing the radiocarbon technique in Australia). By the time of the Davies & Williams (1978) volume, the application of quantitative dating techniques was much more routine and so geochronology is much more a feature of that volume.

Emphases on age(s) of landscapes and landforms and on rates of erosion or denudation are, of course, enduring themes in geomorphology. A key development between the Davies & Williams (1978) volume and the present one has been the emergence of new and exciting geochronological tools. With the advent of these new tools, including cosmogenic nuclide analysis, luminescence dating techniques, accelerator mass spectrometry conventional radiocarbon determinations, and stable isotope-based approaches to landscape dating that exploit Australia's northward drift throughout the Late Mesozoic and Cenozoic, the discipline is much better placed to set landscapes and their evolution in geochemically or geophysically determined quantitative age frameworks. Thus, this volume reports on a range of studies that continue the long-standing theme of the antiquity and stability of the Australia continent. Kaolinite from weathering profiles from the Yilgarn craton have yielded Cenozoic and older ages (**Chivas & Atlhopheng**), pointing to considerable stability of the craton and the slow rates of erosion that characterize much of inland Australia. The extensive dataset of **Heimsath** *et al*. teases out that theme using cosmogenic nuclide analysis. Their data point to a strong dependence between erosion rate and climate, a dependence that is identified in many (but not all) studies of the relationship

From: BISHOP, P. & PILLANS, B. (eds) *Australian Landscapes*. Geological Society, London, Special Publications,
346, 1–6.   DOI: 10.1144/SP346.1   0305-8719/10/$15.00 © The Geological Society of London 2010.

between erosion rate and climate. They also point to interesting (but more enigmatic) relationships between erosion rate and elevation.

Landscape antiquity as a theme in Australian geomorphological research has also been developed in relation to the evolution of cave systems and bedrock rivers. **Osborne** explores the first of these areas in his major review of Australian karst landscapes. His synthesis identifies a series of characteristics that are particular to Australian caves, one of which is a preponderance of complex, multiphase caves in eastern Australia that intersect two or more generations of palaeokarst and display morphological evidence for more than two distinct phases of speleogenesis. Many Australian caves also have great antiquity, with Jenolan Caves, the most complex of these cave systems, containing the oldest dated open caves, formed over 340 Ma ago. Elsewhere, dated Late Mesozoic lava has flowed into caves, likewise confirming their considerable antiquity.

The antiquity of the major river systems of eastern Australia has also been a matter of attention from the late nineteenth century. The late twentieth century 'conventional wisdom' was that these rivers have a long history of stability, but there are dissenters from that view. That debate aside, valley-filling basalts demonstrate that many of the eastern Australian rivers date from the early Neogene and earlier, and so these rivers provide a good test-bed for understanding the evolution of post-orogenic rivers. **Bishop & Goldrick** use bedrock rivers and their associated Early Miocene valley-filling basalts to argue that lithology plays a major role in the development of bedrock river long profiles in post-orogenic settings, principally by slowing the upstream propagation of knickpoints in a way that is consistent with forgotten (or ignored) hypotheses of unequal activity and increasing relief in long-term landscape evolution.

The use of the word 'post-orogenic' to describe the Australian setting may be somewhat contentious given the recent research identifying areas of not-inconsiderable tectonic activity. In parallel with discussions about the antiquity and stability of river systems, long-standing debates about the age of uplift of the SE Australian highlands are being rejoined (e.g. Holdgate *et al.* 2008; Vandenberg 2010). Even the long-abandoned 'Kosciusko Uplift' is being revisited (see Holdgate *et al.* 2008) and is explored further here by **Quigley** *et al*. Thus, the issue of the amount of tectonic activity being experienced by the Australian continent is still ripe for debate. The two editors have been involved in this debate in various ways over several decades and we note that polarized positions may not be especially conducive to good science. It is almost certain that tectonic activity on the Australian continent will be found to be spatially

and temporally complex and discontinuous, and an exciting challenge remains the synthesis of that complexity into a coherent model.

Like the discussion about stability of river systems, the issue of Quaternary climate change in Australia is also long-standing. The lack of extensively glaciated areas on the mainland has historically given a particular flavour to this work, focusing it on the responses of fluvial, aeolian and lacustrine settings to global cooling during Quaternary glacial episodes. A particular emphasis has been the impacts and timing of the glacial periods in terms of aridity. Exploring that vein, which has also been so fruitfully worked by Jim Bowler, **Haberlah** *et al*. present over 120 quantitative age determinations, numbers of determinations that our predecessors could hardly have dreamt of, to assess the responses of the subaerial landscapes of central South Australia to the climate fluctuations about the Last Glacial Maximum (LGM). They find that the area was dry and windy at the LGM, as elsewhere in Australia at that time, and they also identify subtleties in the record, including a short-lived interval of climatic stability about 5 ka before the LGM. Such subtleties are also evident in the isotopic records summarized by **Colhoun** *et al*. for caves in Tasmania, a part of Australia that has not always received the attention for Quaternary studies that its latitudinal position and topography warrant. The record of glaciations now emerging from Tasmania based on cosmogenic nuclide analysis, another technique that our predecessors could scarcely have imagined, is generating challenging new questions.

The long record of glaciation summarized by **Colhoun** *et al*. and the strong association between glacial episodes and aridity in Australia prompt the question as to the antiquity of aridity in Australia. **Fujioka & Chappell** address this issue by reviewing a wide range of literature on the age of aridity in Australia based on multiple dating approaches. Again, cosmogenic nuclide analysis has provided a key technique in dating the onset of Australian aridity, pointing to Late Tertiary and Early Quaternary ages for the stony deserts in central Australia and dunefields in the western Simpson Desert, respectively. As **Fujioka & Chappell** point out, however, these breakthroughs, coupled with those using palaeomagnetism and optically stimulated luminescence, are just a start in unravelling the ages and evolution of large parts of the arid interior. Much remains to be done in that regard.

Our understanding of contemporary arid zone processes has also grown over the last few decades, using advances in remote sensing techniques and new types of digital data on landscape morphology. Using satellite imagery, **Hesse** has compiled a new map of Australian dunefields and concludes that although they cover a large proportion of the continent, most of the dunes receive little new sediment from modern fluvial systems. Rather, they are nourished by reworking of relict sediments. Indeed, large low-relief areas of the arid zone are dune-free because of sediment starvation; the Nullarbor Plain, which is underlain by limestone, is a case in point. On the other hand, luminescence dating confirms that increased aridity during glacial periods allowed the extension of dunefields into currently semi-arid and sub-humid areas where dunes are no longer active. Major dust storms in southeastern Australia (also well documented by satellite imagery), such as the storm that brought enormous amounts of red dust from the arid interior to humid coastal areas during 2009, are a pointed reminder of such past episodes of aridification.

Remotely sensed data (including sub-seabed data) also figure prominently in subsurface and submarine research. **Schmidt *et al*.** synthesize new data from offshore the mouth of the Murray–Darling, the largest drainage system in Australia, which drains one-seventh of the continent (and is the seventeenth largest river basin in the world, by area; Potter 1978). Despite its size, the Murray–Darling often fails to reach the sea and this is not necessarily only due to the 'modern' impact of reservoirs and water abstraction. **Schmidt *et al*.** argue that the Murray's failure to discharge to the sea reflects the low topography and low levels of tectonic activity in Australia (plus presumably the dry interior climates). Indeed, the peculiarities of the Australian continent probably mean that Australian offshore canyons are a special case, which is reflected in the general model that **Schmidt *et al*.** propose for Australian situation.

Thom *et al*. also use remotely sensed submarine data to interpret the evolution of offshore New South Wales. They argue that a planated bedrock surface, the major morphological feature extending out to more than 20 km offshore from the base of the NSW coastal sea cliffs, began forming in the late Palaeogene. It reflects repeated Late Cenozoic transgressions and regressions sweeping the shelf region and progressively planating it. **Thom *et al*.**'s reiteration of an earlier conclusion that the shelf has subsided throughout the Cenozoic raises further interesting and challenging questions concerning the tectonics of the East Australian continental margin, drawing attention again to the issue of the supposed post-orogenic tectonics of the Australian continent.

The link between process and form is an enduring theme in geomorphology, going back to the seminal studies of G. K. Gilbert (e.g. Gilbert 1909) and championed in Australia by Joe Jennings and his students, including John Chappell and Martin Williams. So, although emphasizing the

antiquity of the Australian landscape, we do not overlook the importance of modern process studies, highlighted by two papers concerning beach processes in contrasting coastal settings, from storm-dominated beaches in north Queensland (**Nott**) to low-energy, sheltered sandy beaches in southern Western Australia (**Travers *et al*.**).

In that process context, the relationship between geomorphology and biota has emerged internationally recently as an important and revitalized research area (e.g. Dietrich & Perron 2006; Reinhardt *et al*. 2010). **Dunkerley** explores this theme in the semi-arid zone, using detailed modern process observations, and concludes that plants play a key role in mediating the effects of water scarcity and climate change on geomorphological processes, including overland and ephemeral channel flows.

The interactions between geomorphology, climate change and biota are also central to our understanding of the ways in which indigenous Australians spread across and utilized the continent. As **Holdaway & Fanning** point out, the fleeting nature of much of the human occupation of particular 'sites', plus the very openness of those sites and their susceptibility to erosion and removal of the evidence of occupation, makes it difficult to understand the nature and timing of the occupation. In other words, **Holdaway & Fanning** argue that a different way of thinking about how the preservation of the evidence of occupation in such open settings is required, with careful consideration of the precise meaning(s) of the chronologies obtained from such sites. As in so many other papers in this book, the particularities of the Australian setting add colour and complexity to geomorphological understanding. As was eloquently recognized by Young (1983) in relation to 'normal' rates of landscape evolution, it is essential that Northern Hemisphere approaches, paradigms and interpretations are not adopted uncritically to apply to the whole of the Earth's surface.

Martin Williams and John Chappell both chose to come to Australia for their PhD research, each working on particular aspects of the continent, John on the tectonic activity along the continent's northern margin and Martin on soil processes in different settings across the continent. Both of those themes are evident in this book, along with many others, all dear to the heart and intellect of Martin's and John's PhD supervisor, Joe Jennings. We take the opportunity in the Appendix to this introduction to note Joe's profound and lasting influence on Australian geomorphology.

Australian Landscapes is a statement of current issues in geomorphology in Australia. Not all environments are covered, nor are all viewpoints, but one of our clearest impressions in working with the authors of this volume is the excitement of the questions being asked of the Australian landscape at the end of the first decade of the 21st century. Some of these are long-standing questions, and in other cases researchers are asking the enduring questions in entirely different ways or are asking new questions. It is an exciting time and we have great pleasure in marking the retirement of two key figures, both of whom have played major roles in helping Australian geomorphology reach its dynamic and exciting state.

Australian Landscapes was born out of discussions at the 13th biannual conference of the Australian and New Zealand Geomorphology Group (ANZGG), held in Queenstown, Tasmania, in February 2008. We thank the ANZGG for sponsoring that conference, and both the ANZGG and the Geological Society of London for subsidising the costs of colour plates in this volume.

Appendix on Joe Jennings: Father of modern Australian geomorphology

Joseph Newell Jennings (1916–1984) was appointed to the fledgling Geography Department at the Australian National University (ANU) in 1953, and over the next three decades he had a profound and lasting effect on Australian geomorphology. Joe is best remembered as a karst geomorphologist, but in fact he had wide-ranging research interests and boundless enthusiasm for the entire discipline of geomorphology (Spate & Spate 1985). Indeed, his first published paper on Australian geomorphology was not on karst, but on Lake George (Jennings 1954), the enigmatic lake near Canberra that later was to be the research topic of one of his PhD students, Ross Coventry.

The career of Joe Jennings is detailed in obituaries by Spate & Gillieson (1984), Bowler (1985) and Spate & Spate (1985), as well as the Australian Dictionary of Biography Online (Spate 2006). I wish to demonstrate here Joe's remarkable legacy through construction of his academic family tree, modelled on the family trees used to depict family lineages (http://www.anzgg.org/). In the case of the academic family tree, graduate students are represented as 'children' of their supervisor.

Twelve disciples and a remarkable legacy: Joe Jennings's academic family tree (by B. Pillans)

Joe supervised 12 graduate students to completion at ANU (with year of completion in parentheses): Eric Bird (1959), Nel Caine (1966), Ian Douglas (1966), Martin Williams (1969), Jim Bowler (1970), Bud Frank (1972), John Chappell (1973), Colin Pain (1973), Ross Coventry (1973), Chris Whitaker (1976), Joyce Lundberg (1976) and David Gillieson (1982) (Table A1). In constructing Joe's academic family tree, my focus was on graduate students,

Table A1. *Graduate students supervised by Joe Jennings, their thesis titles and year of completion. All theses completed at ANU, except David Gillieson (University of Queensland)*

Student	Year	Thesis title
BIRD, Eric Charles Frederick	1959	The Gippsland Lakes, Victoria: a geomorphological study
CAINE, Nelson	1966	The blockfields and associated features of northeastern Tasmania
WILLIAMS, Martin Anthony Joseph	1969	Rates of slopewash and soil creep in parts of northern and southeastern Australia: a comparative study
BOWLER, James Maurice	1970	Late Quaternary environments: a study of lakes and associated sediments in south-eastern Australia
FRANK, Ruben Milton (Bud)	1972	Sedimentological and morphological study of selected cave systems in Eastern New South Wales, Australia
CHAPPELL, John Michael Arthur	1973	Geology of coral terraces on Huon Peninsula, New Guinea
COVENTRY, Ross James	1973	Abandoned shorelines and the late Quaternary history of Lake George, New South Wales
PAIN, Colin Frederick	1973	The late Quaternary geomorphic history of the Kaugel Valley, Papua New Guinea
LUNDBERG, Joyce	1976	The geomorphology of Chillagoe Limestones: variations with lithology (MSc)
WHITAKER, Christopher Robert	1976	Pediment form and evolution in the East Kimberleys: granite, basalt and sandstone case-studies
GILLIESON, David	1982	Geomorphology of limestone caves in the Highlands of Papua New Guinea

most of whom I contacted, as well as many of their students. Their inputs are gratefully acknowledged. I have not attempted to extend the tree any further back than Joe himself, because my prime focus here is Joe's legacy, through his students, to Australian geomorphology and related disciplines.

Joe's remarkable legacy lives on through his many academic descendants, who hold (or have held) academic positions at more than half of Australia's 39 universities (Table A2). Many more members of the family tree are employed by government agencies such as CSIRO and Geoscience Australia or work as consultant geomorphologists. Strong international branches of the tree also grew from two of Joe's students who took up university positions overseas, Ian Douglas (University of Manchester) and Nel Caine (University of Colorado), both of whom supervised more than 30 postgraduate students.

This volume is, implicitly, a tribute to Joe Jennings, who spent three decades immersed in Australian geomorphology (Jennings 1983). Not only does it mark the recent retirements of two of Joe's most illustrious students, but also it represents

Table A2. *Academic descendants of Joe Jennings have held teaching and research positions at 24 out of 39 Australian universities*

Australian Catholic University	
Australian National University	John Field, John Chappell, Brad Pillans, John Magee, Pauline Treble, Toshi Fujioka, David Gillieson
Bond University	
Central Queensland University	Robert Miles
Charles Darwin University	Dimity Boggs
Charles Sturt University	Tony Dare-Edwards, Penny Davidson
Curtin University of Technology	Zhongrong Zhu
Deakin University	
Edith Cowan University	
Flinders University	
Griffith University	Ron Neller, Grant McTainsh, Errol Stock, Craig Strong, Andrew Chan
James Cook University	David Gillieson, Ross Coventry
La Trobe University	
Macquarie University	Martin Williams, Paul Hesse, Jim Kohen, Kerrie Tomkins, Tim Ralph, Trish Fanning, Sandy Harrison
Monash University	Martin Williams, David Dunkerley, Nigel Tapper, Paul Bishop, Meredith Orr, Kate Brown, Geoff Goldrick, Dan Penny
Murdoch University	
Queensland University of Technology	
RMIT University	
Southern Cross University	
Swinburne University of Technology	
University of Adelaide	Martin Williams
University of Ballarat	
University of Canberra	Xiang Yang Chen
University of Melbourne	Eric Bird, Jim Bowler, Ian Rutherfurd, Matt Cupper, John Hellstrom
University of New England	Ian Douglas
University of New South Wales	Colin Pain
University of Newcastle	Bob Loughran, Diana Day, Russell Drysdale
University of Notre Dame Australia	Peta Sanderson
University of Queensland	Annie Ross, David Neil
University of South Australia	Chris Whitaker
University of Southern Queensland	Harry Butler
University of Sydney	Diana Day, Paul Bishop, Eleanor Bruce
University of Tasmania	Richard Doyle, Jamie Kirkpatrick, Kate Brown
University of Technology Sydney	
University of the Sunshine Coast	Ron Neller
University of Western Australia	Ian Eliot
University of Western Sydney	
University of Wollongong	Ian Eliot, Marji Puotinen
Victoria University	

the third of a series of edited books on Australian
geomorphology, the first of which, as we have
noted, was co-edited by Joe and the second of
which was a tribute to Joe on his retirement. Furthermore, apart from J. Chappell and M. Williams,
several contributors to this book are also members
of Joe's academic family tree: P. Bishop, J. Bowler,
D. Dunkerley, I. Eliot, P. Fanning, T. Fujioka,
G. Goldrick, D. Haberlah, A. Heimsath, P. Hesse
and B. Pillans.

Joe was a memorable after-dinner speaker at the
first Australian & New Zealand Geomorphology
Group (ANZGG) meeting in 1982 and the text of
that talk was published in the following year
(Jennings 1983). Since 1982, there have been a
further twelve ANZGG conferences, held approximately biennially, with the most recent (13th)
meeting held in Queenstown, Tasmania, in February
2008; the next will be held in Oamaru, New
Zealand, in February 2011. Joe's academic descendants have a strong presence in ANZGG, and will
continue to do so: ANZGG meetings are designed
to encourage the participation of graduate students,
meaning that the tree will continue to grow. The
general ethos of ANZGG conferences is one of
informality, with active mentoring of students by
more senior attendees, most prominent of whom
have been M. Williams and J. Chappell.

References

BOWLER, J. 1985. Joseph Newell Jennings. *Zeitschrift für Geomorphologie, Supplementband*, **55**, v–ix.
DAVIES, J. L. & WILLIAMS, M. A. J. (eds) 1978. *Landform Evolution in Australasia*. Australian National University Press, Canberra.
DIETRICH, W. E. & PERRON, J. T. 2006. The search for a topographic signature of life. *Nature*, **439**, 411–418.
GILBERT, G. K. 1909. The convexity of hilltops. *Journal of Geology*, **17**, 344–350.
HOLDGATE, G. R., WALLACE, M. W., GALLAGHER, S. J., WAGSTAFF, B. E. & MOORE, D. 2008. No mountains to snow on: major post-Eocene uplift of the East

Victoria Highlands; evidence from Cenozoic deposits. *Australian Journal of Earth Sciences*, **55**, 211–234.
JENNINGS, J. N. 1954. Lake George and Lake Bathurst. *Geography*, **39**, 143–144.
JENNINGS, J. N. 1983. Thirty years of geomorphology downunder. *In*: YOUNG, R. W. & NANSON, G. C. (eds) *Aspects of Australian Sandstone Landscapes*. Australian and New Zealand Geomorphology Group Special Publication, **1**, 1–3.
JENNINGS, J. N. & MABBUTT, J. A. (eds) 1967. *Landform Studies from Australia and New Guinea*. Australian National University Press, Canberra.
KING, L. C. 1962. *The Morphology of the Earth*. Oliver & Boyd, Edinburgh.
OLLIER, C. D. 1978. Tectonics and geomorphology of the Eastern Highlands. *In*: DAVIES, J. L. & WILLIAMS, M. A. J. (eds) *Landform Evolution in Australasia*. Australian National University Press, Canberra, 5–47.
POTTER, P. E. 1978. Significance and origin of big rivers. *Journal of Geology*, **86**, 13–33.
REINHARDT, L., JEROLMACK, D., CARDINALE, B. J., VANACKER, V. & WRIGHT, J. 2010. Dynamic inter-actions of life and its landscape: feedbacks at the inter-face of geomorphology and ecology. *Earth Surface Processes and Landforms*, **35**, 78–101.
SPATE, A. 2006. Jennings, Joseph Newell (1916–1984). *Australian Dictionary of Biography Online*. Available online at: http://www.adb.online.anu.edu.au/biogs/A170596b.htm.
SPATE, A. & GILLIESON, D. 1984. Joe Jennings. *Helictite*, **22**, 35–42.
SPATE, O. H. K. & SPATE, A. P. 1985. Joseph Newell Jennings 1916–1984. *Australian Geographical Studies*, **23**, 325–337.
THORNBURY, W. D. 1969. *Principles of Geomorphology*, 2nd edn. Wiley, New York.
TWIDALE, C. R. 1967. Hillslopes and pediments in the Flinders Ranges, South Australia. *In*: JENNINGS, J. N. & MABBUTT, J. A. (eds) *Landform Studies from Australia and New Guinea*. Australian National University Press, Canberra, 95–117.
VANDENBERG, A. H. M. 2010. Paleogene basalts prove early uplift of Victoria's Eastern Uplands. *Australian Journal of Earth Sciences*, **57**, 291–315.
YOUNG, R. W. 1983. The tempo of geomorphological change: evidence from southeastern Australia. *Journal of Geology*, **91**, 221–230.

A theory (involving tropical cyclones) on the formation of coarse-grained sand beach ridges in NE Australia

School of Earth and Environmental Sciences, James Cook University, Cairns, Australia
(e-mail: Jonathan.Nott@jcu.edu.au)

Abstract: A theory is presented that coarse-grained sand beach ridge plains in northeastern Australia have developed their final form (i.e. the height of the ridges above sea level) as a result of marine inundations generated by intense tropical cyclones. Although winds generated during such tempests are of more than sufficient velocity to transport coarse-grained sands the beach is typically inundated by the storm surge and waves during these events and hence there is no viable source for aeolian sand transport. Numerical storm surge and shallow water wave models are run for two sites (Cairns and Cowley Beach) and the results indicate that only wave run-ups generated by category 3 or more intense tropical cyclones can deposit the final units of sediment onto the sand beach ridges at these locations. It is suggested that as a sand beach ridge increases in height, via successive deposition of units of sand, progressively higher marine inundations are required to reach the ridge crest. The final height of the sand beach ridge is dependent upon the interplay of sediment supply rates to the coastal system and the intensity and frequency of tropical cyclones. If tropical cyclones are responsible for the final form of the sand beach ridges then these sequences can be used to assess the long-term climatology of intense tropical cyclones over this broad geographical region.

There are a variety of processes that have been attributed to the formation of sand beach ridges throughout Australia and the world. Aeolian deposition, for example, has been suggested for the ridges in southwestern Australia (Shepherd 1987); Guichen Bay, South Australia (Murray-Wallace *et al.* 2002; Bristow & Pucillo 2006); LeFlevre Peninsula, South Australia (Bowman & Harvey 1986); SE New South Wales (Hesp 1984; Thom & Roy 1985); and Cattle Point, central eastern Queensland (Brooke *et al.* 2008); to name just a few locations within Australia. Globally, ridges at various locations have been attributed to both aeolian and wave processes, and detailed reviews of these have been given by Tanner (1995), Taylor & Stone (1996), Otvos (2000) and Hesp (2006). Essentially both sets of processes (wind and wave) appear to be able to produce 'sand' beach ridges, defined here as convex sand ridges, separated by sand swales paralleling the coast and often forming a ridge plain. These processes can act either in unison or independently depending upon the prevailing environmental conditions. Irrespective of whether aeolian or wave processes dominate at any one site it appears that for ridges to form along a coast there needs to be an abundant supply of sand and a shallow offshore gradient (Woodroffe 2003).

Numerous sand beach ridge plains occur along the coast of NE Queensland. The majority of these between Cape Tribulation and Townsville are composed of poorly to moderately sorted, medium- to coarse-grained sand. Single ridges within these plains typically stand higher than 4 m AHD (Australian Height Datum) or mean spring high tide, with many ridges higher than 5 m AHD. In some locations there are a few ridges that have a capping (0.5–1 m thick) of well-sorted fine-grained sand, presumably of aeolian origin. Other ridge plains throughout this region have single ridges or a group of a few ridges that display these same aeolian sedimentary characteristics, but these are limited in extent and number. Generally, however, the overwhelming characteristic of the sand beach ridges in this region is their height above sea level, their composition of medium- to coarse-grained dominantly quartz sand and their proximity to major streams delivering sediment to the Great Barrier Reef lagoon.

Tropical cyclones are an important meteorological feature of this region and they can produce high-velocity winds capable of transporting coarse-grained sands. However, observation of tropical cyclone impacts along the coast in this region reveals that very little sand is blown from the beach onto an incipient ridge or mature ridges further inland. Sand blown during tropical cyclones travels hundreds of metres inland and is very widely dispersed rather than accumulating at the back of a beach. Also, when the winds are at their greatest velocity (>150 km h^{-1}), and are blowing onshore, the source of that sand, being the beach, is invariably under water because of the associated storm surge

From: Bishop, P. & Pillans, B. (eds) *Australian Landscapes*. Geological Society, London, Special Publications, **346**, 7–22. DOI: 10.1144/SP346.2 0305-8719/10/$15.00 © The Geological Society of London 2010.

and wave action. The vast majority of sand that is transported landward during tropical cyclones in this region (and elsewhere throughout northern tropical Australia) accumulates as a result of wave action. Hence, aeolian processes appear to be an unlikely mechanism for the deposition of the majority of ridges throughout this region of NE Queensland. An alternative hypothesis is presented that tropical cyclone-induced inundations involving storm surge, tide, wave set-up and run-up are the most likely mechanism responsible for depositing these ridges. Here this theory is tested through the use of numerical tropical cyclone wind and storm surge and shallow water wave models to determine which meteorological conditions are capable of generating marine inundations of sufficient magnitude to reach the crests of these sand beach ridges. The critical question of whether a ridge is most probably deposited during a single event or progressively over time is also addressed.

NE Queensland environment

Climate and landscape

The region between Cardwell and Cape Tribulation (Fig. 1), otherwise known as the wet tropics bioregion, experiences a tropical wet climate. Average annual rainfall varies substantially across the region. Cairns receives on average 2200 mm of rain per annum, with rainfall increasing along the eastern edge of the adjacent highlands. Mt Bartle Frere and Mt Bellenden Ker (altitude 1592 m and 1622 m, respectively) to the south of Cairns receive on average over 4000 mm of rain per annum and in some years over 11 000 mm. Tropical rainforest and wet sclerophyll forest dominate the slopes of the region, whereas most of the lowlands have been cleared of forest for agriculture and urban development. The landscape is composed of a plateau c. 400–500 m above sea level, which terminates at a steep east-facing escarpment. A coastal plain extends from the base of the escarpment east to the sea and is contiguous with the continental shelf.

Geology

The geology of the region is dominated by Middle Palaeozoic metasediments, composed principally of interbedded phyllite, schist, quartzite and chert. These rocks strike north–south and are intruded by Carboniferous–Lower Permian granites that crop out irregularly throughout the region and form most of the higher summits (ranging from 1337 m at Black Mountain in the north to 1622 m at Bartle Frere in the south). Devonian–Carboniferous tectonism was responsible for the compression of sediments leading to the present fold belt strata. Early

Triassic upheaval formed the Russell–Mulgrave Shear Zone (248–235 Ma), resulting in deformation and shattering of the metasediments (Willmott et al. 1988; Willmott & Stephenson 1989). North of Cairns, frequent outcrops of foliated and sheared granite and coarse metasediment maintain the steep escarpment of the Macalister Range, which rises abruptly from the coastline. South of Cairns the shear zone is marked by the Mulgrave corridor, which separates the granites of the Malbon–Thompson coastal range from the Tinaroo and Bellenden Ker granites of the Great Dividing Range. Areas of Quaternary basalts occur on the Atherton plateau to the SW of Cairns and at one location (Green Hill volcano) in the Mulgrave corridor, which forms part of the coastal lowland plain. Quaternary sediments occur across the coastal plain and at the base of the ranges. These sediments consist of alluvium, colluvium, coastal sand beach ridges, cheniers, aeolian dunes and fine-grained freshwater lagoon and estuarine sediments. Alluvial fans extend from some of the larger river valleys, such as the Mulgrave River valley near Gordonvale, which is 30 km south of Cairns. Other smaller fan deposits occur in numerous places at the base of the ranges and these merge with marine Pleistocene and Holocene sand beach-ridge sequences to form the coastal plain.

This region has experienced relative tectonic stability throughout much of the Cenozoic, if not the entire Cenozoic. Eastern Australia is a passive tectonic continental margin and in NE Australia it is likely that a short episode of Jurassic rifting may have been responsible for uplift of the eastern Australian highlands (Struckmeyer & Simmonds 1997) although the possibility also remains that these highlands were in existence since before this time (Nott & Horton 2000).

Hydroisostatic flexure of the near-coastal land is likely to have occurred during the latter half of the Holocene. However, there are a number of views on the exact nature of Holocene sea-level variations in this region. Chappell et al. (1982) suggested that sea level had fallen smoothly from 1.5–1.0 m above the present position since c. 5500 years BP as a result of hydroisostasy. Larcombe et al. (1995) suggested that sea level remained c. 1.5 m above present between 6500 and 3000 years BP and then fell sharply. Lewis et al. (2008) advocated two sea-level oscillations with lowstands c. 4500 and c. 2500 years BP before rising again to c. 1 m above present at 2000 years BP and then falling to the present position.

Beach ridge plains throughout the region

There are at least 10 coarse-grained sand beach ridge plains along the nearly 300 km (straight line)

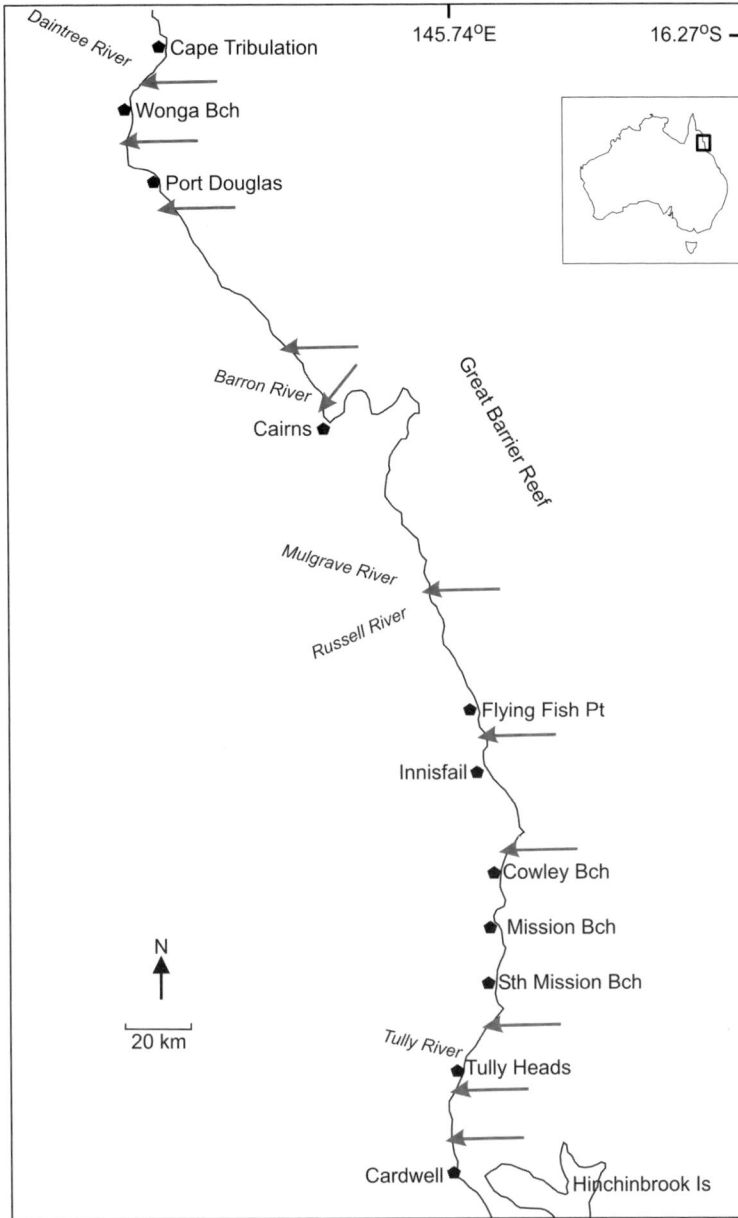

Fig. 1. Location map; arrows show location of named sand beach ridge plains.

length of the wet tropics bioregion coast in NE Queensland (Fig. 1). This area stretches from approximately Cape Tribulation to Cardwell. The majority of these ridge plains have more than 15 ridges paralleling the shore. The most northerly of these is at Wonga Beach immediately south of the mouth of the Daintree River. Here the ridge plain is composed of 12 ridges. At the northern end of the plain the five seaward ridges are composed of

fine-grained well-sorted dominantly quartz sand. The landward seven ridges are composed of moderately sorted medium- to coarse-grained sands. The southern end of the ridge plain contains eight ridges, all of which are composed of moderately sorted, medium- to coarse-grained sands. The southern end of Wonga Beach is also composed of coarser-grained sands than the northern end. It appears that there has been a change in the nature of ridge sedimentation at the northern end of the ridge plain after deposition of the seventh ridge. This does not appear to be the case at the southern end of the Wonga Beach ridge plain, as all of the ridges there are medium to coarse-grained.

The next ridge plain south extends from Newel Beach to nearly Yule Point south of Port Douglas. This plain has been extensively modified by agricultural practices (sugar cane cultivation) over the past century and is clearly visible only in select locations. It varies in the number of ridges along its length. In places it has more than 10 ridges, near Port Douglas these ridges are somewhat indistinct, and further south towards Yule Point there are at least eight ridges. All of these ridges are composed of moderately sorted, medium- to coarse-grained, dominantly quartz sands. This ridge plain is terminated at its southern end by a mangrove-dominated estuary and then a 30 km long section of rocky coast where the east Australian escarpment in places forms sea-cliffs.

A ridge plain containing up to 15 ridges occurs from the northern beaches of Cairns southward to the present-day Central Business District (CBD) of the city of Cairns. Bird (1970, 1971) described

aspects of these ridges and concluded that they were deposited during the mid- to late Holocene when Cairns Bay, the embayment adjacent to Cairns city, contained more sand than it does today. Sand delivery to this embayment diminished after the late 1930s when the mouth of the Barron River avulsed to a new location several kilometres to the north. The ridges were mapped and topographically surveyed in the late 1930s as part of a malaria reduction programme, as local swamps and swales between ridges were ideal locations for mosquito breeding. Many of the ridges in the vicinity of Cairns are now covered by urban and commercial development, and those in the northern beaches are covered by sugar cane. However, aerial photography from the late 1930s and early 1940s along with the mapping undertaken for malarial drainage study show the previous extent of the beach ridges (Figs 2 and 3). Several ridges are also preserved in cemeteries, where the coarse-grained character of the sediments forming these ridges is obvious.

A coarse-grained beach ridge plain containing between 10 and 15 beach ridges occurs at Russell Heads on the northern side of the mouth of the Mulgrave River c. 40 km south of Cairns. Here, as

Fig. 2. Aerial photograph of beach ridges in Cairns prior to major urban and commercial development. Photograph c. 1942. The location is shown as a boxed area in Figure 3.

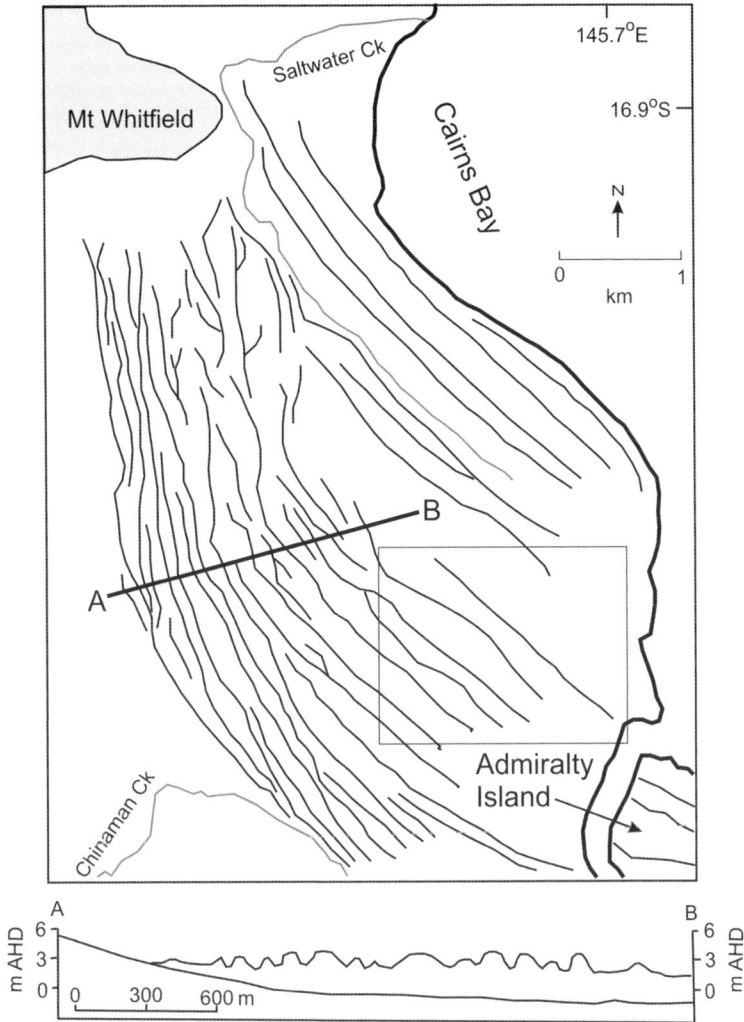

Fig. 3. Map and cross-section of beach ridges at Cairns. Original data were derived from a US military topographic survey of the beach ridge system as part of an anti-malarial drainage study in the early 1940s. Map and cross-section modified from Bird (1971). Boxed area shows approximate location of Figure 2.

at the other locations described, the ridges are composed of moderately sorted, medium- to coarse-grained predominantly quartz sands. The same is true of the other beach ridge plains further south at Flying Fish Point near Innisfail, Cowley Beach, south of Mission Beach and also Tully Heads. Again, each of these locations has multiple shore-parallel ridges ranging in height from 4 to 5 m AHD and containing medium- to coarse-grained sands.

Bathymetry and oceanographic conditions

The continental shelf in this region widens progressively from north to south. Near Cape Tribulation it

is *c.* 20–30 km wide and near Townsville is 80–90 km wide. The Great Barrier Reef (GBR) for its main part is located close to the seaward margin of the continental shelf, hence the outer reef tract increases in distance from shore with distance south. Detailed bathymetric data (<250 m resolution) is not available at present for most of the shelf inside the GBR. Near Cairns, water depths near the GBR are *c.* 50–60 m about 40–50 km offshore. This suggests a shallow gradient across the continental shelf. Three main sedimentary zones exist across the shelf: an inner shelf zone composed of terrigenous sediments found at depths between 0 and 22 m; a middle shelf zone, which experiences

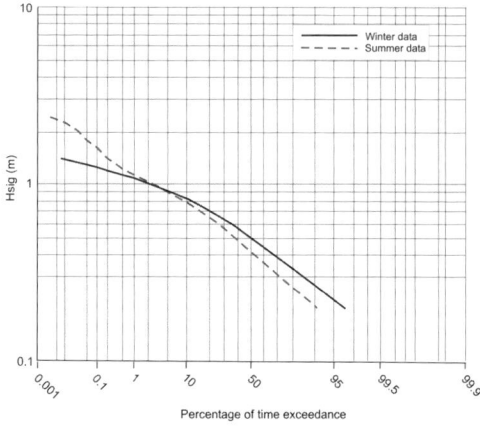

Fig. 4. Offshore wave (significant wave height) data plot for Cairns for 1975–2008. Data measured by wave rider buoy in 15–20 m depth water offshore from Cairns northern beaches. Data supplied by Qld Environmental Protection Agency.

sediment starvation, between 22 and 40 m; and an outer shelf reef tract at depths of greater than or equal to 40 m (Larcombe & Carter 2004). The

mid-shelf zone is thought to be a region of 'cyclone pumping' where predominantly northerly currents generated during these storms transports sediment from this zone into the adjacent inner shelf and outer shelf tracts (Larcombe & Carter 2004).

This region experiences a predominantly low wave energy climate. Figure 4 shows significant wave height (H_{sig}) recorded near Cairns between 1975 and 2007 by a wave rider buoy in c. 20 m depth water. Average H_{sig} varies between 0.4 and 0.6 m with a maximum monthly average of 0.79 m recorded during October 1982. Two main types of waves occur at this location, sea waves generated inside the GBR and swell waves generated some distance away outside the GBR. The majority of the fetch-limited sea waves have a peak period (T_p) of 2.5–5.0 s and swell waves 5.0–9.0 s. Significant wave heights of 2–3 m occur less than 0.1% of the time and this value is even lower for the winter months, when trade winds dominate and there is a lack of tropical cyclone activity. It is the trade winds (southeasterlies) that generate the strongest winds (c. 35 knots) (Fig. 5), except during tropical cyclones, and the longest fetch here is to the SE. At shore, waves rarely exceed 0.5 m height with wave periods of 2–4 s, even during the strongest

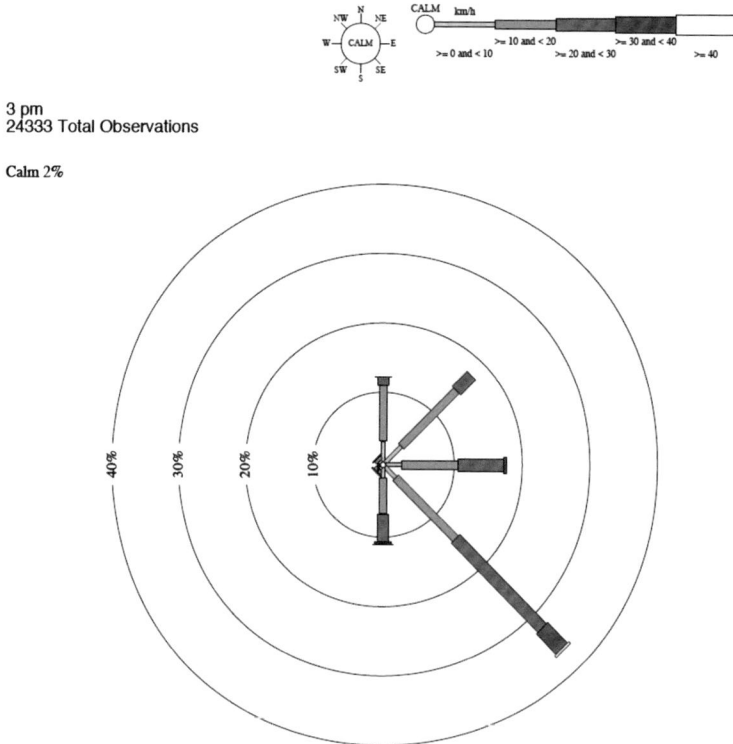

Fig. 5. Wind rose for Cairns.

trade wind conditions. The GBR being here *c.* 40 km offshore provides substantial shelter from larger swells generated by distant storms. However, tropical cyclones close to the mainland coast can generate significant wave heights up to 10 m and wave periods of 12 s between the mainland and the GBR. Such storms can also generate storm surges 4 m or higher above the event tide.

Cairns and Cowley Beach 'sand' beach ridge plain characteristics

The origin of two of the sand beach ridge plains of this region, Cairns and Cowley Beach, *c.* 120 km further south, will be focused on specifically here. These sand beach ridge plains are taken to be representative of many of the sand beach ridge plains throughout the wet tropics bioregion in terms of their sedimentological characteristics and height of the ridges above sea level.

Cairns

Cairns is located on a broad coastal plain covered with Quaternary sediments. These sediments consist of alluvium, colluvium, coastal beach ridges, cheniers, aeolian dunes, and fine-grained freshwater lagoon and estuarine sediments (Jones 1985). Alluvial fans extend from some of the larger river valleys such as the Mulgrave River valley near Gordonvale, 30 km south of Cairns. Other smaller fan deposits occur in numerous places at the base of the ranges, and these merge with marine Pleistocene and Holocene beach and chenier ridge sequences to form the coastal plain. Details of the chronology and sedimentary characteristics of the alluvial fans and fluvial sedimentary sequences throughout this region have been given by Thomas *et al.* (2007).

The city of Cairns and some of its suburbs have been built on top of the beach ridge plain and access to a modern detailed topography and sediment analysis is not possible. The ridges are exposed at a cemetery (Pioneer Cemetry McLeod Street, Cairns) where the original form of two of the ridges has remained largely undisturbed. A detailed topographic survey of the original ridge plain, undertaken as part of the anti-malarial drainage study, has been described and discussed by Bird (1970, 1971). There were 25 ridges paralleling the shore throughout the area now occupied by Cairns and its immediate northern suburbs (Fig. 3). These ridges range in height from 2 to 4 m AHD. A detailed optically stimulated luminescence (OSL) chronology has not been undertaken on the ridges but Bird (1970, 1971) obtained a radiocarbon age of 5530 years BP from shell grit at the base of

the most landward of these ridges (ridge section on Admiralty Island) (Fig. 3).

Where the ridges are exposed in their near original form (such as the Pioneer Cemetery in McLeod Street) they can be seen to be composed of poorly to moderately sorted medium- to coarse-grained sands with clasts up to 3 mm in diameter. The anti-malarial survey shows that the ridges were more closely spaced together at the southern end of the ridge plain (Fig. 3). This may have been due to the nature of sand supply to the region, which was a function of the location of the mouth of the main stream here being the Barron River. Ridge construction at the southern end of the ridge plain (being the area occupied by Cairns city) would have continued to the present day if the sand supply to Cairns Bay had not dramatically diminished following avulsion of the Barron River during the 1930s. Prior to this time Cairns Bay appeared to have an abundant supply of sand, as demonstrated in Figure 6, whereas at the present day it is largely devoid of sand. The Barron River used to enter Cairns Bay near Casuarina Point prior to the 1930s (Fig. 6). After this time it changed course to its present location several kilometres north. This starved Cairns Bay of its sand supply, which eliminated the possibility of further ridge construction in this immediate area.

Cowley Beach

Cowley Beach (Fig. 1) is a 7.5 km long, arcuate-shaped sand beach backed by a beach ridge plain forming an inner and outer barrier sequence (Fig. 7). The beach ridges of the inner barrier are probably Pleistocene in age. The ridges forming the *c.* 2.5 km wide outer barrier are, as shown below, Holocene in age. Salt flats and reworked beach ridge sands also occur landward of the inner barrier and these sediments merge further landward with alluvial deposits and freshwater peat swamps. In all, this section of the coastal sedimentary plain is *c.* 9 km wide. Landward of this, the plain and foothills are dominated by outcrops of Palaeozoic metamorphic rocks and granites, interspersed with Quaternary fluvial and alluvial fan deposits (Thomas *et al.* 2007). The granites of the area supply coarse-grained sands to the nearshore coastal system via a stream (Liverpool Creek) that enters the sea at the southern end of Cowley Beach (Graham 1993). Cowley Beach (mid- to upper tidal beach face slope) itself is composed of these coarse-grained sands below a thin (<0.2 m) layer of medium- to fine-grained sands.

Twenty-nine beach ridges form the outer barrier ridge plain sequence. They vary in height between 3.5 and 5.5 m AHD or mid-high tide level, being approximately mean sea level (Fig. 8). The ridges

Fig. 6. Oblique aerial photograph of north Cairns region showing previous course and mouth of Barron River and beach ridges. This area is now occupied by Cairns International Airport. Photograph *c.* 1938.

are composed of coarse-grained predominantly quartz sands, with minor amounts of feldspars and mica, particularly in the younger ridges closest to the sea where they have not been substantially weathered during pedogenesis. Sand clasts 1–2 mm in diameter are common below *c.* 30 cm depth, and increase in abundance with depth to dominate the deposits below *c.* 1 m. Several of the beach ridges toward the back of the Holocene ridge plain are capped by 1–2 m of fine-grained well-sorted sand that Graham (1993) interpreted to be of aeolian origin. The majority of ridges forming the Holocene outer barrier sequence, however, are largely devoid of an aeolian capping. It is uncertain why this is the case.

The older more landward ridges contain well-developed podzol soils (bleached A2 horizons). Soil development is also occurring within the younger more seaward ridges and this has obscured any sedimentary structures that may have been once present. Trenching of several of the ridges showed no obvious stratigraphic breaks with depth or cyclic fining upwards sequences except for the progressive sediment fining towards the crest of each ridge.

Ridge chronology

Twenty-four of the 29 ridges were dated using OSL analysis (Nott *et al.* 2009) (Fig. 8). Each sample was measured for luminescence using 12-grain aliquots of 180–212 μm quartz grains, adopting a single-aliquot regenerative-dose (SAR) protocol (Murray & Wintle 2000). The samples were assessed initially using a preliminary procedure to determine IR stimulated luminescence (IRSL) contamination, signal sensitivity, approximate equivalent dose (D_e), and an indication of the degree of inter-aliquot variability. Low IRSL values were observed for all samples, suggesting that good quartz separation had been achieved. Preheat conditions adopted for these samples were 220 °C for 10 s for preheat 1 and 200 °C for 10 s for preheat 2. The samples displayed high OSL signal sensitivities and also excellent recycling behaviour; the mean value for all samples was within 3% of unity. Generally low thermal transfer values were observed, although the younger samples showed some aliquots with values up to around 10% of the natural signal. Dose rates were calculated from determinations of U, Th and K concentration based on fusion–dissolution inductively

Fig. 7. Aerial photograph of Cowley Beach beach ridge plain. Outer barrier or Holocene beach ridges can be seen coastward of an estuarine stream network, which partly encompasses part of the inner barrier (Pleistocene) beach ridge system. Transect depicted in Figure 8 is shown.

coupled plasma mass spectrometry (ICPMS) and optical emission spectrometry (OES). These dose rate values were corrected for grain-size attenuation and past water content, using 5% ± 5%. The cosmic dose rate estimation and a soft component were included for shallower samples.

The ridges range in age from 200 to 5740 years BP, giving an average interval for ridge emplacement of *c*. 197 years (Table 1). The ridges increase in age progressively landward except for four age reversals (where the uncertainty margins do not overlap). The first of these is for ridge 14 and the other three form a group between ridges 23 and 26, inclusive. The four age reversals probably are due to the effects of reworking and/or bioturbation.

The three samples collected from ridge 15 returned ages of 3420 ± 120 (75 cm depth),

3480 ± 120 (170 cm depth) and 3400 ± 120 years BP (340 cm depth), suggesting that they are chronologically indistinguishable. A number of ages between ridges also overlap. These include overlaps between ridges 4 and 5, 10 and 12, 16–18, 19 and 20, 19–23, 27 and 28, and 28 and 29. These overlaps could have occurred for a number of reasons including short time periods between depositional events, the nature of the dating process whereby the substantial uncertainty margins (2σ level) are larger than the actual time interval between depositional events, and the possibility that bioturbation or other processes have caused these ages to appear closer together than is actually the case. None the less, despite these age overlaps between ridges there is a clear trend towards each ridge increasing in age with distance inland and with the oldest

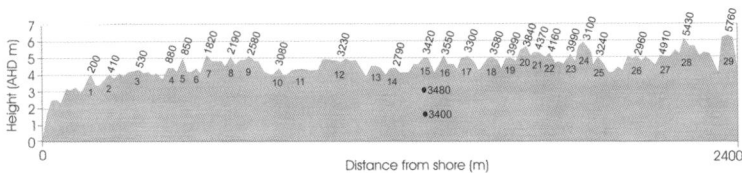

Fig. 8. Cross-section and OSL chronology from Cowley Beach.

Table 1. *OSL age determinations and associated data, Cowley Beach (Data from Nott* et al. *2009)*

Field code	Lab. code	Depth (cm)	D_e (Gy)	Dose rate (mGy a^{-1})	Age (years BP)
Ridge 1	K0451	70	0.35 ± 0.01	1.73 ± 0.11	200 ± 10
Ridge 2	K0442	60	0.65 ± 0.02	1.58 ± 0.10	410 ± 30
Ridge 3	K0443	65	0.79 ± 0.02	1.48 ± 0.09	530 ± 40
Ridge 4	K0444	60	1.37 ± 0.04	1.56 ± 0.10	880 ± 60
Ridge 5	K0390	70	1.26 ± 0.04	1.52 ± 0.12	850 ± 50
Ridge 7	K0445	70	2.57 ± 0.08	1.41 ± 0.09	1820 ± 130
Ridge 8	K0446	60	3.91 ± 0.10	1.78 ± 0.11	2190 ± 150
Ridge 9	K0447	60	4.23 ± 0.10	1.64 ± 0.10	2580 ± 170
Ridge 10	K0448	70	4.86 ± 0.13	1.58 ± 0.10	3080 ± 210
Ridge 12	K0449	40	3.88 ± 0.11	1.20 ± 0.08	3230 ± 240
Ridge 14	K0450	50	3.23 ± 0.08	1.16 ± 0.07	2790 ± 180
Ridge 15	K0391	75	3.98 ± 0.09	1.23 ± 0.06	3420 ± 120
Ridge 15	K0392	170	4.23 ± 0.08	1.21 ± 0.08	3480 ± 120
Ridge 15	K0393	340	4.43 ± 0.08	1.34 ± 0.11	3400 ± 120
Ridge 16	K0452	40	4.81 ± 0.15	1.36 ± 0.09	3550 ± 260
Ridge 17	K0453	40	4.67 ± 0.14	1.42 ± 0.10	3300 ± 240
Ridge 18	K0454	100	5.22 ± 0.14	1.46 ± 0.08	3580 ± 220
Ridge 19	K0455	40	4.59 ± 0.14	1.15 ± 0.09	3990 ± 320
Ridge 20	K0456	40	5.10 ± 0.14	1.33 ± 0.10	3840 ± 290
Ridge 21	K0457	70	7.34 ± 0.28	1.68 ± 0.11	4370 ± 340
Ridge 22	K0458	50	5.05 ± 0.12	1.22 ± 0.08	4160 ± 280
Ridge 23	K0459	70	6.26 ± 0.18	1.57 ± 0.09	3990 ± 250
Ridge 24	K0460	50	5.36 ± 0.15	1.73 ± 0.11	3100 ± 210
Ridge 25	K0461	30	5.02 ± 0.16	1.55 ± 0.12	3240 ± 280
Ridge 26	K0462	30	4.61 ± 0.15	1.56 ± 0.13	2960 ± 260
Ridge 27	K0463	80	3.91 ± 0.10	0.80 ± 0.05	4910 ± 320
Ridge 28	K0464	100	3.54 ± 0.11	0.65 ± 0.04	5430 ± 350
Ridge 29	K0465	70	4.21 ± 0.13	0.73 ± 0.05	5740 ± 400

ridge being deposited relatively soon after the termination of the Holocene marine transgression. The age overlap of the down-profile samples in ridge 15 may represent simultaneous deposition of the entire ridge during one event or it could suggest that the ridge accreted within the time frame (120 years) encapsulated by the uncertainty margins.

Determining the origin of the beach ridge plains

There are two fundamental processes that can contribute to the formation of beach ridges. These are wind or waves and they can act alone or in combination. Determining which process is responsible at any given location requires assessment of the sedimentological and stratigraphical characteristics of the ridges and their height above sea level. Sedimentary characteristics are important for differentiating between wind and wave action, and ridge height is relevant because this will be a function of the magnitude of the marine inundation (including wave run-up) where wind is not the principal process responsible for ridge formation. The coarse-grained sedimentary character of the ridges in this

region suggests that waves may have played a more dominant role than wind; however, tropical cyclones can produce winds of sufficient velocity to easily transport the coarsest sand grains. In such a situation, however, the beach, which would be the obvious source for aeolian transport of sand to a developing ridge, would normally be under a storm surge during episodes of high-velocity onshore winds. Furthermore, observations of sand transport during tropical cyclones in this region (TCs Justin in 1996, Rona in 1999, Steve in 2000, and Larry in 2006) reveal that any sand that is transported inland by wind does not accumulate at the rear of the beach or on a ridge but rather is blown far further inland and is scattered widely. Sediment is transported inland by waves during these tempests and it is deposited as a sand sheet, between 0.5 and 1.5 m thick, generally on the crest of the first sand beach ridge. This latter phenomenon has been observed several times over the past decade, and the volume of sediment transported and deposited is considerably greater than any that may be deposited as a result of aeolian transport onto the sand beach ridges. These observations suggest that wind during tropical cyclones is an unlikely mechanism for ridge construction. Instead, waves and oceanic 'set-up' conditions (the marine inundation) are more probably responsible for the texture of the sands within these ridges along with their height above sea level.

A method to ascertain the nature of the wave processes responsible for depositing the sand beach ridge plains throughout this region is presented here. This method involves the use of numerical storm surge and shallow water wave models to ascertain what type of meteorological and oceanographic conditions are most probably responsible for the deposition of sands onto these ridges at various stages of their development. This technique is applied here to the ridge plains at Cairns and Cowley Beach along with detailed topographic surveys of the ridge plains, texture analysis and chronology, using OSL in the case of Cowley Beach. The method used here accepts from the outset that the ridges were constructed by waves rather than wind because of the coarse textural characteristics of the sediments and the difficulty of transporting such sediments by wind during a tropical cyclone. This approach has been previously applied to coral shingle ridges throughout the region and the details of the method have been discussed by Nott & Hayne (2001) and Nott (2003).

If waves are the most probable depositional mechanism for coarse-grained sand beach ridges it is important to determine the wave conditions responsible. The height of the ridges can provide information in this regard because water levels need to reach an elevation at least equal to the height of the resulting ridge. This water level can either be the vertical extent of wave run-up above the remainder of the marine inundation (storm surge + tide + wave set-up + wave action) or it could be the total marine inundation excluding wave run-up as the flow depth of the inundation would be equal to or above the level of the resulting ridge. It is difficult to ascertain which of these scenarios is likely to be the one responsible and it is possible that either may be dominant at different locations in different circumstances. To be conservative, wave run-up was assumed to play a substantial role in the formation of the beach ridges and it has been further assumed that the marine inundation including wave run-up was equal to the height of the ridges. This is a conservative approach because if flow depth (excluding run-up) was assumed to be equal to or greater than the height of a ridge a more intense (lower central pressure) tropical cyclone generating a larger storm surge is required.

The tropical cyclone wind and storm surge model, otherwise known as the GCOM2D model (Hubbert & McInnes 1999), used here solves a set of mathematical equations over an equally spaced grid to determine water depth, currents, topography and bathymetry, and incorporate wind stresses and atmospheric pressure gradients acting on the ocean surface, and friction on the ocean floor. The model incorporates a movable coastal boundary and so simulates inundation of the coastal terrain. The resolution of the offshore bathymetry used in the model was 250 m. The digital elevation model for the terrestrial land area had a resolution of 100 m. All storm surge model runs were undertaken in a nested grid. This involved generating a grid extending c. 250 km from the coast and at the sites of interest (Cowley Beach and Cairns) a smaller grid area of c. 1 km × 1 km was used. The tropical cyclone wind model was initiated within the larger grid and the tropical cyclone then travels into the smaller grid where surge levels were calculated for at least six separate locations at both Cairns and Cowley Beach. Details of algorithms used within this model and its veracity in estimating surge heights against actual tropical cyclone surge events have been given by Hubbert & McInnes (1999).

The SWAN shallow water wave model was tailored for local bathymetry and run in conjunction with the tropical cyclone wind model incorporated within the GCOM2D storm tide model. The SWAN model used here was able to determine significant wave heights, wave periods, wave set-up and predominant wave travel directions at specified locations during a modelled cyclone event.

The marine inundation during a tropical cyclone is composed of storm surge, tide, wave set-up, wave run-up and wave action. The GCOM2D model estimates storm surge and can be run for any range of tide scenarios. Previously Nott (2003), in his investigation of the origin of coral shingle ridges, estimated wave set-up at 10% of the significant wave height (H_s) in 15–20 m depth water and wave run-up at 30% of H_s. Fortunately, TC Larry, which passed directly over (i.e. the left forward quadrant of the tropical cyclone, which is the zone of maximum wind velocity and surge height) the beach ridge site at Cowley Beach in March 2006, allowed determinations to be made of the wave set-up, run-up and wave action. This was achieved by surveying the marine inundation immediately following landfall of TC Larry. Debris lines composed of grasses, seaweed, marine fauna, sand and pumice accurately recorded the maximum inland extent of the marine inundation for several tens of kilometres along the coast. At Cowley Beach this debris line occurred at 3.5 m AHD. It did not reach the crest of the first major beach ridge. These debris lines represent the maximum marine inundation during the event and so represent a combination of the storm surge plus tide (storm tide) and wave set-up, wave run-up and wave action. Terrestrial run-off during the event was unlikely to have affected the location or height above sea level of the debris lines. This is because

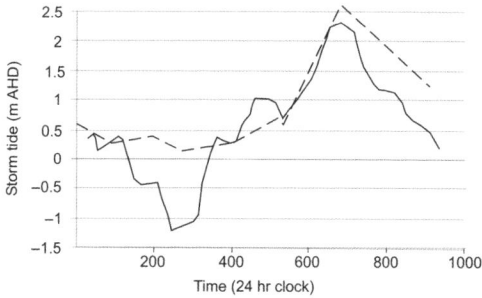

Fig. 9. Comparison of Qld Environmental Protection Agency tide gauge measurements and modelled storm tide generated by TC Larry (March 2006) at Clump Point near North Mission Beach. (Note the similar modelled height and time of the peak storm tide.)

terrestrial floodwaters during the TC Larry peaked c. 8–12 h after the peak in the storm surge. This is a common phenomenon in tropical cyclones, which depends upon the translational speed of the system as it approaches the coast. Often the tropical cyclone makes landfall well before the peak in stream discharge occurs near the coast.

The storm tide generated by TC Larry was modelled using the GCOM2D model and the results were compared with Queensland Environmental Protection Agency storm tide gauge measurements from several locations north and south of Cowley Beach. The results are presented in Figure 9. As can be seen, there is close agreement between both the height and the time of the modelled and measured storm tide. This suggests that the GCOM2D surge model was able to reproduce accurately the inundation generated by TC Larry. The modelled storm tide for Cowley Beach (2.3 m) was subtracted from the height of the surveyed marine inundation (3.5 m) to give the height of the wave set-up and run-up plus the wave action component of the inundation (= 1.2 m). The results suggest that these components combined were equal to 12% of the significant wave height, which was 10 m (at 20 m water depth). This is substantially smaller than previous estimates of the wave run-up and set-up as a percentage of H_s made by Nott (2003).

Apart from the coastal bathymetry and configuration, which are incorporated within the surge model, the height of a storm surge is governed by a number of independent factors including angle of cyclone approach, cyclone translational velocity, radius of maximum winds (R_m), cyclone central pressure and the coastal bathymetry and configuration. A series of sensitivity tests (model runs) were undertaken to determine the variation between these parameters and storm surge for each site. These sensitivity tests examine the range of

possible variation in storm surge height with various cyclone parameters such as forward speed, angle of crossing, distance of landfall from the site in question, radius of maximum wind and central pressure. Storm surge was found to increase substantially with increasing cyclone translational velocity, radius of maximum winds, decreasing central pressure and location of coastal crossing relative to the study sites. The parameters chosen for the final storm surge and shallow water wave model runs were $R_m = 30$ km, tropical cyclone translational velocity of 30 km and a NE–SW crossing angle for Cairns and east–west angle of coastal crossing in the case of Cowley Beach. These approach angles generate the largest possible surges for these sites. A landfall crossing point was also chosen to maximize the surge at each site. In these instances this was c. 30 km north of the study sites. The R_m and cyclone translational velocity of 30 km and 30 km h^{-1} respectively are the historical means for tropical cyclones in this region (McInnes et al. 2003). The values chosen from the sensitivity tests result in a very conservative estimate of the central pressure of the tropical cyclone responsible for generating a marine inundation equivalent to the crest height of the ridges. If values were chosen that resulted in a lower marine inundation then the tropical cyclone responsible would have needed to have a more intense or lower central pressure to generate a marine inundation equivalent to the ridge height. Storm surge relative to central pressure was modelled using these R_m, coastal crossing angle and translational velocity parameters. More detailed information on the modelling approach used here has been given by Nott (2003).

Storm surge and wave modelling results

The results of the storm surge and shallow water wave modelling suggest that only intense tropical cyclones are capable of generating a marine inundation, and specifically wave run-up, capable of reaching the crests of the ridges at both Cairns and Cowley Beach. For the latter a tropical cyclone with 915 hPa central pressure (category 5), an R_m of 30 km, forward velocity of 30 km h^{-1}, tracking in an east–west direction with the zone of maximum winds passing directly over the site and making landfall at mean high tide (0 m AHD) is required to generate wave run-ups capable of reaching the crest of a 5 m (AHD) high ridge.

The tide at the time of emplacement of the final sedimentary units on a ridge is unknown but it can be estimated at the 2σ probability tidal range of the frequency distribution nodal tide curve. For Cowley Beach the 1σ tidal range occurs between −0.582 m

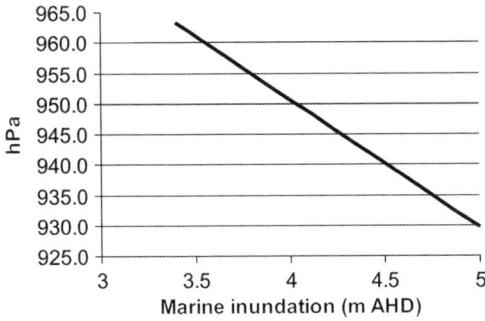

Fig. 10. Cyclone central pressure (hPa) v. marine inundation level for Cowley Beach.

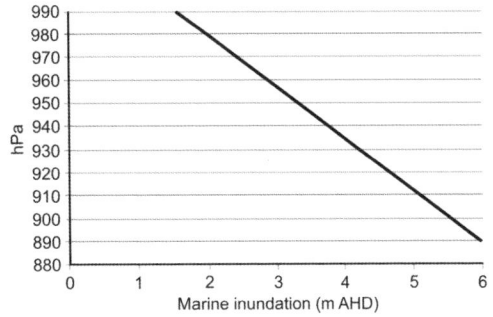

Fig. 11. Cyclone central pressure (hPa) v. marine inundation level for Cairns.

and +0.570 m AHD and the 2σ tidal range between −1.158 and +1.146 m AHD. These tidal ranges effectively form the uncertainty margins associated with the estimated mean central pressure of the cyclone responsible for producing an inundation equal in height to the elevation of a ridge. The 1σ uncertainty margin was determined at ± 12 hPa and the 2σ uncertainty margin ± 25 hPa. This means that the prehistoric tropical cyclone with $R_m = 30$ km responsible for depositing the final sedimentary units on a ridge of 5 m AHD height at Cowley Beach had to have a central pressure between 898 and 922 hPa at the 1σ margin, and 885 and 934 hPa at the 2σ uncertainty margin. Therefore, even if a tropical cyclone crossed at close to the highest possible tide during a full nodal tidal cycle (= 95th tidal quintile) the tropical cyclone central pressure had to be equal to 934 hPa, being a severe category 4 event.

The lower elevation ridges at 4–4.5 m AHD at Cowley Beach still require wave run-ups generated by an intense tropical cyclone to reach their crests. As shown in Figure 10 an inundation of 4 m AHD will be generated by a tropical cyclone with central pressure between 918 and 942 hPa at the 1σ uncertainty level, and 906 and 955 hPa at the 2σ uncertainty level. Hence, the weakest tropical cyclone required to emplace sediments on a ridge of this height is still an intense storm (955 hPa) and is very likely to be considerably stronger than this.

The sand beach ridges at Cairns are not as high as those at Cowley Beach, for the maximum ridge elevation on the Cairns beach ridge plain is 4 m AHD. Also, the different bathymetry and coastal configuration results in different surge characteristics compared with those occurring at Cowley Beach for a given intensity tropical cyclone. A separate storm surge and inundation analysis was therefore undertaken for Cairns. Here a 4 m inundation (including wave run-up) requires a tropical cyclone with a mean central pressure of 935 hPa (1σ

uncertainty = 922–946 hPa; 2σ = 909–959 hPa) tracking from the NE and travelling at 30 km h^{-1} (Fig. 11). A 3 m inundation requires a mean central pressure of 957 hPa (1σ uncertainty = 946–969 hPa; 2σ = 932–982 hPa) and a 2 m inundation requires a mean central pressure of 979 hPa (1σ uncertainty = 967–993 hPa; 2σ = 964–1004 hPa). These results suggest that a 4 m inundation in Cairns requires a weaker tropical cyclone than is required to produce the same size inundation at Cowley Beach. This is probably due to the shallower offshore gradient in Cairns Bay compared with the bathymetry at Cowley Beach. Furthermore, the highest ridges at Cairns are at least 1 m lower than at Cowley Beach. Hence the highest ridges at Cairns can be inundated by a substantially weaker tropical cyclone (mean central pressure = 935 hPa) than that needed for the highest ridges at Cowley Beach (mean central pressure = 910 hPa).

Discussion

The wave and storm surge modelling suggests that only intense tropical cyclones can generate a marine inundation of sufficient magnitude to reach the crests of the sand beach ridges at both Cairns and Cowley Beach. The sand forming the ridges at Cowley Beach is moderately sorted medium- to coarse-grained and dominated by quartz with minor amounts of feldspar (Graham 1993). The sand forming the ridges at Cairns is likewise moderately sorted and medium- to coarse-grained. The grains forming the ridges at both locations appear too coarse to have been blown by trade winds or other non-cyclonic generated wind conditions. High-velocity winds (>200 km h^{-1}) during a tropical cyclone are capable of transporting sand; however, when such conditions occur, the beach, which is the only source of sand, is typically inundated by waves and the storm surge. Therefore there is no source of sand for aeolian processes to construct a ridge during such conditions. Furthermore, regular observations of the impacts of tropical cyclones in this region over the past decade (TCs Justin in 1996, Rona in 1999, Steve in 2000, and Larry in 2006) reveal that wind is not an important factor in transporting and depositing sand onto beach ridges during such events. Instead, any sand blown by wind was transported inland of the beach ridge system and dispersed widely such that no discernible unit or layer of sand was observed to be deposited by this process. The vast majority of sand deposited onto beach ridges during these events was transported by the marine inundation as it overtopped the sand beach ridge. Interestingly, Bird (1971) in his discussion on the origin of the ridges at Cairns stated

> The completion of each ridge was marked by an episode of storm wave action, presumably generated by occasional cyclones, when the sand was piled up at the back of the shore. Subsequently a new ridge would start to develop in front of this, until eventually this too was piled up by storm waves.

Cairns is reasonably protected by Cape Grafton from the SE trade winds and associated wave activity. As a result there is a very limited trade wind wave fetch for Cairns, suggesting that if waves are responsible for deposition of the ridges to their existing height here, then tropical cyclones are the only likely mechanism for generation of these waves.

Hayne & Chappell (2001), Nott & Hayne (2001) and Nott (2003) have suggested that a single coral shingle ridge can be deposited during one storm event. Rhodes *et al.* (1980) have also suggested the same for the sand–shell beach ridges along the western coast of the Gulf of Carpentaria, Queensland. Although there are eyewitness reports of a single (3–4 m high) coral shingle ridge being deposited during one tropical cyclone in this region (Nott 2003), no reliable reports have been published on entire sand beach ridges being deposited during a single event. Rather, as mentioned above, observations here show that units of sand between 0.1 and 1.5 m thick are deposited onto ridge crests during a marine inundation.

Tropical Cyclone Larry deposited a relatively thin unit of sand (2–10 cm tapering landward) onto the incipient ridge at the back of Cowley (ridge before ridge 1 in Fig. 8). Sand units were also deposited on top of the most seaward beach ridge along Cairns' northern beaches during several tropical cyclone induced inundations between 1996 and 2001. One of these units can be seen in Figure 12. This sand unit was deposited during an inundation generated by TC Justin in 1997. The unit varies in thickness from 5 to 40 cm and extends for *c.* 100 m along the crest of an

Fig. 12. Photograph of sand unit deposited onto the first beach ridge at Clifton Beach (Cairns northern beaches) during TC Justin 1999. The photograph shows the sand unit extending down the rear flank of the beach ridge and extending into the swale behind the ridge. This sand deposit was coarse-grained and was over 30 cm thick in places.

existing sand beach ridge that stands only 2 m AHD. TC Justin was only a category 2 tropical cyclone when it made landfall and the inundation generated by this storm was able to overtop this lower-lying section of the beach ridge. Elsewhere, the beach ridge rises to 3 m AHD and the marine inundation did not overtop the ridge in these locations, and as a consequence no sand unit was deposited.

This event and others that have also deposited similar sand units provide a glimpse of how these beach ridges may develop over the longer term; that is, they develop progressively over time when a marine inundation is sufficiently large to overtop or reach the crest of an existing ridge. The inundation results in the deposition of a unit of sand, causing the ridge to grow in height with each successive event. A range of different magnitude inundations can be responsible for depositing a ridge until the ridge approaches a height that is attainable by wave run-ups generated by only very intense tropical cyclones. Hence the final units of sediment deposited on the ridge will be due to the most extreme inundations, whereas the initial units of sediment forming a ridge (those lowest in the ridge stratigraphy) could have been deposited by a range of inundations starting from non-cyclonically induced inundations, such as very high tides and strong trade wind generated wave conditions, to the most intense tropical cyclone generated inundations. Progressively higher inundations are required to deposit sand onto the ridge crest as that ridge grows in height.

A point will be reached where the vast majority of inundations can no longer reach the ridge crest and ridge will cease to increase in height. The time when this terminal point occurs may also be influenced by the rate of growth of the next seaward ridge. This ridge is likely to have already been initiated before the ridge to its landward side has reached its maximum height. Hence the rate and volume of sediment delivery to the coastal system will also probably play a role in the height that ridges can finally attain. Periods when sediment delivery rates and volumes to the coastal system are high may result in a ridge not attaining its maximum height as a function of the intensity of tropical cyclones alone. This is because the ridge on its seaward side has already attained sufficient height to effectively diminish the ability of wave run-up to reach this next inland ridge. Hence most of the sedimentation will occur on the most seaward, probably slightly lower elevation, ridge and the next inland one will become starved of sediment. Ridges may have the opportunity to reach their maximum possible height, as determined by the maximum inundation height for that region, during periods of relatively low sediment delivery rates (and volumes) to the coastal system. In this fashion there will be an interplay between the processes operating to attain the maximum ridge height, these processes being the rate of sediment delivery to the nearshore environment and the maximum height of the inundations able to be generated by tropical cyclones. In the case of the ridge plain at Cowley Beach and Cairns the final inundations responsible for depositing the uppermost units of sediment on the beach ridges were generated by extreme intensity tropical cyclones. Hence the heights of these ridges and the methods presented here for calculating the intensity of the tropical cyclones responsible allow a 5000–6000 year record of intense tropical cyclones in this region to be determined.

Sea-levels have been between 1 and 1.5 m above the present level in this region during the mid- to late Holocene (Chappell *et al.* 1982; Larcombe *et al.* 1995; Lewis *et al.* 2008). This suggests that at Cowley Beach the intensity of a tropical cyclone necessary to reach the ridge crests now at 5–5.5 m AHD may have only needed to be as intense as a present-day tropical cyclone needed to reach a 4 m AHD high ridge crest. Even to reach the crest of a 4 m AHD ridge the tropical cyclone still needs to be at least a category 3–4 event with a central pressure lower (more intense) than 955 hPa and likely to be *c.* 930 hPa. At Cairns a tropical cyclone needs to have a mean central pressure of 957 hPa to reach the crest of a 3 m AHD high ridge that is 1 m lower than the present height of the highest ridges here. Chappell *et al.* (1982) stated that sea level fell smoothly since *c.* 5.5 ka. This suggests that

the deposition of beach ridges since *c.* 3–2 ka would have occurred under conditions close to present-day sea level.

It is likely that other coarse-grained sand beach ridge plains throughout this region of northeastern Australia were also deposited by marine inundations. If this is the case then it may be possible to use these geomorphological features in the assessment of the tropical cyclone hazard in this region. The standard approach to tropical cyclone hazard risk assessment in NE Queensland involves estimating the frequency of high-magnitude events by extrapolating from a short historical time series. There is a substantial difference in the frequency of high-magnitude tropical cyclones suggested by the Cairns and Cowley Beach sand beach ridge records compared with that extrapolated from the short historical time series. The latter suggests that high-intensity tropical cyclone events (Category 4–5 events) occur at millennial time scales (McInnes *et al.* 2003) whereas the geomorphological approach adopted here suggests they occur at centennial scales. This implies that non-stationarity is likely to be an artefact of the longer-term tropical cyclone time series. This is also confirmed by an 800 year long high-resolution (annual) isotope tropical cyclone record presented by Nott *et al.* (2007), which suggests that centennial-scale variations in high-magnitude tropical cyclone frequency have occurred in this region over this time period.

Conclusion

The numerical storm surge and wave modelling, elevation and textural characteristics of the sand beach ridges at Cairns and Cowley Beach suggest that these landfoms, or at least their final form, are due to high-magnitude–low-frequency storm surges and waves generated by high-intensity tropical cyclones. In other coastal environments sand beach ridges have been ascribed to aeolian activity and/or waves generated by undetermined meteorological and oceanographic conditions. Here we have demonstrated that numerical storm surge and shallow water wave models are useful tools with which to determine more precisely the nature of the waves and the meteorological conditions necessary for the formation of these beach ridges.

It is possible that other sand beach ridge sequences throughout NE Australia have also been deposited by tropical cyclone induced surge and waves. If this is the case then these landforms could be used to determine a very broad regional hazard risk assessment, which may provide better insight into the nature of extreme events than that extrapolated from very short historical time series.

References

BIRD, E. C. F. 1970. Coastal evolution in the Cairns district. *Australian Geographer*, **11**, 327–325.

BIRD, E. C. F. 1971. The beach ridge plain at Cairns. *North Queensland Naturalists*, **16**, 4–8.

BOWMAN, G. & HARVEY, N. 1986. Geomorphic evolution of a Holocene beach-ridge complex, Lefevre Peninsula, South Australia. *Journal of Coastal Research*, **2**, 345–362.

BRISTOW, C. S. & PUCILLO, K. 2006. Quantifying rates of coastal progradation from sediment volume using GPR and OSL; the Holocene fill of Guichen Bay, south-east South Australia. *Sedimentology*, **53**, 769–788.

BROOKE, B., RYAN, D. ET AL. 2008. Influence of climate fluctuations and changes in catchment land use on Late Holocene and modern beach-ridge sedimentation on a tropical macrotidal coast: Keppel Bay, Queensland, Australia. *Marine Geology*, **251**, 195–208.

CHAPPELL, J., RHODES, E. G., THOM, B. G. & WALLENSKY, E. 1982. Hydro-isostasy and sea-level isobase of 5500 B.P. in North Queensland, Australia. *Marine Geology*, **49**, 81–90.

GRAHAM, T. 1993. *Geomorphological Response of Continental Shelf and Coastal Environments to the Holocene Transgression—Central Great Barrier Reef*. PhD thesis, Sir George Fischer Centre for Tropical Marine Studies, James Cook University, Townsville.

HAYNE, M. & CHAPPELL, J. 2001. Cyclone frequency during the last 5000 yrs from Curacoa Island, Queensland. *Palaeogeography, Palaeoclimatology, Palaeoecology*, **168**, 201–219.

HESP, P. 1984. Formation of sand 'beach ridges' and foredunes. *Search*, **15**, 289–291.

HESP, P. 2006. Sand beach ridges: Definitions and redefinition. *Journal of Coastal Research, Special Issue*, **39**, 72–75.

HUBBERT, G. & MCINNES, K. 1999. A storm surge inundation model for coastal planning and impact studies. *Journal of Coastal Research*, **15**, 168–185.

JONES, M. R. 1985. *Quaternary Geology and Coastline Evolution of Trinity Bay, North Queensland*. Geological Survey of Queensland Publication, **386**.

LARCOMBE, P. & CARTER, R. M. 2004. Cyclone pumping, sediment partitioning and the development of the Great Barrier Reef shelf system: a review. *Quaternary Science Reviews*, **23**, 107–135.

LARCOMBE, P., CARTER, R. M., DYE, J., GAGAN, M. K. & JOHNSON, D. P. 1995. New evidence for episodic post-glacial sea-level rise, central Great Barrier Reef, Australia. *Marine Geology*, **127**, 1–44.

LEWIS, S. E., WÜST, R. A. J., WEBSTER, J. M. & SHIELDS, G. A. 2008. Mid–late Holocene sea-level variability in eastern Australia. *Terra Nova*, **20**, 74–81.

MCINNES, K., WALSH, K., HUBBERT, G. & BEER, T. 2003. Impact of sea-level rise and storm surges on a coastal community. *Natural Hazards*, **30**, 187–207.

MURRAY, A. S. & WINTLE, A. G. 2000. Luminescence dating of quartz using an improved single-aliquot regenerative-dose protocol. *Radiation Measurements*, **32**, 57–73.

MURRAY-WALLACE, C. V., BANERJEE, D., BOURMAN, R. P., OLLEY, J. M. & BROOKE, B. P. 2002. Optically stimulated luminescence dating of Holocene relict foredunes, Guichen Bay, South Australia. *Quaternary Science Reviews*, **21**, 1077–1086.

NOTT, J. 2003. Intensity of prehistoric tropical cyclones. *Journal of Geophysical Research—Atmospheres*, **108**, 4212–4223.

NOTT, J. & HAYNE, M. 2001. High frequency of 'super-cyclones' along the Great Barrier Reef over the past 5,000 years. *Nature*, **413**, 508–512.

NOTT, J. & HORTON, S. 2000. 180 Ma continental drainage divide in northeastern Australia: The role of passive margin tectonics. *Geology*, **28**, 763–766.

NOTT, J., HAIG, J., NEIL, H. & GILLIESON, D. 2007. Greater frequency variability of landfalling tropical cyclones at centennial compared to seasonal and decadal scales. *Earth and Planetary Science Letters*, **255**, 365–372.

NOTT, J., SMITHERS, S., WALSH, K. & RHODES, E. 2009. Sand beach ridges record 6000 year history of extreme tropical cyclone activity in northeastern Australia. *Quaternary Science Reviews*, **28**, 1511–1520.

OTVOS, E. G. 2000. Beach ridges—definitions and significance. *Geomorphology*, **32**, 83–108.

RHODES, E. G., POLACH, H. A., THOM, B. G. & WILSON, S. R. 1980. Age structure of Holocene coastal sediments, Gulf of Carpentaria, Australia. *Radiocarbon*, **22**, 718–727.

SHEPHERD, M. J. 1987. Sandy beach ridge-system profiles as indicators of changing coastal processes. *New Zealand Geographical Society Conference Series*, **14**, 106–112.

STRUCKMEYER, H. I. A. & SIMMONDS, P. J. 1997. Tectonostratigraphic evolution of the Townsville Basin, Townsville Trough, offshore northeastern Australia. *Australian Journal of Earth Sciences*, **44**, 799–817.

TANNER, W. F. 1995. Origin of beach ridges and swales. *Marine Geology*, **129**, 149–161.

TAYLOR, M. & STONE, G. W. 1996. Beach ridges: a review. *Journal of Coastal Research*, **12**, 612–621.

THOM, B. & ROY, P. 1985. Relative sea-levels and coastal sedimentation in southeastern Australia during the Holocene. *Journal of Sedimentary Research*, **55**, 275–290.

THOMAS, M. J., NOTT, J. F., MURRAY, A. & PRICE, D. M. 2007. Fluvial response to late Quaternary climate changes in northeastern Queensland. *Palaeogeography, Palaeoclimatology, Palaeoecology*, **251**, 119–136.

WILLMOTT, W. F. & STEPHENSON, P. J. 1989. *Rocks and Landscape in the Cairns District*. Queensland Department of Mines, Brisbane.

WILLMOTT, D. L., TREZISE, M. L. & O'FLYNN, M. L. O. 1988. *Cairns Region Sheet 8064 and Part Sheet 8063, Australia 1:100,000 Geological Special*. Queensland Department of Mines, Brisbane.

WOODROFFE, C. D. 2003. *Coasts: Form, Process and Evolution*. Cambridge University Press, Cambridge.

Sheltered sandy beaches of southwestern Australia

A. TRAVERS[1], M. J. ELIOT[2]*, I. G. ELIOT[1] & M. JENDRZEJCZAK[1]

[1]*School of Earth and Environmental Sciences, Building MOO4, University of Western Australia, 35 Stirling Highway, Crawley, WA 6009, Australia*

[2]*School of Environmental Systems Engineering, Building MO15, University of Western Australia, 35 Stirling Highway, Crawley, WA 6009, Australia*

*Corresponding author (e-mail: ian.eliot@bigpond.com)

Abstract: Four beach types were identified from field surveys of beach profiles on low-energy sandy beaches in Cockburn Sound, a micro-tidal, semi-enclosed basin in southwestern Australia. These were delineated by an exposure factor (Ef), which provided a surrogate for the relationship between incident wave energy and attenuation as a result of structures, banks and shoals. The four beach profiles identified were exponential, segmented, concave-curvilinear and convex-curvilinear in form. Whether the different profile shapes occurred in environments subject to different degrees of sheltering and protection under the same tidal regime was open to question. Beach profiles were examined for Como Beach and Princess Royal Harbour in Western Australia and compared with observations from Cockburn Sound to test this proposition. Cockburn Sound and Princess Royal Harbour are subject to minimal penetration of swell waves whereas Como Beach, in the Swan River Estuary, is wholly fetch limited. Profile forms identified in each environment were consistent with those described from Cockburn Sound. Similar profile types had similar exposure values for all sites such that Ef<1 was characteristic of exponential profiles, 1–1.5 of segmented, 1.5–2 of concave-curvilinear, and Ef>2 of convex-curvilinear. Although this approach requires further testing in low-energy environments subject to different tidal ranges and fluctuations in sea level, use of the exposure factor for delineation of low-energy sheltered beach types appears to be a robust procedure.

Sandy beaches are a feature of marine environments in very sheltered locations on the Australian coast. As such they are a significant component of the Australian coastal landscape. These 'low-energy' beaches commonly occur close to the headwaters of deep inlets, bays and gulfs in South Australia, Western Australia and the Northern Territory; within channels on the leeward side of sandy barrier islands in Queensland; and around the shores of tidally attenuated coastal lagoons and estuaries in New South Wales, Victoria and Tasmania. Some examples of low-energy beaches from Western Australia are shown in Figure 1. A shared attribute of them is that there is little if any penetration of open ocean sea and swell, or it is greatly attenuated and is not a significant component of the local wave regime. Additionally, the wave regime is very low, with modal wave heights characteristically less than 25 cm for long periods between short-lasting extreme events.

Research on sandy beaches in environments classified as 'low-energy' has increased dramatically in recent years (Davidson-Arnott & Fisher 1992; Nordstrom 1992; Nordstrom & Jackson 1992; Davidson-Arnott & Conliffe-Reid 1994; Jackson 1995; Hegge *et al.* 1996; Makaske &

Augustinus 1998; Jackson *et al.* 2002; Goodfellow & Stephenson 2005, 2008). Such beaches may be sheltered, fetch limited or a combination of the two. Sheltering implies wave attenuation by refraction, diffraction and shoaling resulting from the aspect of the beach with respect to modal and extreme wave conditions or the presence of offshore structures (Eliot *et al.* 2006). Fetch-limited reaches are characterized by locally generated wind-waves and lack a swell component (Jackson *et al.* 2002). Many sheltered and fetch-restricted beaches, such as estuarine beaches, are markedly affected by non-tidal fluctuation in water level. Hence, for convenience, it is appropriate to refer to them collectively as surge-dominated systems to distinguish them from wave- and tide-dominated environments. In a review of low-energy sandy beaches in marine and estuarine settings Jackson *et al.* (2002) suggested that such environments are subject to minimal non-storm significant wave heights (e.g. <0.25 m), with moderate wave heights (i.e. *c.* 0.5 m) during strong onshore events. This definition has been adopted for the purposes of selecting sites for consideration in this paper.

Investigations in low-energy environments have focused on the discrimination of characteristic

From: BISHOP, P. & PILLANS, B. (eds) *Australian Landscapes.* Geological Society, London, Special Publications, **346**, 23–42. DOI: 10.1144/SP346.3 0305-8719/10/$15.00 © The Geological Society of London 2010.

Fig. 1. Sheltered and fetch-restricted beaches of Western Australia. (**a**) Fetch-restricted, west-facing beach on Vancouver Peninsula, Albany. (**b**) Fetch-restricted, west-facing beach on north Como Beach in the Swan River Estuary, Perth. (**c**) Sheltered, east-facing beach on Garden Island, Cockburn Sound. (**d**) Sheltered, west-facing beach at Gnanga in Shark Bay. (**e**) Fetch-restricted, north-facing beach at Point Walter, Perth. (**f**) Fetch-restricted, east-facing beach at Matilda Bay, Perth.

beach profile forms (Hegge *et al.* 1996) and the nature and mechanism of profile response to changing energy conditions. Nordstrom & Jackson (1992) proposed two types of profile response depending on the dominance of cross- or alongshore sediment transport whereas Makaske & Augustinus (1998) suggested three types of beachface morphology occurring in response to variations in wave height. More recently four distinct profile forms have been discerned on the sandy coasts

of Cockburn Sound (Fig. 2), a sheltered low-energy basin in southwestern Australia (Travers 2007*a*, *b*). The types of beach profile she identified include exponential, segmented, concave-curvilinear and convex-curvilinear profile forms (Fig. 3). The aim of the research reported here was to examine whether the four types of profile reported from Cockburn Sound consistently occurred in similar micro-tidal environments from southwestern Australia.

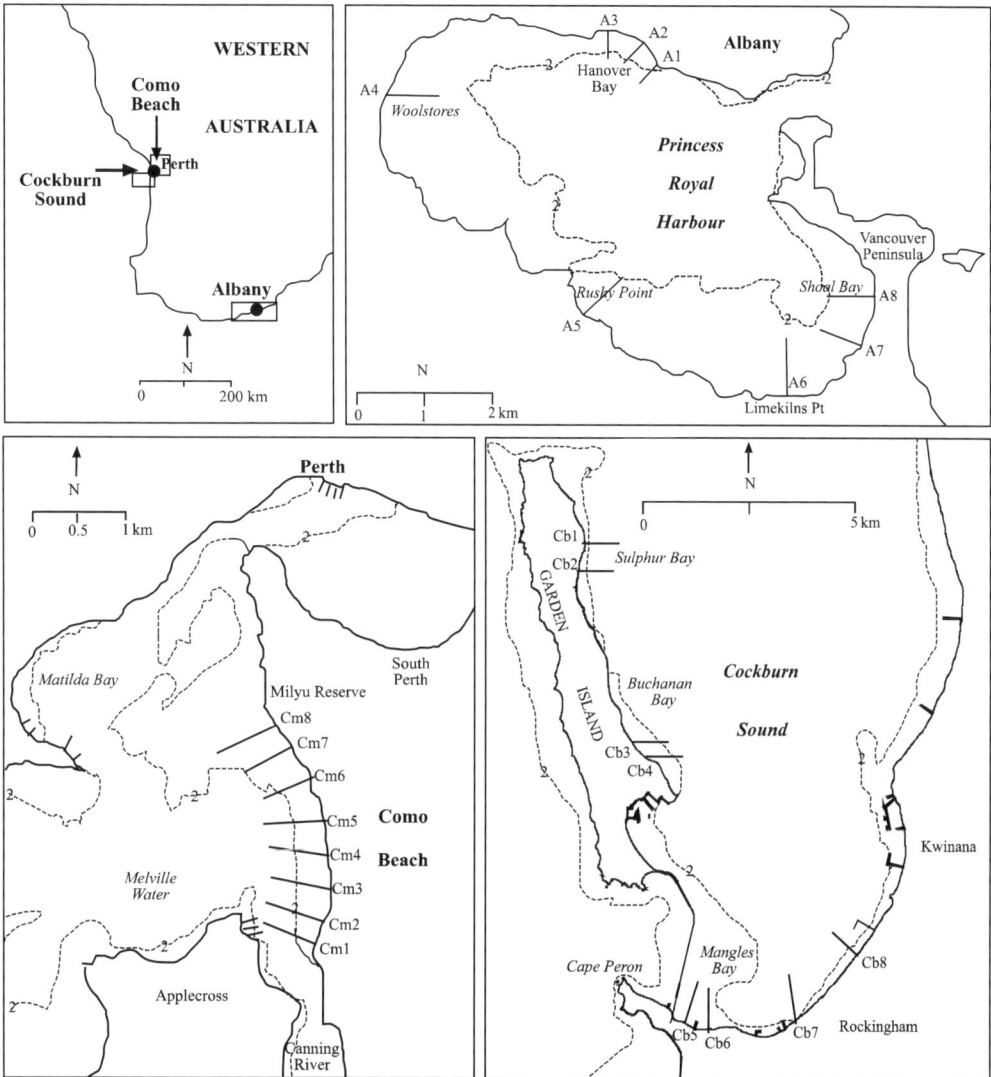

Fig. 2. Survey areas at Como Beach, Cockburn Sound and Princess Royal Harbour in Albany showing the location of profile transects.

Beach profiles in sheltered environments differ markedly from those in high-energy regimes, such as those described by Wright & Short (1984), Sunamura (1989), Lippman & Holman (1990), and Short (2005, 2006). The high-energy beaches display considerable continuity in form ranging from reflective to dissipative beach types (Wright & Short 1984) in response to changes in the wave regime of a wide range of wave and meso-tidal environments (Short 1999, 2006). Additionally, high-energy forms are subject to characteristic levels of beach stability, zones of sediment storage, and modes of beach and dune erosion (Short & Hesp

1982). Through the application of the relative tidal range (Masselink & Short 1993; Short 2005, 2006) the reflective and dissipative model has been applied to wave-dominated beaches under a wide variety of tidal regimes, including macrotidal conditions.

Whether beaches in surge-dominated environments show a similar geographical consistency in a similar range of locations is open to question. Techniques used by Short (2005, 2006) and others to identify open-ocean beach types when applied to very sheltered and fetch-restricted beaches result in a limited range of beach types, which fail to explain the variation observed on low-energy

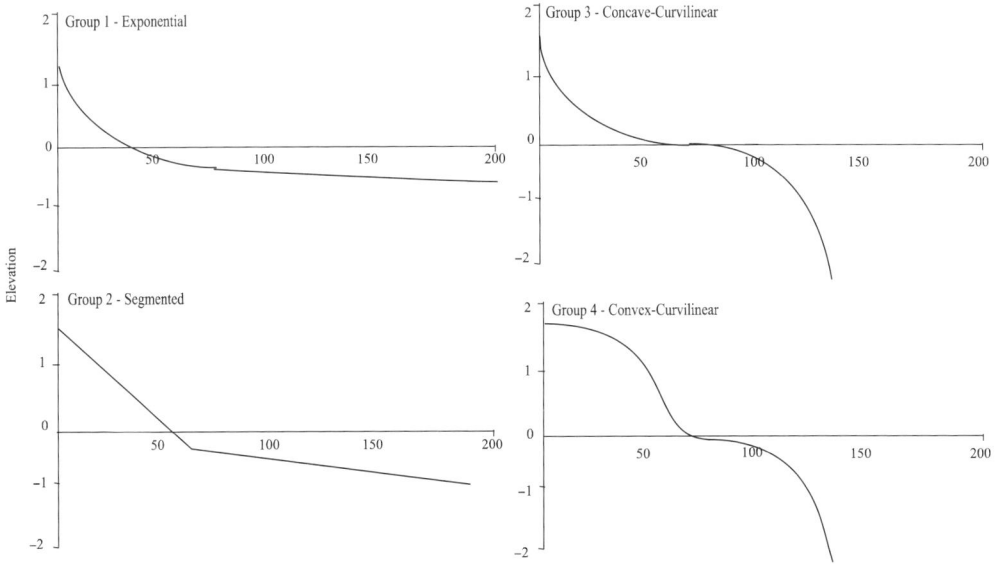

Fig. 3. Profile forms delineated in Cockburn Sound, adapted from Travers (2007*b*).

beaches. It is clear from the definition of 'low-energy' beaches (Jackson *et al.* 2002) that they could only be classified as 'tide-dominated' or 'tide-modified' under the classification schemes used to describe open-ocean beaches. This results from use of extremely low wave heights (particularly annual mean and modal significant wave heights) in the calculation of the dimensionless fall velocity and the relative tide ratio. Irregular and extreme fluctuations in water level, their effects on the upper beachface of low-energy beaches and the potential roles of inherited morphologies are not part of extant classifications of open-ocean sandy beaches. However, they are important attributes of sandy beaches subject to modally very low wave conditions in sheltered and fetch-restricted environments.

Field sites: beaches at Como, Cockburn Sound and Albany

Research reported here investigates a range of low-energy beaches subject to varying degrees of shelter in a common micro-tidal environment. An objective of the research was to establish whether consistent profile configurations occurred in similar low-energy, micro-tidal environments and to assess use of the exposure factor formulated by Travers (2007*b*) in their identification. The three beaches surveyed are in the same diurnal, micro-tidal region, following the terminology of Davies (1980) in which the mean spring tidal range is less than 0.8 m. Differences between the beaches therefore

may relate to their geological setting, including beach sediment characteristics; shoreline aspect; non-tidal variations in sea level, particularly extreme water levels; variation in the locally generated wind-wave regime; and littoral sediment transport. In many circumstances it is difficult to obtain local measurements of these parameters. Hence Eliot *et al.* (2006) and Travers (2007*a, b*) focused on parameters that limit wave growth and relate to the geographical setting of single profiles.

Cockburn Sound is a semi-enclosed micro-tidal basin located on the SW coast of WA *c.* 50 km south of Perth (Fig. 2). It is 16 km long and 9 km wide, with a 17–22 m deep central basin (Steedman 1973). Tides are similar to those at Como Beach in Melville Water, with a marginally greater daily tidal range. In both locations the spring tidal range may be exceeded by fluctuations in ocean water level as a result of storm surge associated with mid-latitude weather systems and other low-frequency water-level fluctuations. Cockburn Sound basin is formed by Garden Island and calcarenite limestone reefs that extend along almost the entire western side of the Sound as well as by Cape Peron and a causeway to Garden Island to the SW. Swell propagating through the northern end of the sound is markedly attenuated by refraction as well as a series of banks to the north. Thus, sandy beaches in this area are extensively protected from offshore wave conditions and non-storm significant wave heights are minimal, especially on the southeastern shores. Different degrees of sheltering create a variety of sandy beaches in the region. Travers (2007*b*) identified four beach types from variation in beach profile shape around the southern shores. The sites selected for analysis in this paper were selected on the basis of her research in Cockburn Sound.

Como Beach lies along the eastern shore of Melville Water, a lagoonal basin of the Swan River Estuary near Perth (Fig. 2). The basin has a complex shape, reflecting limestone rock formations, with a maximum shore-to-shore distance of c. 6 km. Como may be regarded as a 'true' fetch-limited environment because it lacks an open-ocean sea and swell component to its wave field. In this respect, 'fetch-limited' beaches are distinguished from other 'sheltered beaches' where greatly attenuated swell may be a minor component of the local wave regime. Sediment movement and beach change in fetch-limited environments are driven solely by water-level fluctuation and locally generated wind-waves. Wave heights generated by local winds depend principally on wind speed, direction and duration as well as basin width, length and depth (Jackson & Nordstrom 1992). The estuarine location of Como Beach in a fetch-restricted environment implies that different parts of beach are subject to markedly different wave conditions. Hence, eight sites along the 3 km long Como Beach were chosen for consideration according to perceived differences in their profile geometry and similarity to those reported from Cockburn Sound by Travers (2007b).

Princess Royal Harbour is a semi-enclosed, micro-tidal natural basin at Albany on the south coast of WA (Fig. 2). The Harbour is 4 km wide and 8 km long, with an approximate area of 29 km^2. It is oriented in a NW–SE direction and is connected via the Ataturk entrance to the more open basin of King George Sound. The dimensions of the outer sound allied to the narrowness of the Ataturk entrance result in minimal (<5%) open-ocean sea and swell penetration into Princess Royal Harbour. As a result the major source of energy is locally generated wind-waves. Eight profile locations at five beaches around the shores of Princess Royal Harbour were selected for survey on the basis of field reconnaissance and similarity to the forms identified by Travers (2007b). They were chosen to illustrate the range in morphology for different aspect and exposure around the Harbour.

Whereas Como Beach may be considered 'fetch-limited', Cockburn Sound and Princess Royal Harbour are susceptible to some penetration of ocean swell and best described as 'sheltered environments'. However, investigations in both locations have established that swell penetration may be as little as 5% in Cockburn Sound (DEP 1996) and less in Princess Royal Harbour (GEMS 2007). Both environments are sheltered from open-ocean sea and swell inputs to such an extent that sole consideration of wind-generated waves is considered acceptable for the purposes of the current analysis.

Research methods

Evaluation of beach morphology presented here involved collation of data from a series of published and unpublished surveys of low-energy, micro-tidal environments in Western Australia (Table 1). The 30 year record of sandy beach profiles examined from the southern half of Cockburn Sound on the Perth Metropolitan Coast provided primary identification of beach profile shapes by Travers (2007a, b). These results are compared herein with those from separate surveys of beaches at Como, an estuarine shore in the Swan River completed by Eliot et al. (2006) and previously unpublished data collected from Princess Royal Harbour at Albany on the south coast. The locations surveyed are indicated in Figure 2. The time and detail of the surveys is indicated in Table 1.

Beach and shoreface profiles

The number and spacing of survey transects varied. However, for the purposes of the current research, a variety of beach forms from each environment was selected through interpretation of aerial photography and field reconnaissance. All profiles were standardized to the Australian Height Datum (AHD), which is approximately mean sea level, for ease of comparison. The form of profiles at each site was established by determining the function of best fit as straight, segmented or curvilinear (Fig. 3). This approach followed the analysis of beach forms in Cockburn Sound described by Travers (2007b). Subsequent analysis of r^2 values allowed classification of the profiles based on shape.

Profile attributes were recorded to aid in the delineation of beach types within the various environments. These included dimensions of the beachface and sub-tidal terrace (Fig. 4), site-specific sediment grain size, beach aspect in degrees from north, and width of the marginal shoal. For the purposes of this paper the beachface is considered as the area between the +2 m contour (AHD) and the level of the Lowest Astronomical Tide (LAT) (Fig. 4). For sites in Como, Cockburn and Princess Royal Harbour a broad sub-tidal terrace extends for up to a kilometre offshore. Its surface is mainly below spring low-tide level. However, the inner, shoreward part of the terrace may be exposed by combinations of low spring tide, offshore wind and seasonal low mean water level. The low elevation means that wave and current processes affecting surface morphology are strongly linked to the

Table 1. *Timing of profile surveys and site characteristics*

Variable	Beaches		
	Albany	Cockburn	Como
Number of profiles	9	20	14
Period of survey	10 August 2004	1974–2003	18 March 2003
Timing of surveys	Once	Twice a year in 1974–1980 (March and September), once in 1990 and twice in 2003	Once
Tidal range (m)*	0.9	0.6	0.6
Grain size (mm)	0.2–0.34 (fine to medium)	0.19–0.39 (fine to medium)	0.37–0.62 (medium to coarse)
Wave height (m)†	0.12–0.44	0.2–0.78	0.17–0.59
Wave period (s)	1.16–2.49	2.7–5.9	1.3–3.4
Source†	Jendrzejczak (2004)	Travers (2007*a*)	Eliot *et al.* (2006)

*Spring tidal range (MLLW to MHHW).
†Range of significant wave heights for profiles in the study area.

frequency and duration of water-level fluctuations (Eliot *et al.* 2006). In the current project the width of the sub-tidal terrace was measured as the distance offshore from LAT level to a slope break at the 2.0 m isobath, where the profile dropped steeply into a deeper basin. Grain size and sediment settling velocity were determined by settling tube analysis of a sample taken from the central beachface for each profile station.

Winds

The physical setting of each profile site was established from an analysis of local bathymetric maps (DOT 1995; DPI 2003, 2005) in conjunction with meteorological data, particularly wind speed and direction. Local records for each environment were used to create wind frequency distribution plots. Perth Airport data were used to determine the wind climate at Como, whereas records from Rottnest Island and Albany Airport describe the wind regime for Cockburn Sound and Princess Royal Harbour, respectively. All records spanned the 10 year period from 1991 to 2001. The wind

regime for each location was subsequently considered in the context of local aspect and protection to characterize site-specific conditions. Wind speeds over each fetch were established for onshore directions at 22.5° intervals centred on each profile site (Fig. 5). Characteristic wind speeds for hindcasting were defined by the speed that is exceeded for more than $88 \, h \, a^{-1}$ (1% frequency) in the designated directional band. This information was used to determine the prevailing, dominant and longest fetches for the generation of wind waves at each site.

The length of wind fetch length for wave hindcasting was estimated for prevailing and dominant onshore wind directions as well as the longest fetch at each profile location along the beach surveyed. Here, 'prevailing' refers to the most frequent onshore direction from which winds were blowing and 'dominant' to the direction from which the highest wind speeds were recorded.

Water-level ranging

The three field sites were selected for investigation for reasons linked to availability of water-level

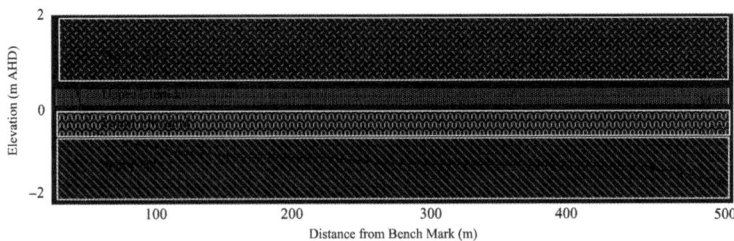

Fig. 4. Profile zones used in morphometric analysis. Example shown applies to Como and Cockburn; HAT = 0.6 m; LAT = −0.6 m. At Albany, HAT and LAT are approximately at +0.7 m and −0.7 m respectively in the figure.

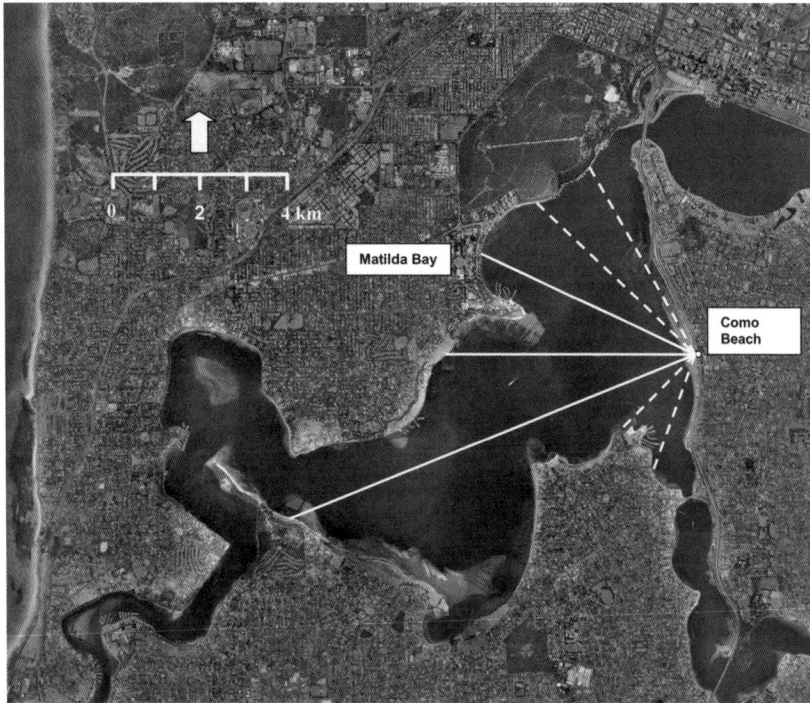

Fig. 5. Onshore winds blowing along fetches between distal and proximal 2.0 m isobaths were used to hindcast parameters for wind-waves immediately offshore from each profile site. Here fetch lines are shown every 22.5° for a profile station at Glyde Street, Como.

records. First, they occur in fetch-restricted environments in which the penetration of oceanic sea and swell is less than 5% of the wave regime, even under extreme wave conditions at the section of coast under investigation. Second, the three sites are close to tide gauging stations at Standard Ports: Fremantle, Albany and Barrack Street. They all share a common micro-tidal environment with a mean daily range of *c*. 0.5 m, a neap range (MHLW to MLHW) of less than 10 cm and a spring range (MLLW to MHHW) of 60 cm (Department of Defence 2009). Under neap conditions the beaches are almost non-tidal.

From neap to mid-tidal (mean) range conditions the tidal range may be exceeded by storm surge. In the context of this paper surge refers to water-level ranging caused by a combination of meteorological and oceanographic process; a full description of surge phenomena and analysis techniques has been given by Pugh (1987, 2004). Surge is approximately described by the residual time series derived when the predicted tide is subtracted from an observed water-level record. The resulting residual time series can be analysed to derive parameters, such as minimum and maximum values and recurrence intervals.

A strong negative surge may be generated by a combination of high barometric pressure and offshore winds at the point of interest on the shore. Conversely, a high positive surge may be generated by low barometric pressure and onshore winds and contribution from wave set-up. Surge is further affected by the geological framework confining the coast and the configuration of the shore (Bode & Hardy 1997). In micro-tidal environments, particularly including those described herein, and elsewhere (Gray 2002), the positive surge may equal or exceed the spring tidal range during extreme events. Low barometric pressure and onshore winds may produce raised water levels up to +0.7 m, whereas high barometric pressure and offshore winds can depress water levels as low as −0.3 m. These effects interact with and may be superimposed on peak tides. Additionally, there are significant seasonal and interannual variations in water level. The seasonal water-level range is *c*. 0.25 m with highest water levels occurring during the winter months (Pariwono *et al.* 1986). Inter-annual variation of the water level corresponding to the El Niño–Southern Oscillation (ENSO) phenomenon may be as large as ±0.2 m (Easton 1970; Pariwono *et al.* 1986; Feng *et al.* 2004).

This raises significant questions concerning the interaction of non-tidal variation in water level with tides, as well as their joint and separate effects on beach morphology. Commonly, in a low-wave, micro-tidal environment there is a high component of inheritance, such that the beach form does not substantially change between extreme events. It also brings into question for the low-energy beach types the validity of classifying sandy beaches on the basis of modal wave and tide conditions, as done in using the dimensionless fall velocity (Wright & Short 1984) and relative tide ratio (Masselink & Short 1993) as descriptive and predictive parameters.

Waves

Wave hindcasting was undertaken with reference to mean sea level as well as for the comparison of extreme conditions during summer and winter. The extreme wave estimates were calculated for a mean summer water level approximated at -0.2 m AHD and mean winter level of $+0.2$ m AHD. Wave conditions were estimated by hindcasting, using the Bretschneider parametric wave equations for shallow water (USACE 1984). Waves passing across the shallows of the broad sub-tidal terrace are subject to considerable depth-effects through wave breaking and bed-friction. Hence wave parameters were calculated by using the depth at 200 m intervals along each fetch line from the 2.0 m isobath on the leeward shore to that on the windward shore subject to survey. Consequently, wave parameters (period and height) have been designated at the fetch crossing of the 2.0 m isobath closest to the profile survey station of interest and seaward of which all water depths are greater than 2.0 m. Wave transformation across the sub-tidal terrace where attenuation took place was not calculated although this may be a useful statistic.

Data analysis

In addition to least-squares identification of beach profile configuration, statistical description of the profile types, each profile was related an exposure factor, $Ef = \log FL/Ms$, which is the proportion of fetch length (FL) to marginal shoal width (Ms), as well as to parameters used to classify open-ocean beaches (Short 2006), including those used in estimation of the dimensionless fall velocity (Wright & Short 1984) and relative tidal range (Masselink & Short 1993). Here, fetch is recorded as the length of the direct fetch between 2 m contours on either side of the basin. The marginal shoal is a terrace in water less than 2 m deep and adjoining the shore. It includes the inter-tidal and sub-tidal environment. The exposure factor delineated profile

form such that $Ef < 1$ was characteristic of exponential profiles, $1-1.5$ of segmented profiles, $1.5-2$ of concave-curvilinear, and $Ef > 2$ of convex-curvilinear profiles.

Once the beach profile configurations had been determined, as indicated in Figure 6, the profile types (exponential, segmented, concave-curvilinear and convex-curvilinear) for each of the three locations (Como, Cockburn and Albany) were plotted against site attributes including wave height, midswash grain size and the exposure factor. A one-way analysis of variance test (ANOVA) carried out in SPSS version 5.1e to further examine potential linkages between profile type and environmental parameters. The profile types identified from each station at the three locations were then plotted for estimates of the exposure factor based on the character of the fetch as the most direct, longest, and associated with the prevailing or dominant onshore wind direction, and relationship between profile type and exposure factor was examined through linear regression.

Results

Plots of mean profile shape for eight sites in each of the three environments under consideration are presented in Figure 6a. The means were estimated as the average elevation for fixed points at 5 m intervals along several profiles surveyed at the same location. Examination of mean shapes for each location indicated the likely presence of four profile forms (groups) consistent with those identified in Cockburn Sound (Travers 2007*b*). Consequently, curve fitting was carried out at a shoreface scale using the functions of best fit described by Travers (2007*b*) to test the whether the forms were consistent. This analysis identified exponential, segmented, convex-curvilinear and concave-curvilinear profile forms at each of the three study sites (Fig. 7). Allied to curve fitting, dimensions of the beachface and shoreface profile were recorded for all sites to aid in the delineation of discrete profile types (Table 2).

Physical setting

Wind roses showing the frequency distribution of hourly wind speeds for Como (Perth Airport, 1991–2001), Albany (1991–2001) and Cockburn Sound (Rottnest Island, 1991–2001) are illustrated in Figure 8. In the context of the physical setting of each site surveyed, the wind distribution helps to define fetches to which each profile site was susceptible. Fetches were established for onshore winds at 22.5° intervals, with each fetch summarized by its length (Table 3). Estimated wave heights and periods indicated spatial variation in

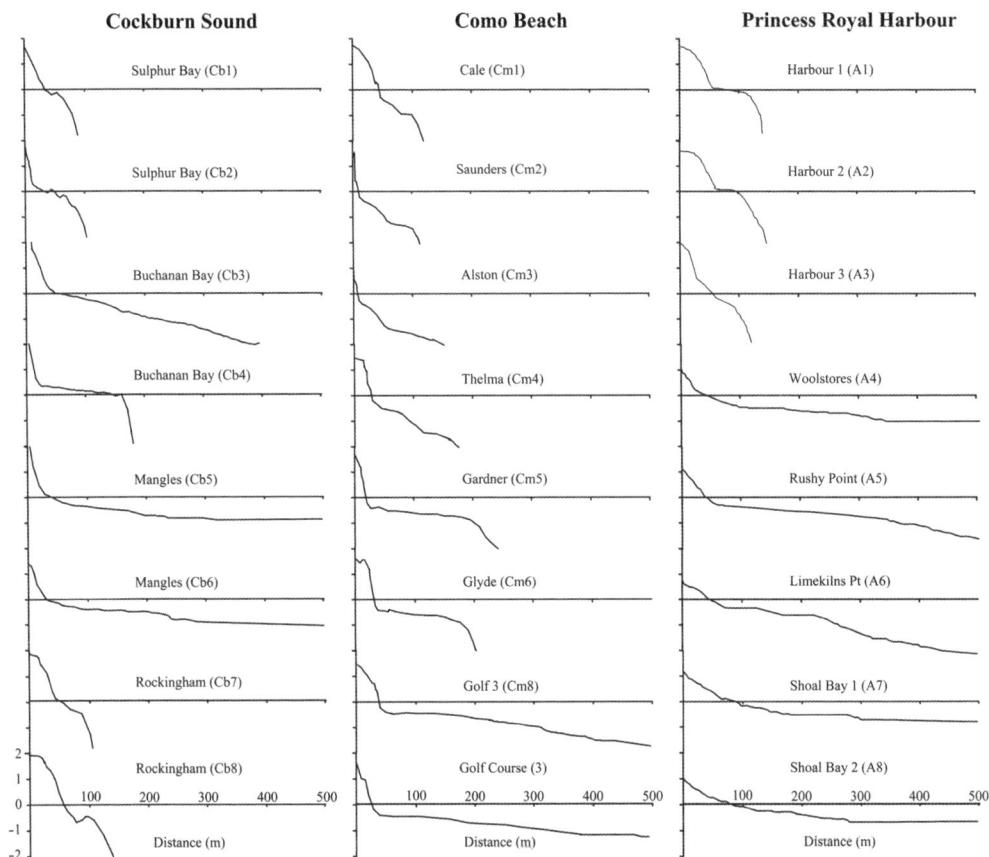

Fig. 6. Mean profile shapes for sites in Cockburn Sound, Como and Princess Royal Harbour.

wave conditions for different fetch directions, wind speeds and water levels at each location. The relative wave climate for winds from the longest fetch and the influence of prevailing and dominant conditions are indicated in Table 4. Incident and attenuated wave heights are listed for all sites.

At Como, wave conditions under dominant (storm) westerly winds are most energetic along the southerly part of the beach, with hindcast waves up to 0.4 m height and 2.5 s period for the 1% exceedance wind speed. Under the same wind conditions, the northern part of the beach experiences waves up to 0.2 m height and 2.2 s period. Wave conditions under prevailing SW winds and from the longest W–SW fetch produce a different pattern of wave exposure, with the most energetic conditions at the seaward edge of the sub-tidal terrace experienced in the central–northern part of the beach, off Glyde Street, with the least energetic waves at Cale Street, in the south. The effect of fetch length on wave conditions is significant, with W–SW winds hindcast to produce up to 0.8 m

height and 4.0 s period waves offshore from the central–northern part of the beach between Eric and Glyde Streets. The effect of the sub-tidal terrace in wave attenuation also is demonstrated for all sites. At the northern end of the beach wave heights were attenuated by c. 50–70% along the prevailing SW fetch line and 30–40% along the dominant westerly direction. At central locations in the vicinity of Eric Street attenuation over the long, low-gradient sub-tidal terrace is also significant with a c. 25–35% decrease in incident wave height. Conversely, for sites in the south, attenuation is minimal, with 5–20% reduction in wave height over all fetches.

Under dominant conditions in Albany, incident wave heights were greatest for sites in the vicinity of Hanover Bay (0.65 m) and Woolstores (0.83 m) and least at Shoal Bay (0.25 m). Similarly, the distribution of wave heights associated with prevailing conditions indicates lowest wave conditions for Shoal Bay whereas highest waves were calculated for the Woolstores site. Waves generated along the longest and direct fetch showed a similar trend; that

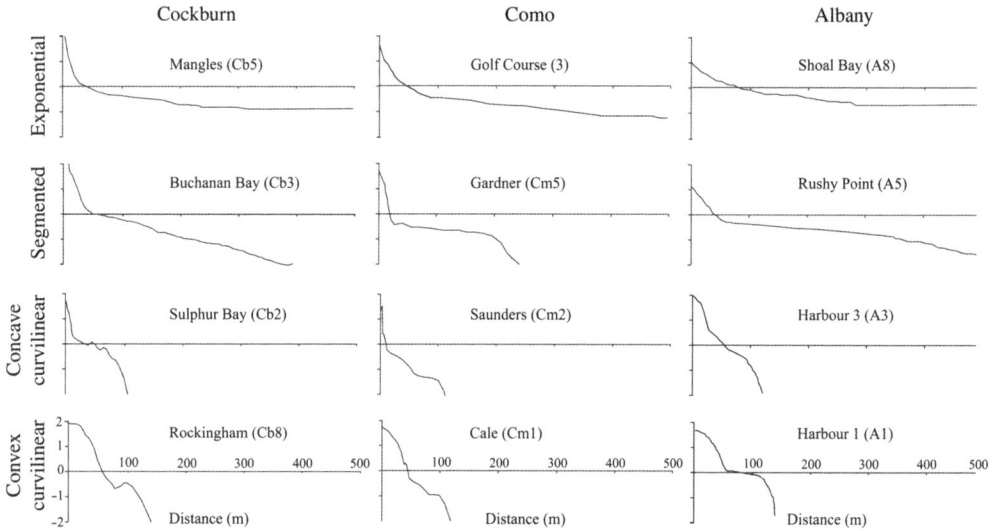

Fig. 7. Examples of the four profile types identified at each survey site.

Table 2. *Morphometric attributes for beach profiles*

Site	W_1	W_2	W_3	W_4	W_5	S_1	S_2	S_3	S_4	S_5	Gs
Albany											
Harbour 1 (A1)	47	22	25	18	29	2.49	3.38	1.60	2.23	2.57	0.30
Harbour 2 (A2)	49	34	15	25	22	2.34	2.13	2.67	1.60	3.38	0.52
Harbour 3 (A3)	38	23	15	12	10	4.09	5.71	2.67	3.34	2.41	0.44
Woolstores (A4)	13	6	7	8	1670	7.59	10.52	5.71	5.00	0.04	0.19
Rushy Point (A5)	22	15	8	11	795	5.19	14.57	5.00	3.64	0.07	0.31
Limekilns (A6)	21	14	7	15	860	4.57	18.00	6.65	2.67	0.06	0.28
Shoal Bay 1 (A7)	13	6	7	8	2010	6.34	23.43	5.71	5.00	0.04	0.32
Shoal Bay 2 (A8)	15	7	8	10	2210	5.71	33.02	5.00	4.00	0.03	0.27
Como											
Cale (Cm1)	36	16	20	11	79	2.07	2.51	1.72	3.12	0.94	0.38
Saunders (Cm2)	22	12	10	19	73	3.38	3.34	3.43	1.81	0.87	0.48
Alston (Cm3)	19	7	12	12	85	3.91	5.71	2.86	2.86	0.53	0.37
Thelma (Cm4)	24	17	7	8	91	3.10	2.36	4.90	4.29	0.65	0.36
Eric (Cm5)	28	19	9	15	175	3.38	2.86	4.29	3.12	0.34	0.43
Gardner (Cm6)	22	14	8	11	220	2.66	2.11	3.81	2.29	0.43	0.56
Glyde (Cm7)	15	8	7	8	1420	4.95	5.00	4.90	4.29	0.05	0.57
Golf (Cm8)	13	7	6	10	1680	5.71	5.71	5.71	3.43	0.04	0.62
Cockburn											
Sulphur 1 (Cb1)	32	21	11	18	62	3.58	1.91	3.12	1.91	1.29	0.22
Sulphur 2 (Cb2)	31	19	12	19	72	3.69	2.11	2.86	1.81	1.11	0.21
Buchanan 1 (Cb3)	21	12	9	11	520	5.44	3.34	3.81	3.12	0.15	0.18
Buchanan 2 (Cb4)	23	15	8	7	425	4.97	2.67	4.29	4.90	0.19	0.19
Mangles 1 (Cb5)	14	6	6	8	960	4.09	6.65	5.71	4.29	0.09	0.22
Mangles 2 (Cb6)	18	10	9	7	910	3.18	4.00	3.81	4.90	0.11	0.23
Rockingham (Cb7)	49	28	21	12	33	2.34	1.43	1.64	2.86	2.43	0.35
Rockingham (Cb8)	53	31	22	18	35	2.16	1.29	1.56	1.91	2.29	0.39

W, length of profile (m); S, slope (degrees): W_1 and S_1, between +2 m AHD and MSL; W_2 and S_2, +2 m AHD to HAT; W_3 and S_3, HAT to MSL; W_4 and S_4, MSL to LAT; W_5 and S_5, LAT to −2 m AHD; Gs, grain size (mm).

Como   **Albany**   **Cockburn**

0.1 1.5 3 5 8 10.8
Wind Speed (Metres Per Second)

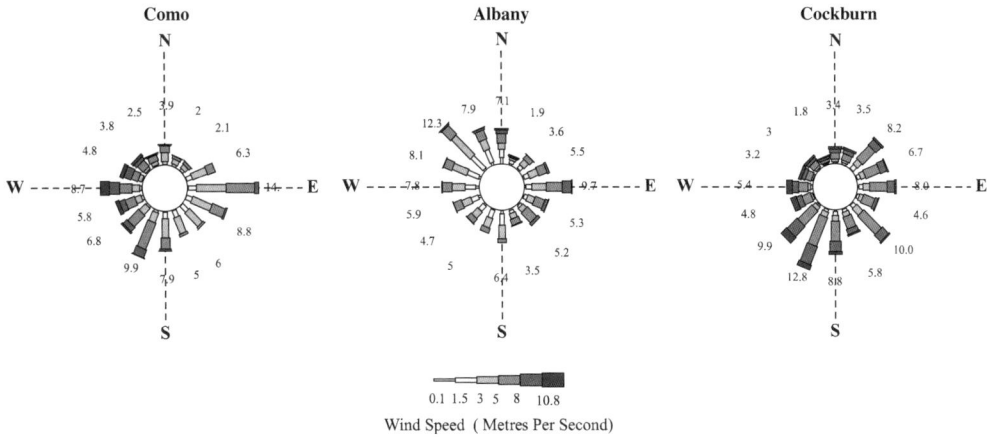

Fig. 8. Frequency distribution of wind speeds for Como (Perth Airport, 1991–2001), Albany (1991–2001) and Cockburn Sound (Rottnest Island, 1991–2001).

Table 3. *Fetch directions and associated lengths for the prevailing, dominant, longest and direct fetch generating wind-waves at each profile site*

	Direction				Length			
	Aspect (deg.)	FL_L (deg.)	FL_P (deg.)	FL_{DOM} (deg.)	FL_{DIR} (m)	FL_L (m)	FL_P (m)	FL_{DOM} (m)
Albany								
Harbour (1)	250	202.5	202.5	202.5	2550	2610	2610	2610
Harbour (2)	217.5	202.5	202.5	202.5	2995	5500	5500	5500
Harbour (3)	212.5	135	202.5	202.5	2610	3750	2420	2420
Woolstores	100	112.5	112.5	112.5	4350	5937	5937	5937
Rushy Point	20	45	45	90	2750	4800	4800	3100
Shoal Bay (1)	340	315	337.5	337.5	3420	4850	3420	3420
Shoal Bay (2)	272.5	270	270	270	5450	5450	5450	5450
Quaranup	270	270	270	270	4275	4275	4275	4275
Como								
Cale (2)	275	292.5	225	315	3510	3520	3120	500
Saunders St	273	289	225	315	3810	3900	3410	680
Alston St	272	225	225	315	3750	4000	4000	740
Preston St	266	272	225	315	3410	3410	3110	650
Eric St	260	247.5	225	315	3100	3450	2350	600
Gardner St	262	270	225	315	2810	2810	2190	950
Glyde St	242	225	225	315	2400	2580	2580	1300
Golf (3)	265	225	225	315	2120	4240	4240	400
Cockburn								
Sulphur (1)	99	135	90	67.5	7210	8910	7210	8320
Sulphur (2)	103	135	90	67.5	8010	9050	8010	8280
Buchanan (1)	39	22.5	90	67.5	9800	9900	6050	14520
Buchanan (2)	38	67.5	90	67.5	9015	9450	5430	9450
Mangles (1)	32	22.5	67.5	67.5	5100	5100	3880	3880
Mangles (2)	353	353	45	45	5810	5810	5450	5450
Rockingham (1)	2	0	22.5	0	23000	23000	12210	23000
Rockingham (2)	332	0	22.5	0	11700	23100	5480	23100

FL_L, longest fetch, FL_P, prevailing fetch; FL_{DOM}, dominant fetch; FL_{DIR}, direct fetch.

Table 4. *Wave heights and periods associated with prevailing, dominant, longest and direct fetch at each profile site*

Location	Hs_{DIR} (m)	Ts_{DIR} (s)	Hm_{DIR} (m)	Tm_{DIR} (s)	Hs_L (m)	Ts_L (s)	Hm_L (m)	Tm_L (s)	Hs_P (m)	Ts_P (s)	Hm_P (m)	Tm_P (s)	Hs_{DOM} (m)	Ts_{DOM} (s)	Hm_{DOM} (m)	Tm_{DOM} (s)
Albany																
Harbour 1 (A1)	0.24	1.2	0.21	1.6	0.32	1.6	0.28	1.8	0.25	1.2	0.22	1.8	0.6	1.9	0.54	2.6
Harbour 2 (A2)	0.25	1.2	0.23	1.8	0.31	1.3	0.26	1.6	0.27	1.2	0.19	1.9	0.65	1.8	0.55	2.4
Harbour 3 (A3)	0.24	1.3	0.12	1.5	0.29	1.7	0.24	1.5	0.24	1.2	0.21	1.7	0.49	1.9	0.53	2.3
Woolstores (A4)	0.36	1.6	0.14	1.1	0.37	1.6	0.14	1.1	0.44	1.7	0.13	1.3	0.83	2.4	0.21	2.4
Rushy Point (A5)	0.19	1.3	0.12	1.3	0.21	1.4	0.14	1.6	0.21	1.2	0.19	1.5	0.49	1.8	0.29	2.1
Limekilns (A6)	0.15	1.5	0.1	1.1	0.22	1.2	0.16	1.8	0.19	1.5	0.14	1.2	0.33	2.6	0.19	1.8
Shoal Bay 1 (A7)	0.11	1.1	0.12	1.0	0.14	1.1	0.11	1.4	0.12	1.2	0.12	1.5	0.25	1.8	0.15	1.6
Shoal Bay 2 (A8)	0.19	1.3	0.1	1.1	0.21	1.2	0.12	1.1	0.19	1.3	0.11	1.6	0.35	2.1	0.21	2.1
Como																
Cale (Cm1)	0.47	2.7	0.42	2.3	0.34	2.4	0.32	2.2	0.17	1.3	0.15	1.4	0.35	2.4	0.32	1.2
Saunders (Cm2)	0.49	2.5	0.43	2.2	0.33	2.4	0.31	2.1	0.18	1.1	0.17	1.8	0.37	2.3	0.31	1.5
Alston (Cm3)	0.46	2.3	0.39	2.0	0.46	2.8	0.41	2.2	0.19	1.3	0.17	1.2	0.38	2.5	0.36	1.8
Thelma (Cm4)	0.32	1.9	0.29	1.7	0.44	2.6	0.37	2.4	0.19	1.3	0.16	1.1	0.36	2.2	0.34	2.1
Eric (Cm5)	0.34	2.2	0.22	2.2	0.77	4.0	0.5	2.5	0.25	1.7	0.18	1.5	0.2	2.1	0.15	1.8
Gardner (Cm6)	0.38	2.4	0.15	1.6	0.6	3.2	0.3	2.1	0.21	1.8	0.13	1.2	0.22	2.3	0.14	1.6
Glyde (Cm7)	0.33	2.1	0.13	1.4	0.62	3.1	0.24	1.8	0.59	3.0	0.11	1.8	0.21	1.8	0.12	1.4
Golf (Cm8)	0.31	2.0	0.1	1.3	0.53	3.5	0.15	1.6	0.49	3.4	0.18	1.7	0.17	1.6	0.1	1.1
Cockburn																
Sulphur 1 (Cb1)	0.53	1.6	0.40	2.3	0.55	2.0	0.44	2.3	0.37	2.6	0.29	1.8	0.71	3.4	0.42	2.5
Sulphur 2 (Cb2)	0.52	1.6	0.42	2.2	0.54	2.0	0.42	2.2	0.46	3.4	0.34	1.5	0.79	3.4	0.47	2.4
Buchanan 1 (Cb3)	0.56	2.0	0.30	2.4	0.56	2.0	0.31	2.0	0.38	3.3	0.19	1.2	0.70	3.3	0.28	2.6
Buchanan 2 (Cb4)	0.57	2.0	0.29	1.3	0.58	2.0	0.03	1.7	0.42	2.7	0.16	2.1	0.71	3.3	0.22	2.1
Mangles 1 (Cb5)	0.43	1.6	0.12	1.4	0.44	2.1	0.12	2.2	0.22	2.4	0.06	1.1	0.45	1.5	0.15	1.8
Mangles 2 (Cb6)	0.46	1.8	0.15	1.6	0.46	2.1	0.15	1.7	0.20	1.9	0.08	1.6	0.58	2.8	0.12	1.6
Rockingham (Cb7)	0.69	2.2	0.52	1.1	0.67	2.4	0.53	2.4	0.38	2.7	0.31	1.8	0.80	4.1	0.62	2.3
Rockingham (Cb8)	0.71	2.4	0.48	1.1	0.66	2.4	0.48	2.3	0.32	2.7	0.28	3.0	0.81	4.1	0.66	2.2

Hs_{DIR} and Ts_{DIR}, significant incident wave height and period generated under direct fetch conditions; Hm_{DIR} and Tm_{DIR}, modified wave height and period generated under direct fetch conditions; Hs_L and Ts_L, longest fetch; Hs_P and Ts_P, prevailing fetch; Hs_{DOM} and Ts_{DOM}, dominant fetch.

is, highest waves were calculated for sites at Hanover Bay and Woolstores and lowest waves were calculated for Shoal Bay. Observations made of attenuated wave height over the sub-tidal terrace altered this pattern of wave distribution around the Harbour. Woolstores, which consistently experienced the highest waves at the 2 m contour underwent up to 75% reduction in wave heights to rank among the lowest wave energy sites for all fetch directions. Comparison of attenuated wave statistics compounded the disparity between highest and lowest wave sites, with Shoal Bay undergoing up to 40% reduction whereas incident waves in Hanover Bay were only subject to up to 15% decrease in height.

In accordance with variation in the wind regime, wave climate varied systematically around Cockburn Sound with aspect and exposure. The most energetic conditions for all fetch directions occurred in the SW around Rockingham, followed by Buchanan Bay and Sulphur Bay. The least energetic conditions for all key wind characteristics under consideration occurred at Mangles Bay on Garden Island. In keeping with results from Albany, attenuation of wave height by the sub-tidal terrace accentuated the disparity between highest and lowest wave sites, with Mangles Bay, with a long, low sub-tidal terrace, experiencing the greatest reduction in wave height and Rockingham the least.

Profile morphology

Exponential forms were identified at Woolstores on the western side of Princess Royal Harbour and at Shoal Bay to the south; at the northern end of Como beach near Glyde Street and the golf course; and in Mangles Bay in the SE reaches of Cockburn Sound. Beaches in this category generally had a narrow concave upper beachface with a long, low-gradient sub-tidal terrace (Table 2).

Segmented forms were located at Rushy Point and Limekilns point in Princess Royal Harbour; Gardner and Eric Street at Como; and in Buchanan Bay on Garden Island in Cockburn Sound. The segmented beaches have two planar components to their profile, with a steep beachface and a flat inshore separated by a distinct break in slope (Fig. 6, Table 2).

Curvilinear profiles had a convex or concave upper beachface with a convex sub-tidal zone dropping off rapidly into deeper water relatively close to shore. Both convex and concave forms were identified at sites in Hanover Bay within Princess Royal Harbour. At Como, convex-curvilinear and concave-curvilinear forms were located at the southern end of the beach. In Cockburn Sound, beaches at Sulphur Bay on Garden Island were concave whereas those in the vicinity of Rockingham on the mainland had a convex profile on the beachface.

Beaches in the exponential category had the narrowest beachface (13–18 m), whereas those in the convex-curvilinear group were widest (36–53 m) for all sites (Table 2). Dimensions of the sub-tidal terrace also varied between profile groups and between study sites. However, for all sites, exponential forms had the most extensive sub-tidal terrace (width 910–2010 m; slope 0.03–0.11°) whereas beaches in the curvilinear groups had a relatively narrow steeply sloping sub-tidal zone (10–91 m; 2.29–3.38°).

Modal sediment size also varied between profile sites for each study environment. In Princess Royal Harbour, sediments ranged from fine to medium (0.19–0.5 mm), with coarsest sediment associated with sites in the vicinity of Hanover Bay and finer grains characterizing beaches in Shoal Bay in the SE of the Harbour. At Como, the segmented and exponential forms at the northern end of the beach are associated with coarser grains (0.43–0.51 mm) whereas the curvilinear forms further south are generally composed of medium-grained sand (0.36–0.48 mm). Modal sediment for beaches on Garden Island in Cockburn Sound were generally categorized as fine (0.19–0.24 mm) whereas mainland beaches, with the exception of those in Mangles Bay, were composed of medium-sized grains (0.31–0.39 mm).

Discussion

The relationship between discrete profile forms and physical attributes on open-ocean coasts traditionally uses descriptors including sediment grain size, wave height and the slope of the beachface profile. Correlations between these descriptors and the low-energy beach profiles were examined. In general, grain size and profile number were positively correlated for all sites except at Como, where finest sediments were associated with the convex-curvilinear profiles (Group 4). Coarsest grains were generally associated with most gently sloping beach faces whereas those in the finer category had steeper slopes (Fig. 9). This was contrary to the results of previous sandy beach research whereby mild slopes occurred in conjunction with fine grains (Hegge *et al.* 1996; Klein & Menezes 2001). The exception was at Como, where beach slope and sediment size were strongly and negatively correlated. Despite these observations, results of the ANOVA test showed that variability in sediment size between profile forms was not statistically significant ($p > 0.05$). Although profile types tended to have an associated range of grain sizes, a clear differentiation between beach forms was lacking. Also, the range of sediment sizes associated with a given profile shape was not consistent across the three study locations (Fig. 9a).

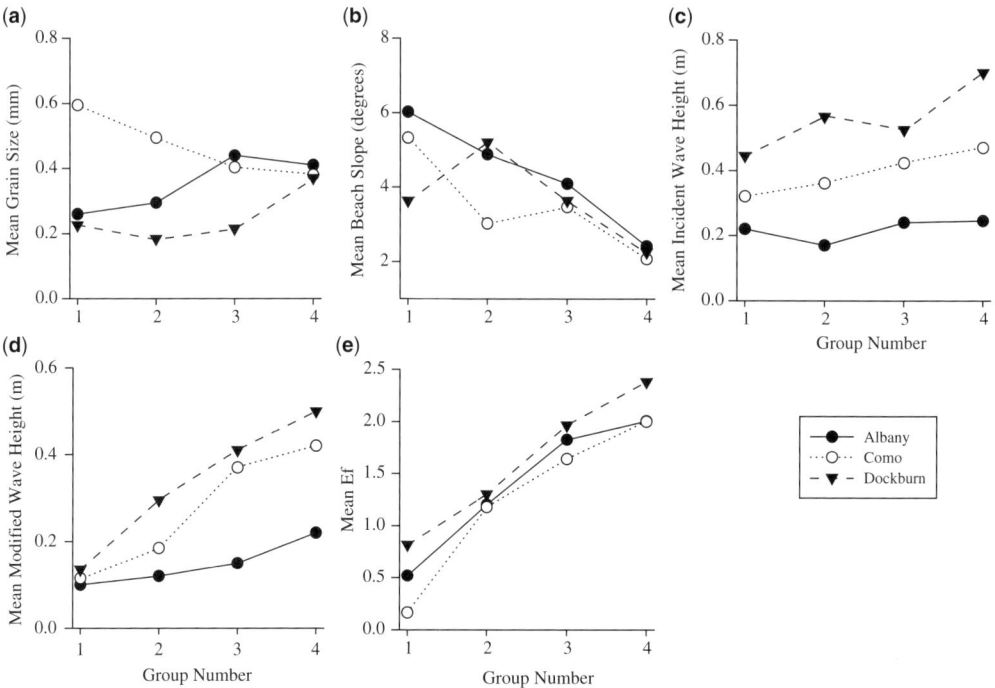

Fig. 9. Means for profile attributes as a function of profile type, indicated by the group of similar profiles, at each of the three survey sites. Group 1, exponential; Group 2, segmented; Group 3, concave-curvilinear; Group 4, convex-curvilinear.

Similarly, values for profile slope and incident wave height were not significantly different between beach types for all environments ($p > 0.05$). Beach-face profiles with comparable slopes, for example Groups 2 and 3, had very different configurations in line with the suggestions of previous researchers (Makaske & Augustinus 1998; Klein & Menezes 2001; Travers 2007*a*, *b*). In keeping with beach slope (Fig. 9b), wave heights associated with profile types were incongruent for the three study environments under consideration (Fig. 9c and d). However, comparison of observed wave heights that had been modified through transformation across the subtidal terrace and profile morphology gave better results (Fig. 9d). Attenuated wave heights were significantly different between profile forms ($p < 0.05$), with exponential forms characterized by lowest waves and curvilinear forms subject to most energetic conditions for all sites. The result was unsurprising in that a long, low-gradient sub-tidal terrace structure typified the exponential forms and resulted in significant attenuation of incident wave energy. This feature appeared to function similarly to the intertidal terrace commonly described in higher tidal range environments, although this is not always clear. Conversely, the

curvilinear forms were characterized by narrow sub-tidal reaches. They tended to be steeply sloping and drop into deeper water. Thus, the degree of wave attenuation at the shoreline appears to be related to systematic variations in the width and depth of the sub-tidal terrace (Eliot *et al.* 2006). Wave attenuation is least and the breaker zone is closest to shore where the sub-tidal terrace is narrowest and attenuation is greatest, whereas breaking occurred furthest from shore at sites where the sub-tidal terrace is extensive. Although a similar trend in results was observed across all three environments, the range of attenuated wave heights associated with morphological groups was not consistent.

The combined relationship of wave, sediment and beach slope properties and profile shape was subsequently assessed through calculation of the omega parameter (Ω) (Short 1999, pp. 177–178) for each site (Table 5). This failed to distinguish between the range of profile shapes in each environment, classifying the majority of sites as dissipative or ultra-dissipative (Table 5). The relationship between the calculated exposure factor (Ef) and beach type was, however, more revealing. An ANOVA test indicated that differences in Ef between profile groups in each environment were statistically significant ($p < 0.001$). The relationship between profile sites, exposure and beach type grouping for each environment is shown in Figure 10a. For all environments, sites with the highest Ef were convex-curvilinear and those with the lowest Ef were segmented. In addition, a comparison of Ef values across environments indicated that profiles for a given beach type had similar exposure values for all sites (Fig. 10b) such that Ef<1 was characteristic of exponential profiles, 1–1.5 of segmented profiles, 1.5–2 of concave-curvilinear profiles, and Ef>2 of convex-curvilinear profiles.

Distinguishing between wave- and surge-dominated sandy beaches

Profile forms described from Cockburn Sound, Como and Princess Royal Harbour are similar to sheltered beaches in meso-tidal sheltered environments described by Nordstrom (1992) and those in macro-tidal environments described by Gray (2002). They also display elements resembling

Table 5. *Predictive morphodynamic indices calculated for each survey site*

Location	RTR	Ω	Ef
Albany			
Harbour 1 (A1)	2.3	6.1	2.01
Harbour 2 (A2)	2.3	5.9	2.00
Harbour 3 (A3)	3.1	4.1	1.83
Woolstores (A4)	3.5	8.5	0.42
Rushy Point (A5)	3.2	5.2	1.11
Limekilns (A6)	3.6	5.3	1.14
Shoal Bay 1 (A7)	7.2	5.5	0.50
Shoal Bay 2 (A8)	6.7	5.6	0.64
Como			
Cale (Cm1)	1.8	15.9	2.00
Saunders (Cm2)	1.8	5.6	1.89
Alston (Cm3)	2.1	5.5	1.55
Thelma (Cm4)	2.4	5.2	1.47
Eric (Cm5)	2.7	4.8	1.25
Gardner (Cm6)	2.5	4.3	1.11
Glyde (Cm7)	3.7	11.9	0.23
Golf (Cm8)	3.9	16.4	0.10
Cockburn			
Sulphur 1 (Cb1)	1.8	7.6	1.95
Sulphur 2 (Cb2)	1.7	11.5	1.97
Buchanan 1 (Cb3)	2.6	16.9	1.28
Buchanan 2 (Cb4)	4.0	6.3	1.33
Mangles 1 (Cb5)	6.2	5.1	0.73
Mangles 2 (Cb6)	5.6	4.8	0.91
Rockingham (Cb7)	1.4	5.8	2.55
Rockingham (Cb8)	1.5	5.1	2.21

RTR, relative tidal range [= mean spring tide range (MSR)/modal breaker height (H_b)]; Ω, dimensionless fall velocity; Ef, exposure factor.

beaches identified through the 'dynamic approach' pioneered by Wright & Short (1984) and described more recently by Short (1999, 2006). For example, the convex-curvilinear profiles resemble the reflective beaches described by Wright & Short (1984) and the exponential forms resemble the tide-modified and tide-dominated forms of Short (2005, 2006). Further, the transition from surge- to wave-dominated beach type is difficult to define, as surge-dominated environments in different areas may be subject to disparate wave and water-level regimes and the two profile types are geographically convergent within deep embayments and along estuary shores. However, the dominant process interactions linked to the profile forms are anticipated to differ, with storm recurrence and inundation frequency playing a more significant role on the low-energy shores. In this respect, Goodfellow & Stephenson (2005, 2008) and Eliot *et al.* (2006) have highlighted the importance of wind direction and speed in governing temporal changes in morphology and dynamics on sheltered beaches, with notable inheritance of morphology following the occurrence of high-energy events.

The observations from southwestern Australia indicate that profile forms from surge-dominated environments may not be predicted by calculation of the dimensionless fall velocity. They also are outside reported boundaries for bar formation although parallel bars may occur as a persistent feature of the shoreface on some curvilinear beaches, as occurs in parts of Mangles Bay. The geometry of the very sheltered, exponential profiles shows similarities to tide-dominated forms described by Short (2005). However, sites in Mangles Bay could be classified only as 'reflective low-tide terrace and rips' through calculation of the surf-scaling parameter or omega factor (Short 1999) and would be termed 'tide-modified' by conventional classification schemes applied to open-ocean wave-dominated coasts such as those reviewed by Short (1999, pp. 177–195) and Masselink & Hughes (2003, pp. 223–228). Travers (2007a) pointed out that neither term is strictly correct as the beaches examined are in extremely sheltered and micro-tidal environments that are near tideless. She argued that sheltered and fetch-restricted beaches in Cockburn Sound are surge dominated and undergo profile adjustment in response to water-level ranging associated with extreme storm events. In some fetch-restricted environments, such as Como, bars are apparent low, transverse ridges and are associated with water-level ranging across the subtidal terrace.

Despite poor prediction of surge-dominated beach forms by indices applicable to wave-dominated coast on open-ocean shores, surge-dominated beaches may be considered as the lower

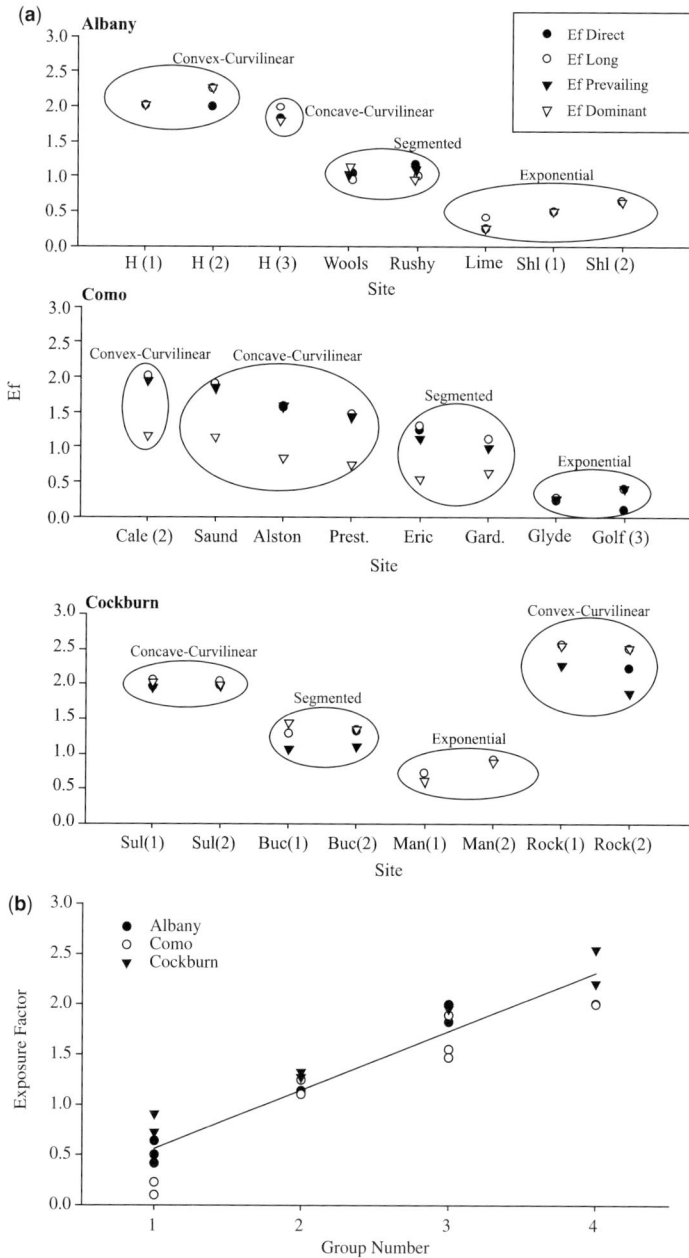

Fig. 10. (**a**) Exposure values associated with each survey site. For each of the three environments under consideration profiles in the convex-curvilinear group are most exposed and those in the exponential group are least exposed. Site names are indicated in Table 2. (**b**) Regression analysis of exposure values and profile types for all sites. Exposure and profile type are strongly and positively correlated, $r^2 = 0.89$, $p < 0.001$.

end of a continuum constituting a hysteresis loop (Fig. 11). The 'highest energy' beaches in surge-dominated environments, the convex-curvilinear profiles, are similar to the lowest energy forms described for open-ocean coasts and may grade into them in an alongshore direction. This was suggested by the work of Goodfellow & Stephenson (2008) at Seaford Beach in Port Phillip Bay,

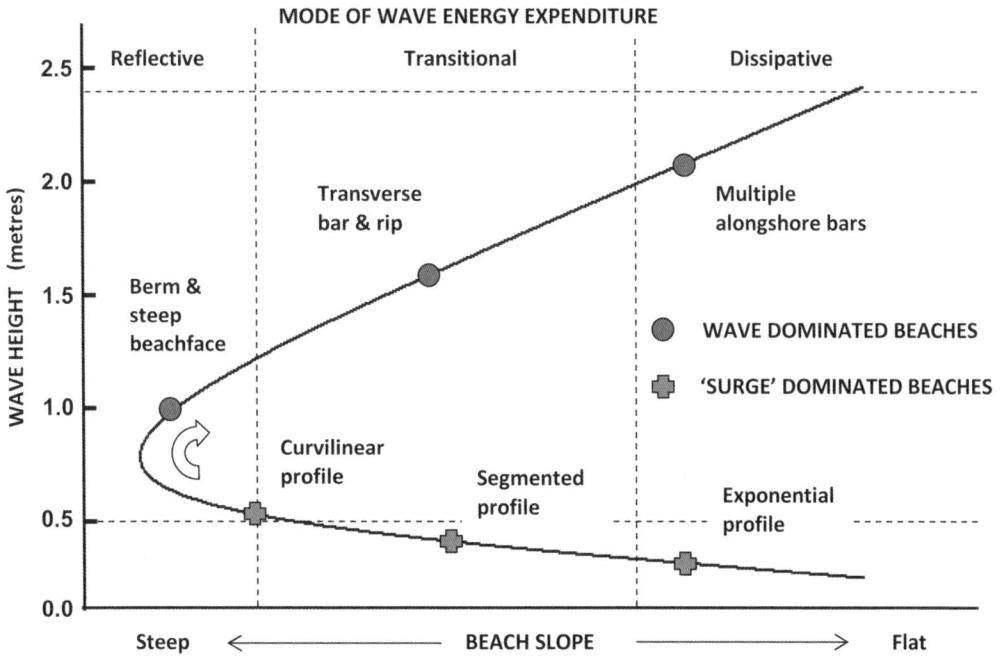

Fig. 11. Continuum of profile shapes on sandy beaches. The continuum forms a hysteresis loop with its end-members being wide and flat. The change from surge to wave domination is marked by wave reflective beaches with a steep beachface and planar inshore morphology.

Victoria. The curvilinear forms could be described as 'low-energy reflective profiles' but this would need to be tested through field measurement of nearshore morphodynamics. However, in the very sheltered and fetch-restricted reaches of the microtidal environments in southwestern Australia these 'low-energy' forms occur at the 'high-energy' end of the scale for surge-dominated systems. Similarly,

the most sheltered and fetch-restricted forms, exponential profiles, could be described as wave-dissipative beaches, especially during storm conditions (Fig. 12).

Sandy beaches in sheltered and fetch-restricted environments where the spring tidal range at least exceeds non-tidal water-level ranging may be 'tide-dominated'. Masselink & Short (1993) and

Fig. 12. View north under low- and high-wave conditions on a fetch-restricted sandy beach at Como, Western Australia. (**a**) Low tide and water level set down on Como beach during November 2002 (note the absence of waves). (**b**) Spring high tide (75 cm), water level set-up (70 cm) and wave action (Hb = 50 cm) on Como Beach during 16 May 2003.

Masselink & Hegge (1995) recognized persistence of 'wave-dominated' profile configurations on beaches in meso- and macro-tidal environments. Whether the profile configurations apparent on 'surge-dominated' beaches in micro-tidal environments also persist under meso- and macro-tidal regimes is open to question. The observations by Nordstrom (1992) from a meso-tidal environment in the USA, and Gray (2002) from macro-tidal beaches in the Northern Territory of Australia suggest a wider geographical spread of beach forms characteristic of fetch-restricted, low-wave environments than that reported for southwestern Australia. However, there are discrepancies between descriptions of the profiles in each region and a more rigorous comparison between surge-dominated sandy beaches under different tidal regimes is required. Potentially, the main effect of an increase in tidal regime, from micro- to macro-tidal, may be an increase in the height and width of the shore profile; that is, the profile is distorted through 'stretching' by tidal processes. This proposition is beyond the scope of the current paper and is a worthwhile subject for further research.

Conclusions

Results of the present research suggest that beach profile types are consistent across a range of surge-dominated micro-tidal environments. Although similar profile forms were repeated at different locations, correlations of profile type and incident wave and sediment statistics were poor. This was unsurprising for surge-dominated reaches, where the spatial distribution of beach forms may be controlled by complex shoreline configurations or where geomorphological contexts, particularly the wave regimes, vary over relatively short reaches. The variation is determined by a site-specific physical setting, which dictates the magnitude of forcing that will lead to sediment mobility and resultant beach morphology.

An exposure factor $Ef = \log FL/Ms$, which is the proportion of fetch length (FL) to marginal shoal width (Ms), potentially provides a useful non-dimensional parameter for the delineation of surge-dominated forms. Whether the relationships elucidated persist for very low-wave environments subject to different tidal ranges and fluctuations in sea level is open to question and should form the basis for further investigations. However, the index accounts for the physical setting of each profile site at the three similar micro-tidal locations examined: Cockburn Sound, Como and Princess Royal Harbour. It does so by including consideration of the fetch length available for generation of local wind-waves and the potential attenuation of wave energy over the width of the sub-tidal terrace or marginal shoal.

Recognition of a hierarchy of sandy beach types in surge-dominated environments at Albany, Cockburn Sound and Como extends our understanding of beach dynamics in two respects. First, in combination the low-energy (fetch-restricted and sheltered) and high-energy (wave-dominated) environments constitute a hysteresis loop with flat dissipative profiles as end-members for extremely low- and extremely high-wave conditions respectively, and the curvilinear forms of the surge-dominated environments grading into reflective beaches as wave height increases. Second, identification of different predictive indices for beaches in high- and low-wave regimes underscores differences in the forcing mechanisms governing beach formation. The regional swell regime and sediment size are critical factors in beach profile formation under open-ocean conditions, and inter-tidal terraces are features of transitional states under low-wave conditions. This contrasts with surge-dominated shores where profile form is tied to local wind conditions and the geometry of the sub-tidal terrace or marginal shoal.

In the immediate context of the paper, our thanks go to L. Nash and J. Collins for unstinting assistance in the collection of field data. The Department of Planning and Infrastructure and Bureau of Meteorology respectively provided water-level and wind data. The principal author was sponsored by an International Postgraduate Research award and additional infrastructural support was provided by the University of Western Australia.

More broadly, the short string of authors of the paper identifies part of the legacy of John Chappell to Western Australian geomorphology. The legacy is an intellectual and practical stream starting with John as a young academic finishing his doctoral studies during which time he accepted Ian Eliot to study coastal processes and landforms as a postgraduate student at ANU. The academic path then passes to Ian's senior students at Honours and postgraduate level from the universities of Wollongong and Western Australia. Some now occupy positions in government and academia; the latter continuing the legacy at Sydney, Wollongong, UWA and Notre Dame (Fremantle). The range of employment followed by them is diverse. For example, A.T. is now an environmental consultant, as are people with whom she works. M.J. is a town planner.

Additionally, John's supervision and mentoring of Ian through his postgraduate studies and internship as a lecturer brought their young families into close contact and friendship. Social discussions ranging across a very wide variety of subjects from art to science have had a profound influence on Ian's children; something his younger son Matt, a coastal engineer, acknowledges as forming his interest in the application of metocean processes and geomorphology to his engineering roles.

Beyond John's legacy there is highly valued friendship and cultural awareness. In the small, lonely hours of a very dark, rain-drizzling night Ian stood quietly counting people working in the surf on a beach survey he was supervising. He became aware of company. John had come to share the task, discuss the inevitable problems of working in an unfriendly environment and provide some humour to the situation. That was not an unusual event. John has always been there when needed, always provided intellectual rigour, practical advice, a strong sense of environment and good humour at dark times.

References

BODE, L. & HARDY, T. 1997. Progress and recent developments in storm surge modelling. *Journal of Hydraulic Engineering*, **123**, 315–331.

DAVIDSON-ARNOTT, R. G. D. & CONLIFFE-REID, H. E. 1994. Sedimentary processes and the evolution of the distal bayside of Long Point, Lake Erie. *Canadian Journal of Earth Science*, **31**, 1461–1473.

DAVIDSON-ARNOTT, R. G. D. & FISHER, J. D. 1992. Spatial and temporal controls on overwash occurrence on a Great Lakes barrier spit. *Canadian Journal of Earth Sciences*, **29**, 102–117.

DAVIES, J. L. 1980. *Geographical Variation in Coastal Development*, 2nd edn. Longman, Harlow.

DEP. 1996. *Southern Metropolitan Coastal Waters Study 1991–1994: Final Report (1)*. WA Department of Environmental Protection, Perth.

DEPARTMENT OF DEFENCE. 2009. *Australian National Tide Tables*. Australian Oceanographic Publication, **11**.

DOT. 1995. *Ocean Reef to Cape Peron, Western Australia*, Scale 1:75 000, Nautical Chart WA 001, Edition 7. Department of Transport, WA Government, Perth.

DPI. 2003. *Swan and Canning Rivers, Western Australia*, Scale 1:25 000, Nautical Chart WA 001, Edition 898. Department for Planning and Infrastructure, WA Government, Perth.

DPI. 2005. *Albany, Western Australia, Scale 1:25 000, Nautical Chart WA 1083*. 1st edn. Department for Planning and Infrastructure, WA Government, Perth.

EASTON, A. 1970. *The Tides of the Continent of Australia*. Horace Lamb Centre of Oceanographical Research, Flinders University, Research Paper, **37**.

ELIOT, M. J., TRAVERS, A. & ELIOT, I. 2006. Morphology of a low-energy beach; Como Beach, Western Australia. *Journal of Coastal Research*, **22**, 63–77.

FENG, M., LI, J. & MEYERS, G. 2004. Multidecadal variations of Fremantle sea level: footprint of climate variability in the tropical Pacific. *Geophysical Research Letters*, **31**, L16302.

GEMS. 2007. *Grange Resources and Albany Water Authority Port Development: Oceanographic Studies and Dredging Program Simulation Studies*. Global Environmental Modelling Systems Report, **376/06**.

GOODFELLOW, B. W. & STEPHENSON, W. J. 2005. Beach morphodynamics in a strong-wind bay: a low-energy environment? *Marine Geology*, **214**, 101–116.

GOODFELLOW, B. W. & STEPHENSON, W. J. 2008. Role of infragravity energy in bar formation in a strong-wind bay: observations from Seaford, Port Phillip Bay, Australia. *Geographical Research*, **46**, 208–223.

GRAY, P. 2002. *Beach Profile Change on the Eastern Beaches of Port Darwin, 1996 to 2001*. MSc thesis, Northern Territory University, Darwin, NT.

HEGGE, B. J., ELIOT, I. & HSU, J. 1996. Sheltered sandy beaches of Southwestern Australia. *Journal of Coastal Research*, **12**, 748–760.

JACKSON, N. L. 1995. Wind and waves: influence of local and non-local waves on mesoscale beach behaviour in estuarine environments. *Annals of the Association of American Geographers*, **85**, 21–37.

JACKSON, N. L., NORDSTROM, K. F., ELIOT, I. & MASSELINK, G. 2002. Low energy sandy beaches in marine and estuarine environments: a review. *Geomorphology*, **48**, 147–162.

JENDRZEJCZAK, M. 2004. *Sandy beaches of Princess Royal Harbour, on the south coast of Western Australia*. BSc Honours Thesis, The University of Western Australia, Nedlands, WA.

KLEIN, A. H. & MENEZES, J. T. 2001. Beach morphodynamics and profile sequence for a headland bay coast. *Journal of Coastal Research*, **17**, 812–834.

LIPPMAN, T. C. & HOLMAN, R. A. 1990. The spatial and temporal variability of sand bar morphology. *Journal of Geophysical Research*, **95**, 11575–11590.

MAKASKE, B. & AUGUSTINUS, G. E. F. 1998. Morphologic changes of a micro-tidal, low wave energy beach face during a spring–neap cycle, Rhône Delta, France. *Journal of Coastal Research*, **14**, 632–645.

MASSELINK, G. & HEGGE, M. G. 1995. Morphodynamics of meso- and macrotidal beaches: examples from Central Queensland, Australia. *Marine Geology*, **129**, 1–23.

MASSELINK, G. & HUGHES, M. G. 2003. *Introduction to Coastal Processes and Geomorphology*. Hodder–Arnold, London.

MASSELINK, G. & SHORT, A. D. 1993. The effect of tide range on beach morphodynamics and morphology: a conceptual beach model. *Journal of Coastal Research*, **9**, 785–800.

NORDSTROM, K. F. 1992. *Estuarine Beaches*. Elsevier, Amsterdam.

NORDSTROM, K. F. & JACKSON, N. L. 1992. Two-dimensional change on sandy beaches in meso-tidal estuaries. *Zeitschrift für Geomorphologie*, **36**, 465–478.

PARIWONO, J., BYE, J. & LENNON, G. 1986. Long-period variations of sea-level in Australasia. *Geophysical Journal of the Royal Astronomical Society*, **87**, 43–54.

PUGH, D. 1987. *Tides, Surges and Mean Sea-Level*. Wiley, Chichester.

PUGH, D. 2004. *Changing Sea Levels. Effects of Tides, Weather and Climate*. Cambridge University Press, Cambridge.

SHORT, A. D. (ed.) 1999. *Handbook of Beach and Shoreface Morphodynamics*. Wiley, Chichester.

SHORT, A. D. 2005. *Beaches of the Western Australian Coast: Eucla to Roebuck Bay. A Guide to their Nature, Characteristics, Surf and Safety*. Sydney University Press, Sydney.

SHORT, A. D. 2006. Australian beach systems—nature and distribution. *Journal of Coastal Research*, **22**, 11–27.

SHORT, A. D. & HESP, P. A. 1982. Wave, beach and dune interactions in S.E. Australia. *Marine Geology*, **48**, 259–284.

STEEDMAN, R. K. 1973. *Sulphur Bay Investigations—Garden Island WA: Long Period Waves from Jervoise*

Bay Wave Records—Cockburn Sound. Report No. 1. R. K. Steedman and Associates, Perth.

SUNAMURA, T. 1989. Sandy beach geomorphology elucidated by laboratory modeling. *In*: LAKAN, V. C. & TRENHAILE, A. S. (eds) *Applications in Coastal Modelling.* Elsevier Oceanography Series, **23**, 159–213.

TRAVERS, A. 2007*a. Sediment Exchange Between the Inshore and Beachface of Sheltered, Low-energy beaches.* PhD thesis, University of Western Australia, Crawley, WA.

TRAVERS, A. 2007*b.* Low-energy beach morphology with respect to physical setting: a case study from Cockburn Sound, Southwestern Australia. *Journal of Coastal Research*, **23**, 429–444.

USACE. 1984. *Shore Protection Manual.* 3rd edn. US Army Corps of Engineers, Coastal Engineering Research Center (CERC), US Government Printing Office, Washington, DC.

WRIGHT, L. D. & SHORT, A. D. 1984. Morphodynamic variability of surf zones and beaches: a synthesis. *Marine Geology*, **56**, 93–118.

Are the Murray Canyons offshore southern Australia still active for sediment transport?

S. SCHMIDT[1,2], P. DE DECKKER[3]*, H. ETCHEBER[1,2] & S. CARADEC[2]

[1]CNRS, UMR5805 EPOC, Avenue des Facultés, F-33405 Talence Cedex, France

[2]Université de Bordeaux, UMR5805 EPOC, F-33405 Talence Cedex, France

[3]Research School of Earth Sciences, The Australian National University, Canberra, ACT 0200, Australia

*Corresponding author (e-mail: patrick.dedeckker@anu.edu.au)

Abstract: The Australian continental margin hosts numerous canyons. Some of the most spectacular canyons are located offshore Kangaroo Island, and these are linked to ancient courses of the River Murray, which would have flowed across the very wide Lacepede Shelf during periods of low sea level. During the AUSCAN-1 project, modern sedimentation was assessed using a multi-tracer approach on interface sediments from 350 to 2500 m water depth. The presence of freshly deposited particles, tagged by ^{234}Th in excess, ^{210}Pb-based sediment accumulation (0.03–0.13 cm a^{-1}) and ^{230}Th-based focusing ratios, supports the occurrence of significant advection of marine sediments within these canyons. In the absence of direct riverine inputs, the shelf, being the site of intensive carbonate production, is the main supplier of material. The presence of incised channels in the eastern portion of the Murray Canyons Group (MCG) indicates recent to sub-recent activity along the canyons. The presence of underwater slides in the western side of the MCG confirms that sediment transport to the abyssal plain does occur. Based on our preliminary investigation and by synthesizing previous work on other canyons, we provide a conceptual model for sediment focusing and transfer within the canyons offshore Australia.

Canyons are often presented as natural conduits for the transfer of particulate matter from the shelf to the deep ocean (Carson *et al.* 1986; Gardner 1989). Although the Holocene sea-level rise drastically reduced the supply of coarse-grained sediments to abyssal depths, canyons continue to be favourable sites for the concentration and accumulation of fine-grained sediments (Biscaye & Olsen 1976; Hickey *et al.* 1986). Sediment is probably not produced locally within the canyon but transported from shallow coastal areas (Granata *et al.* 1999). Canyons seem to act as sediment traps for material transported along slope by oceanic currents (Liu & Lin 2004; Palanques *et al.* 2005). Some active downslope transport of sediments, such as in turbidity events, also occurs (Mulder *et al.* 2001; Khripounoff *et al.* 2003; Puig *et al.* 2003). However, the age and the frequency of sedimentary processes transporting sediments through canyons, and to the overbank deposits on the flanks of associated channels, are usually unknown. Nevertheless, this knowledge is important for understanding the reason for canyons being 'hotspots' for biological activity, and how changes in the quality and intensity of particle supply could affect a canyon and its associated productivity. It is known that additional supply of particulate matter, especially

organic in origin, can enhance biological productivity and diversity (Ruhl & Smith 2004).

The Australian landmass is bordered by numerous canyons, some of which are not even charted on currently available maps. Along the southern margin of Australia, there are over 50 canyons, principally offshore the southern portion of Western Australia and offshore South Australia. It appears that some of the most spectacular canyons thus far known in the Australia region are located offshore Kangaroo Island (Fig. 1). Some of these appear to be linked to ancient courses of the River Murray, which would have flowed across the Lacepede Shelf during periods of low sea level (Fig. 2).

The objective of this investigation is to characterize sediment transport and deposition within the Murray Canyons Group on seasonal to century time scales using a multi-tracer approach. We refer here to the Murray Canyons Group as a series of canyons that coalesce from the continental rise down to 5000 m (Fig. 1). The largest canyon, called Sprigg Canyon (Figs 2 and 3), has two arms (West and East Sprigg Canyons) with, on its western side, the du Couedic Canyon and with the narrow Murray Canyon located further west, and the Gantheaume Canyon, located east of Sprigg Canyon (Fig. 1). The AUSCAN project (cruise MD131, R.V.

From: Bishop, P. & Pillans, B. (eds) *Australian Landscapes.* Geological Society, London, Special Publications, **346**, 43–55. DOI: 10.1144/SP346.4 0305-8719/10/$15.00 © The Geological Society of London 2010.

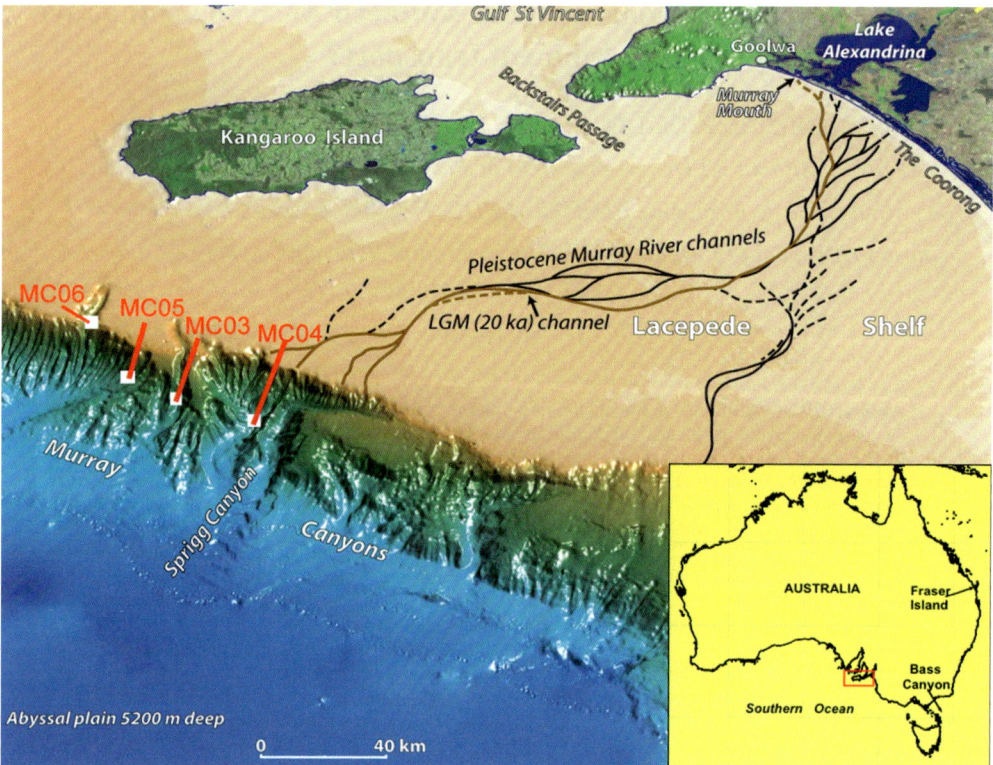

Fig. 1. Enlarged image of the Murray Canyons Group offshore Kangaroo Island in South Australia showing the location of the multicores studied here taken at various depths in the canyons area. The location of the Lacepede Shelf, and the various courses of the palaeo-Murray River for periods of low sea level (after Hill *et al.* 2009), should be noted. The image was generated by P. J. Hill. The inset shows the location of the Murray Canyons and the Lacepede Shelf in Australian waters.

Marion Dufresne, in February–March 2003) permitted the first investigation of these Australian canyons to assess if they are active today or have been in the recent past. We report here detailed depth profiles of the particle-reactive radionuclides ^{234}Th (half-life 24.1 days), ^{210}Pb (22.3 years), and ^{230}Th (75.3 ka) and of $CaCO_3$ and organic carbon (C_{org}) content in interface sediments collected at different depths within the canyons during the AUSCAN cruise (Hill & De Deckker 2004). These results are discussed to characterize the present sedimentation framework (bioturbation, sediment accumulation, focusing) of the Murray Canyons Group.

Physical setting

It is through the principal efforts of Francis Sheppard and colleagues in the USA (see Sheppard *et al.* 1979), starting in the 1950s, that oceanographers and marine geologists became interested in deep-sea canyons. It is only during the last two

decades that marine biologists and ecologists commenced investigating the biodiversity, productivity and uniqueness of deep-sea canyons (see Vetter 1994; Vetter & Dayton 1998). Deep-sea canyons have been extensively studied offshore California, parts of Canada and along the east coast of the USA (Eittreim *et al.* 1982; Mullenbach *et al.* 2004; Mullenbach & Nittrouer 2006). Other canyons have received attention offshore the European margin (Radakovitch & Heussner 1999; van Weering *et al.* 2002; Palanques *et al.* 2005), as well as New Zealand (Lewis & Barnes 1999; Orpin 2004; Orpin *et al.* 2006), Taiwan (Liu & Lin 2004) and Japan (Nakajima *et al.* 1998; Noda *et al.* 2008).

Surprisingly, since the pioneering work of the eminent South Australian geologist R. C. Sprigg in 1947 when the canyons offshore South Australia became known, and the subsequent investigations of J. Conolly and C. von der Borch up to 1968 (von der Borch 1968; von der Borch *et al.* 1970), little work has been done except for the recent investigations of Bass Canyon on the eastern edge

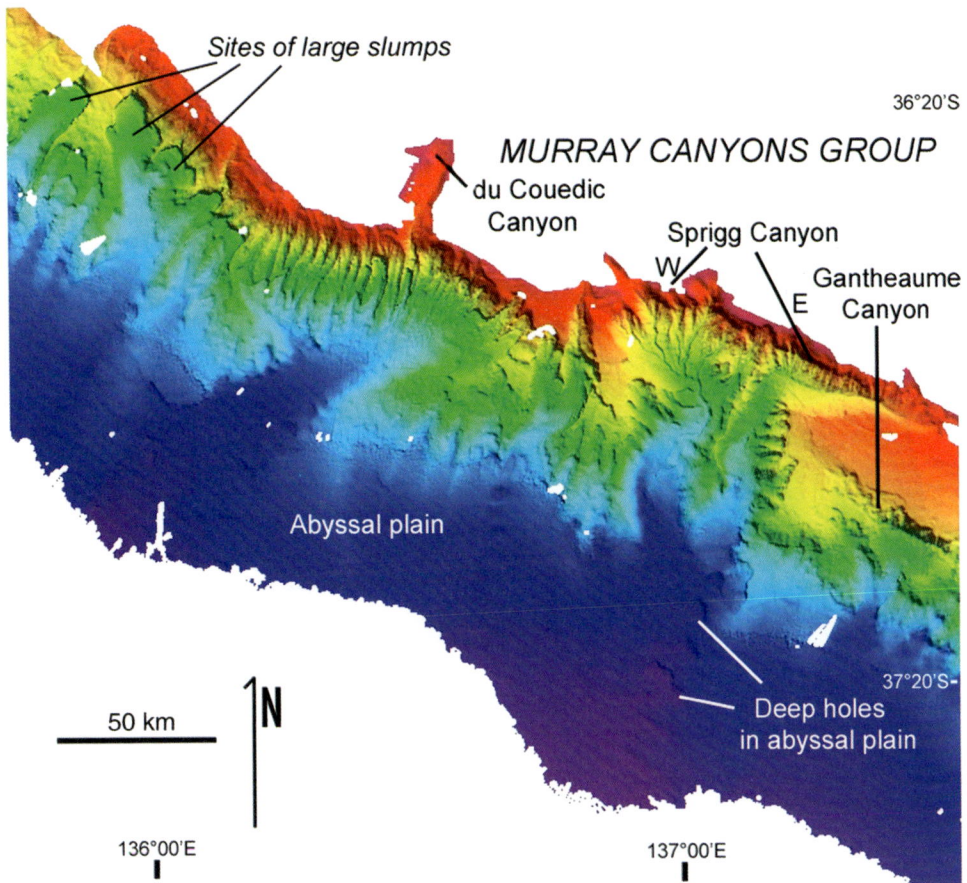

Fig. 2. Image generated from swath mapping data gathered during several cruises [for more details refer to Hill & De Deckker (2004) and updated with additional information gathered during the R.V. *Southern Surveyor* cruise SS02/06]. The white areas are those for which no information is currently available. A noteworthy feature is the presence of deep holes in the abyssal plain that appear to be lined up. Recent investigations by M. Mojtahid at the Australian National University have revealed obvious turbiditic sequences in a 2.71 m long gravity core (GC2), thus indicating that at least one hole can be considered to be a large 'plunge pool' formed by erosive turbidity flow down a canyon, as postulated by Hill *et al.* (2009). Their origin is still controversial. On the left side of the image can be seen the scars left by very large underwater slides, resulting in slope failure, which potentially could have generated large tsunamis. Such features are common along the continental slope of southern Australia (see also Exon *et al.* 2005) and have been mapped as far as the Victorian border during cruise SS02/06. They may be caused by water seepage along the shelf and be the trigger for the earlier formation of deep-sea canyons, prior to being incised further by large rivers, especially during periods of low sea level. Depths: red <200 m; yellow >200 m and <1000 m; green >1000 m and <4000 m; blue >4000 m.

of Bass Strait (Mitchell *et al.* 2007*a*, *b*). Hill *et al.* (2001) were the first to adequately document the nature and morphology of several of the deep canyons along the margin of southeastern Australia, including the Murray Canyons (see also the cruise report of Hill & De Deckker 2004). In addition, Exon *et al.* (2005) explored the Albany canyons offshore the SW of Western Australia. Subsequent studies on the sedimentological and geochemical aspects of two long cores taken in the vicinity of the

Murray Canyons Group were published by Gingele *et al.* (2001, 2004, 2007; Gingele & De Deckker 2005).

Material and methods

Analytical method

Our investigation deals with sediments at the interface between the sea floor and the overlying water.

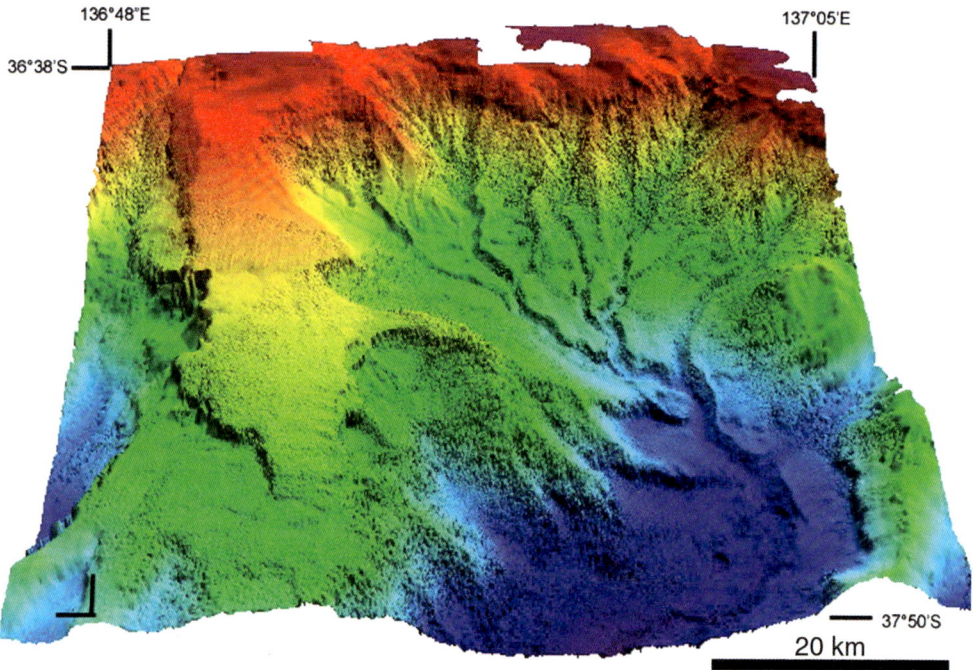

Fig. 3. Detailed view, tilted by *c.* 15°, of the western side of Sprigg Canyon showing the entrenched underwater 'fluvial' channels that display evidence of erosion and underwater transport, which therefore must still be effective today otherwise the channels would have filled up. The Sprigg Canyon is named after Reginald Sprigg, who was the first to document these underwater features (Sprigg 1947). Multicore MUC04 was taken on the 'peninsula' on the western side of the canyon. The colour scheme is similar to that used in Figure 2.

The sediments were obtained from the head of du Couedic Canyon and along the West Sprigg Canyon as well as the spur that separates the two arms of Sprigg Canyon (Figs 1–3; Table 1). Immediately after core retrieval from a multicorer, tubes were carefully extruded with sediment taken at 0.5 cm intervals from 0 to 5 cm, and 1 cm intervals below that depth.

In the laboratory, dry bulk density was measured by determining the weight after drying (60 °C) of a known volume of wet sediment. Following this procedure, ^{234}Th, ^{210}Pb and ^{226}Ra activities were measured using a low background, high-efficiency, well-shaped γ detector (Schmidt *et al.* 2001). Error on radionuclide activities are based on 1 SD counting statistics. Excess ^{234}Th and ^{210}Pb data were calculated by subtracting the activity supported by their parent isotope, ^{238}U and ^{226}Ra respectively, from the total activity in the sediment, and then by correcting ^{234}Th values for radioactive decay that occurred between sample collection and counting (this correction is not necessary for ^{210}Pb because of its longer half-life). Errors on ^{234}Th$_{xs}$ and ^{210}Pb$_{xs}$ are calculated by propagation of errors in the corresponding pair, ^{234}Th and ^{238}U, or ^{210}Pb and ^{226}Ra.

The determination of U and Th isotopes relies on a complete dissolution of the sediment by a mixture of $HF-HNO_3-HClO_4$. The radionuclides of interest were purified by ion exchanges on anionic resins (Schmidt 2006). A known amount of a calibrated ^{228}Th/^{232}U spike was added at the beginning of the digestion to determine chemical yield. ^{230}Th, ^{234}U and ^{238}U activities were determined by α counting as previously explained by Schmidt (2006).

The organic carbon (C_{org}) and carbonate ($CaCO_3$) content were determined on dry weight sediment by combustion in an LECO CS 125 analyzer (Etcheber *et al.* 1999). Samples were acidified in crucibles with 2N HCl to destroy carbonates, then dried at 60 °C to remove inorganic C and most of the remaining acid and water. The analyses were performed on bulk and decarbonated sediments by direct combustion in an induction furnace, and the CO_2 formed was determined quantitatively by IR absorption.

Data treatment

Sediment accumulation rates derived from ^{210}Pb. The ^{210}Pb method is based on the measurement of

Table 1. *Site location, sedimentation and mass accumulation rates derived from $^{210}Pb_{xs}$ profiles, $^{234}Th_{xs}$ inventories and bioturbation rates for the multicores from the Murray Canyons Group (AUSCAN cruise)*

Site	Latitude (S)	Longitude (E)	Depth (m)	$^{210}Pb_{xs}$		$^{234}Th_{xs}$		CaCO$_3$ (%)	C$_{org}$ (%)
				S (cm a^{-1})	MAR (mg cm^{-2} a^{-1})	I (dpm cm^{-2})	D_b (cm^{-2} a^{-1})		
Sprigg canyon, west side									
MD131-MC03	36°43.40′	136°47.32′	949	0.119	100	2.0	0.15	82	0.93
MD131-MC04	36°48.80′	136°48.93′	1619	0.025	33	0.7	<0.10	87	0.35
du Couedic canyon									
MD131-MC06	36°31.43′	136°25.96′	354	0.088	66	12.1	2.88	96	0.74
MD131-MC05	36°43.72′	136°32.87′	2476	0.030	23	1.5	0.48	87	1.12

C$_{org}$ and CaCO$_3$ contents (%) in surficial sediment (0–0.5 cm).

the excess or unsupported activity of ^{210}Pb ($^{210}Pb_{xs}$) which is incorporated rapidly into the sediment from atmospheric fallout and water column scavenging (Appleby & Oldfield 1992, and references therein). Once incorporated into the sediment, unsupported ^{210}Pb decays with depth, equivalent to time, in the sediment column according to its known half-life. Sediment accumulation rate can be derived from ^{210}Pb, based on two assumptions: constant flux and constant sediment accumulation rates (referred to as the CF:CS method) (Robbins & Edgington 1975). Then, the decrease of $^{210}Pb_{xs}$ activities with depth is described by the following relation:

$$[^{210}Pb_{xs}]_z = [^{210}Pb_0]_0 \exp\left(-z\frac{\lambda}{S}\right) \quad (1)$$

where $[^{210}Pb_{xs}]_{0,z}$ is the activitiy of excess ^{210}Pb at surface, or the base of the mixed layer, and depth z, λ is the decay constant of the nuclide, and S is the sediment accumulation rate. In this model, the compaction effect is not considered, and the sediment accumulation rates correspond to maximum values. An alternative method is to plot the regression of $^{210}Pb_{xs}$ against cumulative mass to calculate a mass accumulation rate (MAR), which integrates the compaction effect. Both estimates are given, and the first one is the more commonly used.

Bioturbation rates. Taking into account its very short half-life and the sedimentation rates (which are far less than 1 cm a^{-1}), $^{234}Th_{xs}$ should be present only at the water–sediment interface. Its penetration to variable depths indicates efficient mixing of the upper sediments, usually by bioturbation. The simplest way to derive bioturbation rates (D_b) from radionuclide profiles is to assume bioturbation as a diffusive process occurring at a constant rate within a surface mixed layer under steady state (Schmidt *et al.* 2001, 2002; Lecroart *et al.* 2007). The steady-state approximation is often used to

derive bioturbation rates from radionuclide profiles and introduces only limited errors (Lecroart *et al.* 2007). The latter simplification is supposed to respect the inequality $S^2 \ll 4D_b\lambda$ (Wheatcroft 2006). Sedimentation accumulation rates obtained in this study are always <0.2 cm a^{-1} (Table 1); therefore one can assume that sediment accumulation rates are not likely to affect $^{234}Th_{xs}$ profiles. These simplifications allow the determination of bioturbation rates from a simple plot of radionuclide activity as a function of depth, using the equation

$$[^{234}Th_{xs}]_z = [^{234}Th_0]_0 \exp\left(-z\sqrt{\frac{\lambda}{D_b}}\right) \quad (2)$$

where $[^{234}Th_{xs}]_{0,z}$ is the activity (dpm g^{-1}) of excess ^{234}Th at the water–sediment interface (Schmidt *et al.* 2001). We present ^{234}Th-derived bioturbation rates and inventories as an indication of particle input over the last few months. Such bioturbation rates must be considered as an instantaneous signal (Aller & Demaster 1984; Schmidt *et al.* 2002).

Sediment focusing

Sediment focusing is a process whereby water turbulence or other processes transfer sediment from shallow to deeper zones in the ocean. The focusing factor ϕ, based on ^{230}Th, has been proposed to distinguish between the contribution from vertical fluxes originating from the overlying waters, and lateral fluxes resulting from sediment redistribution by bottom currents (François *et al.* 2004, and references therein) using the relationship

$$\phi = \frac{F_s}{F_p} \quad (3)$$

where F_p, the flux of scavenged ^{230}Th reaching the sea floor with particles settling through the water column, is close to the rate of ^{230}Th production from the decay of ^{234}U over the depth (z in metres) of the overlying water column: $F_p = 0.0267 \times z$ dpm m^{-2} a^{-1}.

F_s is the accumulation rate of ^{230}Th$_{xs}$ (dpm m^{-2} a^{-1}) in sediment: $F_s = [^{230}$Th$_{xs}]_0 \times S \times$ DBD, where $[^{230}$Th$_{xs}]_0$ is the concentration (decay-corrected) of ^{230}Th$_{xs}$, S is the sedimentation rate (cm a^{-1}), and DBD is the dry bulk density (g cm^{-3}).

^{230}Th$_{xs}$ is calculated by subtracting from the measured total ^{230}Th the detrital and authigenic contributions (for details, see Veeh *et al.* 2000) to calculate the amount of ^{230}Th originating from scavenging in the water column, referred to as 'in excess' (in contrast to the supported detrital and authigenic fractions of ^{230}Th in sediment). Independent chronology was obtained for interface sediments using sedimentation rates derived from ^{210}Pb. Detrital uranium was derived from the measured total U and Th concentrations by assuming that all Th resides in detrital phases. Authigenic uranium was calculated as the difference of total and detrital uranium. Despite the occurrence of a significant fraction of authigenic U in our cores, the corrections for *in situ* growth of ^{230}Th are minor, with respect to the half-life of ^{230}Th, on such young sediments at the sediment–water interface.

Results

In the Murray Canyons Group, surface excess ^{210}Pb activities range from 14 to 27 dpm g^{-1} (Fig. 4). Profiles of ^{210}Pb$_{xs}$ present an upper mixed layer, more pronounced in the du Couedic Canyon, followed by an exponential decrease, with ^{210}Pb activities reaching supported activity levels at about 5 cm depth. The absence of disturbance in ^{210}Pb$_{xs}$ profiles is appropriate to the determination of sediment and mass accumulation rates (MAR) (Table 1). The MAR are high in the West Sprigg Canyon, with the highest value (100 mg cm^{-2} a^{-1}) registered at the site MC03 located at 949 m water depth (Fig. 4). In contrast, the shallowest site, MC06 (354 m) at the head of the du Couedic Canyon, registers a lower value (66 mg cm^{-2} a^{-1}). The deepest sites in the Murray Canyons Group have MAR ranging from 33 mg cm^{-2} a^{-1} (MC04; depth 1619 m) to 23 mg cm^{-2} a^{-1} (MC05; depth 2476 m). Sediment accumulation rates and MAR exhibit decreasing values with depth, as usually observed in margin and canyon sediments (Sanchez-Cabeza *et al.* 1999; Matthai *et al.* 2001; van Weering *et al.* 2002).

Excess ^{234}Th was always detected at substantial levels (4 and 22 dpm g^{-1}) in surface samples, in the range of values reported for continental margins elsewhere (Aller & Demaster 1984; Schmidt *et al.*

2002). Such an occurrence of ^{234}Th, with its short half-life (24.1 days), shows the presence of freshly deposited particles. The changes in activities are associated, to a lesser extent, with variations in penetration depth. The deepest penetration of ^{234}Th$_{xs}$ does not exceed the mixed layer of ^{210}Pb, this latter registering a mixing event on a longer time scale. The sea-bed inventory of excess ^{234}Th can be used as an alternative tool to sediment traps for the investigation of particle dynamics at the water–sediment interface on the 100 day time scale. Sea-bed ^{234}Th$_{xs}$ inventories range from 0.7 to 12.1 dpm cm^{-2} (Table 1). Sediments in the upper canyons are rather enriched in C_{org} compared with usual margin sediments (Etcheber *et al.* 1999; Fig. 4).

U and Th contents range between 1.5 and 2.6 ppm and 0.7 and 1.6 ppm respectively (Table 2), in the lower range of values for margin sediments (McManus *et al.* 2006; Yamada & Aono 2006; Yamada *et al.* 2006). The low Th concentration indicates a low fraction of detrital sediment, thus resulting in low calculated detrital uranium. This is corroborated by the high carbonate content (>80%; Fig. 4). Focusing factors, derived from ^{230}Th, are high, from 84 to 6, indicating a net lateral input of resuspended or advected sediment components containing excess ^{230}Th in addition to the vertical flux of particulate ^{230}Th out of the water column.

Discussion

Present-day sedimentation within the Murray Canyons Group

Bioturbation rates derived from ^{234}Th profiles (0.1–2.9 cm^2 a^{-1}) decrease with depth in the Murray Canyons Group. Mixing rates are in the range of reported values for margin environments (Soetaert *et al.* 1996), with the highest rates observed in the heads of the canyons. These values are higher than those reported for the Nazaré Canyon (Schmidt *et al.* 2001), an example of a canyon deeply incised into the continental shelf (the canyon head opens 500 m from Nazaré beach). In the Murray Canyons Group, we note significant organic inputs at the sediment–water interface, and this concurs with the observation of abundant faecal pellets on the sea floor and of significant C_{org} content (0.9–1.2%). Sediment accumulation rates based on ^{210}Pb (0.03–0.13 cm a^{-1}) exhibit the same decreasing trend with depth.

In the upper head of du Couedic Canyon (west side of the Murray Canyons Group), core MC06 displays evidence of bioturbation, which explains why ^{234}Th values are found down to 3 cm in depth

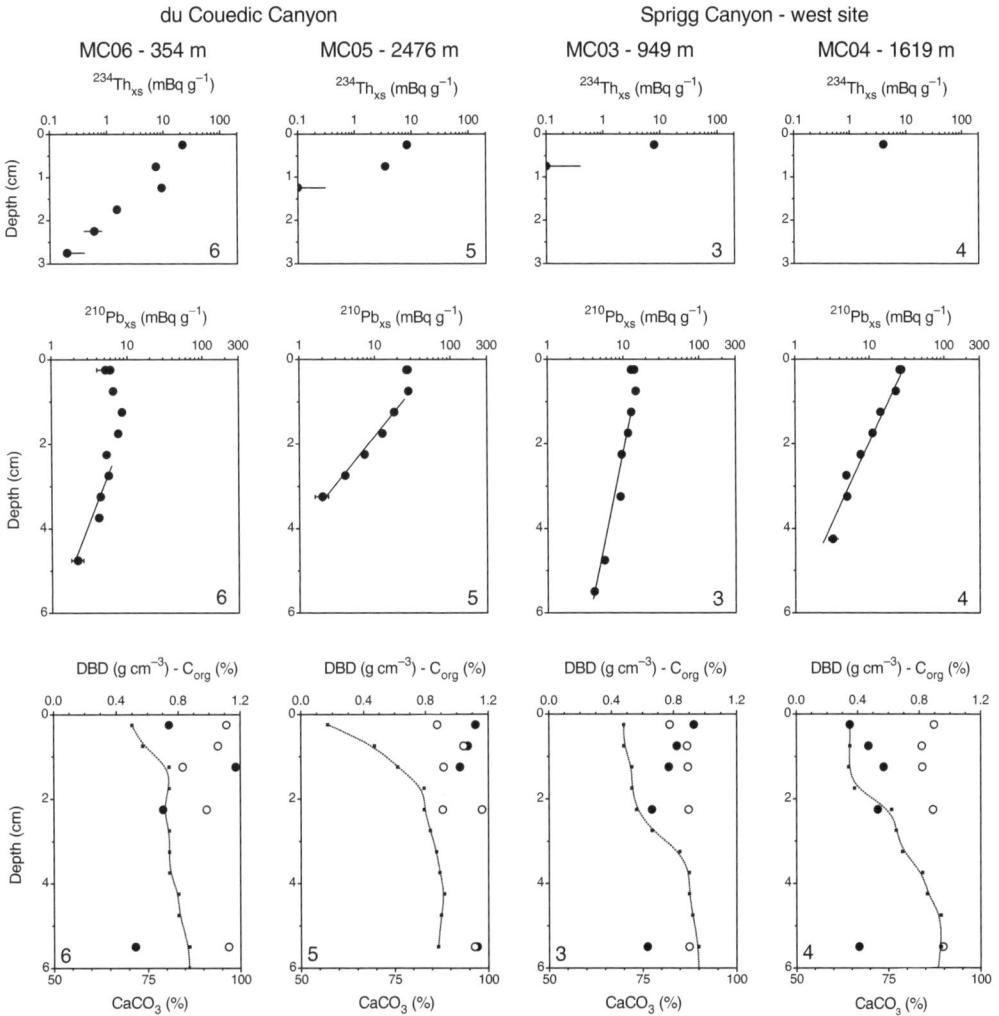

Fig. 4. $^{234}Th_{xs}$ and $^{210}Pb_{xs}$ profiles carried out on the four multicores from the two sides of the Murray Canyons Group. Dry bulk density (■, dotted line), C_{org} (●) and $CaCO_3$ (○) contents are also plotted with depth in the bottom row of panels. Error bars on radionuclides profiles correspond to 1 SD. For cores 6, 5 and 3, it is noticeable that the deepest penetration of the short-lived radionuclide corresponds to the base of the mixed layer of $^{210}Pb_{xs}$ profiles.

(Fig. 4). $^{234}Th_{xs}$-derived bioturbation rates are similar to those observed in other continental margins (Table 1; DeMaster *et al.* 1994; Schmidt *et al.* 2002). These highest signals are associated with highest $^{234}Th_{xs}$ inventories and the deepest mixed layer observed on $^{210}Pb_{xs}$ profiles. On the other hand, $^{234}Th_{xs}$-derived bioturbation rates in the west Sprigg Canyon are rather low, in agreement with a negligible mixed layer for $^{210}Pb_{xs}$ profiles. In particular, in the middle reaches of the Sprigg Canyon (MC03), there is the highest sediment deposition rate reported in this study, but the $^{234}Th_{xs}$-derived bioturbation rate and inventory are low (Table 1). Although we cannot exclude a possible loss of the uppermost layer, we interpret this trend based on the difference in particle supply to these two canyons. The decoupling between ^{234}Th and ^{210}Pb in the upper sediment can be explained by the nature and the intensity of sediment supply (Schmidt *et al.* 2001). The du Couedic Canyon appears to be a preferential conduit for fresh and organic material, although the Sprigg Canyon is active for particle transfer to the deep ocean. Dredging of the sea floor during AUSCAN-1 cruise supports this hypothesis, indicating the occurrence of abundant organisms and of faecal pellets in the Sprigg Canyon (Hill & De Deckker 2004).

Table 2. *U and Th concentrations (ppm), detrital uranium fraction ($^{238}U_d$, % of total U), ^{230}Th activities and focusing factors, ϕ, for the multicores from the Murray Canyons Group (AUSCAN cruise)*

Label	Layer (cm)	^{238}U (ppm)	$^{238}U_d$ (%)	^{232}Th (ppm)	^{230}Th (dpm g^{-1})	ϕ
Sprigg Canyon, west side						
MD131-MC03	1.25	1.55 ± 0.06	9	0.77 ± 0.08	0.72 ± 0.04	30
	5.50	1.88 ± 0.09	9	0.90 ± 0.08	0.80 ± 0.03	33
	15	2.31 ± 0.09	5	0.63 ± 0.06	0.89 ± 0.03	40
MD131-MC04	1.25	1.55 ± 0.06	11	0.97 ± 0.08	1.26 ± 0.05	6
	5.5	1.98 ± 0.08	8	0.86 ± 0.07	1.15 ± 0.04	6
du Couedic Canyon						
MD131-MC06	1.25	1.96 ± 0.05	7	0.75 ± 0.08	0.98 ± 0.04	67
	5.5	2.61 ± 0.09	5	0.70 ± 0.06	1.21 ± 0.04	84
MD131-MC05	0.75	1.69 ± 0.07	12	1.15 ± 0.08	1.58 ± 0.05	6
	5.5	2.44 ± 0.10	6	0.87 ± 0.09	1.46 ± 0.06	5

^{230}Th activities (dpm g^{-1}, number of disintegrations per minute per gram of dry sediment). ϕ, the ratio of the lateral fluxes resulting from sediment redistribution by bottom currents to the vertical particulate fluxes originating from the overlying waters.

In the Murray Canyons Group, Th and detrital U concentrations are low, indicating a limited contribution of detrital phase (Table 2). These values are very low compared with those obtained by Dosseto *et al.* (2006) for colloids and suspended sediments in the River Murray system, which have values that are at least one order of magnitude higher compared with the canyons' sediment. The findings by Dosseto *et al.* (2006) therefore indicate that contributions from the River Murray to the canyons are minimal. This result is consistent with the observation of high carbonate contents in these sediments and in long cores recovered at the same sites (Gingele *et al.* 2004); most of the sediments consist of carbonates, with a little terrigenous detritic component.

We can observe, as for the short-lived radionuclides, a decoupling of long-lived radionuclide data (Table 2) from the cores MC03 and MC06. MC03 exhibits the highest U content and increase in authigenic fraction. It has been previously suggested that the U accumulation rate is sensitive to organic carbon delivery to the sea-bed (Anderson *et al.* 1998) and therefore may indicate an enhanced input of fresh material at this site (MC03), which is consistent with high bioturbation rates and $^{234}Th_{xs}$ inventories, and C_{org} content. Focusing factors, derived from $^{230}Th_{xs}$, are between 80 and 6 (Table 2). As for the other parameters, the highest values are observed in the head of the du Couedic Canyon (MC06), thus supporting the occurrence of a significant horizontal advection of sediments within the canyon. Both tracers confirm that the upper part of Murray Canyons Group acts as an active locus for sedimentation.

Specificity of the Murray Canyons Group

In comparison with most canyons, which usually indent the coastline (Mulder *et al.* 2001; Schmidt *et al.* 2001; Palanques *et al.* 2005; Mullenbach & Nittrouer 2006), the Murray Canyons Group lies far away from the coastline and may present different dynamics of sediment particle supply. Today, fluvial inputs are likely to be negligible because of the high sea level, as well as the dams built during the last century across many locations along the River Murray (Gingele *et al.* 2004, 2007). The Murray Canyons Group multicore samples, with their extremely high carbonate contents, indicate that there is little supply of River Murray sediments.

The Lacepede Shelf is very wide, reaching 200 km width in places (Fig. 2). Along the eastern part of the Lacepede Shelf, seasonal upwellings have been recorded (Lewis 1981; Schalinger 1987), thus adding to the potential biological productivity on the shelf. This phenomenon potentially causes particulate matter to become recycled in this shallow area, and it may eventually become transported down canyons via their shallow conduits. The high percentage of carbonate sediments in the Murray Canyons multicores confirms the findings of James *et al.* (1992) that a large amount of biogenic carbonate is produced on the Lacepede Shelf.

Figure 3 clearly shows meandering channels that would have been formed by gravity flows of sediment-laden currents. The material would have obviously originated from the Lacepede Shelf as mentioned above, and the reader should refer to the description by James *et al.* (1992) of the nature of the biogenic sediment for more information. The fact that the channels are still clearly incised and empty indicates that either they must still be active or they were so in the recent past as they are not filled with sediments. Recent investigations of gravity core GC2 taken during the R.V. *Southern Surveyor* cruise SS02/06 (37°07.98'S, 136°29.753'E, 4978 m depth; 271 cm long), obtained from one of the deep holes in the abyssal plain, showed evidence of several superimposed turbiditic sequences. The unpublished work of M. Mojtahid at the Australian National University on core GC2 found that the sediment is mostly sandy and consists of calcareous biogenic remains, predominantly composed of benthic foraminifers and bryozoans, some of which would have lived in shallow water such as on the Lacepede Shelf. Of interest is the presence in some samples of fragile pteropod skeletons composed of aragonite, thus indicating a rapid burial and a shallow origin (<1000 m), as otherwise they would have been dissolved as the core location is well below the aragonite compensation depth.

The concept that underwater sediment drifts can occur today, even in the absence of onshore rivers, was elegantly illustrated by Boyd *et al.* (2008) who showed that deep-water sands that originate near the shelf edge can be delivered to great oceanic depths. In that case, Boyd *et al.* (2008) demonstrated that longshore transport occurs offshore Fraser

Island along the east coast of Australia. Those workers documented obvious channels that converge towards submarine canyons. Similar channels had already been illustrated by Hill *et al.* (1998) at the head of the large Bass Canyon located on the eastern side of Bass Strait between the Australian mainland and Tasmania.

The presence of underwater slides as shown on the west of the de Couedic Canyon (Fig. 1) illustrates the presence of underwater slope failure. Jenkins & Keene (1992) found evidence of structures associated with large-scale slope failure along the southeastern Australia continental slope. Similar features were seen during R.V. *Southern Surveyor* cruise SS02/06 offshore the coast of Victoria, west of Portland. In addition, Exon *et al.* (2005) provided spectacular images of widespread and large slides on the southern tip of Western Australia, pointing to the commonness of such features. We argue here that perhaps these slides originate as a result of water seepage, of continental origin, that

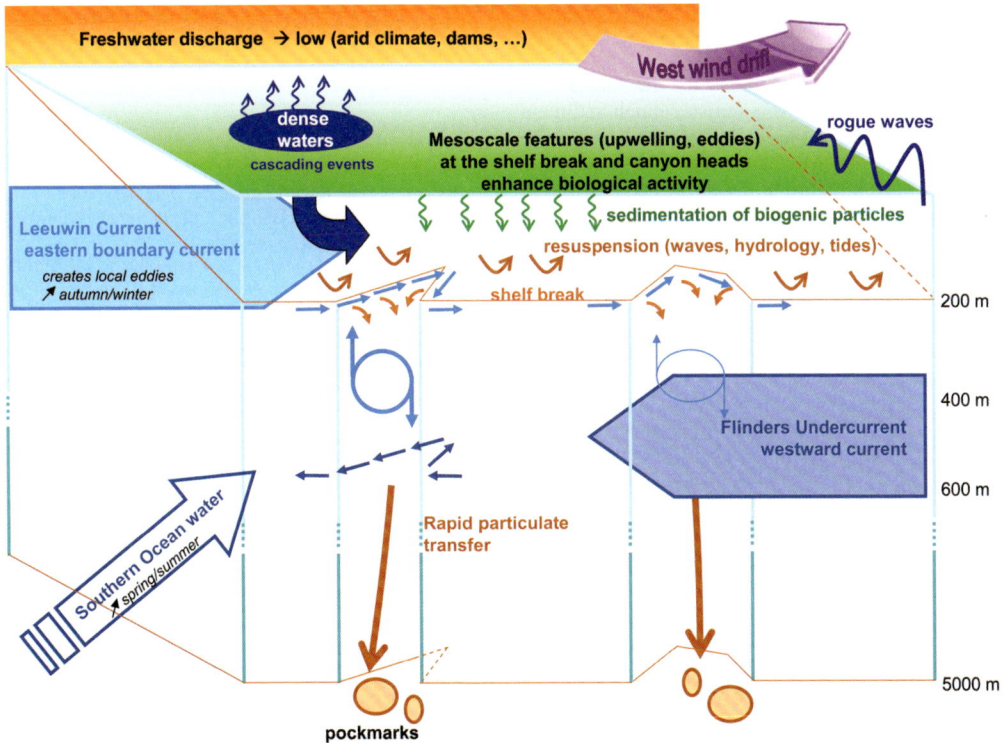

Fig. 5. Simplified model of sediment focusing and trapping mechanisms of the blind canyons off the southern margin of Australia. This model combines the various processes that are likely to contribute positively to particulate transfer towards the canyons (water fluxes in blue; particle production and transfer in green and brown). The main processes identified from the literature (Li *et al.* 1999; Ogston *et al.* 2008; Puig *et al.* 2008; Rennie *et al.* 2009a, b) principally involve hydrology, in particular the influence of the Leeuwin Current and the Flinders Undercurrent, and the occurrence of dense water formed on the Lacepede Shelf that cascades down the slope [for more information, refer to Lennon *et al.* (1987)].

'crops out' on the continental shelf and upper slope, engendering slope instability. The possible groundwater transfer along ancient courses of large rivers that are found on the continental shelf (such as documented by Hill *et al.* (2009) for palaeo-channels of the River Murray, and that would have been active during periods of low sea levels) and that contain coarse (and porous) fluvial sediments, could be the cause of underwater 'erosion' and gullying formation. This would eventually trigger the commencement of canyon formation. In addition, it is possible that such large underwater slides, as mentioned above, may potentially be able to generate tsunamis.

If the hypothesis of groundwater seepage and cropping out on the shelf and upper slope is accepted, this would explain the presence of canyons or channels that are not necessarily linked to rivers on land.

Conclusions

The Murray Canyons Group acts as an 'amplifier' of sediment supply through the conduits down to the deep ocean despite the fact that the River Murray today sheds little sediment to the ocean. In general, deep-sea canyons register much sediment deposition because they are directly linked with large rivers. Our Australian data on the Murray Canyons Group highlight, therefore, how different they are compared with many other deep-sea canyons. One plausible explanation is that Australia, as a continent, is not affected by rapid and substantial erosion, as a result of its low topography and reduced tectonic activity. The other important factor is the extensive breadth of the Lacepede Shelf.

Based on this preliminary investigation of the canyons in the MCG and by synthesizing previous work on equivalent features in deep-sea canyons (Li *et al.* 1999; Ogston *et al.* 2008; Puig *et al.* 2008; Rennie *et al.* 2009a, b), we provide a conceptual model for sediment focusing and transfer in the canyons along the southern margin of Australia (Fig. 5). Canyons do not necessarily need to be linked to specific fluvial sources to be active (Mullenbach *et al.* 2004). Our investigations of canyons offshore South Australia point to the importance of the interaction between currents and wind and the overall canyon bathymetry and configuration. Regarding circulation, the canyon topography influences the circulation dynamics of the Leeuwin Current and, subsequently, eddy development and vertical transport, which in turn affect upwelling and productivity (Li *et al.* 1999; Rennie *et al.* 2009a, b). As reported for the submarine canyons of the Gulf of Lions in the Mediterranean Sea, storm-induced downwelling and dense water cascading events efficiently transport sediment down-canyon (Ogston *et al.* 2008; Puig *et al.* 2008). The combination of these processes, acting at various degrees through different seasons, is assumed to act in favour of an efficient sediment transfer through canyons along the southern margin of Australia down to the deep ocean.

This work was supported by the CNRS, the Institut Polaire Français Paul Emile Victor and the Centre d'Energie Atomique. The Australian National Oceans Office funded part of this project, and an ARC grant awarded to P. D. D. helped facilitate the study of the canyons and cores. Y. Balut is particularly thanked for all his help in obtaining the cores, and A. Rathburn helped with the retrieval of the multiple cores. We are also grateful to M. Spinoccia of Geoscience Australia, who generated the two swath maps while on the R. V. *Southern Surveyor* cruise SS02/06. P. J. Hill and J. Rogers helped collect the swath data during that cruise. We wish to thank the two reviewers A. Heap and B. Thom, for their pertinent comments that helped improve this paper, and J. Shelley for proofreading. Our thanks also go to editor P. Bishop for his final editorial comments that helped clarify our text.

The authors thank the Australian & New Zealand Geomorphology Group for contributing to the colour plates production costs.

References

ALLER, R. C. & DeMASTER, D. J. 1984. Estimates of particle flux and reworking at the deep-sea floor using ^{234}Th/^{238}U disequilibrium. *Earth and Planetary Science Letters*, **67**, 308–318.

ANDERSON, R. F., KUMAR, N., MORTLOCK, R. A., FROELICH, P. N., KUBIK, P., DITTRICH-HANNEN, B. & SUTTER, M. 1998. Late-Quaternary changes in productivity of the Southern Ocean. *Journal of Marine Systems*, **17**, 497–514.

APPLEBY, P. G. & OLDFIELD, F. 1992. Application of lead-210 to sedimentation studies. *In*: IVANOVICH, M. & HARMON, R. S. (eds) *Uranium-Series Disequilibrium: Applications to Earth, Marine, and Environmental Sciences*. Clarendon Press, Oxford, 731–778.

BISCAYE, P. E. & OLSEN, C. R. 1976. Suspended particulate concentrations and compositions in the New York Bight. *In*: GROSS, M. G. (ed.) *Middle Atlantic Continental Shelf and the New York Bight*. American Society of Limnology and Oceanography, Special Symposium 2, 124–137.

BOYD, R., RUMING, K., GOODWIN, I., SANDSROM, M. & SCHRÖDER-ADAMS, C. 2008. Highstand transport of coastal sand to the deep ocean: A case study from Fraser Island. *Geology*, **36**, 15–18.

CARSON, B., BAKER, E. T., HICKEY, B. M., NITTROUER, C. A., DeMASTER, D. J., THORBJARNARSON, K. W. & SNYDER, G. W. 1986. Modern sediment dispersal and accumulation in Quinault submarine canyon—a summary. *Marine Geology*, **71**, 1–13.

DeMASTER, D. J., POPE, R. H., LEVIN, L. A. & BLAIR, N. E. 1994. Biological mixing intensity and rates of organic carbon accumulation in North Carolina slope sediments. *Deep-Sea Research II*, **41**, 735–753.

DOSSETO, A., TURNER, S. P. & DOUGLAS, G. B. 2006. Uranium-series isotopes in colloids and suspended sediments: Timescale for sediment production and transport in the Murray–Darling River system. *Earth and Planetary Science Letters*, **246**, 418–431.

EITTREIM, S., GRANTZ, A. & GREENBERG, J. 1982. Active geologic processes in Barrow Canyon, northeast Chukchi Sea. *Marine Geology*, **50**, 61–76.

ETCHEBER, H., RELEXANS, J.-C., BELIARD, M., WEBER, O., BUSCAIL, R. & HEUSSNER, S. 1999. Distribution and quality of sedimentary organic matter on the Aquitanian margin (Bay of Biscay). *Deep-Sea Research II*, **46**, 2249–2288.

EXON, N. F., HILL, P. J., MITCHELL, C. & POST, A. 2005. Nature and origin of the submarine Albany canyons offshore southwest Australia. *Australian Journal of Earth Sciences*, **52**, 101–115.

FRANÇOIS, R., FRANK, M., RUTGERS, M. M., VAN DER LOEFF, M. R. & BACON, M. P. 2004. ^{230}Th normalization: An essential tool for interpreting sedimentary fluxes during the late Quaternary. *Paleoceanography*, **19**, PA1018, doi: 10.1029/PA000939.

GARDNER, W. D. 1989. Baltimore Canyon as a modern conduit of sediment to the deep sea. *Deep-Sea Research*, **36**, 323–358.

GINGELE, F. X. & DE DECKKER, P. 2005. Late Quaternary fluctuations of palaeoproductivity in the Murray Canyons area, South Australian continental margin. *Palaeogeography, Palaeoclimatology, Palaeoecology*, **220**, 361–373.

GINGELE, F. X., DE DECKKER, P. & HILLENBRAND, C.-D. 2001. Clay mineral distribution in surface sediments between Indonesia and NW Australia—source and transport by ocean currents. *Marine Geology*, **179**, 135–146.

GINGELE, F. X., DE DECKKER, P. & HILLENBRAND, C.-D. 2004. Late Quaternary terrigenous sediments from the Murray Canyons area, offshore South Australia and their implications for sea level change, palaeoclimate and palaeodrainage of the Murray–Darling Basin. *Marine Geology*, **212**, 183–197.

GINGELE, F. X., DE DECKKER, P. & NORMAN, M. 2007. Late Pleistocene and Holocene climate of SE Australia reconstructed from dust and river loads deposited offshore the River Murray Mouth. *Earth and Planetary Science Letters*, **255**, 257–272.

GRANATA, T. C., VIDONDO, B., DUARTE, C. M., SATTA, M. P. & GARCIA, M. 1999. Hydrodynamics and particle transport associated with a submarine canyon off Blanes (Spain), NW Mediterranean Sea. *Continental Shelf Research*, **19**, 1249–1263.

HICKEY, B., BAKER, E. & KACHEL, N. 1986. Suspended particle movement in and around Quinault submarine canyon. *Marine Geology*, **71**, 35–83.

HILL, P. J. & DE DECKKER, P. 2004. *AUSCAN Seafloor mapping and geological sampling survey on the Australian southern margin by RV* Marion Dufresne *in 2003*. Geoscience Australia Record, **2004/04**.

HILL, P. J., EXON, N. F., KEENE, J. B. & SMITH, S. M. 1998. The continental margin off east Tasmania and Gippsland: structure and development using new multibeam sonar data. *Exploration Geophysics*, **29**, 410–419.

HILL, P. J., ROLLET, N. & SYMONDS, P. 2001. *Seafloor mapping of the South-east Marine Region and adjacent waters—AUSTREA final report: Lord Howe Island, south-east Australian margin (includes Tasmania and South Tasman Rise) and central Great Australian Bight*. Australian Geological Survey Organisation Record, **8**.

HILL, P. J., DE DECKKER, P. & EXON, F. 2005. Geomorphology and evolution of the gigantic Murray canyons on the Australian southern margin. *Australian Journal of Earth Sciences*, **52**, 117–136.

HILL, P. J., DE DECKKER, P., VON DER BORCH, C. C. & MURRAY-WALLACE, C. V. 2009. Ancestral Murray River on the Lacepede Shelf, southern Australia: Late Quaternary migrations of a major river outlet and strandline development. *Australian Journal of Earth Sciences*, **56**, 135–157.

JAMES, N. P., BONE, Y., VON DER BORCH, C. C. & GOSTIN, V. A. 1992. Modern carbonate and terrigenous clastic sediments on a cool-water, high-energy, mid-latitude shelf: Lacepede, southern Australia. *Sedimentology*, **39**, 877–903.

JENKINS, C. J. & KEENE, J. B. 1992. Submarine slope failures of the southeast Australian continental slope: a thinly sedimented margin. *Deep-Sea Research A*, **39**, 121–136.

KHRIPOUNOFF, A., VANGRIESHEIM, N., BABONNEAU, PH., CRASSOUS, B., DENNIELOU, B. & SAVOYE, B. 2003. Direct observation of intense turbidity current activity in the Zaire submarine valley at 4000 m water depth. *Marine Geology*, **194**, 151–158.

LECROART, P., SCHMIDT, S. & JOUANNEAU, J.-M. 2007. A numerical estimation of the error on the bioturbation coefficient in coastal environments by short-lived radioisotopes modelling. *Estuarine and Coastal Shelf Science*, **72**, 543–555.

LENNON, G. W., BOWERS, D. G. ET AL. 1987. Gravity currents and the release of salt from an inverse estuary. *Nature*, **327**, 695–697.

LEWIS, R. K. 1981. Seasonal upwelling along the southeastern coastline of South Australia. *Australian Journal of Marine and Freshwater Research*, **32**, 843–854.

LEWIS, K. B. & BARNES, P. M. 1999. Kaikoura Canyon, New Zealand: active conduit from nearshore sediment zones to trench-axis channel. *Marine Geology*, **162**, 39–69.

LI, Q., JAMES, N. P., BONE, Y. & MCGOWRAN, B. 1999. Palaeoceanographic significance of recent foraminiferal biofacies on the southern shelf of Western Australia: a preliminary study. *Palaeogeography, Palaeoclimatology, Palaeoecology*, **147**, 101–120.

LIU, J. T. & LIN, H.-L. 2004. Sediment dynamics in a submarine canyon: a case of river–sea interaction. *Marine Geology*, **207**, 55–81.

MATTHAI, C., BIRCH, G. F., JENKINSON, A. & HEIJNIS, H. 2001. Physical resuspension and vertical mixing of sediments on a high energy continental margin (Sydney, Australia). *Journal of Environmental Radioactivity*, **52**, 67–89.

MCMANUS, J., BERELSON, W. M., SEVERMANN, S., POULSON, R. L., HAMMOND, D. E., KLINKHAMMER, G. P. & HOLM, C. 2006. Molybdenum and uranium geochemistry in continental margin sediments: Paleoproxy potential. *Geochimica et Cosmochimica Acta*, **70**, 4643–4662.

MITCHELL, J. K., HOLDGATE, G. R., WALLACE, M. W. &
    GALLAGHER, S. J. 2007*a*. Marine geology of the
    Quaternary Bass Canyon system, southeast Australia:
    a cool-water carbonate system. *Marine Geology*, **237**,
    71–96.

MITCHELL, J. K., HOLDGATE, G. R. & WALLACE, M. W.
    2007*b*. Pliocene–Pleistocene history of the Gippsland
    Basin outer shelf and canyon heads, southeast
    Australia. *Australian Journal of Earth Sciences*, **54**,
    49–64.

MULDER, T., WEBER, O., ANSCHUTZ, P., JORISSEN, F. J. &
    JOUANNEAU, J.-M. 2001. A few months old storm-
    generated turbidite deposited in the Capbreton
    Canyon (Bay of Biscay, SW France). *Geo-Marine
    Letters*, **21**, 149–156.

MULLENBACH, B. L. & NITTROUER, C. A. 2006. Decadal
    record of sediment export to the deep sea via Eel
    Canyon. *Continental Shelf Research*, **26**, 2157–2177.

MULLENBACH, B. L., NITTROUER, C. A., PUIG, P. &
    ORANGE, D. L. 2004. Sediment deposition in a
    modern submarine canyon: Eel Canyon, northern
    California. *Marine Geology*, **211**, 101–119.

NAKAJIMA, T., SATOH, M. & OKAMURA, Y. 1998. Channel-
    levee complexes, terminal deep-sea fan and sediment
    wave fields associated with the Toyama Deep-Sea
    Channel system in the Japan Sea. *Marine Geology*,
    147, 25–41.

NODA, A., TUZINO, T., FURUKAWA, R., JOSHIMA, M. &
    UCHIDA, J. I. 2008. Physiographical and sedimentolo-
    gical characteristics of submarine canyons developed
    upon an active forearc slope: The Kushiro Submarine
    Canyon, northern Japan. *Geological Society of
    America Bulletin*, **120**, 750–767.

OGSTON, A. S., DREXLER, T. M. & PUIG, P. 2008. Sedi-
    ment delivery, resuspension, and transport in two
    contrasting canyon environments in the southwest
    Gulf of Lions. *Continental Shelf Research*, **28**,
    2000–2016.

ORPIN, A. R. 2004. Holocene sediment deposition on the
    Poverty-slope margin by the muddy Waipara River,
    East Coast New Zealand. *Marine Geology*, **209**,
    69–90.

ORPIN, A. R., ALEXANDER, C., CARTER, L., KUEHL, S. &
    WALSH, J. P. 2006. Temporal and spatial complexity in
    post-glacial sedimentation on the tectonically active,
    Poverty Bay continental margin of New Zealand.
    Continental Shelf Research, **26**, 2205–2224.

PALANQUES, A., EL KHATAB, M., PUIG, P., MASQUE, P.,
    SANCHEZ-CABEZA, J. A. & ISLA, E. 2005. Downward
    particle fluxes in the Guadiaro submarine canyon
    depositional system (northwestern Alboran Sea), a
    river flood dominated system. *Marine Geology*, **220**,
    23–40.

PUIG, P., OGSTON, A. S., MULLENBACH, B. L., NITTROUER,
    C. A. & STERNBERG, R. W. 2003. Shelf-to-canyon
    sediment-transport processes on the Eel continental
    margin (northern California). *Marine Geology*, **193**,
    129–149.

PUIG, P., PALANQUES, A., ORANGE, D. L., LASTRAS, G. &
    CANALS, M. 2008. Dense shelf water cascades and
    sedimentary furrow formation in the Cap de Creus
    Canyon, northwestern Mediterranean Sea. *Continental
    Shelf Research*, **28**, 2017–2030.

RADAKOVITCH, O. & HEUSSNER, S. 1999. Fluxes and
    budget of ^{210}Pb on the continental margin of the Bay
    of Biscay (northeastern Atlantic). *Deep-Sea Research
    II*, **46**, 2175–2203.

RENNIE, S., HANSON, C. E. *ET AL.* 2009*a*. Physical proper-
    ties and processes in the Perth Canyon, Western
    Australia: Links to water column production and seaso-
    nal pygmy blue whale abundance. *Journal of Marine
    Systems*, **77**, 21–44.

RENNIE, S. J., PATTIARATCHI, C. B. & MCCAULEY, R. D.
    2009*b*. Numerical simulation of the circulation within
    the Perth Submarine Canyon, Western Australia.
    Continental Shelf Research, **29**, 2020–2036.

ROBBINS, J. & EDGINGTON, D. N. 1975. Determination of
    recent sedimentation rates in Lake Michigan using
    ^{210}Pb and ^{137}Cs. *Geochimica et Cosmochimica Acta*,
    39, 285–304.

RUHL, H. A. & SMITH, K. L., JR. 2004. Shifts in deep-sea
    community structure linked to climate and food supply.
    Science **305**, 513–515.

SANCHEZ-CABEZA, J. A., MASQUÉ, P. *ET AL.* 1999. Sedi-
    ment accumulation rates in the southern Barcelona
    continental margin (NW Mediterranean Sea) derived
    from ^{210}Pb and ^{137}Cs chronology. *Progress in Ocean-
    ography*, **44**, 313–332.

SCHALINGER, R. B. 1987. Structure of coastal upwelling
    events observed off the south-east coast of South
    Australia during February 1983–April 1984. *Austra-
    lian Journal of Marine and Freshwater Research*, **38**,
    439–459.

SCHMIDT, S. 2006. Impact of the Mediterranean Outflow
    Water on particle dynamics in intermediate waters of
    the North-East Atlantic, as revealed by ^{234}Th and
    ^{228}Th. *Marine Chemistry*, **100**, 289–298.

SCHMIDT, S., DE STIGTER, H. C. & VAN WEERING, TJ. C. E.
    2001. Enhanced short-term sediment deposition within
    the Nazare Canyon, North-East Atlantic. *Marine
    Geology*, **173**, 55–67.

SCHMIDT, S., VAN WEERING, TJ. C. E., REYSS, J.-L. & VAN
    BEEK, P. 2002. Seasonal deposition and reworking at
    the sediment–water interface on the north-western Ibe-
    rian Margin. *Progress in Oceanography*, **52**, 331–348.

SHEPPARD, F. P., MARSHALL, N. F., MCLOGHLIN, P. A. &
    SULLICAN, G. G. 1979. Currents in submarine canyons
    and other sea valleys. *AAPG Studies in Geology*, **8**,
    1–172.

SOETAERT, K., HERMAN, P. M. J., MIDDELBURG, J. J., DE
    STIGTER, H. S., VAN WEERING, TJ. C. E., EPPING, E.
    & HELDER, W. 1996. Modelling ^{210}Pb-derived
    mixing activity in ocean margin sediments: diffusive
    v. non-local mixing. *Journal of Marine Research*, **54**,
    1207–1227.

SPRIGG, R. C. 1947. Submarine canyons of the New Guinea
    and South Australian coasts. *Transactions of the Royal
    Society of South Australia*, **71**, 296–310.

VAN WEERING, TJ. C. E., DE STIGTER, H. C., BOER, W. &
    DE HAAS, H. 2002. Recent sediment transport and
    accumulation on the NW Iberian margin. *Progress in
    Oceanography*, **52**, 349–371.

VEEH, H. H., MCCORKLE, D. C. & HEGGIE, D. T. 2000.
    Glacial/interglacial variations of sedimentation on
    the West Australian continental margin: constraints
    from excess ^{230}Th. *Marine Geology*, **166**, 11–30.

VETTER, E. W. 1994. Hotspots of benthic production. *Nature*, **372**, 47.

VETTER, E. W. & DAYTON, P. K. 1998. Macrofaunal communities within and adjacent to a detritus-rich submarine canyon system. *Deep-Sea Research II*, **45**, 5–54.

VON DER BORCH, C. C. 1968. Southern Australian submarine canyons: their distribution and ages. *Marine Geology*, **6**, 267–279.

VON DER BORCH, C. C., CONOLLY, J. R. & DIETZ, R. S. 1970. Sedimentation and structure of the continental margin in the vicinity of the Otway Basin, Southern Australia. *Marine Geology*, **8**, 59–83.

WHEATCROFT, R. A. 2006. Time-series measurements of macrobenthos abundance and sediment bioturbation intensity on a flood-dominated shelf. *Progress in Oceanography*, **71**, 88–122.

YAMADA, M. & AONO, T. 2006. ^{238}U, Th isotopes, ^{210}Pb and $^{239+240}$Pu in settling particles on the continental margin of the East China Sea: Fluxes and particle transport processes. *Marine Geology*, **227**, 1–12.

YAMADA, M., WANG, Z.-L. & KATO, Y. 2006. Precipitation of authigenic uranium in suboxic continental margin sediments from the Okinawa Trough. *Progress in Oceanography*, **66**, 570–579.

East Australian marine abrasion surface

BRUCE G. THOM, JOCK B. KEENE, PETER J. COWELL* & MARC DALEY

School of Geosciences, University of Sydney, Sydney, NSW 2006, Australia

Corresponding author (e-mail: cowell@usyd.edu.au)

Abstract: Almost one-third of the seabed off the coastline north and south of Sydney comprises a planated bedrock surface, evident from sidescan surveys over the inner continental shelf. In seismic records, this rock surface extends up to 23 km offshore from the sea cliffs along 300 km of the coast. The rock surface dips offshore to as much as 180 m below sea level, where it merges with a major unconformity in the shelf sediment wedge. The surface is eroded into Mesozoic and Palaeozoic rocks and is heavily dissected by sediment-filled, palaeo-valley incision and structural jointing. The sediment-fills comprise sand wedges that thicken landwards to form beaches and estuarine flood-tide deltas, respectively, in smaller and larger palaeo-valleys incised to below present sea level. At the base of the cliffs, the planated surface is buried by shelf sand bodies up to 30 m thick in places. The seaward edge of the surface is everywhere buried by the onlapping continental-shelf sediment wedge. The contiguity of the abrasion surface with the unconformity in the shelf sediment wedge suggests that marine planation began in the Mid-Oligocene, indicating time-average rates of gross cliff retreat at about 1 mm a^{-1}.

Coastal cliffs are one the most striking geomorphological features in southeastern Australia (Fig. 1a). The submarine abrasion surface created by retreat of these cliffs is even more striking (Fig. 1b). The cliffs are fringed by rock platforms and provide a scenic backdrop to the extensive headland and bay coastline of southeastern Australia. These cliffs range in height from 20 to 110 m above sea level and are cut into both horizontal and folded sedimentary rocks of Palaeozoic and Mesozoic age. Whereas considerable attention has been given to developing models for the evolution of Quaternary depositional systems and landform origins of embayments and offshore (Roy *et al.* 1994), there has been less effort in understanding the age and formation of the rocky coast and rock-bound offshore areas (Langford-Smith & Thom 1969). Over the last decade or so, seismic and other data have provided insights into the geological and geomorphological character of the continental shelf. These data offer new opportunities for deciphering the evolutionary history of the cliffs and offshore bedrock surface formed in rocks of varying lithology, structure and age that predate the formation of the continental margin.

Various studies have noted the division of the continental shelf of this region into an outer-shelf sediment wedge and an inner–middle shelf, which appears to have a veneer of mobile sediments over a rocky substrate (Davies 1975; Jones & Kudrass 1982; Jones *et al.* 1982; Roy *et al.* 1994). The shelf break off the Sydney coast occurs at depths of 150–160 m and is 30–50 km offshore. More detailed divisions of shelf morphologies and sediment

features have been outlined by Boyd *et al.* (2004). However, the distinction between the outer prograding wedge and the inner and more gently sloping mid-shelf surfaces is fundamental to the focus of this paper.

In his recent textbook *Coasts*, Woodroffe (2002, p. 151) stated:

> Erosion of rocky coasts over sufficient time can form a near-horizontal submarine platform. For example, there is an abrasion surface, which is of the order of 1000 km long and 10 km wide along the east Australian coast, reflecting the concentration of wave energy along this coastline since the Miocene.

Thom and Cowell (2005, p. 252), in commenting further on this apparent feature, noted:

> On the east coast, the narrow continental shelf has been subjected to slow marginal subsidence. This has facilitated two geomorphic outcomes: shelf edge accretion of a sediment wedge ca. 500 m in thickness, and an inner-shelf abrasion surface ... (The East Australian Marine Abrasion Surface) ... [which] is now veneered with Late Quaternary sediments and is entrenched by infilled paleo-valleys carved by rivers draining to lower sea levels.

The abrasion surface referred to by these researchers has been revealed in sidescan sonar, multibeam and bathymetric surveys, together with some bottom sampling and shallow coring at selected locations (Fig. 2). The surveys show considerable geomorphological variability on the continental shelf, but include extensive evidence that a planar bedrock surface underlies or crops out on many parts of the inner shelf down to depths of generally more than

From: BISHOP, P. & PILLANS, B. (eds) *Australian Landscapes*. Geological Society, London, Special Publications, **346**, 57–69. DOI: 10.1144/SP346.5 0305-8719/10/$15.00 © The Geological Society of London 2010.

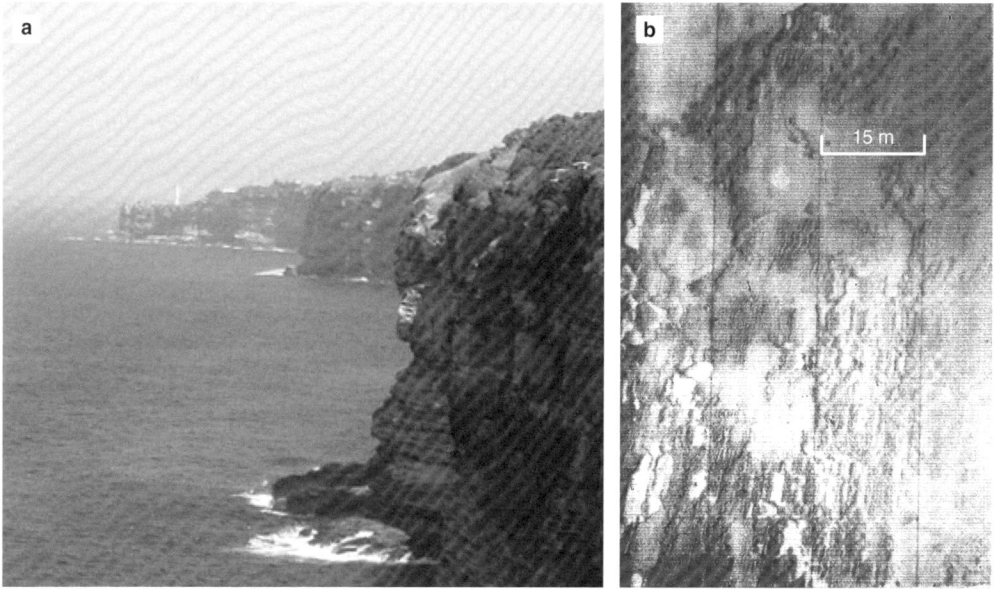

Fig. 1. Marine erosion features on the Sydney coast: (**a**) cliffs and modern rock platforms along the coast immediately south of Sydney Harbour; (**b**) sidescan-sonar image of abrasion surface cut into block jointed bedrock offshore from the cliffs shown in (**a**), in 40 m water depth.

120 m, and to almost 190 m in some places. This was tantalizingly suggested in a seismic section published by Roy (1998, fig. 25.12), but also indicated in earlier reports of the Bureau of Mineral Resources (see Davies 1975, 1979). The purpose of this paper is to show evidence for the abrasion surface that lies seaward of the cliffs and embayments of the coast of central NSW (Fig. 1) and to discuss its likely origin.

Continental margin and coastal setting

The principal geomorphological features in the east Australian margin of the Tasman Sea were formed during rifting and thinning of the continental crust in the Late Cretaceous between 110 and 80 Ma ago (Gaina *et al.* 1998; Persano *et al.* 2005). This was followed by a period of sea-floor spreading with the formation of new basaltic ocean crust, which continued until *c.* 52 Ma, as determined by magnetic anomalies (Ringis 1972; Hayes & Ringis 1973; Gaina *et al.* 1998). Sea-floor spreading created the Tasman Sea ocean basin and associated failed-rift troughs, ridges and plateaux. The basic first-order geomorphology of the present continental margin, including the development of drainage basins, shelf-edge sedimentation and shelf-slope erosion, began to develop after the cessation of sea-floor spreading and subsidence, although its progressive evolution has been punctuated by periodic

intraplate volcanism (Keene *et al.* 2008). It is possible that the continental-shelf edge in its present configuration could be as young as early Eocene, although magnetic anomalies date the age of the oceanic–continental crust boundary, at the base of the continental slope, at 80–60 Ma offshore of southern and central New South Wales (Gaina *et al.* 1998).

The eastern Australian continental margin and the Lord Howe Rise can be regarded as passive margins, which, we hypothesize, have been subsiding to current depths since rifting ceased, perhaps since the early Eocene. The rates and timing of subsidence must be viewed in the context of global eustatic sea-level movements. Drilling on the Rise and dredging of sediments suggest open marine shelf conditions in deeper parts of these margins with little terrigenous input for the last 60 Ma or so (Burns *et al.* 1973; Quilty *et al.* 1997).

The terrestrial landscape of the eastern continental margin was once viewed as reflecting the imprint of river incision following uplift in the Late Tertiary (Browne 1969). Over the past 30 years detailed geomorphological studies complemented by thermochronometry and radiometric dating of basalts occupying valley floors have revealed a much more ancient landscape, which may be traced back to the Mesozoic (Bishop & Goldrick 2000; Persano *et al.* 2005). There should be a linkage between terrestrial tectonics, landform genesis, opening of the Tasman

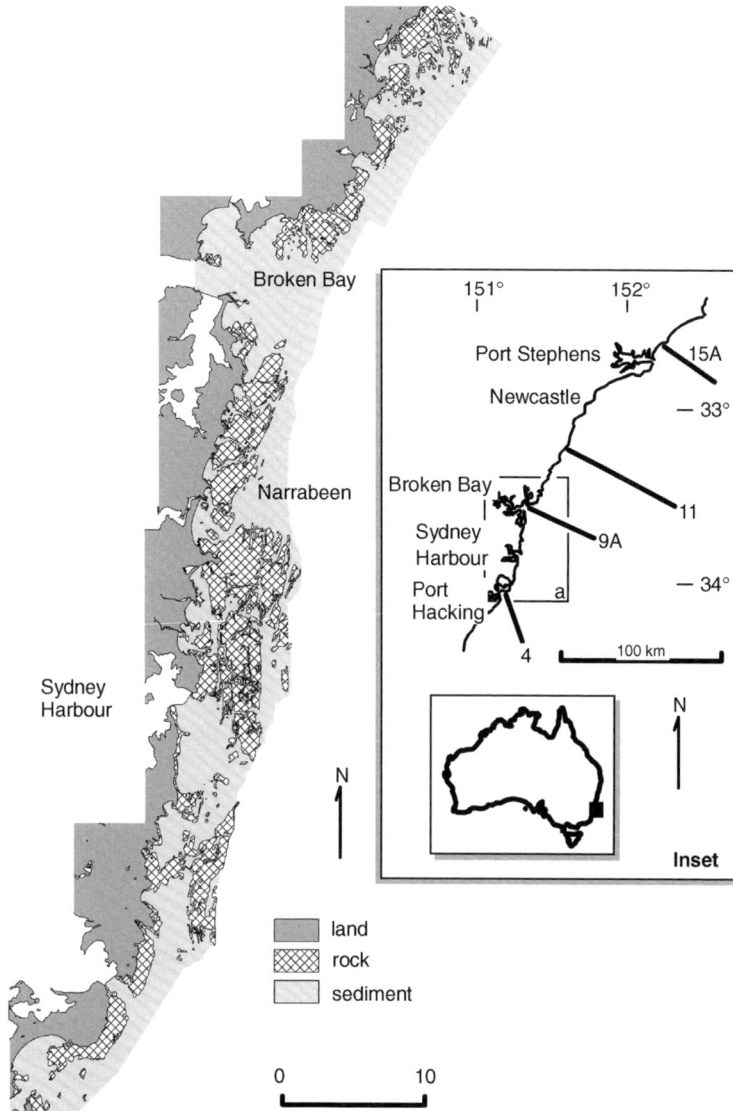

Fig. 2. Seabed rock and sediment surfaces in sidescan-sonar surveys (digitized from maps documented by Gordon & Hoffman 1986), locations of places referred to in the text, and (inset) seismic lines from surveys conducted by Geoscience Australia, The University of Sydney, and the NSW Department of Mineral Resources (Heggie *et al.* 1992).

Sea, asymmetric rifting processes, and subsequent volcanism, in relation to evolution of the continental shelf and slope (Roy & Thom 1991). We show below that incision of river valleys into this ancient landscape continues onto the continental shelf, with truncated and cliffed interfluves of ancient drainage lines forming many of the head-lands and cliffs of this coast (Bishop & Cowell 1997). Is it possible to go one step further and demonstrate a continuum between fluvial denuda-tion on the subaerially exposed rock surfaces

of the eastern highlands, with the formation of valleys and ridges extending seawards, and the marine denudation of rock surfaces that are sub-jected to progressive submergence on the subsiding continental margin? This paper aims to answer that question.

Marine abrasion surfaces: an enigma?

The theory of marine planation resulting from pro-cesses of marine abrasion has a long and chequered history. Ramsay (1846) contested the marine-valley hypothesis of Lyell in his study of planation surfaces in Wales, and established a paradigm that was readily accepted by many coastal geomorphologists, especially in Great Britain (see Chorley *et al.* 1964, Chapter 16). Many of the classical studies in 19th and early 20th century geomorphology accepted marine planation as a powerful process in carving surfaces across rocks of different lithologies. Aspects of marine and subaerial denudation were discussed by Davis (1896), and various studies were reviewed by Cotton (1974*a*) and Dietz (1963). Woodroffe (2002, Chapter 4) has recently reviewed the literature on the polycyclic and polygenetic character of rocky coasts. Trenhaile (1989) more specifically reviewed ideas on marine planation in the context of models for polycyclic development of abrasion surfaces. He differentiated earlier theories as involving either submarine erosion or shore platform development near and at sea level. Significantly for our paper, Trenhaile's own work emphasized the diachronous formation of a residual surface as a polycyclic process during repeated fluctuations in relative sea level commensurate with Pleistocene eustatic cycles. Trenhaile's polycyclic processes relate to erosion of friable bedrock on the Pleistocene time scale (of the order of 10^5– 10^6 years), whereas we consider the evidence for similar processes in southeastern Australia operating in more resistant sedimentary rocks on time scales of 10^7 years.

Arguably the most powerful proponent for marine planation in the classical literature was Johnson (1919). He developed theoretical models for submarine planation based on the concept of a profile of equilibrium under conditions of stable sea level. From his perspective, marine erosion, given sufficient time, will eventually destroy an island or even a land mass of considerable size (see Johnson 1919, figs 35 and 40; amended by Cotton 1974*a*). The result will be a submarine abraded surface little above local level of wave base. He stated that 'careful analysis of the process of marine erosion must lead to the conclusion that marine planation is possible without coastal subsidence' (Johnson 1919, p. 235). He cited examples where he thought this condition might apply. Cotton (1974*a*, *b*) was critical of Johnson's reasoning, especially in the light of what he considered to be the submergence history of submerged surfaces and sea-level oscillations.

Although some marine-planation surfaces on land have been reinterpreted as the product of subaerial processes, there remains the likelihood that rock-bound surfaces surrounding or capping basaltic tropical and subtropical islands are the result of marine abrasion (Woodroffe 2002, pp. 151–152). Along the Lord Howe Rise there are truncated seamounts forming flat-topped guyots and volcanic island peaks, remnants of once sizeable, subsiding island masses. Fraser Guyot in the northern Tasman Sea has significant terraces on its flanks formed in basalt now at depths of 1030 and 1350 m (Exon *et al.* 2005, 2006). There are also examples of apparently 'wave-cut' terraces on uplifted coasts (see Ota 1986). At higher latitudes, including areas of glacial rebound, there are many instances of shore platforms cutting across various rock types (Hansom 1983).

On passive continental margins, the existence of marine planation surfaces is less well established. Even Johnson (1925), in his regional study of the New England coast, was sceptical. However, Zenkovich (1967) and others have cited evidence from subsurface data. Widespread seismic and drilling programmes have subsequently given more emphasis to the constructional nature of shelves. Where a passive margin has received relatively little sediment in the Cenozoic, and is subsiding, it may be possible that, given sufficient time, marine abrasion processes may truncate bedrock. Oscillations of sea level should enhance these processes, as demonstrated numerically by Trenhaile (1989) for Pleistocene eustatic cycles (see Roy & Thom 1991, fig. 5, for sea-level oscillations during the Cenozoic in relation to different rates of margin subsidence). Also, where such processes are continuing to truncate submerging interfluves, the resulting coastal morphology of today will be the cliffed headlands and infilled embayments of the SE coast of Australia (Bishop & Cowell 1997).

SE coast inner and mid-continental shelf

Although inner and outer sections of the continental shelf of SE Australia have been subjected to seismic surveys over the last 30 years, very little of this information has been published. However, some reports have indicated the existence of a shallow basement surface underlying the inner shelf (Davies 1975; Jones *et al.* 1982; Albani *et al.* 1988; Heggie *et al.* 1992; Roy 1998). There is no deep drilling on the continental shelf between latitude 28 and 40 °S. For the purposes of this study, seismic and sidescan-sonar data on the inner shelf were analysed over a 300 km stretch of shelf from south of Sydney to north of Newcastle (Fig. 2). Over this stretch of coast considerable work has been done, and is continuing, on the sediment history of surficial 'drowned' barrier deposits and lobate shelf sand bodies (Albani *et al.* 1988; Roy *et al.* 1994; Boyd *et al.* 2004). These features are Late Quaternary in age and constitute a record of sea-level change through the glacial–interglacial cycles. They are being variously reworked by contemporary inner and mid-shelf wave and current processes.

In 1992 a cruise was conducted as a joint project with Geoscience Australia, The University of Sydney, and the NSW Department of Mineral Resources (Heggie *et al.* 1992). Several seismic lines from that cruise will be illustrated here. The seismic records were collected as single channel using a 120 cubic inch airgun. Detailed information is lacking on surficial sediments because of the resolution of the system.

Profile line 4 starts in 40 m of water offshore Port Hacking. It shows clearly that the outer-shelf sediment wedge thins landwards to 80 m water depth (Fig. 3). Bedrock with a thin veneer of sediment extends towards the shore with an average slope of 0.7° between −80 and −40 m. This forms the more steeply sloping inner shelf. Flat basement rocks are truncated to form an erosion surface that extends beneath the onlapping sediments at −80 m. Buried valleys exist seaward of this and beneath two prominent planar erosion surfaces. These surfaces are seen as reflectors in the profile and were called S1 (upper) and S2 (lower) by Davies (1975). The floor of the most seaward and largest valley is *c.* 250 m below present sea level.

Profile line 9A runs SE from Broken Bay (Fig. 4). It starts in a water depth of 40 m with a thin veneer of sediment overlying valley fills with at least 50 m relief. These are the palaeo-Hawkesbury channels, which extend seaward from the cliffed headlands at Broken Bay (see Albani *et al.* 1988) and include incised Late Pleistocene and, perhaps, Late Tertiary fill in the incised bedrock valley, as documented further upstream by Roy (1983). Figure 4 shows that fill in some of these channels has been truncated between 65 and 45 m water depths. On the mid-shelf, the basement-erosion surface is relatively flat and inclined seaward. The exact thickness of the sediment cover on the mid-shelf between 60 and 120 m is difficult to determine from the relatively low resolution of the airgun data. The outer shelf forms a prograding sediment wedge at −120 m below present sea level and extends over the shelf break at −150 m and wedges out between −1000 and −1500 m.

Profile line 11 starts at −38 m off Norah Head (Fig. 5). Landward tilted rocks of the Triassic–Permian Sydney Basin are truncated by a relatively flat unconformity that either crops out on the inner and mid-shelf or is covered with only a thin veneer of sediment. Here there are no valleys cut into this surface on the inner or mid-shelf, but they are present buried beneath the sediment wedge on the outer shelf below 120 m water depth.

Profile line 15A runs SE and shore normal to the headlands north of Port Stephens starting in 27 m of water (Fig. 6). The eroded basement surface has low relief of 10 m or less over a distance of *c.* 10 km along the length of the profile. It is seaward dipping until 100 m water depth, where it is onlapped by the outer shelf wedge. Beneath the wedge the basement reflector outlines eroded valleys. Off Newcastle and the Hunter River to the south of line 15A, other seismic data from Heggie *et al.* (1992) show the sediment wedge extending further onto the inner shelf but still overlying an eroded basement that slopes seaward. A major broad valley is buried beneath the sediment wedge at 122 m water depth. The floor of this valley is *c.* 250 m below present sea level.

Sidescan sonar, bathymetric, shallow-sampling and seismic data, collected by the then NSW Public Works Department (Gordon & Hoffman 1989) between Port Hacking in the south to Broken Bay (Fig. 2), reveal extensive outcrops of rock on the inner continental shelf. These outcrops extend to water depths of 50–80 m and are most pronounced and continuous off cliffed headlands (Gordon & Hoffman 1986; Albani *et al.* 1988; Albani & Rickwood 2000). The data were assembled into

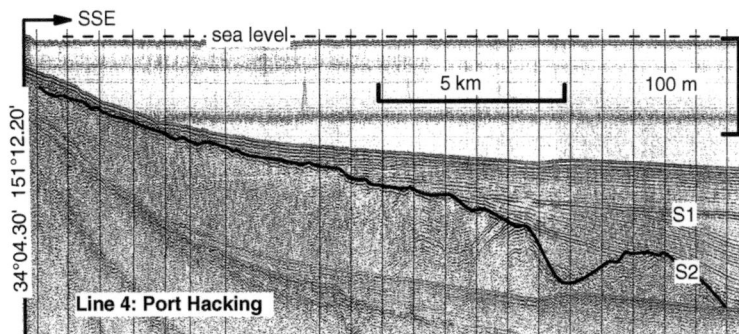

Fig. 3. Seismic profile off Port Hacking (part of Line 4, Fig. 2b).

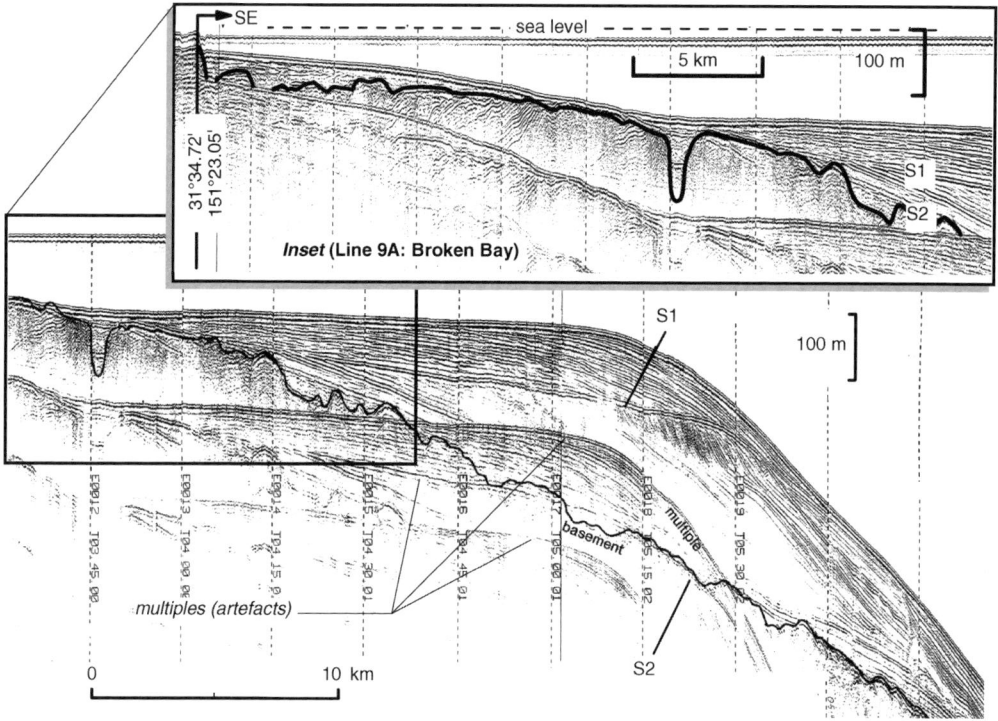

Fig. 4. Continental-shelf and continental-slope sediment wedge in seismic profile off Broken Bay (Line 9A, Fig. 2b).

a series of map sheets at a scale of 1:25 000 (Gordon & Hoffman 1989), which we digitized and integrated in ArcGIS (Fig. 2) for quantitative analysis. This analysis shows a total of 468 regions of rock outcrop on the inner shelf, collectively covering 150 km² within the region covered by the dataset. The sand surface is divided by the exposed rock into 251 separate compartments, with a combined coverage of 315 km². The rock outcrops thus make up 32% of the total area covered by the sidescan-sonar surveys. This proportion of rock is less than might be expected from the subaerial dominance of sea cliffs, which form roughly 70% of the coastline. The lower proportion of rock area on the inner shelf is attributable to the existence of overlying sand bodies off the headlands (Roy *et al.* 1994), especially along the south Sydney coast (Cowell 1986).

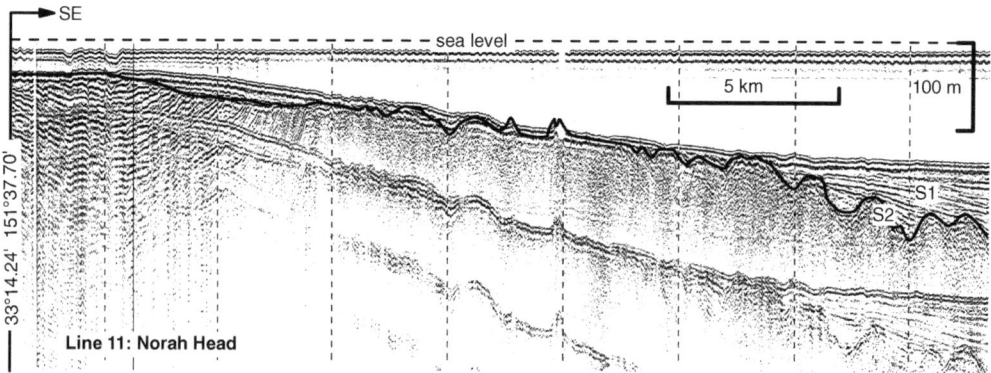

Fig. 5. Seismic profile off Norah Head (part of Line 11, Fig. 2b).

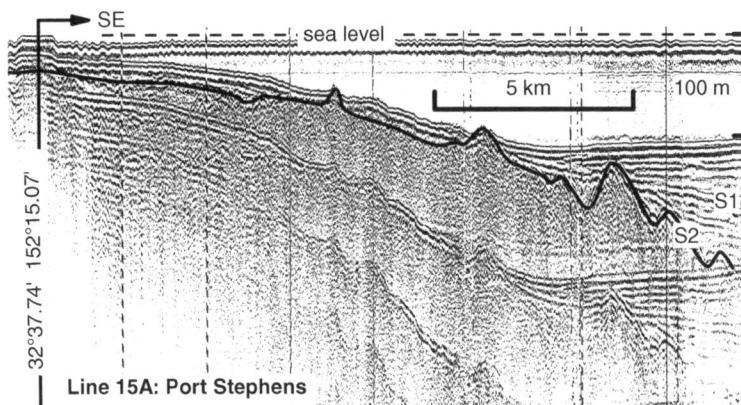

Fig. 6. Seismic profile off headlands north of Port Stephens (part of Line 15A, Fig. 2b).

Albani *et al.* (1988) used the data to define palaeo-drainage channels on the continental shelf as well as to describe the surface sediment characteristics (see also Gordon & Hoffman 1986). They also noted the existence of dissected plateaux carved into bedrock on the inner shelf, but did not discuss the geomorphological significance of these surfaces. These features are evident in the surface pattern of rock outcrops (Fig. 2), more clearly illustrated when focusing a more limited stretch of this coast (Fig. 7). The sediment fills evident in this illustration expand in proximity to the beaches, not because the width of the channels themselves increases, but because the sediment fill thickens to form bay barriers on which the beaches occur. Elsewhere, the landward thickening sand wedges form flood-tide deltas in entrance of the larger, drowned valley estuaries (Fig. 2). A network of palaeo-valley tributaries trending seaward is apparent between rock outcrops off Narrabeen Beach (Fig. 7). Connected sand filaments roughly perpendicular to the channel fills are probably indicative of joint planes enlarged through weathering and erosion during periods of subaerial exposure. Rectilinear structural patterns are typical of the regional lithology, and location of the palaeo-valleys is probably controlled by this structural grain (Bishop & Cowell 1997).

The depth contours derived from the sidescan-sonar dataset generally show a lack of marked deflection where they cross from regions of rock to sand, indicating that the rock is, like the sand, fairly planar in its surface geometry (see 40 m contour in Fig. 7). The general conformity of the sediment and rock surfaces is also evident in the presence of sand patches within the regions of rock outcrop (Fig. 1b). Patches of coarse sand are visible on the rock surface where wave-rippled bedforms are evident, and fine sand is evident in

the regions of the image containing light homogeneous tones. These sand inclusions demonstrate that the rock outcrops are so low that inner shelf sand is transported over them without topographic impediment.

The general conformity between rock and sand surface elevations across the inner shelf is evident from analysis of depths in the entire sidescan-sonar dataset (Fig. 8). The dataset was partitioned into horizontal bands across the inner shelf, spaced at 100 m increments either side of the 40 m depth contour. Depth data were averaged separately for rock and sediment zones within each band. The difference between mean water depths over rock and sediment regions within each band is less than 3 m almost everywhere, as indicated by the dashed bands in Figure 8. In shallower water (toward the top left of the diagram), the rock regions tend to be higher because of proximity effects of the headlands. In deeper water (lower right), mean sand elevations are higher because of proximity of the onlapping shelf sediment wedge that buries the abrasion surface further offshore (Figs 3–6). These comparative elevations are put into perspective when considering that the abrasion surface is the residual topography left as a result of landward retreat of the eroding cliffs and associated features (e.g. shore platforms). Given that the relief associated with these cliffs is 30–100 m, variability in elevation of the order of 1 m along the inner shelf is negligible by comparison. This subdued relief demonstrates that the inner shelf can be regarded regionally as having a planar, seaward dipping morphology.

The rock surface on the inner shelf forms a seaward dipping surface ranging in width from more than 11 km to 23 km. The elevation of this surface ranges from 30 to 180 m below present sea level, where it passes into the S1 unconformity in

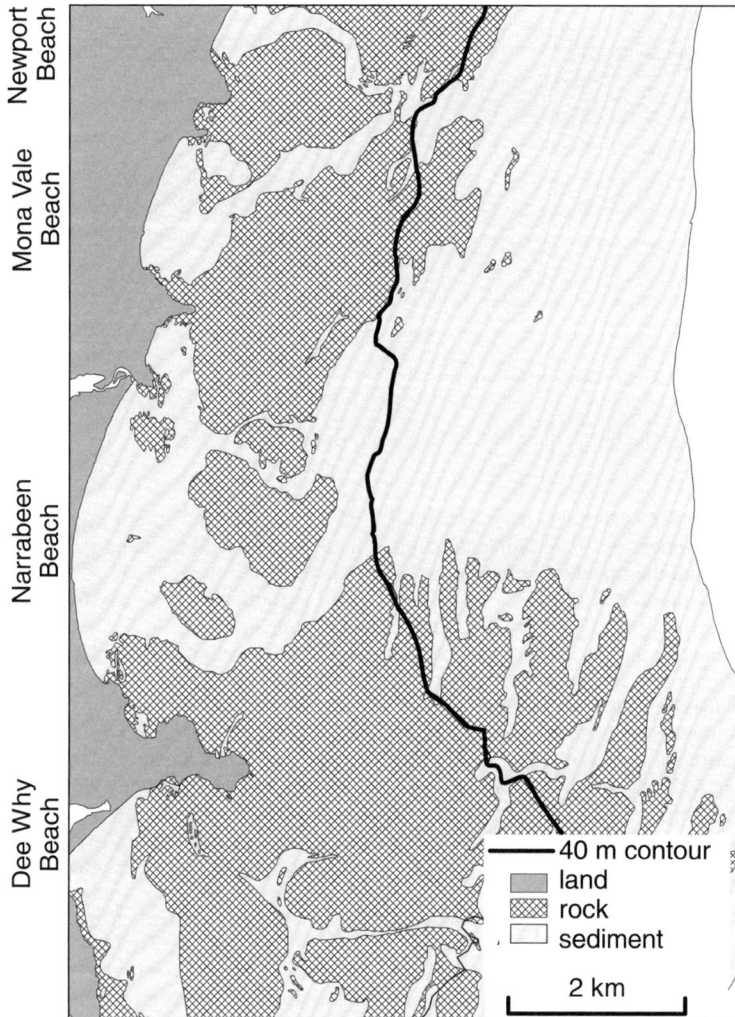

Fig. 7. Detail of sidescan-survey data off Narrabeen and adjacent beach, 10 km north of Sydney Harbour (Fig. 2), showing 40 m depth contour, and distribution of rock and sediments on the inner shelf, including the palaeo-valleys and structural depressions filled with sediments.

the shelf sediment wedge (Figs 3–6). The outer portion of the surface generally has the same slope as the unconformity. The seaward edge of the surface is everywhere buried by the onlapping continental-shelf sediment wedge.

The rock surface slope decreases landward, resulting in mild convexity in the across-shelf profile. At the base of the cliffs, shelf sand bodies, up to 30 m thick, bury the planated surface in parts of the region. Field and Roy (1984) and Ferland (1990) have described and discussed these remarkable lobate shelf sand bodies off the toe of cliffs along the south Sydney coast (also see Roy *et al.* 1994). These features are of Holocene age

and together with the other more planar sediment veneers on the inner and mid-shelf sections provide a patchy cover over the Triassic rocks of this region. The mobility of the sand and gravel fractions of this veneer has been documented by Cowell (1986) and Gordon and Hoffman (1986). Terraces and nick-points are common down to 160 m and were formed by erosion during times of lower sea level (Jones *et al.* 1975).

Process implications and conclusions

Six key features of the southeastern Australia continental margin must be explained to provide a model

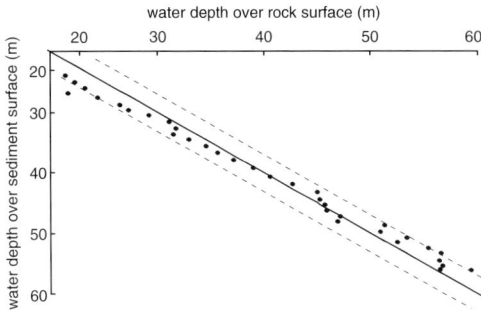

Fig. 8. Mean water depths over regions of rock (horizontal axis) and sediment (vertical axis) partitioned in horizontal bands at 100 m increments across the inner shelf. The continuous line is 1:1 correspondence between sand and rock elevations, and the dashed lines define the ± 3 m envelope.

for marine planation: (1) an ancient terrestrial landscape with incised valleys extending offshore; (2) truncated interfluves of an embayed coast forming cliffs; (3) rock platforms at base of cliffs, with polycyclic and polygenetic origin; (4) division of the continental shelf into an outer sediment wedge and inner–mid-shelf in which basement rocks are at or near the sea floor; (5) an outer sediment wedge onlapping the truncated basement surface; (6) buried and infilled river valleys cut into inner, mid-, and outer shelf regions indicative of fluvial incision and fill on a subsiding margin.

Possibly two phases of 'interidal' marine abrasion exist, both associated with transgressive and fluctuating relative sea levels. The earlier, lower surface in seismic data of Davies (1975) and Jones *et al.* (1975) was described by Glenn *et al.* (2008) as being a flat seaward dipping surface underlying the entire sediment wedge of early Tertiary age off central NSW. Davies (1975) considered this S2 surface to be an extension offshore of a Cretaceous peneplain, but Jones *et al.* (1975) interpreted this surface as a possible marine abrasion platform formed during the initial transgression following the opening of the Tasman Sea. Dredge samples described by Quilty *et al.* (1997) that are of shallow marine origin and Palaeocene age may have been deposited on this surface.

The sediment wedge on the outer continental shelf extending over and forming the shelf break in central NSW has been interpreted as being post-Eocene in age. The younger portion of the wedge above the distinct S1 unconformity was regarded by Davies (1975) as Pliocene and younger, although Roy and Thom (1991) correlated the S1 erosion surface with relative changes in sea level during the Miocene. Based on the seismic data, continued sedimentation during the Late Cenozoic appears to

have extended the sediment wedge landward as the margin subsided (Fig. 4).

The incised valleys can be traced across the shelf and beneath the sediment wedge and cut into the bedrock basement (Fig. 4; see also Roy & Thom 1991). These valleys are infilled with sediment and have been discussed in detail by Albani *et al.* (1988) where they occur on the inner and mid-sections of the shelf. Their seismic stratigraphy suggests that the valleys have been subjected to complex erosional and depositional episodes that have yet to be deciphered. However, on the inner and mid-shelf, their upper surfaces are truncated to levels generally comparable with the adjacent sea-floor bedrock, the dissected plateaux of Albani *et al.* (1988).

Rock platforms at the base of the cliffs along the central NSW coast have long been the subject of debate as to age and origin since first described by Dana (1863). He and others attributed them to wave erosion, although other marine processes have sometimes been invoked (Langford-Smith & Thom 1969; Stephenson 2000; Woodroffe 2002). Radiometric dating on surface crusts on the platforms has demonstrated that they are pre-Holocene (Young & Bryant 1993) and buried by 5000–6000 year old sediments, indicating an age older than the Holocene stillstand. This implies that the platforms are polycyclic and, by inference, so would be the cliffs that they front.

Overall the distinctive character of the cliffs, platforms, buried valleys, bedrock-bound inner–mid-shelf and the outer-shelf wedge can be best interpreted through a time-transgressive model of a subsiding margin dominated by polycyclic marine erosion. Trenhaile (1989) has simulated polycyclic shore platform retreat in weak rock with rising and falling sea levels during the Quaternary. On the NSW coast we have extended the time scale to include the effect of oscillating sea levels across basement surfaces formed in more resistant Palaeozoic and Mesozoic strata since the Eocene. These changes in levels should be combined with an uneven subsidence history of the continental margin (Roy & Thom 1991) giving rise to cyclic sweeping of the basement rocks leading to their planation. Subsidence has allowed for accommodation space on the outer shelf for a wedge of sediment to accumulate as a back-stepping deposit that onlaps the truncated bedrock strata of the mid-shelf. Continued landward erosion, following concepts put forward by Dana, Davies, Johnson, Cotton, Trenhaile and others, highlights the capacity for marine abrasion processes to develop planation surfaces. The more prolonged exposure of the abrasion surface on the inner shelf, compared with further offshore where the surface has been buried by the onlapping deposits, is consistent with the inshore convexity of the surface referred to above.

An evolutionary model is proposed for the origin of the truncated bedrock of the shelf and the cliffs along the coast north and south of Sydney. Over a period of at least 30 Ma, the mid- and inner-shelf sections have been peeled landward. Figure 9 schematically shows how the shelf abrasion surface might have evolved. The contiguity of the abrasion surface with the unconformity in the shelf sediment wedge suggests that marine planation began in the Mid-Oligocene, based on previous speculation about the age of the unconformity; although these proposed ages range widely from Pliocene to Palaeocene. The reconstructed palaeo-land surface (dashed line) is hypothetical. Strata in the present-day cliffs tend to dip gently upward in the seaward direction, suggesting a higher palaeo-surface further to the east (Persano *et al.* 2005). Whatever the initial elevation and trend of this surface, subaerial lowering of the surface is likely to have continued. Although the initial surface in Figure 9 merely

serves an illustrative purpose, the time lines of indicative retreat were computed from a uniform volume-rate of erosion (0.02 m^3 per m of coastline per year). For the hypothesized initial surface, rates of retreat decrease through time as the elevation of the initial land surface rises with distance from the sea. The need remains, however, to reconcile the erosion history of the dissected coastal plain seaward of the Eastern Highlands with the geomorphological evolution of the continental shelf and slope.

If margin subsidence is assumed to have brought sea level to a point where the bedrock was exposed landward of the shelf sediment wedge during the Mid-Oligocene, then the gross, time-average rate of cliff retreat can be estimated from the width of the abrasion surface evident today. The estimate also requires a hypothesized initial topography, which could be varied through sensitivity analysis. For the initial terrain shown in Figure 9, the gross, time-average rate of cliff retreat is estimated to be *c*. 1 mm a^{-1} (0.4–2 mm a^{-1}). These gross rates

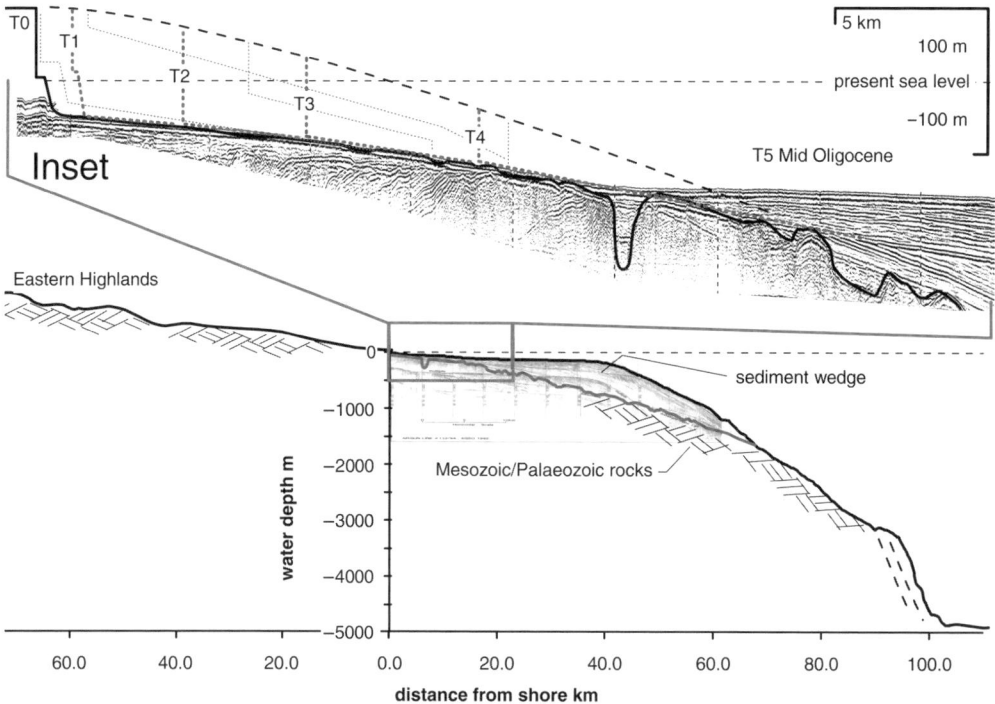

Fig. 9. Schematic representation (illustrated using stylized seismic section based on Fig. 4) showing disposition of the shelf sediment wedge relative to the continental margin. Inset shows detail of abrasion surface and its hypothesized evolution (bold dotted lines) initiated in the Mid-Oligocene (T5), then stepping landward to the start of the Miocene (T4), Mid-Miocene (T3), Early Pliocene (T2), and Early Pleistocene (T1), to the present (T0). The faint dotted lines schematically emphasize the likelihood of polycyclic back-cutting and lowering during successive irregular cycles of rising and falling sea level, rather than through the simplified representation of nick-point retreat along a single trajectory.

compare with published measurements of modern cliff retreat ranging from 9 to 20 mm a^{-1} in similar rock types (Sunamura 1992). The comparison supports the feasibility of gross marine planation rates inferred for the Sydney abrasion platform during the past 30 Ma (± 20 Ma) through polycyclic platform lowering and cliff retreat. That is, the order of magnitude lower rates on the geological time scale seem feasible considering that the retreat is active at the landward-most cutting face during sea-level highstands.

Nevertheless, down- and back-cutting is not limited to the highest nick-point at any given time (Trenhaile 1989). Shore platform development is likely to have been working away at the rock mass to be removed at all times, except during periods of lowest sea levels. In this sense, the older shore platforms are effaced during subsequent eustatic cycles: shore-platform development can thereby be regarded simply as a transient, sub-grid process on the Cenozoic scale. The abrasion surface is a residual morphology, ultimately resulting from the aggregate process proposed by Trenhaile (1989) for a multitude of sea-level fluctuations through many tens of metres (polycyclic erosion). This point is illustrated by the faint timelines in Figure 9, which represent the irregular episodes of erosive retreat that ultimately result in the surface development plotted as the bold timelines (compare Trenhaile's figures 9.4 and 9.5).

The concepts underlying Figure 9 depart from Trenhaile's formalism in two respects. First, Trenhaile considered platform development in friable bedrock during Pleistocene sea-level fluctuations, whereas here abrasion-platform evolution involves harder sedimentary lithologies acted upon over a much longer time (i.e. throughout much of the Cenozoic). Second, it is necessary on this longer time scale to consider the erosion as a gross (lumped) process of polygenetic surface development. The gross evolution is the aggregate effect of intertidal weathering and physical erosion responsible for shore-platform growth and decay, consequent nick-point retreat that causes undermining and failure of cliffs, and, necessarily, removal of detritus produced by cliff failure and shore-platform disintegration. The Trenhaile model does not include this last aspect of the overall process.

Without this detritus removal, talus blanketing the bedrock impedes further erosion of the bedrock. For example, accumulation of the shelf sediment wedge (Fig. 4) isolated the S2 surface from erosion processes, preventing further planation. Conversely, the onset of wholesale erosion of the shelf that produced the S1 unconformity would have re-exposed the landward portion of the bedrock surface previously buried by the pre-S1 shelf-wedge onlap. This broad-scale detritus removal

was therefore a fundamental requirement before development of the abrasion surface could proceed, just as was the case subsequently regarding removal of detritus produced during shore-platform development.

Detritus removal was likely to have involved weathering and abrasion, plus transport offshore, alongshore, and landward to form valley fills. Temporary burial even may have enhanced weathering through groundwater effects, at the base of cliffs on platform surfaces and within the detritus itself, during periods of subaerial exposure. Progressive breakdown of the detritus to sand, silt and clay would have served to make its removal possible through sediment-transport processes. Much of the detritus can be expected to have weathered and abraded to finer grades, which are readily removed by diffusive transport to the mid- and outer continental shelf, or to the adjacent ocean basin. Bedload transport of the sand fraction offshore to the shelf sediment wedge, although feasible, is likely to have been much less important than alongshore transport in the surf zone. Sand dispersed alongshore would have eventually been sequestered by drowned-valley fills to landward, under transgressive conditions, or in downdrift sinks, such as Fraser Island, or bypassed to the continental slope where the shelf is sufficiently narrow, immediately north of Fraser Island (Boyd et al. 2004).

The feasibility of sand-detritus removal through alongshore transport alone was evaluated by estimating the indicative capacity of erosive sand fluxes, assuming that the cliff-detritus volume was all sand (a maximum sand-volume assumption). Erosive fluxes require positive transport gradients along the entire coastal tract from which the detritus is to be removed. Net alongshore transport of sand in southeastern Australia is to the north, unless impeded by headlands, which were far less prominent at lower sea level than at present (Roy & Thom 1981). The long-term removal of sand detritus along some parts of the inner shelf may have required offshore transport (by streams) during periods of lower sea levels to allow the sand to enter the unimpeded littoral transport stream.

The length of the coastline from Sydney to Fraser Island is roughly 1000 km. The estimated net transport rate northward past Fraser Island is 500 000 m^3 a^{-1} (Schröder-Adams et al. 2008). Assuming this transport rate increases progressively to the north from zero at Sydney, the transport gradient, and thus the potential capacity for detritus removal, is $+0.5$ m^3 a^{-1} m^{-1} of coastline. Removal of detritus required an estimated gradient of only $+0.02$ m^3 m^{-1} a^{-1} to maintain a bare abrasion surface, susceptible to continuing platform development. The actual requirement averaged along the coastal tract was less because a significant proportion of the total coastline consists of incised valleys. These are potential sinks for detritus, rather than a source.

The marine-abrasion surface may extend the full length of the coastal from Wilsons Promontory in the south to Fraser Island in the north. If so, the length of coastline from which removal of sand detritus was necessary increases by up to 1900 km, again ignoring the proportion of coastline consisting of incised valleys. In these circumstances, the potential alongshore-transport gradient averages c. $+0.26 \, \mathrm{m}^3 \, \mathrm{a}^{-1}$. Even this weaker gradient is an order of magnitude greater than that required for detritus removal, once talus is abraded to sand (i.e. at the time-averaged rate of roughly $0.02 \, \mathrm{m}^3 \, \mathrm{a}^{-1} \, \mathrm{m}^{-1}$ of coastline). Therefore, erosion and transport processes are more than sufficient to allow formation of an abrasion surface, 20 km wide, after the continental shelf was flooded, while the margin subsided relative to sea level.

Overall, the data seem to support the general conclusion that, on the Cenozoic time scale, the non-friable bedrock forming the cliffs along the Sydney–Newcastle coast was peeled back with no more resistance than inherent in the unconsolidated shelf sediment wedge located immediately offshore. This conclusion can be inferred from the coplanar disposition of the bedrock abrasion surface, the S1 unconformity in the shelf sediment wedge, and the truncated surface of valley fills immediately offshore the present coast. The aggregate processes tend to plane off high points, and fill low points with sediments, along a diachronous shoreline trajectory that probably reflects the long-term transgressive trend associated with combined passive-margin subsidence and slow eustatic rise throughout the Cenozoic, regardless of superimposed eustatic fluctuations. These fluctuations are, however, responsible for the submarine terraces and nickpoints that occur down to 160 m below present sea level. The scale of these features, and other higher residuals, seems negligible compared with that of the marine-abrasion surface and retreating sea cliffs.

References

ALBANI, A. D & RICKWOOD, P. C. 2000. Marine aggregates near Sydney. *In*: McNALLY, G. H. & FRANKLIN, B. J. (eds) *Sandstone City—Sydney's Dimension Stone and other Sandstone Geomaterials*. Geological Society of Australia, EEHSG Monograph, **5**, 260–266.

ALBANI, A. D., TAYTON, J. W., RICKWOOD, P. C., GORDON, A. D. & HOFFMAN, J. G. 1988. Cainozoic morphology of the inner continental shelf near Sydney. *Journal and Proceedings of the Royal Society of NSW*, **121**, 11–28.

BISHOP, P. & COWELL, P. J. 1997. Lithological and drainage network determinants of the character of drowned, embayed coastlines. *Journal of Geology*, **105**, 685–699.

BISHOP, P. & GOLDRICK, G. 2000. Geomorphological evolution of the East Australian continental margin. *In*: SUMMERFIELD, M. A. (ed.) *Geomorphology and Global Tectonics*. Wiley, New York, 227–235.

BOYD, R., RUMING, K. & ROBERTS, J. J. 2004. Geomorphology and surficial sediments on the southeast Australian continental margin. *Australian Journal of Earth Sciences*, **51**, 743–764.

BROWNE, W. R. 1969. Geomorphology. *Journal of the Geological Society of Australia*, **16**, 559–569.

BURNS, R. E., ANDREWS, J. E. & SHIPBOARD SCIENTISTS. 1973. *Initial Reports of the Deep Sea Drilling Project, 21*. US Government Printing Office, Washington, DC.

CHORLEY, R. J., DUNN, A. J. & BECKINSALE, R. P. 1964. *The History of the Study of Landforms. Volume 1, Geomorphology before Davis*. Methuen, London.

COTTON, C. 1974a. The theory of secular marine planation. *In*: COLLINS, B. W. (ed.) *Bold Coasts*. Reed, Wellington (republished from *American Journal of Science*, **253**, 1955), 164–174.

COTTON, C. 1974b. Plunging cliffs and Pleistocene coastal cliffing in the southern hemisphere. *In*: COLLINS, B. W. (ed.) *Bold Coasts*. Reed, Wellington, 245–266.

COWELL, P. J. 1986. Wave-induced sand mobility and deposition on the south Sydney inner-continental shelf. *In*: FRANKEL, E., KEENE, J. B. & WALTHO, A. E. (eds) *Recent Sediments in Eastern Australia – Marine Through Terrestrial*. Geological Society of Australia Special Publication, **2**, 1–28.

DANA, J. D. 1863. *Manual of Geology*. 1st edn. Bliss, Philadelphia, PA.

DAVIES, P. J. 1975. Shallow seismic structure of the continental shelf, southeast Australia. *Journal of the Geological Society of Australia*, **22**, 345–359.

DAVIES, P. J. 1979. *Marine geology of the continental shelf off southeastern Australia*. Bureau of Mineral Resources Bulletin, **195**.

DAVIS, W. M. 1896. Plains of marine and sub-aerial denudation. *Geological Society of America Bulletin*, **7**, 377–398.

DIETZ, R. S. 1963. Wave base, marine profile of equilibrium, and wave-built terraces: a critical appraisal. *Geological Society of America Bulletin*, **74**, 971–990.

EXON, N., HILL, P. *ET AL.* 2005. *The geology of the Kenn Plateau off northeast Australia: results of* Southern Surveyor *Cruise SS5/2004 (Geoscience Australia Cruise 270)*. Geoscience Australia Record, 2005/04.

EXON, N., HILL, P., LAFOY, Y., HEINE, C. & BERNARDEL, G. 2006. Kenn Plateau off northeast Australia: a continental fragment in the southwest Pacific jigsaw. *Australian Journal of Earth Sciences*, **53**, 541–564.

FERLAND, M. A. 1990. *Shelf Sand Bodies in Southeastern Australia*. PhD thesis, University of Sydney, Sydney.

FIELD, M. E. & ROY, P. S. 1984. Offshore transport and sand body formation: evidence from a steep, high energy shoreface, southeastern Australia. *Journal of Sedimentary Petrology*, **54**, 1292–1302.

GAINA, C., MULLER, D. R., ROYER, J.-Y., STOCK, J., HARDEBECK, J. & SYMONDS, P. 1998. The tectonic history of the Tasman Sea: a puzzle with 13 pieces. *Journal of Geophysical Research*, **103**, 12413–12433.

GLENN, K. C., POST, A. *ET AL.* 2008. *Geoscience Australia Marine Survey Post-Cruise Report—NSW Continental Slope Survey*. Geoscience Australia, Record, **2008/14**.

GORDON, A. D. & HOFFMAN, J. G. 1986. Sediment features and processes of the Sydney continental shelf. *In*: FRANKEL, E., KEENE, J. B. & WALTHO, A. E. (eds) *Recent Sediments in Eastern Australia – Marine Through Terrestrial*. Geological Society of Australia Special Publication, **2**, 29–51.

GORDON, A. D. & HOFFMAN, J. G. 1989. *Seabed Information, 1:25,000 Sheets: Bate Bay, Sydney Heads, Broken Bay, Gosford*. Public Works Department New South Wales Coast and Rivers Branch, Sydney.

HANSOM, J. 1983. Shore platform development in the South Shetland Islands, Antarctica. *Marine Geology*, **53**, 211–229.

HAYES, D. E. & RINGIS, J. 1973. Seafloor spreading in the Tasman Sea. *Nature*, **243**, 454–458.

HEGGIE, D. & SHIPBOARD SCIENTISTS. 1992. *Preliminary results of AGSO RV Rig Seismic Survey 112: Offshore Sydney Basin continental shelf and slope geochemistry, sedimentology and geology*. Australian Geological Survey Organisation Record, **1993/5**.

JOHNSON, D. W. 1919. *Shore Processes and Shoreline Development*. Wiley, New York (reprinted Hafner, New York, 1965).

JOHNSON, D. W. 1925. *The New England–Arcadian Shoreline*. John Wiley, New York.

JONES, H. A., DAVIES, P. J. & MARSHALL, J. M. 1975. Origin of the shelf-break off southeast Australia. *Journal of the Geological Society of Australia*, **22**, 71–78.

JONES, H. A. & KUDRASS, H. R. 1982. *Sonne* cruise (SO-15 1980) off the east coast of Australia bathymetry and sea floor morphology. *Geologisches Jahrbuch*, **Reihe D56**, 55–68.

JONES, H. A., LEAN, J. & SCHLÜTER, H.-U. 1982. Seismic reflection profiling off the east coast of Australia, Newcastle to Cape Hawke. *Geologisches Jahrbuch*, **Reihe D56**, 69–75.

KEENE, J., BAKER, C., TRAN, M. & POTTER, A. 2008. *Sedimentology and Geomorphology of the East Marine region of Australia: A spatial analysis*. Geoscience Australia Record, **2008/10**.

LANGFORD-SMITH, T. & THOM, B. G. 1969. New South Wales coastal morphology. *Journal of the Geological Society of Australia*, **16**, 572–580.

OTA, Y. 1986. Marine terraces as reference surfaces in late Quaternary tectonics studies: examples from the Pacific Rim. *Bulletin of the Royal Society of New Zealand*, **24**, 357–375.

PERSANO, C., STUART, F. M., BISHOP, P. & DEMPSTER, T. J. 2005. Deciphering continental breakup in Eastern Australia using low-temperature thermochronometers. *Journal of Geophsyical Research*, **110**, B12405.

QUILTY, P. G., SHAFIK, S., JENKINS, C. J. & KEENE, J. B. 1997. An Early Cainozoic (Paleocene) foraminiferal fauna with *Fabiania* from offshore eastern Australia. *Alcheringa*, **21**, 299–315.

RAMSAY, A. C. 1846. *The Denudation of South Wales*. Memoir of the Geological Survey of Great Britain, 1. HMSO, London, 297–335.

RINGIS, J. 1972. *The Structure and History of the Tasman Sea and Southwest Australian Margin*. PhD thesis, University of NSW, Sydney.

ROY, P. S. 1983. Quaternary geology. *In*: HERBERT, C. (ed.) *Geology of the Sydney 1:100,000 Sheet 9130*. Geological Survey of New South Wales, Sydney, 41–91.

ROY, P. S. 1998. Cainozoic geology of the New South Wales coast and shelf. *In*: SCHEIBNER, E. & BASDEN, H. (eds) *Geology of New South Wales: Synthesis, Volume 2 Geological Evolution*. Geological Survey of New South Wales, Memoir, **13**, 361–385.

ROY, P. S. & THOM, B. G. 1981. Late Quaternary marine deposition in New South Wales and southern Queensland—an evolutionary model. *Journal of the Geological Society of Australia*, **28**, 471–189.

ROY, P. S. & THOM, B. G. 1991. Cainozoic shelf sedimentation model for the Tasman Sea margin of southeastern Australia. *In*: WILLIAMS, M. A. J., KERSHAW, A. P. & DE DECKKER, P. (eds) *The Cainozoic in Australia: A Reappraisal of the Evidence*. Geological Society of Australia, Special Publication, **18**, 119–136.

ROY, P. S., COWELL, P. J., FERLAND, M. A. & THOM, B. G. 1994. Wave-dominated coasts. *In*: CARTER, R. W. G. & WOODROFFE, C. D. (eds) *Coastal Evolution*. Cambridge University Press, Cambridge, 121–186.

SCHRÖDER-ADAMS, C. J., BOYD, R., RUMING, K. & SANDSTROM, M. 2008. Influence of sediment transport dynamics and ocean floor morphology on benthic foraminifera, offshore Fraser Island, Australia. *Marine Geology*, **254**, 47–61.

STEPHENSON, W. J. 2000. Shore platforms: remain a neglected coastal feature? *Progress in Physical Geography*, **23**, 311–327.

SUNAMURA, T. 1992. *Geomorphology of Rocky Coasts*. Wiley, Chichester.

THOM, B. G. & COWELL, P. J. 2005. Coastal changes, gradual. *In*: SCHWARTZ, M. L. (ed.) *Encyclopedia of Coastal Science*. Springer, Dordrecht, 251–253.

TRENHAILE, A. S. 1989. Sea level oscillations and the development of rock coasts. *In*: LAKNAN, V. C. & TRENHAILE, A. S. (eds) *Applications in Coastal Modelling*. Elsevier Oceanography Series, **49**, 271–295.

WOODROFFE, C. D. 2002. *Coasts*. Cambridge University Press, Cambridge.

YOUNG, R. W. & BRYANT, E. A. 1993. Coastal rock platforms and ramps Plcistocene and Tertiary age, southern New South Wales, Australia. *Zeitschrift für Geomorphologie*, **37**, 257–272.

ZENKOVICH, V. P. 1967. *Processes of Coastal Development*. Oliver & Boyd, Edinburgh.

Geoarchaeology in Australia: understanding human–environment interactions

SIMON J. HOLDAWAY[1]* & PATRICIA C. FANNING[2]

[1]*Department of Anthropology, University of Auckland, Private Bag, Auckland, New Zealand*

[2]*Graduate School of the Environment, Macquarie University, Sydney, NSW 2109, Australia*

*Corresponding author (e-mail: sj.holdaway@auckland.ac.nz)

Abstract: This paper reviews the long history of interaction between scientists working in geomorphology, stratigraphy, sedimentology and chronology and those working in archaeology to understand past human–environment interactions in Australia. Despite this close collaboration, differentiating environmental impacts from the influence of human behaviour has proven difficult in research on key topics such as the causes of megafauna extinction, the significance of fire, and the impact of climatic shifts such as the El Niño–Southern Oscillation. Geoarchaeological research focused on depositional environments and post-depositional change in western New South Wales, Australia, provides important examples of how processes acting over different temporal scales affect archaeological deposits. The archaeological record is in some places discontinuous in time because geomorphological activity has removed the record of particular time periods, and it is discontinuous in space because it is preserved only in places that are geomorphologically relatively inactive. Important inferences concerning past human behaviours may be drawn from the record, but the processes responsible for both the presence and absence of the record must be considered. More attention needs to be given to ensuring that datasets with a similar temporal resolution are compared if the causes for behavioural changes in the past are to be correctly understood.

Geoarchaeology refers to the application of the geosciences to solve research problems in archaeology (Butzer 1982; Pollard 1999). Although the term was first used relatively recently (Renfrew in Davidson & Shackley 1976), interaction between the geosciences and archaeology goes back to the early nineteenth century, when geology and prehistoric archaeology developed essentially in parallel (Pollard 1999). Other terms such as archaeogeology, archaeological geology and archaeometry have also been used in the same context, although the differences in meaning between them are considered to be trivial (Herz & Garrison 1998). In this paper, the term geoarchaeology refers to archaeological studies concerned with geomorphology, pedology, stratigraphy, sedimentology and chronology (e.g. Gladfelter 1977; Pollard 1999).

Strong interdisciplinary activities have characterized Australian archaeology since its foundation, with geologists, geomorphologists and pedologists making important site-specific contributions to a number of 'classic' studies of archaeological research (Shawcross & Kaye 1980; Hughes & Sullivan 1982). However, geoarchaeology has had little recognition as a distinct sub-discipline of either archaeology or geology in Australian universities, with most archaeology being taught in social, rather than natural, science faculties. Despite this, many contemporary archaeological studies in Australia incorporate geological and geomorphological investigations of archaeological materials and their landscape settings.

Hughes & Sullivan (1982) reviewed Australian geoarchaeology in the proceedings of the first Australasian archaeometry conference held at the Australian Museum in January 1982 (Ambrose & Duerden 1982). The majority of papers at this and later conferences focused on the geoscientific applications to archaeology more commonly recognized as archaeometry, such as the chemistry, provenance and dating of artefacts and archaeological sediments, and conservation of archaeological materials (e.g. Ambrose & Mummery 1987; Fankhauser & Bird 1993). However, there were sufficient papers with a geoarchaeological theme for Hughes & Sullivan (1982) to identify two main groupings: those drawing on the methods and theories of geomorphology, geology and pedology that were largely site-specific in approach, and those drawing additionally on the geographical sciences with a regional focus, utilizing a wide range of environmental and spatial approaches. From the first group, they reviewed investigations of the stratigraphy and chronology of rockshelter sites, the nature and sources of raw materials for stone artefact manufacture, and rock art conservation studies. Two further areas of investigation characterized the second group, namely palaeoenvironmental and

From: BISHOP, P. & PILLANS, B. (eds) *Australian Landscapes*. Geological Society, London, Special Publications, **346**, 71–85. DOI: 10.1144/SP346.6 0305-8719/10/$15.00 © The Geological Society of London 2010.

chronological investigations of places such as Lake Mungo and river terraces in Victoria, and the impact of Pleistocene and Holocene environmental changes, more closely reflecting the way in which Butzer's landmark volume on geoarchaeology, *Environment and Archaeology* (Butzer 1971), was interpreted in Australia.

Published in the same year as Butzer's work, Mulvaney & Golson (1971) brought together geologists, geomorphologists, biologists and archaeologists in the volume *Aboriginal Man and Environment in Australia*, to discuss the relationship between humans and the environment. The result was a series of papers that discussed aspects of palaeoenvironmental reconstruction and chronology. Few of the papers, however, dealt specifically with the integration of environmental and archaeological data in ways consistent with geoarchaeology. As Allen (1992) commented some 20 years later, most of the participants limited themselves to discussion of their own discipline, grappling with the combination of a human presence in Australia that was then thought to be at least 20 000 years old and a range of new techniques that were changing the nature of their own specializations as well as bringing their subject matter closer to that of interest to archaeologists.

In *Naïve Lands*, published 20 years later (Dodson 1992), theoretical changes in archaeology had moved the more ecologically minded archaeologists closer to the concerns of Quaternarists. Levels of integration between archaeology and natural science had increased dramatically, as had the body of knowledge concerning the Quaternary. Pancontinental models of cultural change were replaced by a concern with more local sequences, evidenced most clearly in archaeology by the rejection of universal typological schemes for stone artefact classification. The debate concerning chronology was no longer about a 20 000 years sequence but one at least twice that length. Yet, in summing up the *Naïve Lands* volume, Allen (1992) noted that the various disciplines still continued to maintain their own theoretical positions, methods and intellectual agendas. Whereas archaeologists were more interested in environmental explanations than was apparent in the Mulvaney & Golson (1971) volume, Allen felt that natural scientists were less inclined to seek cultural explanations for their data, maintaining an interest in the environment as the locus of change.

In 2010, there is ample evidence that natural scientists seek out human, as well as natural, explanations for environmental changes. Both of the important debates about human–environment interactions in Australia, the cause(s) of megafaunal extinction (e.g. Gillespie *et al.* 1978; Dodson *et al.* 1992, 1993; Baynes 1999; Field & Dodson 1999; Miller *et al.* 1999; Roberts *et al.* 2001; Gillespie 2002; Johnson 2005; Trueman *et al.* 2005; Wroe & Field 2006) and the human use of fire to modify the environment (e.g. Jones 1968, 1969; Horton 1982; Head 1994, 1996, 2000*a*; Bowman *et al.* 2001; Kershaw *et al.* 2002), continue to receive attention. To these debates may be added a renewed interest in the effect of global environmental change on regional human adaptations, seen most clearly in the work on climate drivers such as the El Niño–Southern Oscillation (ENSO) (e.g. Haberle & David 2004; Turney & Hobbs 2006; Turney & Palmer 2006; Bourke *et al.* 2007; Cosgrove *et al.* 2007). However, disciplinary differences between explanations of human behaviour derived from natural sciences (i.e. environmental impacts on human behaviour) and from social sciences (i.e. cultural change) remain, and although explanations now incorporate new data, disciplinary differences would probably still be recognizable to the authors of papers in both of the previously mentioned volumes from the early 1970s and early 1990s.

Here, we argue that one of the key difficulties faced by all Quaternary scientists interested in the interaction between humans and their environment is the complex array of processes acting over different spatial and temporal scales that are responsible for the formation of the archaeological record (e.g. Haberle & Chepstow-Lusty 2000), a problem identified by Dodson *et al.* (1992). Given that complex causation is increasingly recognized in both archaeological and environmental research, accounts of human–environmental interactions in the past must take account of the complexity of both records.

The archaeological record

Writing in the volume by Mulvaney & Golson (1971), Peterson presaged many of the problems that interest us today by discussing the different temporal scales at which the ethnographic and archaeological data are constructed. He suggested that archaeologists should change the nature of the questions they pose of the archaeological record. They should ignore the 'idiosyncrasies' of human behaviour that concerned anthropologists, and instead consider long-term aspects of the environment that acted to 'hold fast' basic regularities of human life. They should incorporate space by investigating areas greater in extent than those occupied by ethnographically defined bands. Long-term continuity in human–environment relationships might, Peterson (1971) argued, be determined by replacing time with space.

At the time that they were written, Peterson's proposals fitted well with contemporaneous developments in archaeological theory. Theoretical developments outside Australia emphasized artefact function over ethnically correlated style, and settlement patterns rather than single-site analyses as a way of understanding landscape use (summarized by Wandsnider 1996). In Australia, Allen's (1972) doctoral research applied these new ideas. He outlined a method whereby stratified rockshelter deposits provide the means to order surface artefact scatters in time on the basis of stone artefact assemblage composition. Using this technique, surface scatters of artefacts can be used to document consistency in human landscape use across prolonged time periods. His approach was applied in area studies (e.g. Veth 1993), and became the default method for much of the Cultural Resource Management (CRM) archaeology that was then in its infancy, but that has now become the primary vehicle for documenting the archaeological record in Australia.

Although widely adopted, the settlement pattern approach that effectively replaced time with space as Peterson had proposed was flawed because, as a number of archaeologists commented, the archaeological record is subject to the actions of different processes operating at different temporal and spatial scales that are not easily resolved (e.g. Hiscock 1983; Bird & Frankel 1991; Head 2000a, p. 128; Head 2000b, chapter 3; Allen & Holdaway 2009). Archaeological sites distributed across the landscape in different concentrations and with different densities and types of stone artefacts were initially interpreted as a record produced by Aboriginal people practising a regional adaptation to an environment largely unchanging, or at least one that showed only a few shifts over thousands of years. The temptation was always to make the connection between the distribution of sites (often represented as dots on maps) and an interpretation of groups of Aboriginal people moving from place to place exploiting seasonally available resources or perhaps coming together for socially important events. Peterson (1971) had earlier recognized that archaeologists were not able to resolve the archaeological record at the level of an ethnographic band that this type of interpretation requires. Like other archaeologists, Peterson was aware of the impossibility of dating surface sites, or indeed stratified deposits, at the resolution needed for drawing inferences about day-to-day behaviour. However, few archaeologists embraced Peterson's alternative to inferences based on ethnographic analogy. Implications of the difference in the scale of observation versus the scale of interpretation were underplayed.

The interpretative difficulties are well illustrated in the mismatch between the geomorphological and geochronological interpretations provided by the Willandra Lakes research of Bowler and others

(Bowler 1998; Bowler et al. 2003), and the level of inference adopted by the archaeologists interpreting the contents and distribution of the associated shell and stone artefact deposits (Allen et al. 2008; Allen & Holdaway 2009). At the Willandra Lakes (see Fig. 1), inferences about the existence of ethnographically derived units, such as dinnertime camps, have been proposed for deposits situated within stratigraphic units that took millennia to accumulate. The stratigraphic evidence has been used to support arguments for the contemporaneity of a large number of small, dispersed sites leading to an interpretation closely tying mobility and large-scale environmental change. As a consequence, changes in behaviour are apparently limited to long-term shifts in the appearance of people on the landscape. The overall impression is one of groups of people with a stable adaptation shifting only in response to major changes in Late Pleistocene environments. Unfortunately, it is difficult to determine whether this stability truly reflects past human behaviour, or whether it is an artefact of the temporal scale of the stratigraphic and palaeoenvironmental record in which the archaeology is situated. As Allen et al. (2008) commented, much of the archaeological record in the Willandra Lakes consists of small, isolated, shell middens (McIntyre & Hope 1978; Johnston 1993). Even when multiple middens occur in close proximity, there is no reason why these should be considered contemporaneous. The differences in the temporal scale at which geomorphological processes are described versus the 'fragile and fleeting evidence of human occupation' have not so far been adequately addressed (Allen & Holdaway 2009, p. 101).

Issues of temporal scale are also apparent in the interpretation of human–environment interactions during long stretches of the Holocene in different parts of Australia. The Holocene archaeological record is also commonly interpreted as demonstrating stable adaptations, or, following various views on demographic increases and the development of social complexity, cultural changes unrelated to environmental variability (e.g. Lourandos 1983; Lourandos & David 1998). Like the situation at the Willandra Lakes, however, sets of evidence with markedly different chronological resolution are commonly tied together, promoting either interpretations that emphasize stability or, alternatively, slow accumulative directional change (Frankel 1988). Where chronologies with different precisions are mixed, all must default to the chronology with the coarsest age estimates. The perception of either stable adaptations or of cumulative progressive change in such situations may well be a function of the coarseness of the measurements.

Dodson et al. (1992) reviewed the archaeological and palaeoenvironmental evidence from

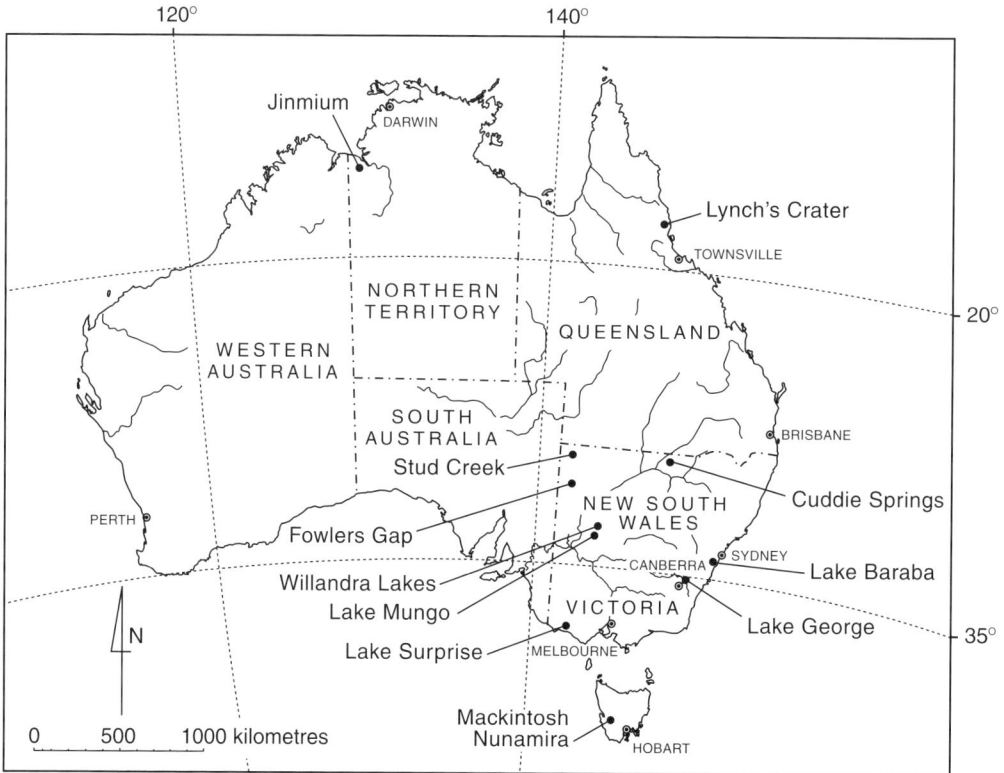

Fig. 1. Map of Australia showing locations of places mentioned in the text.

the SE of Australia using a multi-scale approach, contrasting large-scale changes that took place over thousands of years, and therefore many human generations, with local and shorter-term variability. The latter varies in temporal scale locally, as different systems responded to change in different ways. To some degree, archaeologists and environmental scientists have adopted the multi-scale approach in more recent studies. Tibby *et al.* (2006), for instance, reviewed the palaeoenvironmental record from Lake Surprise in southwestern Victoria. They noted that although Period 3 (3750–1500 cal. years BP) shows evidence for considerable environmental variability, it is difficult to determine if increases in the rate of environmental change are the result of climate, human activity, or some combination of both. Environmental changes evidenced in the Lake Surprise record were possibly associated with social changes including systematic eel fishing (Williams 1988), but confirmation using precise dating is required.

In other studies, general trends rather than precise correlations continued to be emphasized. In the same volume as the Tibby *et al.* (2006) paper, Veth (2006) summarized occupation evidence in

the Western Desert, arguing largely on the basis of an increase in the number of sites dated to the late Holocene for a decrease in residential mobility and the greater scheduling of resource use. In our own work in western New South Wales, such late Holocene increases can be shown to be due to differential preservation of the surfaces on which the archaeological record is deposited. Older surfaces are rare and more recent surfaces are relatively common, reflecting the cumulative nature of erosion through time, leading to the impression of an abundant late Holocene record simply because older deposits have more frequently eroded away (Holdaway *et al.* 2008*a*).

The theoretical significance of the recognition of a mismatch between the temporal resolution of archaeological data and the scale of interpretation has been discussed only recently (Bailey 2007). One of the best documented cases concerns the study of the spatial distribution of artefacts. In the 1950s and 1960s, archaeologists working in Africa and Europe were intrigued by the idea of defining 'living floors', places where the distribution of artefacts was taken to reflect past activities (Holdaway & Stern 2001). Living floors were at first thought to offer the possibility of detailed behavioural interpretations and their presence featured in debates about the behaviour of humanity's earliest ancestors (e.g. Isaac 1981). The definition of living floors also fitted well with artefact analyses that emphasized the definition of functional sets: groups of artefacts used to perform one or a limited number of functions (Wandsnider 1996).

Much research on artefact accumulation occurred from the early 1970s, helped by Schiffer's Behavioral Archaeology program directed at understanding how archaeological sites were formed (Schiffer 1972). However, the results of these studies, although providing much detail on how artefacts accumulated, provided little support for the idea that archaeological deposits represent the outcome of short-term or isolated events. Focus shifted from defining living floors and functional artefact sets to the definition of activity areas: less precisely defined locations where a variety of artefacts were used and deposited as a result of a range of activities. Attention was focused on assemblage formation rather than artefact function. Once it was demonstrated that artefacts were used in multiple ways with little clear relationship between form and function, and that spatial proximity did not necessarily imply contemporaneity, archaeologists began to discuss artefact life-histories (e.g. Shott 1989). That is, groups of artefacts found together do not represent toolkits intended to be used together to accomplish a particular task. Rather, they represent artefacts that had accumulated together at a particular location as a result of a variety of different processes, both behavioural and taphonomic, through time.

In Australia, this change in emphasis from function to formation was slow in coming. Late twentieth century studies of Australian stone artefacts emphasized function, obtainable by considering usewear (i.e. damage on the edge of artefacts indicating how artefact edges were used), rather than correlations between activity and artefact form (Holdaway & Stern 2004). Where function could not be directly determined, tool technology (i.e. how the artefacts were made) seemingly provided an alternative route to determining the nature of activities represented by the accumulation of artefacts at single locations. Most recently, studies that demonstrate tool reuse have been emphasized as a means of investigating technological organization (e.g. papers in the book by Clarkson & Lamb 2005).

Artefacts obviously functioned in the past but, as Peterson (1971) recognized many years ago, this function may only be indirectly related to what went on at a particular place. Peterson noted that much of the apparent conservatism in Australian stone artefacts might reflect the fact that they were used to work wood and bone. As there are only a

few techniques available to work these materials, one would not expect a particularly varied toolkit. Artefacts deposited at a particular location might have been used to form other artefacts used at a variety of different places. It might also be true that the bulk of the use-life of an artefact might have occurred somewhere other than the place where it was deposited. Therefore, artefacts found at a particular location do not necessarily inform on what went on there. What they can tell us about is the amount of artefact deposition at a particular location. Analysis in ways that take account of the life histories of different artefact forms allows inferences about the length of occupations at different locations (Holdaway et al. 2008b).

Understanding the scale at which artefact deposits can be resolved, and developing behavioural explanations to match that scale, was a primary aim of the Southern Forests Archaeological Project that investigated rockshelter deposits in Tasmania (Allen 1996). A number of limestone rockshelters with archaeological deposits spanning the Last Glacial Maximum (LGM) were excavated. Stern (2008) reviewed the mechanisms responsible for the accumulation of archaeological deposits at two of the rockshelters, Mackintosh and Nunamira. Both have deposits that took thousands of years to accumulate and both were repeatedly occupied by small groups of hunter–gatherers during that time. Stern argued that the human occupation of Mackintosh cave created a mixed deposit of sands and clays from the rockshelter floor, with organic materials, stone artefacts and gravels brought in from outside the shelter. Human activity homogenized these deposits, meaning that there is no sedimentological evidence for breaks in occupation. Features such as hearths are preserved only on the surface of the deposit, not within the sediment layer itself. A carbonate deposit is present in the sediment sequence, to some extent dividing the deposit into two parts. However, even this stratigraphic division is sometimes obscured by human activities such as trampling and natural processes such as carbonate impregnation. Because of the degree of homogenization, radiocarbon determinations show little consistency across the site. Mixing across layers means that even when lenses of non-cultural sediments do occur, they cannot be used to separate different occupations. Stern (2008) concluded that the Mackintosh deposit contains two 'minimum-archaeological stratigraphic units', one that accumulated between 17 000 and 15 000 years BP and a second that took a further 10 000 years to accumulate. Separating the archaeological materials contained in these units into finer chronological periods has, Stern argued, no stratigraphic basis.

Understanding the nature of time averaging at archaeological sites such as Mackintosh offers the opportunity to develop comparative analyses aimed at determining whether different types of behavioural information are preserved in different types of deposit. As more time elapses and the number and/or duration of occupations increase, the greater the probability that assemblages will contain material relating to rare activities (Shott 1997). In Tasmania, this argument can be used to explain the presence of artefacts made from Darwin Glass (a locally available impactite used to manufacture stone artefacts) in levels dating to periods after the LGM, and their absence in deposits dating before this time (Holdaway 2004), as considerably greater occupation occurs after the LGM than before.

To summarize, whereas Peterson and his contemporaries hoped that equating time and space might simplify the archaeological interpretation, recent studies such as those carried out on the Tasmanian deposits indicate that it is the complex interplay between processes operating at different temporal and spatial scales that provide a more interesting set of behavioural inferences. Geoarchaeological studies have provided important examples of how processes acting over different temporal scales affect archaeological deposits, through studies of the depositional and post-depositional environments. Through these studies, concerns with the different scales of interpretation have also extended to considerations of the environment.

Environment and time

Peterson (1971) characterized human–environment relationships as stable for long periods between major incidents of change. He criticized studies where economies were thought to be highly localized, arguing instead that, because hunter–gatherers could not survive by exploiting a single plant or animal community and because they had to take into account seasonal availability, territories exploited by hunter–gatherers had to encompass a full range of plant communities. When combined with a functionally based, settlement-pattern approach to interpreting the archaeological record, the application of Peterson's characterization of hunter–gatherer environmental relationships led to two forms of study.

First, great attention was given to chronology. It was clear even in the early 1970s that Australian environments had, in the past, differed from their current state (see, e.g. Bowler 1971; Costin 1971; Galloway 1971; Pels 1971). Chronology provided the means to associate archaeological materials with one or other of these past environments (Smith & Sharp 1993). Efforts to establish the earliest age of settlement concentrated on sites that might indicate movement into the continent from island South East Asia, and on human modification

of the environment (O'Connell & Allen 2004). Charcoal peaks within pollen deposits allowed Singh et al. (1981) to propose a 130 ka settlement age, with subsequent evidence used to propose dates as early as 185 ka (Kershaw et al. 1993; Wang et al. 1998). Apparently pre-LGM dates from the site of Jinmium lent support for such an early arrival; however, further study showed that the archaeological record at the site was no older than 20 ka (Roberts et al. 1998) and the charcoal peaks and changes in pollen spectra at Lake George and Lynch's Crater were subsequently shown to be the result of environmental changes including the prevalence of natural fires (O'Connell & Allen 2004, p. 837).

The oldest age estimates from archaeological deposits (i.e. where datable material is associated with artefacts) also proved problematic because the ages obtained pushed radiocarbon techniques to the limit (e.g. Chappell et al. 1996; Roberts 1997) and required the use of alternative techniques (e.g. thermoluminescence (TL) and optically stimulated luminescence (OSL); Roberts et al. 1990). Ancient radiocarbon age estimates face the dual problem of sample contamination and calibration. Nevertheless, most recent assessments place the age of initial settlement in Australia at 42–45 ka, with widespread occupation of the continent by 35 ka (e.g. O'Connell & Allen 2004, 2007).

Second, past environments provided different resources to Aboriginal people. Therefore it was necessary to estimate what resources were available at different times in the past, or at least to determine how past environments affected access to these resources. Smith (2005) outlined earlier views that deserts in general, and the Australian arid zone in particular, were not settled until well after the LGM, a conclusion now refuted by recent evidence of settlement from around 35 ka (Hiscock & Wallis 2005). Although lakes existed in northern and eastern Australia, these were largely dry or saline by the time of human settlement, apart from the Darling and Willandra lakes (Smith 2005). Whether the pre-LGM arid zone environment encouraged settlement and subsequently became drier, requiring cultural changes to groups already inhabiting arid regions (Hiscock & Wallis 2005), or whether desert regions were colonized during arid periods with dispersal during relatively short periods of climatic amelioration (Smith 2005), is debated. Evidence indicates a range of habitats in use by 30 ka from the central Australian ranges through to the arid littoral regions of the North West Cape, but the imprecision of the dating techniques makes it difficult to determine if inland regions were occupied at the same time or sequentially (Hiscock & Wallis 2005). Dating imprecision, along with the nature of palaeoenvironmental records, also make it hard to precisely interpret the nature of human settlement in the face of environmental change. Given the relative poverty of the archaeological record, a result of a limited number of well-preserved deposits, large geographical regions and research by only a handful of archaeologists, there is a tendency to search for general explanations based on only a few sets of data. Hiscock and Wallis (2005, fig. 3.1), for instance, described eight sites in the arid or semi-arid regions of Australia and used these to propose a 'desert transformation' model where generalist foragers colonized the interior, concentrating on riverine and lacustrine environments. According to this model, desert adaptations emerged later during subsequent environmental desiccation. There are obvious sampling issues when trying to construct such a model on so small a dataset. Although a better understanding of environmental change is now available than in Peterson's time, and few would characterize past environments as unchanging, there are very few case studies (beyond those in the Willandra region and SW Tasmania, both discussed above) in which the archaeological and palaeoenvironmental records are well matched.

Similar problems occur when trying to assess the causes of megafaunal extinctions in Australia. The fossil bones of large extinct marsupials were first recorded for the Australian continent by Owens in 1870 (cited by Williams *et al.* 1998). As described by Williams *et al.* (1998, p. 237), the initial reaction of scientists was to invoke climate change as the sole cause. However, the apparent synchronicity of the arrival of humans on the Australian continent with the timing of megafaunal extinctions led many researchers to abandon climate change as an explanation in favour of human predation (e.g. Flannery 1994, 2007; Miller *et al.* 1999; Roberts *et al.* 2001). Sampling issues plague debate about megafaunal extinction, together with the inability to establish a definitive set of criteria with which to differentiate human activity from the effect of environmental change. Many species became extinct well before human settlement of the continent. Without a precise chronology, differentiating between human impact during initial settlement and environmental change is extremely difficult (Hiscock & Wallis 2005). Although evidence from a single site can never be considered definitive, the existence of megafauna and of artefacts at Cuddie Springs from 30–20 ka suggests that human–megafaunal interactions were both drawn out and likely to be regionally variable (Field & Dodson 1999; Wroe *et al.* 2002).

Global environmental change since the LGM is sometimes considered to be a primary driver of human societal change, particularly population increases supported by the change from a hunter–gatherer mode to settled agriculture. In Australia, Rowlands (1999) proposed that Holocene variability should be given more consideration in accounting for change in Aboriginal society, particularly changes related to ENSO. Subsequently, Hiscock (2002) suggested that the onset of heightened ENSO variability beginning in the mid-Holocene might help explain changes in stone artefact technology. Cosgrove *et al.* (2007) have argued that occupation of the Queensland rainforest involved movement into a zone with relatively high-cost resources and that this movement is best understood as a response to increased uncertainty brought on by ENSO-induced resource unpredictability. Working in far north Queensland, Haberle & David (2004) related climatic and environmental transformations in the early to mid-Holocene to notable increases in population across north Queensland by 6000 cal. years BP (the so-called mid-Holocene climatic optimum). When natural levels of bioproduction began to decrease and climatic variability increased in the late Holocene, under the influence of increasingly intense and frequent ENSO activity, it is suggested that heightened regional populations began to split into smaller, land-owning and land-using groups, evidenced by a regionalization of rock art styles around 3700 cal. years BP (Haberle & David 2004, p. 177). Bourke *et al.* (2007) summarized the evidence for substantial changes in shell fishing practices in three sites from Arnhem Land after 1000 cal. years BP related to ENSO activity as well as the Little Climatic Optimum (LCO; 1200–700 years BP) and the Little Ice Age (600–100 years BP). These environmental changes were also related to shifts in the culture and society of people inhabiting the Torres Strait 800–600 cal. years BP (McNiven 2006).

Despite greater attention to the relationship between human activity and environmental change, however, direct correlations between the two are not easy to achieve. Smith *et al.* (2008) have used spectral analysis to demonstrate a periodicity of 1340 years in a dataset of radiocarbon determinations from Australian dryland sites. They reported a positive cross-correlation with the number of ENSO events per 100 years. The 1340 year periodicity may reflect a 1470 year periodicity reported for ENSO or may be correlated with shorter 210 year de Vries or Suess solar cycles (Smith *et al.* 2008). However, these relationships are apparent only when radiocarbon samples from very large geographical regions are combined. It may be hard therefore to determine local variation in the behavioural significance of these regional correlations.

Fire as a driver of vegetation change in Australia is another subject where the role of humans is debated, but difficult to resolve, mainly because of the different temporal and spatial scales at which the evidence has accumulated. The pollen and charcoal records from Lake George and Lynch's Crater (Singh *et al.* 1981) were the first to be interpreted as evidence that Aboriginal burning had dramatically changed the Australian vegetation, in each case because a pollen and charcoal signature that cannot be explained from the glacial–interglacial cycles could be discerned (Head 2000*a*, p. 123). These records have since been substantially reinterpreted, but debate about the role of Aboriginal activity in modifying the environment at the local scale remains. Head (2000*a*, p. 125), for example, cited a number of examples of plant management strategies and practices by Aboriginal people that must, she claimed, have considerably affected vegetation communities at the local scale. Although such evidence is regularly discussed within anthropology and archaeology, Head maintained that there are very few ecologists who analyse Australian vegetation with the assumption of an inbuilt legacy of thousands of years of Aboriginal interaction. Differences in temporal scale are a major problem: ethnographic and historical records cover a mere 200 years, pollen records may span thousands of years, and archaeological chronologies often fall somewhere between. Impacts that are obvious at the scale of human lifetimes might not be apparent at the centuries to millennia scales of pollen and charcoal records. A good example of the mismatch between archaeological and ecological data is the comparison of Attenbrow's (2004) data on the number of habitations occupied by Aboriginal people in the Sydney region and the abundance of charcoal in a sediment core spanning the late Pleistocene and Holocene from Lake Baraba, SW of Sydney (Black *et al.* 2006). Black and colleagues were attempting to determine whether or not the pattern of charcoal deposition in the core sediments could be related to Aboriginal burning. Although the charcoal data are high resolution and site-specific, the archaeological occupancy data against which they are compared are regionally based and pooled from archaeological site records of variable resolution. This variability in resolution makes it very difficult to draw firm conclusions.

To summarize, human–environment relationships are a key aspect of human presence in Australia but issues of temporal and spatial scale continue to pose challenges for geoarchaeologists in much the same way that scale poses challenges for those who attempt to derive behavioural inferences from the contexts in which artefacts are found. Low-resolution temporal and spatial correlations may be hard to interpret in ways that are relevant to the activities of past peoples. Differentiating environmental impacts from the influence of human behaviour has proven difficult in research

on megafaunal extinction, the impact of climatic shifts such as ENSO and the significance of fire. One solution advanced in recent studies is to stop trying to separate human and environmental effects and concentrate on the complex interrelated set of processes operating at different temporal and spatial scales that characterize this relationship.

Integrating geomorphology and archaeology: decoupling space and time

As discussed above, Peterson (1971) proposed that time could be subsumed by space if archaeologists ignored what he termed the 'idiosyncrasies' of the archaeological record and instead focused on the use of large areas where environment might be held constant. Application of Peterson's suggestion involved use of the distribution of archaeological materials (normally identified as sites) across landscapes typically defined relative to a drainage system. The distribution of sites was then interpreted as the archaeological manifestation of a settlement pattern, the long-term consequence of accessing the different resources available within a single environment.

At the time it was written, Peterson's (1971) approach matched the functional notions of how spatial distributions of artefacts should be assessed. If groups of artefacts represented toolkits and if surfaces could be thought of as living floors (or at least approximations of living floors), then sites might be thought of as the material remains from sets of activities, themselves identified with particular functions. Following this line of reasoning, sites with different sets of artefacts distributed across space should provide the means to understand how space was used in the past. One might expect patterns of largely consistent use of landscape, separated by periods of marked change when either the environment was in transition or the relationship of people to that environment shifted; for example, during the mid- to late Holocene according to the 'intensification theory' of Lourandos (1983, 1985, 1997). However, the studies that emphasize how artefacts accumulate in sites rather than how they functioned has forced a review of the chain of inference that leads from artefact to site function (Wandsnider 1996). If single artefact life histories are considered without the assumption that spatial proximity equates with functional homogeneity, then the warrant for identifying site function becomes much less secure. Single artefact life histories also require us to acknowledge that groups of artefacts found at one location most probably accumulated as the result of multiple occupations. Although always implicitly understood, the effects of multiple occupations were effectively ignored while interpretation was focused on a general understanding of function. The impact of multiple occupations has become much harder to ignore when artefact deposition is itself understood to be an outcome of the life history of the uses of artefacts, and the composition of groups of artefacts is understood to reflect the duration of artefact accumulation (e.g. Holdaway *et al.* 2008*b*).

As archaeologists have better understood patterns in groups of artefacts, the range of variables involved has greatly increased. For stone artefacts, for example, the location and shape of the raw material together with its flaking properties are important. So is the technology of manufacture. These variables are in turn influenced by the relative mobility of the people who made the artefacts as well as the constraints imposed by the uses for which the artefacts were intended (e.g. Douglass *et al.* 2008). Added to this, the units used to group artefacts together (e.g. sediment strata, or the length of time a surface was potentially available for artefact accumulation) have a bearing on the composition of resulting groups of artefacts and therefore the nature of the inferences drawn on the basis of these groups. Our own research in western NSW has demonstrated that geomorphological processes of erosion and deposition, operating over a range of time scales from the immediate to the long term, have removed or covered up some of the surface artefacts discarded by Aboriginal people in the past (Fanning *et al.* 2007). Thus, the archaeological record is in some places discontinuous in time because geomorphological events have removed the record of particular time periods (Fanning *et al.* 2008), and it is discontinuous in space because it is preserved only in places that are geomorphologically relatively inactive (Fanning *et al.* 2009). Important inferences concerning the past may still be drawn from the record but the processes responsible for both the presence and absence of the record must be considered.

The interactions among the variables discussed above have the potential to create many different patterns. Like geomorphologists, archaeologists face a problem of equifinality of assemblage form, in that groups of artefacts that look similar in composition may reflect different depositional histories. We can no longer be confident that the patterns in artefacts from different locations simply represent differences in the types of activities undertaken at those localities. Function is certainly still implicated, but it is better thought of as 'functioning', a set of processes responding to a host of changing variables that themselves are subject to change though time. The last 20 years of artefact studies have provided a very good appreciation of the wealth of processes that contribute to variability in the archaeological record, but this appreciation has not always been matched by our ability to interpret what this variability might mean in a behavioural sense.

It is not at all clear how a formational view of artefacts is to be integrated with past environments. Allen (1992), for instance, commented on the movement away from pan-continental explanations of the archaeological record to a more regional approach, citing the decline of the use of the terminology 'Australian core tool and scraper industry' and 'Australian small tool tradition' as good examples. A variety of regional sequences of artefact change might prove to be better suited to describing human–environment relationships across the diverse Australian continent. However, with comparisons between groups of artefacts found at different locations requiring the type of contextual information suggested by the above discussion, it remains unclear how such regions might be defined. It remains to be seen whether the level of variability between regions is in fact greater than the level of variability among groups of artefacts that are spatial neighbours. Despite repeated calls for a regional approach, very few studies have succeeded in defining what is distinctive about a particular region in comparison with others. The validity of regional approaches cannot be assumed just because different researchers each work in different regions.

The simplest way to decouple space and time is to pay attention to the age of surfaces within a particular study region. Dealing with surface archaeological deposits necessitates an emphasis on chronology because, up to now, chronology has been seen as limiting the utility of these deposits compared with those that are stratified. In western NSW, a combination of OSL and radiocarbon age estimates has provided chronologies for a range of surface deposits (Fanning & Holdaway 2001; Holdaway *et al.* 2002, 2005; Fanning *et al.* 2008; Rhodes *et al.* 2009, 2010). These chronologies demonstrate that surfaces separated by, at most, a few kilometres can differ in age by thousands of years. In a relatively few locations, such as stream terraces, surfaces have survived since the beginning of the Holocene, whereas in the majority of cases erosion has removed sediment deposits with ages greater than 1000–2000 years. As a consequence, the archaeological records preserved on these surfaces have variable chronologies. Thus the current distribution and age of groups of artefacts on the surface reflect a long history of differential erosion, and not simply the use of the landscape during one period.

There is evidence to suggest that surface erosion was initiated by climatic events; in particular, intense rainstorms that occur from time to time in western NSW. At the Nundooka (ND) study site at Fowlers Gap, for example, there is a good correlation between the age estimates previously determined by Jansen & Brierley (2004) for a series of palaeofloods and OSL age estimates for sediments deposited by those floods. In turn, ages for these flood events correlate with radiocarbon age estimates we obtained for the remains of heat-retainer hearths excavated into those sediments (Fanning *et al.* 2007). Our interpretation is that floods removed all but remnants of the sediments that once lined the valley in which the ND site is located, removing any trace of earlier archaeological deposits. Some centuries after these floods Aboriginal people returned and, over a period of about 600 years, constructed hearths and deposited artefacts. It may be less that the floods drove Aboriginal people away than that the immediate aftermath of the floods changed the local environment sufficiently to make it less attractive for reoccupation. Regardless of the reasons for the delay in post-flood reoccupation of the location, the effect of a localized flood event was a change in the chronology of human occupation at this site. The variability that this produces in archaeological chronology and artefact assemblage composition will need to be assessed before regional comparisons are made.

At the same time, radiocarbon determinations from hearths on relatively stable surfaces indicate a discontinuous pattern of radiocarbon ages through time (Holdaway *et al.* 2005). Hearth ages tend to group together, each group separated from the others by periods in which no hearth ages fall. This pattern is unlikely to be due to sampling error, as we have acquired nearly 200 radiocarbon determinations from multiple locations across western NSW. With no field evidence for differential preservation of these hearths, nor differential preservation of the surfaces in which they rest (groups of hearths with different ages are separated by a few tens of metres on the same depositional surfaces (Holdaway *et al.* 2005, 2008*a*)), then the gaps in the record from these geomorphologically stable locations indicate discontinuities in occupation. We conclude that Aboriginal occupation of at least some parts of western NSW was intermittent, with periods of abandonment measured in centuries.

An environmental correlation with these periods of abandonment (or, at least, of no hearth construction) is seen at Stud Creek in the far northwestern corner of NSW. Here, the gap in the distribution of hearth ages correlates with the global climatic change known as the Little Climatic Optimum (LCO) or the Medieval Climatic Anomaly (MCA) (Holdaway *et al.* 2002). There is little evidence from which to argue that the MCA caused people to abandon the Stud Creek location: we lack the palaeoclimatic data from local sites that would indicate the local effects of the MCA. However, the coincidence does suggest that gaps in the radiocarbon record may be a measure of human–environment interaction visible at a temporal scale measured in centuries. Humans interacted successfully with the environment over periods spanning several centuries. Almost certainly, Aboriginal people abandoned areas, and subsequently returned, for periods of time not discernible with current radiocarbon resolution. However, at Stud Creek between *c.* AD 800 and 1100, something clearly changed and people stopped returning. Using Head's (2000*b*) terminology, gaps in the radiocarbon record may represent the sum of small-scale changes in human–environment interaction that are difficult to detect using normal palaeoecological approaches. Although at first it may seem paradoxical, gaps in the chronological record provide just the type of temporal detail that is lacking in accounts that describe a static human adaptation to the environment.

It is not possible to tie groups of stone artefacts directly to the hearth chronologies. Artefacts are present where hearths are found and it is probably true that at least some were deposited at the same time as hearth construction, but precise correlation is not possible. Continuities in the forms of artefacts found at all of the sampling locations we have studied suggest a form of environmental interaction that persisted at some level for many centuries. However, there are also variations in the composition of groups of artefacts found at different locations that suggest a more complex occupation history. At the previously discussed ND location, for instance, flakes, cores and tools, made from combinations of materials that are local and materials that must have been brought from elsewhere, show patterns of core reduction, raw material use and tool deposition that are somewhat inconsistent with each other (Shiner *et al.* 2005; Holdaway *et al.* 2008*b*). The expected intensity of core reduction of locally available quartz is not sufficient to account for the number of long-uselife tools present. Without multiple patterns of core reduction and tool production reinforcing one another it becomes more difficult to suggest a single functional interpretation for the groups of artefacts found at ND. The alternative, following the view that emphasizes the processes by which artefacts accumulate, is that the archaeological record is in fact the result of separate occupations each producing a different group of artefacts. The lack of consistent patterns in core reduction and tool production resulting from these occupations is the outcome of history in action. Rather than locations in the landscape that each represent a single function, variability within groups of artefacts and variability between groups of artefacts reflect the complex history of human–environment relationships, not in the static sense that humans undertook a series of tasks at different points in the landscape, but in a dynamic sense where environmental opportunities were in a state of flux and Aboriginal people reacted in different ways at different times.

Culturally, there is much interest in the degree of variability in the way places were used at different times in the past, just as there is interest in variation in when different places were occupied. Variability within the record accrues as the result of times passing. Instants of behaviour, even if they could be defined, will produce no pattern. The gaps in the records of hearth construction, for example, are definable only in relation to clusters of age determinations at both ends of their span. Environmentally, the challenge is to introduce a similar consideration of variability into explanations of the interaction between people and the environment so as to move beyond the static view of human adaptation to a single environment (such as 'the arid zone', for example).

Conclusions

In spite of the intervening four decades of research, some of the problems foreseen by Peterson (1971), with marrying archaeological records of human behaviour with geoscientific records of environmental change, remain. The papers published 20 years later in *Naïve Lands* (Dodson 1992) demonstrated not only the vast increase in knowledge of the Quaternary in the intervening two decades, but also the increasing interdisciplinarity of much of the research (Allen 1992). However, until now, the limitations of existing dating methods and the related inability to construct robust chronologies of Aboriginal occupation patterns across space have meant that there have been very few attempts to fully integrate geomorphological and archaeological survey and analysis to investigate human–environment interactions on the same spatial and temporal scales.

There are now good Australian examples of such integration to be found alongside other studies that require greater attention to the temporal scales with which different datasets are constructed. One of the challenges for scientists from all disciplines is how to relate the effects of large-scale environmental processes to the actions and perceptions of the people who inhabited different parts of Australia in the past. Answering this challenge will require careful consideration of the scales at which both archaeological and palaeoenvironmental records can be resolved. One profitable research direction is to consider how records resolvable at different scales interact, thereby producing a more nuanced view of causation (Brown 2008). To be successful, such studies will require that archaeologists and environmental scientists develop a good understanding of both the potential and the limitations of each others' datasets. This is a direction that could be profitably followed over the next 20 years of geoarchaeological studies.

References

ALLEN, H. 1972. *Where the Crow Flies Backwards: Man and Land in the Darling Basin*. PhD thesis, Australian National University, Canberra.

ALLEN, H. & HOLDAWAY, S. J. 2009. The archaeology of Mungo and the Willandra Lakes: looking back, looking forward. *Archaeology in Oceania*, **44**, 96–106.

ALLEN, H., HOLDAWAY, S. J., FANNING, P. C. & LITTLETON, J. 2008. Footprints in the sand: appraising the archaeology of the Willandra Lakes. *Antiquity*, **82**, 11–24.

ALLEN, J. 1992. Conclusion: prolonged applause and stamping of feet? *In*: DODSON, J. (ed.) *The Naïve Lands: Prehistory and Environmental Change in Australia and the SouthWest Pacific*. Longman, Harlow, 242–247.

ALLEN, J. 1996. Report of the Southern Forests Archaeological Project Volume 1 Site Descriptions, Stratigraphies and Chronologies. School of Archaeology, La Trobe University, Bundoora, Vic.

AMBROSE, W. & DUERDEN, P. 1982. *Archaeometry: An Australasian Perspective*. Department of Prehistory, Research School of Pacific Studies, Australian National University, Canberra.

AMBROSE, W. & MUMMERY, J. M. J. 1987. *Archaeometry: Further Australasian Studies*. Department of Prehistory, Research School of Pacific Studies, Australian National University, Canberra.

ATTENBROW, V. 2004. *What's changing? Population size or land use patterns? The Archaeology of Upper Mangrove Creek*. Pandanus Books, Sydney Basin.

BAILEY, G. 2007. Time perspectives, palimpsests and the archaeology of time. *Journal of Anthropological Archaeology*, **26**, 198–223.

BAYNES, A. 1999. The absolutely last remake of Beau Geste: yet another review of the Australian megafaunal radiocarbon dates. *In*: BAYNES, A. & LONG, J. A. (eds) *Papers in Vertebrate Palaeontology*. Records of the West Australian Museum, Supplement, **57**, 391.

BIRD, C. F. M. & FRANKEL, D. 1991. Problems in constructing a prehistoric regional sequence: Holocene southeast Australia. *World Archaeology*, **23**, 179–192.

BLACK, M. P., MOONEY, S. D. & MARTIN, H. A. 2006. A >43,000 year vegetation and fire history from Lake Baraba, New South Wales, Australia. *Quaternary Science Reviews*, **25**, 3005–3016.

BOURKE, P., BROCKWELL, S., FAULKNER, P. & MEEHAN, B. 2007. Climate variability in the mid to late Holocene Arnhem Land region, north Australia: Archaeological archives of environmental and cultural change. *Archaeology in Oceania*, **42**, 91–101.

BOWLER, J. M. 1971. Pleistocene salinities and climate change: evidence from lakes and lunettes in Southeastern Australia. *In*: MULVANEY, D. J. & GOLSON, J. (eds) *Aboriginal Man and Environment in Australia*. ANU Press, Canberra, 47–65.

BOWLER, J. M. 1998. Willandra Lakes revisited: environmental framework for human occupation. *Archaeology in Oceania*, **33**, 120–155.

BOWLER, J. M., JOHNSTON, H., OLLEY, J. M., PRESCOTT, J. R., ROBERTS, R. G., SHAWCROSS, W. & SPOONER, N. A. 2003. New ages for human occupation and climatic change at Lake Mungo, Australia. *Nature*, **421**, 837–840.

BOWMAN, D. M. J. S., GARDE, M. & SAULWICK, A. 2001. *Kunj-ken makka man-wurrk* Fire is for kangaroos: interpreting Aboriginal accounts of landscape burning in Central Arnhem Land. *In*: ANDERSON, A., LILLEY, I. & O'CONNOR, S. (eds) *Essays in Honour of Rhys Jones*. ANH Publications. Australian National University, Canberra, 61–78.

BROWN, A. G. 2008. Geoarchaeology, the four dimensional (4D) fluvial matrix and climatic causality. *Geomorphology*, **101**, 278–297.

BUTZER, K. W. 1971. *Environment and Archaeology: An Ecological Approach to Prehistory*. 2nd edn. Methuen, London.

BUTZER, K. W. 1982. *Archaeology as Human Ecology: Method and Theory for a Contextual Approach*. Cambridge University Press, Cambridge.

CHAPPELL, J., HEAD, J. & MAGEE, J. 1996. Beyond the radiocarbon limit in Australian archaeology and Quaternary research. *Antiquity*, **70**, 543–552.

CLARKSON, C. & LAMB, L. 2005. *Lithics 'Down Under': Australian Perpectives on Lithic Reduction, Use and Classification*. British Archaeological Research Series, **1408**.

COSGROVE, R., FIELD, J. & FERRIER, A. 2007. The archaeology of Australia's tropical rainforests. *Palaeogeography, Palaeoclimatology, Palaeoecology*, **251**, 150–173.

COSTIN, A. B. 1971. Vegetation, soils and climate in Late Quaternary Southeastern Australia. *In*: MULVANEY, D. J. & GOLSON, J. (eds) *Aboriginal Man and Environment in Australia*. ANU Press, Canberra, 26–37.

DAVIDSON, D. A. & SHACKLEY, M. L. 1976. *Geoarchaeology: Earth Science and the Past*. Duckworth, London.

DODSON, J. 1992. *The Naïve Lands: Prehistory and Environmental Change in Australia and the Southwest Pacific*. Longman, Harlow.

DODSON, J., FULLAGAR, R. & HEAD, L. 1992. Dynamics of environment and people in the forested crescents of temperate Australia. *In*: DODSON, J. (ed.) *The Naïve Lands: Prehistory and Environmental Change in Australia and the Southwest Pacific*. Longman, Harlow, 115–159.

DODSON, J., FULLAGAR, R., FURBY, J., JONES, R. & PROSSER, I. 1993. Humans and megafauna at Cuddie Springs, NSW. *Archaeology in Oceania*, **28**, 93–99.

DOUGLASS, M. J., HOLDAWAY, S. J., FANNING, P. C. & SHINER, J. I. 2008. An assessment and archaeological application of cortex measurement in lithic assemblages. *American Antiquity*, **73**, 513–526.

FANKHAUSER, B. L. & BIRD, J. R. 1993. *Archaeometry: Current Australasian Research*. Department of Prehistory, Research School of Pacific Studies, Australian National University, Occasional Papers in Prehistory, **22**.

FANNING, P. C. & HOLDAWAY, S. J. 2001. Temporal limits to the archaeological record in arid western NSW, Australia: lessons from OSL and radiocarbon dating of hearths and sediments. *In*: JONES, M. & SHEPPARD, P. (eds) *Australasian Connections and New Directions: Proceedings of the 7th Australasian Archaeometry Conference*. Research in Anthropology and Linguistics, **5**, 85–104.

FANNING, P. C., HOLDAWAY, S. J. & RHODES, E. J. 2007. A geomorphic framework for understanding the surface archaeological record in arid environments. *Geodinamica Acta*, **20**, 275–286.

FANNING, P. C., HOLDAWAY, S. J. & RHODES, E. J. 2008. A new geoarchaeology of Aboriginal artifact deposits in western NSW, Australia: establishing spatial and temporal geomorphic controls on the surface archaeological record. *Geomorphology*, **101**, 526–532.

FANNING, P. C., HOLDAWAY, S. J., RHODES, E. J. & BRYANT, T. G. 2009. The surface archaeological record in Australia: geomorphic controls on preservation, exposure and visibility. *Geoarchaeology*, **24**, 121–146.

FIELD, J. & DODSON, J. 1999. Late Pleistocene megafauna and archaeology from Cuddie Springs, southeastern Australia. *Proceedings of the Prehistoric Society*, **65**, 275–301.

FLANNERY, T. F. 1994. *The Future Eaters: An Ecological History of the Australasian Lands and People*. Reed Books, Chatswood.

FLANNERY, T. F. 2007. The trajectory of human evolution in Australia 10,000 BP to the present. *In*: COSTANZA, R., GRAUMLICH, L. J. & STEFFEN, W. (eds) *Sustainability of Collapse? An Integrated History and Future of People on Earth. Report of the 96th Dahlem Workshop on Integrated History and Future of People on Earth (IHOPE) Berlin, 12–17 June 2005*. MIT and Freie Universität Berlin, Cambridge, MA, and Berlin, 89–94.

FRANKEL, D. 1988. Characterising change in prehistoric sequences: a view from Australia. *Archaeology in Oceania*, **23**, 41–48.

GALLOWAY, R. W. 1971. Evidence for Late Quaternary climates. *In*: MULVANEY, D. J. & GOLSON, J. (eds) *Aboriginal Man and Environment in Australia*. ANU Press, Canberra, 14–25.

GILLESPIE, R. 2002. Dating the first Australians. *Radiocarbon*, **44**, 455–472.

GILLESPIE, R., HORTON, D. R., LADD, P., MACUMBER, P. M., RICH, T. H., THORNE, A. R. & WRIGHT, R. V. S. 1978. Lancefield Swamp and the extinction of the Australian megafauna. *Science*, **200**, 1044–1048.

GLADFELTER, B. G. 1977. Geoarchaeology: the geomorphologist and archaeology. *American Antiquity*, **42**, 519–538.

HABERLE, S. G. & CHEPSTOW-LUSTY, A. 2000. Can climate influence cultural development? A view through time. *Environment and History*, **6**, 349–369.

HABERLE, S. G. & DAVID, B. 2004. Climates of change: human dimensions of Holocene environmental change in low latitudes of the PEPII transect. *Quaternary International*, **118–119**, 165–179.

HEAD, L. 1994. Landscapes socialised by fire: post-contact changes in Aboriginal fire use in northern Australia, and implications for prehistory. *Archaeology in Oceania*, **29**, 172–181.

HEAD, L. 1996. Rethinking the prehistory of hunter–gatherers, fire, and vegetation change in northern Australia. *Holocene*, **6**, 481–487.

HEAD, L. 2000a. *Cultural Landscapes and Environmental Change.* Arnold, London.

HEAD, L. 2000b. *Second Nature: the History and Implications of Australia as Aboriginal Landscape.* Syracuse University Press, Syracuse, NY.

HERZ, N. & GARRISON, E. G. 1998. *Geological Methods for Archaeology.* Oxford University Press, Oxford.

HISCOCK, P. 1983. Stone tools as cultural markers? *Australian Archaeology*, **16**, 48–57.

HISCOCK, P. 2002. Pattern and context in the Holocene proliferation of backed artefacts in Australia. *In*: ELSTON, R. G. & KUHN, S. L. (eds) *Thinking Small: Global Perspectives on Microlithization.* Archaeological Papers of the American Anthropological Association, **12**, 163–177.

HISCOCK, P. & WALLIS, L. 2005. Pleistocene settlement of deserts from an Australian perspective. *In*: VETH, P. & HISCOCK, P. (eds) *Desert Peoples: Archaeological Perspectives.* Blackwell, Oxford, 34–57.

HOLDAWAY, S. J. 2004. *Continuity and Change: An Investigation of the Flaked Stone Artefacts from the Pleistocene Deposits at Bone Cave South West Tasmania, Australia, Bundoora.* Archaeology Program, School of Historical and European Studies, La Trobe University, Bundoora, Vic.

HOLDAWAY, S. J. & STERN, N. 2001. Paleolithic archaeology. *In*: MURRAY, T. (ed.) *The Encyclopedia of Archaeology, History and Discoveries.* ABC-CLIO, Santa Barbara, CA, 974–984.

HOLDAWAY, S. J. & STERN, N. 2004. *A Record in Stone: Methods and Theories for Analysing Australian Flaked Stone Artefacts.* Museum Victoria and Australian Institute of Aboriginal and Torres Strait Islander Studies, Melbourne.

HOLDAWAY, S. J., FANNING, P. C., WITTE, D. C., JONES, M., NICHOLLS, G., REEVES, J. & SHINER, J. 2002. Variability in the chronology of Late Holocene Aboriginal occupation on the arid margin of southeastern Australia. *Journal of Archaeological Science*, **29**, 351–363.

HOLDAWAY, S. J., FANNING, P. C. & SHINER, J. 2005. Absence of evidence or evidence of absence? Understanding the chronology of indigenous occupation of western New South Wales, Australia. *Archaeology in Oceania*, **40**, 33–49.

HOLDAWAY, S. J., FANNING, P. C. & RHODES, E. J. 2008a. Challenging intensification: human–environment interactions in the Holocene geoarchaeological record from western New South Wales, Australia. *Holocene*, **18**, 411–420.

HOLDAWAY, S. J., FANNING, P. C. & RHODES, E. J. 2008b. Assemblage accumulation as a time dependent process in the arid zone of western New South Wales, Australia. *In*: HOLDAWAY, S. J. & WANDSNIDER, L. A. (eds) *Time in Archaeology: Time Perspectivism Revisited.* University of Utah Press, Salt Lake City, 110–133.

HORTON, D. R. 1982. The burning question: Aborigines, fire and Australian ecosystems. *Mankind*, **13**, 237–251.

HUGHES, P. J. & SULLIVAN, M. E. 1982. Geoarchaeology in Australia: a review. *In*: AMBROSE, W. &

DUERDEN, P. (eds) *Archaeometry: an Australasian Perspective.* Australian National University Press, Canberra, 100–111.

ISAAC, G. 1981. Stone age visiting cards: approaches to the study of early land use patterns. *In*: HODDER, I., ISAAC, G. & HAMMOND, N. (eds) *Pattern of the Past: Studies in Honour of David Clarke.* Cambridge University Press, Cambridge, 131–155.

JANSEN, J. D. & BRIERLEY, G. J. 2004. Pool-fills: a window to palaeoflood history and response in bedrock-confined rivers. *Sedimentology*, **51**, 901–925.

JOHNSON, C. N. 2005. What can the data on late survival of Australian megafauna tell us about the cause of their extinction? *Quaternary Science Reviews*, **24**, 2167–2172.

JOHNSTON, H. 1993. Pleistocene shell middens of the Willandra Lakes. *In*: SMITH, M. A., SPRIGGS, M. & FANKHAUSER, B. (eds) *Sahul in Review: Pleistocene Archaeology in Australia, New Guinea and Island Melanesia.* Department of Prehistory, Research School of Pacific Studies, Australian National University, Occasional Papers in Prehistory, **24**, 197–203.

JONES, R. 1968. The geographical background to the arrival of man in Australia and Tasmania. *Archaeology and Physical Anthropology in Oceania*, **3**, 186–215.

JONES, R. 1969. Fire-stick farming. *Australian Natural History*, **16**, 224–228.

KERSHAW, A. P., MCKENZIE, G. M. & MCMINN, A. 1993. Quaternary vegetation history of northeastern Queensland from pollen analysis of ODP site 820. *Proceedings of the Ocean Drilling Program, Scientific Results*, **133**, 107–114.

KERSHAW, A. P., CLARK, J. S., GILL, A. M. & D'COSTA, D. M. 2002. A history of fire on Australia. *In*: BRADSTOCK, R., WILLIAMS, J. & GILL, M. (eds) *A History of Fire in Australia.* Cambridge University Press, Cambridge, 3–25.

LOURANDOS, H. 1983. Intensification: a late Pleistocene–Holocene archaeological sequence from south-western Victoria. *Archaeology in Oceania*, **18**, 81–94.

LOURANDOS, H. 1985. Intensification and Australian prehistory. *In*: PRICE, T. D. & BROWN, J. A. (eds) *Prehistoric Hunter–Gatherers the Emergence of Cultural Complexity.* Academic Press, New York, 385–423.

LOURANDOS, H. 1997. *Continent of Hunter–Gatherers: New Perspectives in Australian Prehistory.* Cambridge University Press, Cambridge.

LOURANDOS, H. & DAVID, B. 1998. Comparing long-term archaeological and environmental trends: north Queensland, arid and semi-arid Australia. *Artefact*, **21**, 105–114.

MCINTYRE, M. L. & HOPE, J. H. 1978. *Procoptodon* fossils from the Willandra Lakes, Western New South Wales. *Artefact*, **3**, 117–132.

MCNIVEN, I. 2006. Dauan 4 and the emergence of ethnographically known social arrangements across Torres Strait during the last 600–800 years. *Australian Archaeology*, **62**, 1–12.

MILLER, G. H., MAGEE, J., JOHNSON, B. M., FOGEL, M. L., SPOONER, N. A., MCCULLOCH, M. T. & AYLIFFE, L. K. 1999. Pleistocene extinction of *Genyornis newtoni*: human impact on Australian megafauna. *Science*, **283**, 205–208.

MULVANEY, D. J. & GOLSON, J. 1971. *Aboriginal Man and Environment in Australia*. ANU Press, Canberra.

O'CONNELL, J. F. & ALLEN, J. 2004. Dating the colonization of Sahul (Pleistocene Australia–New Guinea): a review of recent research. *Journal of Archaeological Science*, **31**, 835–853.

O'CONNELL, J. F. & ALLEN, J. 2007. Pre-LGM Sahul (Pleistocene Australia–New Guinea) and the archaeology of early modern humans. *In*: MELLARS, P., BOYLE, K., BAR-YOSEF, O. & STRINGER, C. (eds) *Rethinking the Human Revolution*. McDonald Institute for Archaeological Research, Cambridge, 395–410.

PELS, S. 1971. River systems and climate changes in Southeastern Australia. *In*: MULVANEY, D. J. & GOLSON, J. (eds) *Aboriginal Man and Environment in Australia*. ANU Press, Canberra, 38–46.

PETERSON, N. 1971. Open sites and the ethnographic approach to the archaeology of hunter–gatherers. *In*: MULVANEY, D. J. & GOLSON, J. (eds) *Aboriginal Man and Environment in Australia*. ANU Press, Canberra, 239–248.

POLLARD, A. M. (ed.) 1999. *Geoarchaeology: Exploration, Environments, Resources*. Geological Society, London, Special Publications, **165**.

RHODES, E. J., FANNING, P. C., HOLDAWAY, S. J. & BOLTON, C. 2009. Ancient surfaces? Dating archaeological surfaces in western NSW using OSL. *In*: FAIRBAIRN, A. & O'CONNOR, S. (eds) *New Directions in Archaeological Science*. Terra Australis, **28**, 189–200.

RHODES, E. J., FANNING, P. C. & HOLDAWAY, S. J. 2010. Developments in optically stimulated luminescence age control for geoarchaeological sediments and hearths in western New South Wales, Australia. *Quaternary Geochronology*, **5**, 348–352.

ROBERTS, R. G. 1997. Luminescence dating in archaeology: from origins to optical. *Radiation Measurements*, **27**, 819–892.

ROBERTS, R. G., JONES, R. & SMITH, M. A. 1990. Thermoluminescence dating of a 50,000 year old human occupation site in northern Australia. *Nature*, **345**, 153–156.

ROBERTS, R. G., BIRD, M. *ET AL*. 1998. Optical and radiocarbon dating at Jinmium rock shelter in northern Australia. *Nature*, **395**, 358–362.

ROBERTS, R. G., FLANNERY, T. F. *ET AL*. 2001. New ages for the last Australian megafauna: continent-wide extinction about 46,000 years ago. *Science*, **292**, 1888–1892.

ROWLANDS, M. 1999. Holocene environmental variability: have its impacts been underestimated in Australian prehistory? *Artefact*, **22**, 11–40.

SCHIFFER, M. B. 1972. Archaeological context and systemic context. *American Antiquity*, **37**, 156–165.

SHAWCROSS, F. W. & KAYE, M. 1980. Australian archaeology: implications of current interdisciplinary research. *Interdisciplinary Science Reviews*, **5**, 112–128.

SHINER, J., HOLDAWAY, S. J., ALLEN, H. A. & FANNING, P. C. 2005. Understanding stone artefact assemblage variability in Late Holocene contexts in western New South Wales, Australia: Burkes Cave, Stud Creek and Fowlers Gap. *In*: CLARKSON, C. & LAMB, L. (eds) *Lithics Down Under: Australian Perspectives on Lithic Reduction, Use and Classification*. British

Archaeological Reports, International Monograph Series, **S1408**, 67–80.

SHOTT, M. J. 1989. Diversity, organization, and behavior in the material record. *Current Anthropology*, **30**, 283–315.

SHOTT, M. 1997. Activity and formation as sources of variation in Great Lakes Paleoindian assemblages. *Midcontinental Journal of Archaeology*, **22**, 197–236.

SINGH, G., KERSHAW, A. P. & CLARK, R. 1981. Quaternary vegetation and fire history in Australia. *In*: GILL, A. M., GROVES, R. H. & NOBLE, I. R. (eds) *Fire and the Australian Biota*. Australian Academy of Science, Canberra, 23–54.

SMITH, M. A. 2005. Desert archaeology, linguistic stratigraphy, and the spread of Western Desert language. *In*: VETH, P., SMITH, M. A. & HISCOCK, P. (eds) *Desert Peoples: Archaeological Perspectives*. Blackwell, Oxford, 222–242.

SMITH, M. A. & SHARP, N. 1993. Pleistocene sites in Australia, New Guinea and island Melanesia: geographic and temporal structure of the archaeological record. *In*: SMITH, M. A., SPRIGGS, M. & FANKHAUSER, B. (eds) *Sahul in Review: Pleistocene Archaeology in Australia, New Guinea and Island Melanesia*. Department of Prehistory, Research School of Pacific Studies, Australian National University, Occasional Papers in Prehistory, **24**, 37–59.

SMITH, M. A., WILLIAMS, A. N., TURNEY, C. S. M. & CUPPER, M. L. 2008. Human–environment interactions in Australian drylands: exploratory time-series analysis of archaeological records. *Holocene*, **18**, 389–401.

STERN, N. 2008. Time averaging and the structure of Late Pleistocene archaeological deposits in southwest Tasmania. *In*: HOLDAWAY, S. J. & WANDSNIDER, L. A. (eds) *Time in Archaeology: Time Perspectivism Revisited*. University of Utah Press, Salt Lake City, 134–148.

TIBBY, J., KERSHAW, A. P., BULITH, H., PHILIBERT, A., WHITE, C. & HOPE, G. S. 2006. Environmental change and variability in southwestern Victoria: changing constraints and opportunities for occupation and land use. *In*: DAVID, B., BARKER, B. & McNIVEN, I. (eds) *The Social Archaeology of Indigenous Societies: Essays on Aboriginal History in Honour of Harry Lourandos*. Aboriginal Studies Press, Canberra, 254–269.

TRUEMAN, C. N. G., FIELD, J. H., DORTCH, J., CHARLES, B. & WROE, S. 2005. Prolonged coexistence of humans and megafauna in Pleistocene Australia. *Proceedings of the National Academy of Sciences of the USA*, **102**, 8381–8385.

TURNEY, C. S. M. & HOBBS, D. 2006. ENSO influence on Holocene Aboriginal populations in Queensland, Australia. *Journal of Archaeological Science*, **33**, 1744–1748.

TURNEY, C. S. M. & PALMER, J. G. 2007. Does El Niño–Southern Oscillation control the interhemispheric radiocarbon offset? *Quaternary Research*, **67**, 174–180.

VETH, P. 1993. *Islands in the Interior: the Dynamics of Prehistoric Adaptations within the Arid Zone of Australia*. International Monographs in Prehistory. Archaeological Series, **3**.

VETH, P. 2006. Social dynamism in the archaeology of the Western Desert. *In*: DAVID, B., BARKER, B. &

McNiven, I. (eds) *The Social Archaeology of Australian Indigenous Societies*. Aboriginal Studies Press, Canberra, 242–253.

Wandsnider, L. A. 1996. Describing and comparing archaeological spatial structures. *Journal of Archaeological Method and Theory*, **3**, 319–384.

Wang, X., van der Kaars, S., Kershaw, A. P., Bird, M. & Jansen, F. 1998. A record of fire, vegetation and climate through the last three glacial cycles from Lombok Ridge core G6-4, eastern Indian Ocean, Indonesia. *Palaeogeography, Palaeoclimatology, Palaeoecology*, **147**, 241–256.

Williams, E. 1988. *Complex Hunter–gatherers: A Late Holocene Example from Temperate Australia*. British Archaeological Reports International Series, **423**.

Williams, M. A. J., Dunkerley, D. L., De Deckker, P., Kershaw, P. & Chappell, J. 1998. *Quaternary Environments*. 2nd edn. Arnold, London.

Wroe, S. & Field, J. H. 2006. A review of the evidence for a human role in the extinction of Australian megafauna and an alternative interpretation. *Quaternary Science Reviews*, **25**, 2692–2703.

Wroe, S., Field, J. H., Fullagar, R. & Jermiin, L. 2002. *Lost giants. Nature Australia*, **25**, 54.

Ecogeomorphology in the Australian drylands and the role of biota in mediating the effects of climate change on landscape processes and evolution

DAVID DUNKERLEY

School of Geography and Environmental Science, Monash University, Clayton, Victoria 3800, Australia (e-mail: David.dunkerley@arts.monash.edu.au)

Abstract: Australian dryland landscape has developed under the influence of aridity, low relief, tectonic stability and biota adapted to nutrient and water scarcity. The biota in general, but notably the plants, mediates the impact of water scarcity and of climate change on ecohydrological and geomorphological processes. It reduce the partitioning of rain into overland flow, and so limit soil erosion, notably through the development of patch structures that partition the landscape into local runoff sources and runon sinks. In large rain events, when flow does reach ephemeral streams, channel-associated plants again modify flow conditions, reducing flow speeds and flow competence. Given the diverse influences of biota on landscape processes, it is argued that it likewise moderated the effects of Quaternary and Holocene climate change. Field evidence from Australian and other drylands suggests that the effect of changing land surface properties on runoff and erosion may exceed the effect of moderate climate change. Knowledge of the role of dryland biota and its role in land surface change is therefore a prerequisite to understanding the responses of landscapes to climate change, to understanding the complex spatio-temporal variability in landscape development, and to developing the ability to correctly interpret the alluvial record of changing geomorphological processes in terms of changes in climate and other external drivers.

The Australian mainland contains extensive and diverse dryland landscapes (for overviews, see Croke 1997; Goudie 2002). Practical reasons for research into these landscapes and environments include assessments of their vulnerability or resilience in the face of grazing pressure and recreational activity, the management and rehabilitation of mining-related impacts (e.g. Read *et al.* 2005), the identification and management of continental dust sources (McGowan & Clark 2008), and the understanding of runoff, infiltration, and groundwater recharge that support both natural ecosystems and dryland agriculture and pastoralism (Gifford 1978; Morgan 1999). Impacts of primary production in the drylands include important off-site impacts, such as the arrival of fluvial suspended sediments in the Great Barrier Reef Lagoon (Bartley *et al.* 2006). Many of these issues are of increasing importance as extractive industry and population grow, and as regional climatic change continues to exert an influence on the hydroclimate at all scales from hourly rainfall to multi-year fluctuations related to El Niño–Southern Oscillation (ENSO) and other atmosphere–ocean processes (Holmgren *et al.* 2006). The understanding of land degradation or 'desertification' likewise requires, amongst the perspectives of other discipline areas, a knowledge of dryland ecohydrology and geomorphology, and of their interactions (see Pickup 1998). A potentially very significant issue in the Australian drylands is the

longevity of human occupation, with the associated use of fire. The antiquity of aboriginal occupation in Australia is not resolved with certainty, but the dating of archaeological sites in northern Australia suggests that occupation has certainly extended for 40 ka, and perhaps 50–60 ka (e.g. O'Connell & Allen 2004, 2007; David *et al.* 2007) and archaeological sites in the arid continental interior certainly date from before 20 ka BP (e.g. see Hiscock 2008). Manipulation of fire regimes by aboriginal people has been implicated in driving land surface change of a magnitude sufficient to affect the summer monsoon circulation (Johnson *et al.* 1999; Miller *et al.* 2005), and hence, potentially, other aspects of ecohydrology, ecosystem structure, runoff and erosion.

However, the diverse Australian drylands also provide a laboratory for the investigation of geomorphological and ecohydrological processes generally. They stand somewhat apart from other widely studied global drylands when judged against a number of criteria. For instance, they display less severe aridity than the widely researched desert environments of the Negev or those of northern Africa, much of the Australian drylands receiving at least 200 mm of rain annually. The Australian drylands occur at much lower elevations (generally less than 250 m above sea level) than the interior deserts of the western USA, and receive no snow and, over much of their area, only occasional rain

From: BISHOP, P. & PILLANS, B. (eds) *Australian Landscapes*. Geological Society, London, Special Publications, **346**, 87–120. DOI: 10.1144/SP346.7 0305-8719/10/$15.00 © The Geological Society of London 2010.

from tropical storms. The low relief limits stream gradients to very low values, except locally where channels debouch from the limited areas of rocky uplands. Stream channel collapse, with the dispersal of floodwaters across alluvial plains or into dune-fields, commonly follows the entry of flow onto the flat desert floors, where no further tributary inflows occur. Thus, drainage networks are often not fully integrated (i.e. low-order tributary streams may rarely contribute flow to the apparent trunk channel). No Quaternary ice caps served as physical barriers to shift atmospheric circulations and modify the tracks of rain-bearing systems across Australia, as occurred in the western USA. This resulted in a different history of pluvial and arid phases of the late Quaternary, and correspondingly, a different history of expansion and retreat of plant cover, of erosion and sediment accumulation, and of lake drying or filling. Large areas of the Australian dry-lands drain endorheically to lake basins, and there fine-grained transported sediments accumulate, often accompanied by salts precipitated during dry-ing phases. These lake-bed materials appear to have been released during phases of active lake deflation (efflorescence of salts related to episodic wetting probably being involved) and transported eastward over the continent, to accumulate where rainout occurred or where plant cover or topographic bar-riers existed that could promote deposition. In this way the erosional and depositional cycle of the dry-lands exerted an influence on soil development in the arid, subhumid and humid zones of the continent (e.g. Chartres 1982; Walker et al. 1988). Holocene dust fluxes from the interior drylands appear to reflect the strength of the dry phases of the ENSO phenomenon (e.g. Marx et al. 2009) and, indeed, it is speculated that cooling related to atmospheric dust loading may tend to suppress convection and further reinforce rainfall variability through the ENSO cycle (e.g. Rotstayn et al. 2009). Feedbacks from dust to reduced rainfall have been reported from other global drylands also (e.g. Hui et al. 2008).

However, in terms of landscape processes, a striking feature of the Australian drylands is their biota. In some of the most arid areas, vegetation is locally absent in severe drought years, or where grazing pressure is intense, but over large areas, there is a partial but permanent cover of grasses, shrubs or trees. Although the vegetation is dispersed across the landscape, rather than being limited to contracted distributions along drainage lines and other relatively benign locations, there is a notable increase in tree growth along many ephemeral stream channels. There, various Eucalyptus species such as E. camaldulensis (the river red gum) and E. coolabah (the coolibah) grow in riparian corri-dors as well as on sandy and gravelly stream beds. The Australian dryland flora survives in the face of a rainfall environment of marked unreliability, and characteristics such as sclerophylly are wide-spread. Furthermore, in the eastern part of the drylands particularly, the rainfall climate exhibits periodic fluctuations from dry to wet, on a time scale of several years, related to the ENSO mechan-isms and to other longer-term oscillations within the ocean–atmosphere system (Nicholls 1991; Nicholls & Kariko 1993; Power et al. 1999; Rakich et al. 2008). Droughts may be widespread and long-lasting. Statistical investigation of river flows has confirmed higher levels of variability in the Australian environment than in drylands elsewhere that relate to these features of the Australian climatic environment (McMahon 1982; Finlayson & McMahon 1988).

Australian dryland fauna is also to some extent distinctive. The late Quaternary megafauna is extinct, its disappearance possibly the result of a trend of progressive continental aridification and linked decline in the extent of suitable habitat, although perhaps with some hunting pressure (e.g. see Webb 2008). The modern fauna includes many species of burrowing marsupials, and a great diver-sity in groups such as the ants. On the other hand, although native ungulates are completely absent, the macropods (kangaroos, euros, etc.) do cause soil disturbance locally when scraping away the hot surface soil materials to expose a resting place on the cooler subsurface (Eldridge & Rath 2002). As noted above, the period through which indigen-ous people modified the Australian biota, notably through the use of fire, remains unclear, but cer-tainly extends beyond 40 millennia. The widespread use of fire has resulted in changes to land surface characteristics including the nature and distribution of flora and fauna, as well as in the hydroclimate. Certainly, the longevity of human occupation in the Australian drylands is another feature that makes these areas different from many other global dry-lands. Some of the distinctive aspects of dryland ecology in Australia were reviewed by Stafford Smith & Morton (1990). The extent to which human use of fire has modified the Australian drylands remains an area of debate, but there is little doubt that differences in the occurrence and nature of fire, including the abundance of fuel, provide one of the mechanisms through which cli-mate change may result in diverse outcomes across the extensive drylands.

Taken together, the distinctive features of the climate, relief, elevation, tectonic stability and the biota (and, perhaps, the human use of fire) suggest that there might also be some degree of distinctive-ness to Australian geomorphological and ecohydro-logical processes, and perhaps to the rates and evolutionary pathways of landscape development. It is the objective of this paper to review some aspects of this idea, with particular attention to the role of biota in driving or modifying landscape processes and, importantly, in mediating the effects of climate change. This is an area of increasing debate and exploration, and has variously been identified as 'biomorphodynamics' and 'ecogeomorphology' (e.g. Murray *et al.* 2008). The paper is divided into sections that follow the path of precipitation from first arrival at the land surface to its return to the atmosphere via evapotranspiration or its percolation into the groundwater system. Thus, in turn, accounts are presented of the rainfall climate, of the ecohydrological and geomorphological processes (infiltration, runoff production) on hillslopes, and of ephemeral stream channel processes. In all cases, the objective is to explore and illustrate some of the roles played by biota. The treatment is necessarily both brief and selective, being weighted toward parts of the arid zone better known to the author. The listed objectives have been adopted for several reasons. The first is to contribute an Australian perspective to the literature dealing with the relative roles and 'ecosystem engineering' role of biota in the global drylands, which is a very active area of research but where there are major knowledge gaps. Biota is and will continue to be affected by global and regional climate change, and may provide important feedbacks to landscape process. Some of these are likely to be non-linear (e.g. Ludwig *et al.* 2007), and perhaps will exhibit thresholds beyond which the rate of change will accelerate. Exploration of some of these feedbacks is a second objective here. Finally, a critical reason for focusing on the role of biota is that environmental changes of the Quaternary caused changes in dryland ecosystems that in turn drove shifts in geomorphological processes. Given the non-linearity of the relationships between, say, plant cover and overland flow generated on hillslopes (to be explored below), it is apparent that the sedimentary record of the Quaternary cannot be interpreted properly without knowing by how much or how little key landscape processes were modified under the altered vegetation cover and hydroclimate of particular phases of the late glacial or Holocene. Thus, a goal here is to highlight some of the ways in which changes in the role of biota may indeed have modified landscape processes. Aspects of the sedimentary record that might initially be interpreted as reflecting changes in climate may then be seen to be partially the result of climate change modulated by changes in flora and fauna. With this perspective, it may be possible to draw understanding from the sedimentary record that will be relevant to the ecosystem responses to contemporary environmental changes. Fundamentally, therefore, this paper explores the extent to which biota and ecosystem functions mediate the effects

of imposed external climate changes (e.g. astronomically driven changes, changes resulting from changed atmospheric composition, changes in sea surface temperatures, or global atmospheric circulation). This highlights the importance of considering landscape change in terms of climatic and ecosystem factors more diverse than temperature and effective precipitation as used in many attempts to understand landscape responses to climate change (e.g. see Bull 1991, and references therein).

This paper does not seek to provide systematic solutions to questions of the path and rate of landscape evolution, but rather to present arguments for the systematic inclusion of biota in research on these topics. Theoretical models provide a different path into some of these issues that is not pursued here. For example, Tucker & Slingerland (1997) explored some of the responses of drainage basins to climate change using the GOLEM landscape model. Their results highlighted a number of complex aspects of landscape response to climate change. They showed that a response to an increase in runoff intensity can occur more rapidly, and along a different pathway, from the response to a decrease in runoff intensity. They also showed how divergent landscape responses can arise from differing time scales of climate change and landscape relaxation. Tucker & Slingerland (1997) raised but did not pursue some issues, such as the linkages between hillslope and channel processes, and the outcome for the landscape and its developmental pathway if (as seems likely) erosion thresholds differ between hillslopes and stream channels. When the kinds of feedback processes related to biota that are reviewed here are added to conceptual models of landscape response to climate change, an even more complex view of landscape change emerges. Explorations of this using various modelling approaches can be found in many publications (e.g. Collins *et al.* 2004; Istanbulluoglu & Bras 2006, and studies cited therein).

We begin by considering briefly the nature of the hydroclimate of the Australian drylands, whose spatio-temporal complexity is often sidestepped in studies of ecology, hydrology and geomorphology. The goal of this short review is to emphasize that there is much more to dryland rainfall climates than merely low annual totals and high inter-annual variability.

The nature of the hydroclimate over the Australian drylands

The Australian drylands extend over a very broad geographical range, covering more than 5×10^6 km^2, and spanning more than $30°$ of longitude and $15°$ of latitude. The Köppen–Geiger 'B' (arid) climate zones cover 77.8% of the continent, according to the revised climate maps of Peel *et al.* (2007), which are based upon 1807 rainfall and 351 temperature stations. The basic climatology of the drylands is well known, and includes a summer rainfall dominance in the north and a winter rainfall season in the south, whereas much of the interior lacks strong seasonality in rainfall. Clear skies are generated by atmospheric subsidence within the Hadley cell circulation (the core of the Australian drylands lying at $25-30°S$), and resulting high surface net radiation and very large potential evapotranspiration result in persistent dryness. Rainfall and streamflow both exhibit temporal variability that is high when compared with other global drylands (McMahon 1982). Semi-arid drylands are additionally known to generate higher unit discharges for a given recurrence interval than do basins in wetter climates (e.g. Osterkamp & Friedman 2000). The continental core is arid, with annual rainfall declining with distance from the coastline at rates of around 1 mm a^{-1} km^{-1}. Toward the interior, potential evapotranspiration rises at a considerably faster rate and in the interior is an order of magnitude larger than the mean rainfall. Finally, tectonic stability has left the continent flat (and of generally low elevation) so that topographic gradients are slight, and over large areas of ancient basement rocks, large-scale organized surface drainage is absent except locally following large rain events. For example, across the Yilgarn craton in WA, the general gradient from Kalgoorlie to Wiluna (*c.* 500 km) is only about 0.0003 (Morgan 1999). Large areas of the major dunefields are also arheic. Elsewhere, basins of internal drainage occur, including the vast Lake Eyre basin that alone covers more than 1.1×10^6 km^2 (Alley 1998; McMahon *et al.* 2008), or more than 20% of the Australian drylands.

In such an environment, surface runoff and erosion occur only intermittently in response to rainfall. This may be frontal rain that is spatially widespread and relatively long-lasting, or local convective rain, occurring primarily in the late afternoon. At this time of day, temperatures are high and relative humidity low, conditions that favour significant losses of moisture from interception stores on foliage and from the moist soil surface (Dunkerley 2008a). These processes are enhanced by the limited spatial extent of convective rain, and the advection of warm dry air from the surrounding rainless areas, especially once rain ends. To build an understanding of the controls on infiltration and runoff, a more detailed picture of the arrival of rain at the land surface is needed than is contained in conventional climatologies at daily, seasonal or annual time scales. In particular, the erratic arrival of rain in relatively short events, which themselves may include many short rainless intervals when evaporation can proceed, influences infiltration and runoff behaviour in ways that cannot be understood from rainfall data even at daily resolution. Dunkerley (2008b) presented rain event analyses derived from a 6 year pluviograph record collected in a chenopod shrubland at the Fowlers Gap Arid Zone Research Station, located in far western NSW. The tipping bucket gauge used had a sensitivity of 0.5 mm per tip, and tip events were logged with a time resolution of 1 s. From the pluviograph record, rain events were defined using the minimum inter-event time (MIT) approach, which delineates separate rain events if there is a minimum rainless time separating them. Results derived using an MIT of 6 h are presented here. This is an interval that allows virtually complete drying of shrub canopies following rain, and that therefore represents the commencement of a new rain event on vegetation canopies manifesting their full canopy interception storage capacity. Using the 6 h criterion, the rainfall record at Fowlers Gap revealed an average of 34 storms per year, typically lasting for 1 h 40 min, and delivering 2 mm of rain at an average intensity of 1.5 mm h^{-1}. Such small rain event depths, low rainfall rates and relatively short storm durations in an environment of low relative humidity mean that the drying of wet foliage, and drying of the damp soil surface, will be important processes both intrastorm and post-storm. Evaporation during brief gaps in rainfall would be of reduced significance in the humid zone, owing to higher rainfall depths and lower vapour pressure deficits there. However, in drylands, the effective rainfall, the part that infiltrates the soil and remains as a moisture store that can be withdrawn by plants later, will be significantly reduced in comparison with the meteorological or gauge rainfall. Changes in rain event properties in the drylands will undoubtedly develop as global environmental changes occur. For instance, rain events of reduced rain rate and event depth would tend to diminish the yield of functionally available moisture in the root zone. This in turn may drive floristic and structural changes as moisture stress increases. Baudena *et al.* (2007) explored the effects of rainfall intermittency on dryland vegetation by means of an ecohydrological model. They concluded that changes in intermittency, even without a change in annual or seasonal total, can drive significant changes in plant cover. Further modelling by Köchy (2008) has suggested that the effects of a change in mean annual rainfall overlap with those of a change in mean daily rainfall, if the annual total is held fixed. A change to less frequent but larger rain events was found to recharge soil moisture to 30 cm depth more effectively than the same annual total rainfall delivered in more frequent but smaller events. Köchy (2008) explored how such outcomes vary along a Mediterranean climatic gradient, and showed that drylands exhibited a greater sensitivity than more humid areas. The particular sensitivity to change in meteorological rainfall among Australian dryland ecosystems is, however, virtually unknown, as are the likely changes in sub-daily (event) rainfall properties that will be associated with the forecast changes in annual totals (e.g. Murphy & Timbal 2007). Quaternary changes in rainfall climate undoubtedly involved altered rain event properties, and this provides a first feedback capable of amplifying or reducing the change in effective rainfall arising from a change in meteorological rainfall.

It is worth observing that different ecohydrological processes may exhibit distinct sensitivities to changed rainfall climate. For example, significant recharge of groundwater may arise principally in very large rainfall events ('episodic recharge'; Wischusen 1994; Harrington et al. 2002; Lewis & Walker 2002). For the Wiluna area on the Yilgarn craton, Morgan (1999) reported that successive rain days delivering an aggregate of >50 mm provided significant recharge. He emphasized that rain rate was important, not merely total depth. In large recharge events, the fraction of rainfall entering the groundwater typically lay in the range 0.75–5% or so. Similarly, for some aquifers of the NT drylands, Harrington et al. (1999) reported that >100 mm in a month is needed for significant recharge. Important contributions to recharge are made by percolation losses to ephemeral stream beds, in the transmission loss process described below (e.g. Osterkamp & Lane 2003). In comparison, dryland vegetation can receive significant growth stimulus from much smaller rain events. According to Martin (2006), small falls (<6 mm) are ineffective within dry periods, but can be beneficial during wet intervals. Falls of rain that are effective for vascular vegetation typically occur on only about 9 days per year (Martin 2006). On the other hand, it has been hypothesized that small rain events may have a relatively larger impact (ecosystem effect per millimetre) than large events, because a larger fraction of the water is available in near-surface locations where nutrient cycling is focused (e.g. Sala & Lauenroth 1982). Rapid pulses of microbial activity, important for the nutrient status of many dryland soils (e.g. Goberna et al. 2007), can follow from even minor rain events. This kind of argument leads to the idea that organisms able to utilize small events will be important in many drylands, and that the ecosystem structure partly driven by these events will relate to a different property of the rainfall climate than hillslope erosion, the latter perhaps linked more strongly to the larger but rarer events. Thus, for hillslope erosion

and the generation of streamflow, there may be a larger threshold rain event magnitude than for the maintenance of plant community composition and structure. For Fowlers Creek in western NSW, Cordery et al. (1983) reported that flows resulted from all rain events delivering >16 mm or >5 mm in 1 h. From data such as these, it can be hypothesized that thresholds for significant ecohydrological events or change differ between plants, hillslopes, streams and the groundwater system. Thus, as the nature and occurrence of rain events change (or changed in the late glacial and Holocene), there will be differing changes in event frequencies between these other components of the dryland environment. Event spacing, for example, which can reach many months in the case of streamflow, affects the degree to which moisture stores in the sandy beds of ephemeral channels are depleted prior to the next flow event. Greater event spacing or diminished event depth may result in greater transmission losses into the stream bed, with a reduced downstream flow length and diminished capacity to transport bedload materials. The fundamental significance of these ideas for geomorphological response is as follows. If the rainfall climate changes, including shifts in rain event properties such as duration, depth and rain rate, but also the period between rain days, then the fraction lost to canopy interception and soil surface drying will also change. If the trends are to smaller events, then proportionally less rain will reach the land surface. But if plant cover then declines because the soils are drier, then the fraction of the incident rain that is converted to overland flow may in fact increase. The more widespread development of surface seals and crusts on the enlarged area of exposed soil would amplify this effect. It might then be anticipated that runoff ratios would increase, speeds of overland flow would rise, and therefore, stream hydrograph shape (risetime, peak discharge, etc.) would also change. Bed sediment transport and bank erosion (including perhaps more lateral cutting) might then also become more active, at least in events at the high-magnitude, low-frequency end of the event spectrum. Therefore, it is not a straightforward exercise to deduce what geomorphological consequences would follow from a change in the rainfall climate of the kind involved in the climate fluctuations of the late glacial and Holocene. It also follows that the consequences of contemporary climate change for processes and landscape stability in dryland environments are not simple to envisage. It has been argued that land surface erosion exhibits a marked sensitivity to rainfall change (Angel et al. 2005). Nearing et al. (2004), for example, estimated that soil loss rates could increase by 1.7% for each 1% increase in annual rainfall total. However, the bulk of our understanding of such issues comes from agricultural soils of the humid zone, and not from the drylands. Moreover, it is important to recall that the effects of rainfall event properties such as depth or rainfall rate drive outcomes that vary with spatial scale, from patch to hillslope to sub-catchment and catchment scales (e.g. see discussion by Kirkby et al. 2005).

Importance of rain event structure in the drylands

The separation of rain events (storms) in time is clearly likely to be important for the production of overland flow and surface erosion, probably via effects on the antecedent soil water, but has been investigated very little for locations in the Australian arid zone. Ring (1988) highlighted this parameter in a study of storm runoff in the Macintyre River valley in NSW in the period 1966–1981. In 1950–1965, storms of >80 mm depth were separated by about 40 days on average, but this declined to 19 days in 1966–1981. He suggested that this decrease of more than 50% in the inter-storm period accounted for part of an observed increase in storm runoff ratio, the land surface having greater antecedent wetness when storms occurred more closely in time. Rain events tend to be widely spaced in the drylands, and most rain days deliver very small depths. In his work on mulga water balance near Alice Springs, Slatyer (1965) reported 26 rain days per year, of which about 42% (about 11 rain days) were >5 mm but only two per year >25 mm. For Wiluna, Morgan (1999) reported a mean of 13.2% of days with rain (records 1898–1997) or about 40 days per year. For Fowlers Gap in western NSW, Cordery et al. (1983) reported about nine events of >5 mm depth per year, but only about 2.4–3 events per year >20 mm. From a 6 year pluviograph record from Fowlers Gap, Dunkerley (2008b) reported that only 8.9% of days recorded rain, which is the equivalent of about 33 rain days per year on average, somewhat less than the value for Wiluna. Other dryland locations record even fewer rain days per year (e.g. Alice Springs averages <30 days having ≥1.0 mm) whereas in the humid zone, Sydney and Melbourne record about three times as many rain days per year, and locations in western Tasmania six times as many. The issue of rain event spacing has been addressed for the purposes of arid watershed streamflow modelling (e.g. see Shamir et al. 2007), but more analyses of dryland rainfall climates are needed to deepen our knowledge of the phenomena involved. As noted above, rain event properties are likely to change as annual rainfall totals change, but may not change at the same rate or even in the same direction.

Spatial variability of rainfall climate in the drylands

Maps of the number of rain days by depth (Fig. 1) reveal different aspects of the rainfall climate of the Australian drylands. In terms of rain days with >1 mm, the area with the fewest annual events is centred on the NE corner of South Australia, between Oodnadatta and Birdsville in Queensland. For days >5 mm, the region with the smallest occurrence frequencies is a larger zone that extends across SA and into WA. For days with >25 mm, the region of lowest frequency of occurrence shifts southward, extending inland from the southern coastlines of WA and SA. This pattern presumably reflects the southward penetration of summer monsoon rainfall in the more northerly parts of those states. However, in terms of the differing responses of landscape processes to event size discussed above, the differing portrayals of the drylands seen in Figure 1 suggest, for example, that the region 25–27°S (spanning approximately the latitudes of Oodnadatta and Maree) would experience the fewest rain days delivering falls capable of generating overland flow or of delivering useful pulses of soil moisture, but that the wetter areas to the south experience fewer large events likely to drive groundwater recharge.

Rain events in western NSW are typically small. Dunkerley (2008b) found that using a 6 h MIT, the mean event depth was 2.06 mm. Published data are seemingly not available for other Australian dryland sites. For a study site in Arizona, Shaw & Cooper (2008) reported a mean rain day depth of 6 mm (summer) and 4 mm (winter), but these daily totals might represent the rain from more than a single rain event as defined using the MIT approach to event separation. An analysis of 27 years of rainfall records from Oman showed that up to 95% of rain occurred on days of <10 mm (Kwarteng et al. 2008). Here it is relevant to recall that the canopies of dryland shrubs and grasses often have a canopy storage capacity comparable with the mean rain event depth (Dunkerley & Booth 1999; Dunkerley 2008c). Litter lying beneath dryland plants provides an additional potential interception store and surface area for intrastorm evaporation. Therefore, it is likely that many rain events fail to generate functionally effective rainfall in sub-canopy locations. The arrival of runon water from upslope (described below) may thus become an important moisture source. This is a point of great importance in the Australian drylands: many ecohydrological and geomorphological processes are driven by water arriving from upslope or from the upper parts of the drainage basin. Because, as just noted, contemporary dryland rain events often deliver depths comparable with the interception

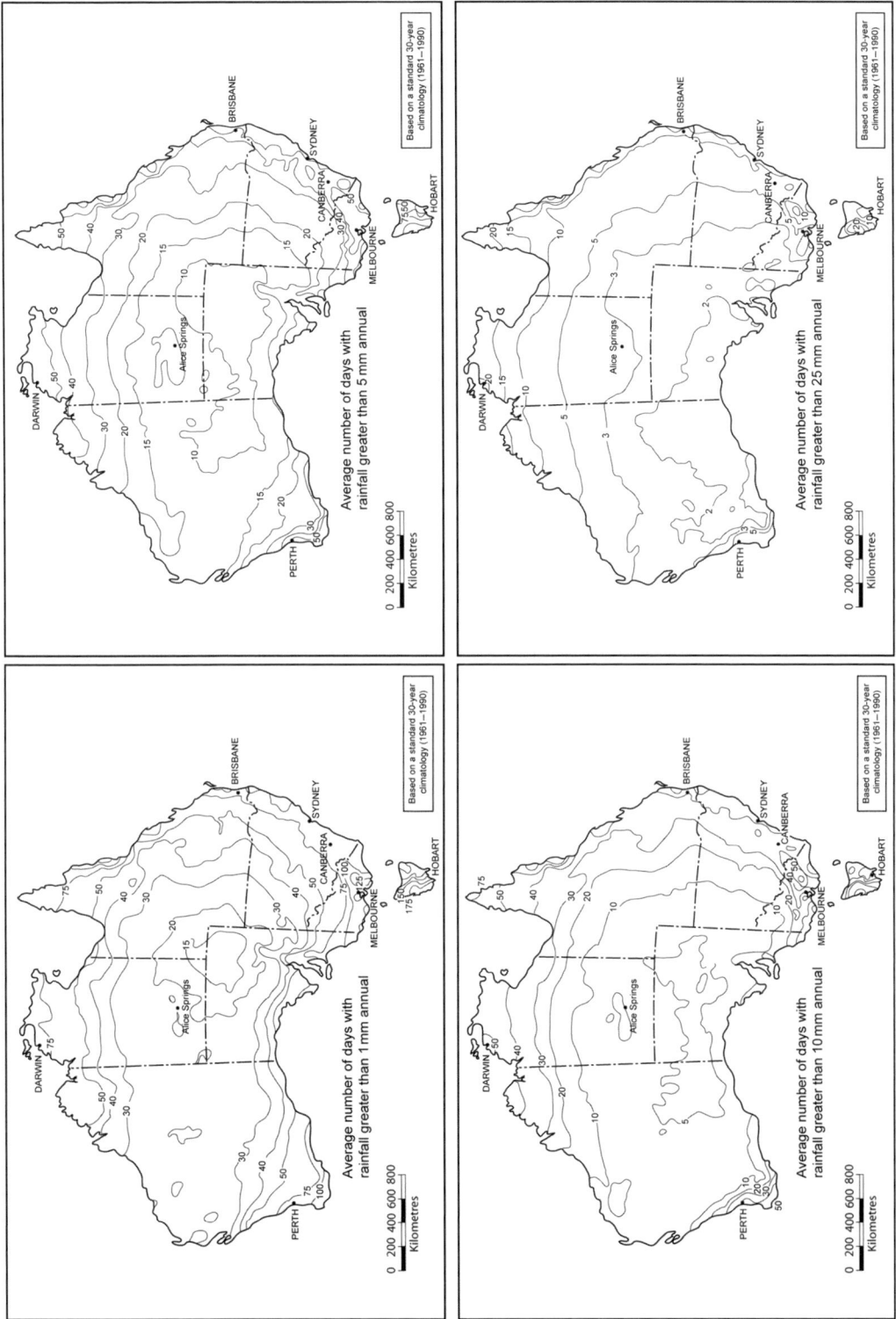

Fig. 1. The spatial pattern of rain day amounts across the Australian drylands, for days receiving >1 mm, >5 mm, >10 mm and >25 mm. Modified after online data derived from the Bureau of Meteorology (http://www.bom.gov.au).

capacity of shrubs and their associated litter, there is a non-linear relationship between event depth and potential infiltration excess.

To further explore the phenomenon of rain arrival, the detailed pluviograph records from four Australian dryland locations (Alice Springs, NT; Giles, WA; Woomera and Oodnadatta, SA) were obtained from the Bureau of Meteorology, and the data for 1970–2007 processed to yield rain events defined using an MIT of 6 h. The key event characteristics of duration, depth, and rain rate are summarized in Table 1. Distributions of rain event depths (Fig. 2) show that for all stations, 50–60% of rain events are >4 mm. Events of <30 mm make up less than 10% of the events. The low rain rates are noteworthy, given that medians are typically only about 1.5 mm h^{-1}, so that half of the rain events exhibit even lower rates. These are likely to be of little functional significance to vascular plants, but important for the widespread biological crusts that stabilize soil surfaces (mentioned below). At all four sites, occasional events do reach high rain rates (30–40 mm h^{-1} is typical, but even higher at Giles), and large event depths of 150–300 mm. These large events would drive widespread runoff and groundwater recharge.

As noted above, the arrival of events in time affects runoff production via antecedent wetness and the vigour or otherwise of plant growth. For the same four dryland locations, the pluviograph data were used to identify the inter-event times, based upon an MIT of 6 h. The results (Table 2 and Fig. 3) show that at all sites, 30–40% of events arrive within 2 days of a previous rain event. However, the distributions are highly positively skewed, and 5–15% of events arrive after >40 days without rain. Moreover, there are evident differences between the sites in terms of several key characteristics of the rainfall climate. These cannot be explored in detail here, but it is interesting to note that the inter-event waiting time is much longer in relation to event duration at the drier sites (Oodnadatta & Woomera). At those sites, the median waiting time is 73–77 times the median event duration, whereas at the sites having higher rainfalls (Alice Springs and Giles), the waiting times only exceed the event duration by 34–38 times. Median event depths differ less, so that water scarcity is increased by the long waiting time at the drier locations.

The importance of soil properties for ecohydrological processes

Having highlighted some of the key characteristics of rainfall events in the Australian drylands, we now turn to consider aspects of the ecological and landscape roles for rain. Interaction of rain with plants and soil begins the series of transformations of this water resource into surface ponding, soil water, overland flow, and channelized flow. Dryland soils provide key influences on these transformations, and we now turn to consider these.

The lower frequency of rain occurrence in the drylands than in the humid zone means that soil moisture stores become more critical for plant

Table 1. *Statistical properties of the rain events for four Australian dryland sites*

Location	Alice Springs, NT	Oodnadatta, SA	Woomera, SA	Giles, WA
Period of record analysed	1970–2007	1970–2006	1970–2006	1970–2006
Number of rain events	1141	392	969	1320
Median duration (h)	2.30	2.00	1.90	1.90
Median depth (mm)	3.10	4.80	2.80	2.80
Median rain rate (mm h^{-1})	1.30	2.10	1.50	1.40
Event maximum duration (h)	80.4	52.3	37.0	70.6
Event maximum depth (mm)	305.1	210.2	148.7	259.8
Event maximum rain rate (mm h^{-1})	41.0	27.8	27.6	164.6
Standard deviation of event durations (h)	7.03	5.26	4.75	5.73
Standard deviation of event depths (mm)	20.6	18.5	10.8	17.1
Standard deviation of event rain rates (mm h^{-1})	3.15	3.44	2.64	6.46

Data processed by the author from pluviograph records at 6 min resolution, and aggregated into events using MIT = 6.0 h. For each station, the data from 1970 to the latest available Bureau of Meteorology data were analysed. It should be noted that because the data relate to events, longest durations extend to more than 3 days. However, breaks in rainfall of up to 6 h are by definition possible within long events.

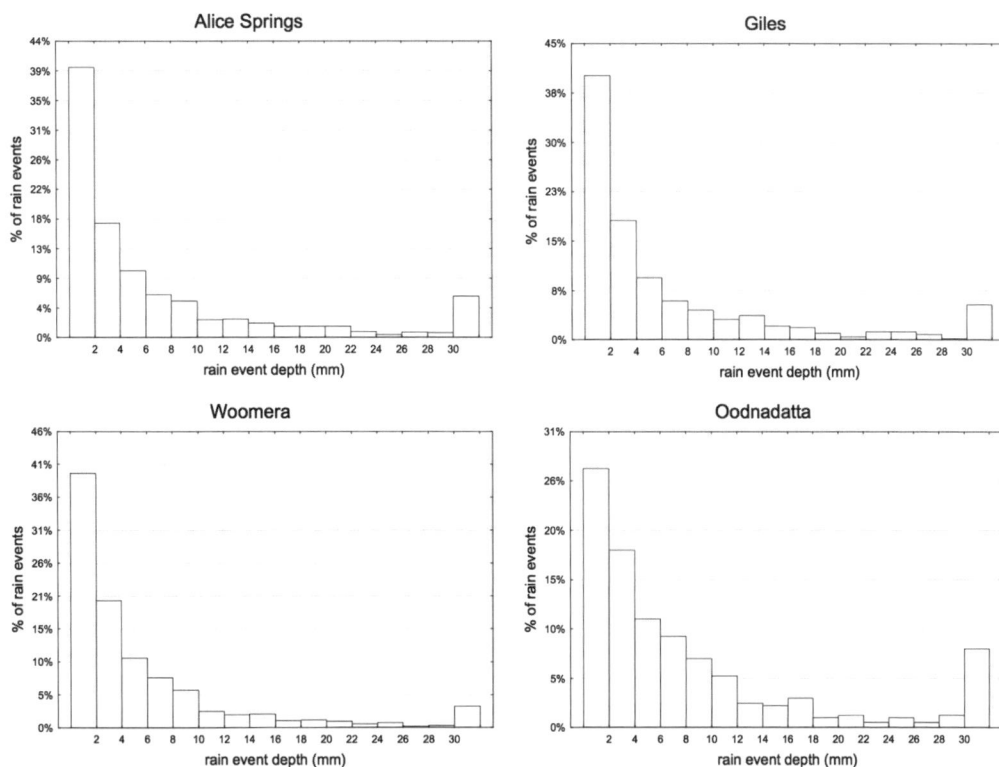

Fig. 2. Distributions of rain event depths for rain events defined using a minimum inter-event time (MIT) of 6 h, at four locations in the Australian drylands. Depths exceeding 30 mm are grouped into the rightmost column of each graph. Derived from 6 min pluviograph data for 1970–2007 supplied by the Bureau of Meteorology, Australia.

survival. Here, two key characteristics of dryland soils may mitigate against efficient moisture storage. First, surface seals may limit the infiltrability of exposed soils to <15 mm h^{-1}, so that in intense rain events, considerable water is partitioned into overland flow. Second, over rocky uplands soils are shallow and provide only limited capacity for water storage. An analysis of storms that produced overland flow and/or streamflow in the arid Fowlers Gap area of western NSW led Cordery *et al.* (1983) to conclude that most runoff events were produced by rain events of low intensity but

Table 2. *Statistical properties of the rainless interval between events for four locations in the Australian drylands*

Location of rainfall station (Bureau of Meteorology station number)	Period of record analysed	Mean annual rainfall in period (mm)	Number of inter-event rainless intervals	Median rainless interval (days)	Standard deviation (days)	Skewness
Alice Springs, NT (15590)	1970–2007	280	1139	3.24	25.2	8.39
Oodnadatta, SA (17043)	1970–2006	194	389	6.11	33.6	4.59
Woomera, SA (16001)	1970–2006	181	967	6.17	21.6	3.75
Giles, WA (13017)	1970–2006	303	1319	3.03	18.6	5.64

Data processed by the author from pluviograph records at 6 min resolution, and aggregated into events using MIT = 6.0 h. For each station, the data from 1970 to the latest available Bureau of Meteorology data were analysed. It should be noted that *c.* 15 years of record (1985–2000) were missing from the Oodnadatta pluviograph data. These years were excluded from the statistical analysis. The reduced sample size for this station reflects the missing data. Summary climatic data for these stations can be found at *http://www. bom.gov.au/climate/averages/*

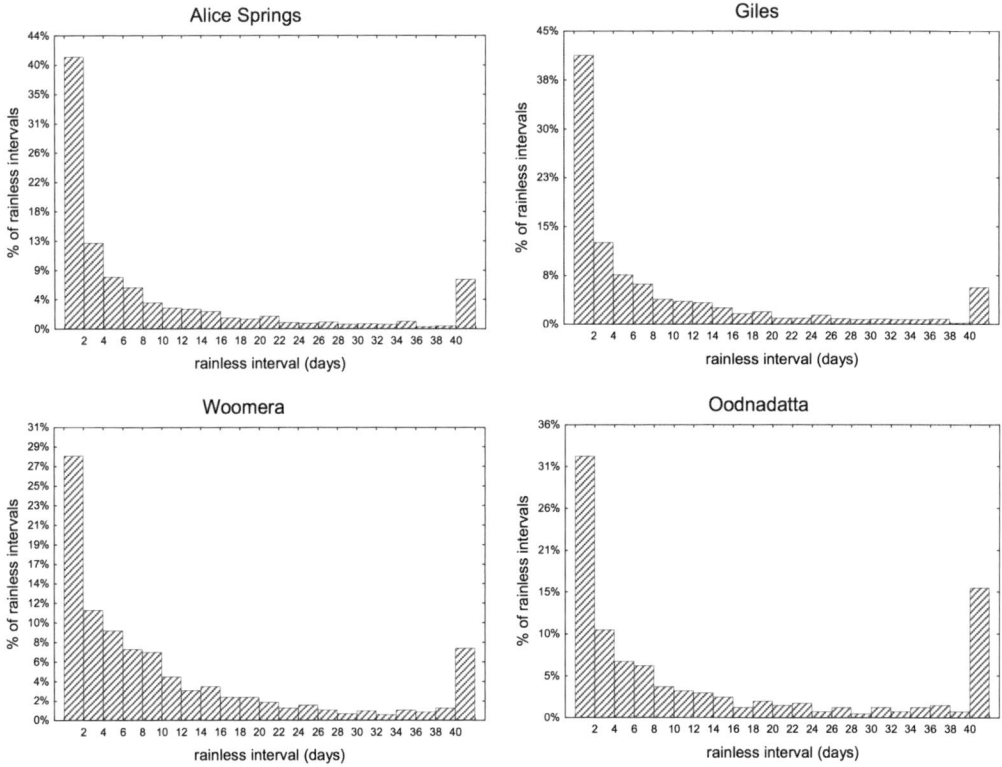

Fig. 3. Distributions of inter-event times for the same four dryland locations as in Figure 2, in classes of 2 day duration. It should be noted that all durations exceeding 40 days are grouped into the rightmost column of each graph. Derived from 6 min pluviograph data for 1970–2007 supplied by the Bureau of Meteorology, Australia.

long duration. On rocky uplands in this area, it is likely therefore to be primarily the restricted soil depth that most often drives the partitioning of rain into overland flow. Downslope, however, hillslope runoff may combine with rainfall to exceed the surface infiltrability even where soils are deep and more permeable, so that runoff may continue over larger spatial scales. It has to be remembered that rain day depths provide insufficient resolution of rainfall to adequately understand the generation of overland flow and the occurrence of channel-forming floods. In a study of floods in Nahal Arava in Israel, Jacoby *et al.* (2008) noted the influence of both long-duration, low-intensity rain events from winter depressions, and autumn and spring convective storms of high intensity but short duration. The latter generate most of the floods in Nahal Arava. For many Australian dryland streams, comparable analyses have not yet been carried out. It is clear that the delivery of water to streams from many dryland hillslopes depends in a complex way on the conversion of rain to overland flow (a local or patch-scale phenomenon, at least initially) and the subsequent integration of hillslope flow as rainfall

continues. Thus, rainfall rates in relation to soil properties (infiltrability and depth) combine with temporal aspects (rainfall rate profile during the event, overall event duration) to influence the extent to which overland flow connectivity is achieved and integrated hillslope flow is delivered to streams (e.g. see Puigdefábregas *et al.* 1998, 1999; Mueller *et al.* 2007; Reaney *et al.* 2007; Mayor *et al.* 2008).

It is thus apparent that multiple factors influence the partitioning of the open-field rainfall into re-evaporated water lost to the landscape, functionally available water within the soil, or into surface ponding and overland flow. The latter two pathways are of relevance to geomorphological processes. Among the external, hydroclimate controls are the storm properties: duration, rain rate, event depth and rain intermittency within the event, and meteorological variables including wind speed, air temperature and vapour deficit. The other group of controls on water partitioning are those related to the land surface: plant cover, soil type and properties, slope gradient, soil surface microtopography, litter cover, presence of organic and inorganic surface crusts, soil depth, surface and subsurface stones, etc. Many of these are directly or indirectly related to the nature of biota, and it can therefore be hypothesized that they exhibited responses to Quaternary and Holocene changes in climate.

A note on the use of contemporary landscape process observations and historical climatic data

A significant limitation of the data and analysis so far presented has to be acknowledged. This is the limitation arising with most contemporary studies of ecosystem and landscape processes: they have been carried out within a very restricted timeframe. This may not have adequately revealed processes in low-frequency events of higher magnitudes than have been directly observed in the historical period. Exploration of dryland palaeoflood hydrology, for instance, points to the occurrence of Holocene floods larger than any known from the historical period (e.g. Jansen & Brierley 2004). These in turn suggest larger rainfall events and considerably altered processes of hillslope and channel sediment transport. However, it is not possible to resolve the fine details of these events from the sedimentary record, but it is certainly possible to use the palaeohydrological evidence to emphasize the uncertainty that must be attached to interpretations based upon 'contemporary' observations and data. Studies of Holocene floods have highlighted some general conclusions relating to the occurrence of floods, links to the ENSO phenomena (e.g. see Ely 1997; Baker 2006, 2008). Although in particular locations there may be evidence of floods larger than have been recorded instrumentally, Enzel et al. (1993) used an analysis of contemporary and palaeoflood magnitudes from the SW USA to suggest that in fact the palaeofloods did not exceed the envelope curve linking peak discharge to catchment area defined from modern observational data. This they took to reflect the limitations imposed by atmospheric delivery of moisture, movement of frontal systems, etc., which are likely to be relatively unchanging. Whether this same conclusion applies to the past flood history of the Australian drylands remains unclear.

Role of biota in affecting hillslope runoff mechanisms

We have briefly considered some fundamental principles relating to rainfall climate, as a conceptual framework for environmental change in the drylands, and something of the nature of rainfall partitioning at and near the soil surface. It is now appropriate to consider in a little more detail some of the roles of biota in overland flow production and fluvial erosion and deposition. We begin by exploring more fully the role of biota in hillslope water partitioning.

How efficiently do upland surfaces convert open-field rainfall into overland flow and streamflow? The answer is very much dependent upon the spatial scale over which the accounting is done. There are also feedback links that operate via the mediation of plant cover in runoff production. At the m^2 scale on rocky uplands, overland flow is produced efficiently by the processes already described, such as the Hortonian runoff mechanism. As water balance is evaluated over increasingly large spatial scales, the efficiency of runoff production is reduced owing to the inclusion of sites showing greater soil depth or infiltrability, or groves of plants (discussed below) where runon water is re-absorbed, and where less runoff originates. Finally, once flow enters organized stream channels, transmission loss into dry sandy sediments can progressively consume the flow, leaving a peak in runoff occurrence and volume at some intermediate spatial scale. At locations increasingly far from the sources of hillslope runoff that feed streamflow, transmission loss may totally consume flood flows, so that flow events, and opportunities for the shaping of the channel, become progressively less common away from the uplands.

The behaviour of runoff of course must be viewed in the context of the temporal magnitude–frequency spectrum of event sizes. The occurrence of a large rainfall event may modify the landscape response to subsequent smaller events. Thus, Cordery et al. (1983) suggested from limited field data that runoff events in the rocky uplands at Fowlers Gap were less common following the record high rainfall year of 1974, which delivered rain over large areas of the Australian inland, resulting in one of the episodes of filling of Lake Eyre. In particular, although there were multiple rainfall events in the following years (1975–1977) of sufficient depth to have produced runoff, most failed to do so. Runoff events appeared to resume by 1978–1980, by which time the plant cover had once again become sparse. Data from 25 m^2 runoff plots, larger basins and small stream catchments up to about 20 km^2 showed that plots generated a runoff depth an order of magnitude larger than the larger catchments. In this way, processes in the upland catchments, undoubtedly related to the ENSO phenomena, ensure that there are linked changes in channel processes downstream.

We can identify many distinct and interacting roles for biota in this spatial hierarchy. These arise primarily from the effects of plants, but secondarily in many cases from soil fauna associated with the sheltered and resource-rich micro-environment associated with plants. A selection of key mechanisms is considered here, although necessarily the treatment is selective.

Effects of plants on local-scale soil hydraulic properties

It is known from field monitoring that the areas occupied by isolated dryland plants, or small clusters of plants, shed less overland flow and eroded soil than do bare, inter-plant surfaces (Dunkerley 2000). Indeed, shrubs are considered to concentrate resources into 'fertile islands' (Vetaas 1992; Schlesinger *et al.* 1996; Wezel *et al.* 2000; Thompson *et al.* 2005; Arnau-Rosalén *et al.* 2008; Bedford & Small 2008). Several features contribute to this. The plant canopy intercepts rain and the fraction that escapes interception and evaporative loss from the foliage is delivered to the soil surface with reduced kinetic energy, owing to the reduced fall height. Only the free throughfall drops that pass through canopy gaps can reach the soil at terminal velocity, and even then, they are likely to strike materials in the litter layer where, again, erosive energy is dissipated (Fig. 4). Soils beneath plants are also differentiated from those of interspace sites, owing to more abundant organic matter, sandier texture, and more abundant biopores produced by roots or by soil fauna, including termites. Sub-canopy soils may therefore exhibit greater macroporosity, larger intergranular porosity, and considerably greater overall infiltrability. These effects are not restricted to the sub-canopy soils, but extend into the interspace where roots spread and where litter is dispersed by wind and water. Dunkerley (2000) has presented data on this for both chenopod shrubs and mulga. Table 3 compares these Australian data with those from drylands of the USA, and this shows that the local influence of plants is comparable across a wide range of locations. These properties mean that vegetation tends to do two significant things in relation to hillslope hydrology: first, it reduces the volume of overland flow by creating sites where runoff production is minimized and, second, plants may act as sinks for overland flow arriving from upslope (that is, they promote the infiltration of runon water; e.g. Howes & Abrahams 2003). This role will be mentioned again in connection with patterned vegetation, which is discussed below.

A significant role for plant litter

Many dryland plants shed abundant litter of leaves, flowers, or other parts, and these accumulate beneath the canopy. They may also be dislodged

Fig. 4. Examples of plant communities that generate abundant foliar cover and ground litter in the Australian drylands. (**a**) Dryland landscape south of Broken Hill in NSW, where the groundcover is a spinifex–mallee association. Both groups of plants produce abundant litter, and offer significant canopy interception capacity. (**b**) Large dryland shrub (*Cassia* sp.) near Lake Torrens, South Australia. The extensive canopy and corresponding thick litter accumulation on the sub-canopy soil surface should be noted. Both provide significant interception capacity.

by wind, flowing water, or splash processes (Geddes & Dunkerley 1999) and move across the landscape. In shallow laminar overland flow, floating litter commonly accumulates in sinuous litter dams or barriers, perhaps where flow divergence occurs, or where small obstacles promote deposition (Fig. 5). These barriers may reach heights of 50–60 mm. Given the gentle slopes of many landscapes in the Australian drylands, sets of such dams (often tens to hundreds in a cascade) act to reduce flow speeds, and to drive the deposition of eroded soil materials, seeds, and organic matter. They consequently tend to reduce the erosivity of overland flow, and by holding back ephemeral ponds of water, they tend to increase the opportunity time for infiltration into soils of low infiltrability (Mitchell & Humphreys 1987; Cerda & Doerr 2008). These effects have not been widely investigated in the Australian drylands, but are likely

Table 3. *Selected data on the relationship of soil infiltrability beneath plant canopies and in the adjacent open interspace*

Location	Relation of sub-canopy to interspace infiltration	Reference
Chihuahuan Desert, New Mexico, USA	Up to 1.54 × faster beneath canopy	Bhark & Small (2003)
Idaho, USA	>6.6 × faster beneath canopy	Eldridge & Rosentreter (2004)
Nevada, USA	3.4 × faster beneath canopy	Blackburn (1975)
	2.6–3.0 × faster beneath canopy	Lyford & Qashu (1969)
Mojave Desert, USA	Up to 10 × faster beneath shrub canopies on old surfaces	Caldwell *et al.* (2008)
New Mexico, USA	2–3 × higher beneath shrub canopies	Perkins & McDaniel (2005)
New Mexico, USA	35% higher beneath canopy (but depends on landscape position)	Bedford & Small (2008)
Western NSW, Australia	4–8 × faster beneath canopy of chenopod shrubs	Dunkerley (2000)
Alice Springs, Australia	10–20 × faster beneath mulga canopy	Dunkerley (2002)

The published source for each result is listed. All data are from drylands in the USA, with the exception of the last two studies, which are from the author's work in the Australian drylands.

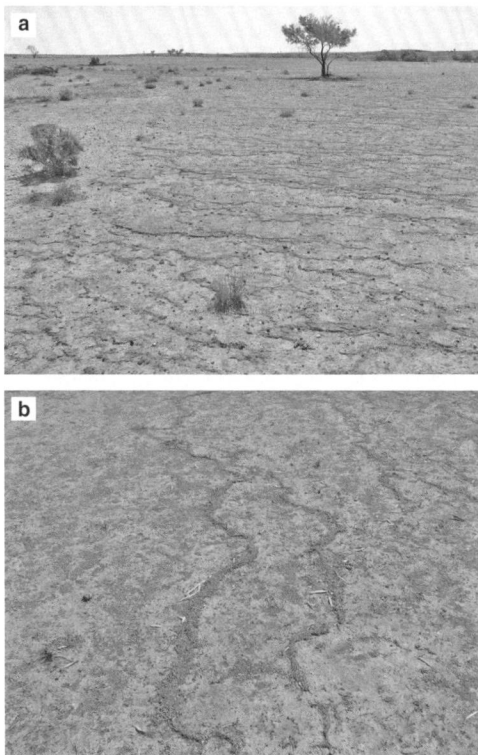

Fig. 5. Typical litter barriers on low-gradient dryland hillslopes. (**a**) Tiers of litter barriers on a drought-affected dryland hillslope north of Broken Hill in western NSW, Australia. Height of the barriers here is only about 1 cm. (**b**) Detailed view of some litter barriers seen in (a). Flow direction from left to right.

to take on particular importance there owing to the low gradients and laminar flow conditions in overland flows.

Patterned vegetation and runoff-runon systems

The two effects of plants and litter just described act at small spatial scales but can be distributed widely across the landscape. Over very large areas of the Australian drylands the runoff–runon system that operates in the vicinity of single plants also functions on a far more extensive scale, involving the structuring of entire vegetation communities into mosaics (Fig. 6). Indeed, a variety of regular and quasi-regular structural arrangements of plants has been described from drylands, generally on relatively gentle slopes. There has been considerable interest in the nature and development of vegetation distribution and patterned vegetation in drylands, including a considerable diversity of approaches through modelling (e.g. Dunkerley 1997a; Lefever & Lejeune 1997; Rietkerk *et al.* 2004; Boer & Puigdefábregas 2005; Barbier *et al.* 2006; Gilad *et al.* 2007; Sherratt & Lord 2007; Liu *et al.* 2008). Only a brief description can be included here. Essentially, in vegetation banding, which occurs very extensively in grassland, chenopod shrubland and in mulga (*Acacia aneura*) woodlands in the Australian drylands, plants are clustered in groves separated by areas virtually devoid of vascular plants. The groves and intergroves are precisely aligned along the contour (Fig. 7b, d) and subdivide the landscape

Fig. 6. Four instances of vegetation mosaics from the Australian drylands. (**a**) View along a grassy intergrove between mulga groves. NW of Alice Springs, NT, Australia. Runoff direction is from left to right. (**b**) Aerial view of the area of groves seen in (a). The groves are strongly contour-aligned. Slope direction is from bottom right to top left. (**c**) View downslope across multiple intergrove–grove cycles in a banded vegetation mosaic near Menindee, NSW, Australia. (**d**) Aerial view of the area of groves seen in (c). Runoff direction is from left to right. A highway crosses the upper part of the photograph.

into runoff sources (where the soil surface is bare, smooth, of low porosity, and carrying raindrop impact and biological crusts) and runon sinks in the groves (where, in contrast, litter is abundant, soils are porous, the formation of surface seals is reduced by plant canopies, and infiltrability is enhanced in ways already described in proximity to dryland plants, their litter, and associated soil fauna). Soil erodibility is very low in the runoff sources (Dunkerley & Brown 1999a) but localized erosion and deposition of soil materials do yield a subtle microtopography, which assists in the formation of ephemeral surface ponding forming large depression storages in the lowermost intergroves, and from which water trickles into the groves. These systems inhibit the occurrence of integrated hillslope runoff, and limit runoff to the local runoff–runon system of adjacent groves (Fig. 7). That is, during most rain events, overland flow is likely to be restricted to the intergrove where it is produced, without the flow passing downslope to combine with that generated on the next downslope intergrove. It is therefore likely

that integrated runoff is rare, and limited to the largest (longest duration, highest rain rate, largest rain depth, etc.) events. For a contrasting form of banded vegetation in the Chihuahuan Desert, McDonald *et al.* (2009) reported that overland flow was able to pass through groves in rain events delivering >16 mm. There are no observational data on the occurrence of integrated overland flow from any area of patterned vegetation in the Australian drylands. Modelling suggests that the compartmentalization of the hillslope allows a greater biomass to be supported than would be the case if the scarce water resources were distributed uniformly over the landscape. In terms of sediment transport, this suggests that patterned vegetation should be associated locally with reduced streamflow volumes, and reduced stream sediment transport capacity.

The origin and antiquity of these widespread ecosystems has not been resolved. One of the few hypotheses that specifically addressed the antiquity of vegetation patterning is that proposed by Dunkerley & Brown (1995), in a field study of

Fig. 7. Two examples of runon effects in patterned vegetation communities of the Australian drylands. (**a**) An irregular mosaic consisting of gibber patches (upper left) and shrub patches (lower right). The darkening of the soil surface along the junction, and among the plants, is the effect of just 2.8 mm of rain falling at low intensity on 22 August 2008. The photograph was taken the following day, and demonstrates the efficiency of bare intergroves in generating runoff in very small rain events, and delivering useful moisture to associated plants. (**b**) A contour-aligned banded vegetation mosaic near Menindee, NSW (seen also in Fig. 6c, d). The bare intergrove on the left of the photograph has recently shed runoff to the grove of Mitchell grass to the right. The edge of the grove has captured the runon water efficiently, and has responded by renewed growth, whereas the grasses further downslope (to the right-hand edge of the photograph) have consequently been denied runon water, and remain dormant.

contour-parallel banding in a chenopod shrubland in western NSW. There, the soil surface is mantled by gibbers (stones) whose size declines steadily downslope from outcrop sources near the slope crest. However, active transport and sorting of these stones is not occurring today, as the landscape is strongly banded, and the groves are pockmarked with collapse structures (crabholes) that would trap any stones moving across the grove.

Dunkerley & Brown (1995) therefore reasoned that the stone sorting predated the development of vegetation banding. This development was likely to be related to increasing rainfall as climate ameliorated after the marked dryness of the last glacial period. If this is correct, then integrated hillslope runoff would have been possible in the colder and drier conditions of the late glacial, before more extensive plant cover and contour-aligned vegetation banding developed, but not in the modern environment. Consequently, the volume of runoff passed to the streams, and the rate of hillslope erosion, were probably both larger (at least in low-magnitude events) during dry conditions than during the relatively higher rainfall conditions of today. Across the Australian drylands, vegetation banding is very widespread, especially in the form of mulga groves on broad, gently sloping landscapes. If much of this area lacked vegetation banding in the late glacial or early Holocene, and subsequently developed it, then this has the potential to be a major driver of ecohydrological and geomorphological change. The signature of this event might, however, be reflected in the sedimentary record by a reduced rate of sedimentation or even a cessation in sedimentation. This is because sediment delivery from the slopes upon which patterned vegetation developed would have declined significantly, despite the increase in rainfall. However, the effect of pattern development needs to be weighed with other potential drivers of landscape change, such as changing temperature and rain event properties, and aboriginal use of fire, in an ecologically complete interpretation of landscape record. These ideas largely remain to be tested. However, microfossil and charcoal analyses from stratigraphic analysis of some dryland archaeological sites provide some supporting data. For example, charcoal analyses have shown that mulga (*Acacia aneura*), which provides high-quality fuel wood, became abundant in central Australian rock shelter charcoals only after about 13 ka BP (Smith *et al.* 1995; Smith 2009). If this tall shrub was not abundant prior to this time, then the strongly banded mulga woodlands that dominate low-gradient landscapes over large areas of the Australian drylands could not have been present. This would presumably have allowed freer drainage of these landscapes, and higher rates of hillslope sediment transport, even though conditions were drier prior to the early Holocene. Analyses of stable isotopes of carbon and nitrogen in the soils of mulga grove landscapes suggest that broad areas occupied by mulga have been stable for about 1000 years (Bowman *et al.* 2007) but there are as yet no data specifically bearing on the antiquity of the mulga grove and intergrove system.

Widespread biological soil crusts

Although of course the ground cover provided by vascular plants may be low in drylands (typically 5–30%), the uppermost soil materials are extensively occupied by microphytic organisms, including cyanobacteria, mosses, and lichens (Eldridge & Greene, 1994; Belnap & Lange 2001). There is a considerable literature on these organisms and their influence on soil surface stability and erosion resistance, on soil nutrient and moisture status, vulnerability to grazing impacts, etc. For drylands, it is very likely that biological soil crusts are more significant than vascular plants in hardening soil surfaces against splash and water erosion (e.g. Xie *et al.* 2007; Bowker *et al.* 2008). This arises from mechanisms including the production of exopolysaccharides, which bind soil particles. The soil surface is generally roughened by the presence of biological crusts, and the surface undulations increase surface depression storage and reduce the

speed of overland flow, providing more opportunity time for infiltration (Fig. 8).

The role of biological soil crusts is not explored in more detail here because the organisms involved are able to colonize soil and rock surfaces even in extremely arid locations. It therefore seems unlikely that these crusts would have been eliminated from the drylands during glacial phase aridity, but would rather have persisted through even the harshest times. There is consequently little likelihood that significant changes in ecohydrological and geomorphological processes in the Australian drylands were driven by climate-related changes in biological soil crusts.

Some ecohydrological and geomorphological effects of Australian dryland fauna

Both vertebrate and invertebrate fauna affect the processes in dryland landscapes. Among the faunal

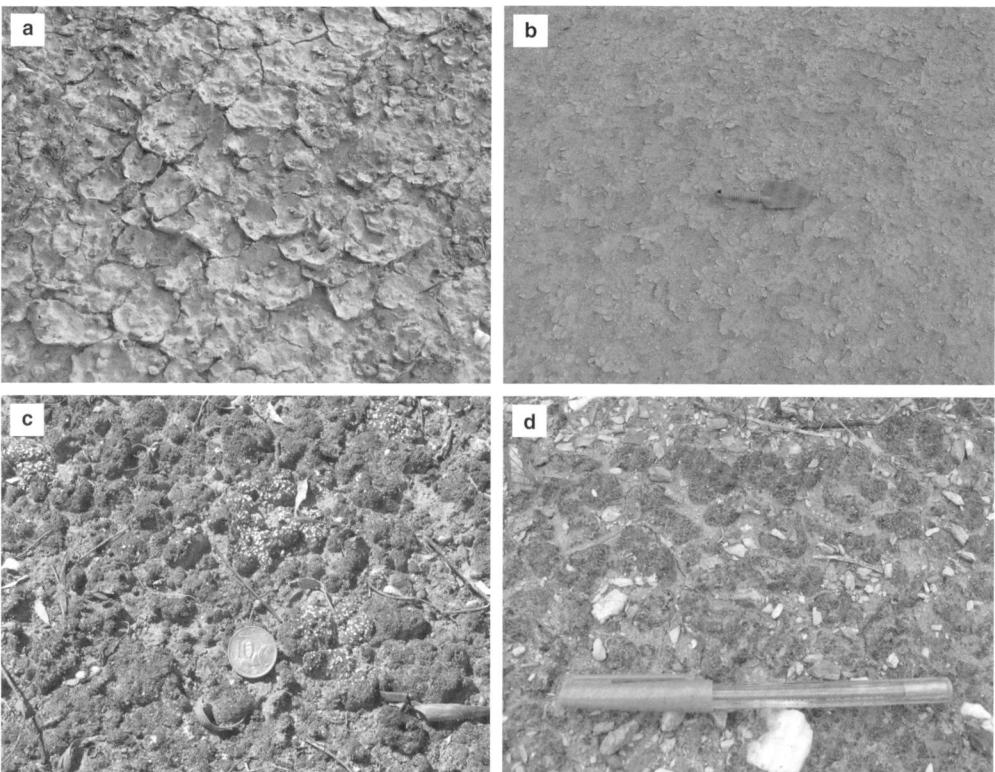

Fig. 8. Biological soil crusts from the Australian drylands. (**a**) Typical pedicled cyanobacterial crust on calcareous soils, western NSW. (**b**) Cyanobacterial crust abraded by blowing sand during severe drought conditions, western NSW. (**c**) Lichen crust exhibiting significant surface microrelief and depression storage capacity, northwestern Victoria. (**d**) Lichen crust with significant surface microrelief and depression strorage capacity, in a long-term grazing exclosure, far western NSW.

effects in the drylands are those arising from the excavation of soil pits by foraging, for instance by fossorial animals such as goannas (*Varanus gouldii*) (Whitford 1998) and the echidna (*Tachyglossus aculeatus*) (Eldridge & Mensinga 2007). From a semi-arid site in NSW, Eldridge & Mensinga (2007) reported infiltrability to be about twice as rapid in echidna foraging pits than at sites not disturbed (75.9 ± 17 mm h^{-1} v. 38.0 ± 6.3 mm h^{-1}). Given the decline or extinction of many Australian dryland mammal taxa, their reintroduction for landscape management has been considered by James & Eldridge (2007). Research in drylands elsewhere has highlighted the effects of faunal activity. For example, Neave & Abrahams (2001) showed that at a site in New Mexico, sediment concentration in overland flow could be related to the extent of small mammal disturbance of the soil surface. A discussion dealing more generally with the links between biota, soils, geomorphology and landscape processes in drylands has been presented by Monger & Bestelmeyer (2006).

The bettong (*Bettongia lesueur)*, a small marsupial, formerly occupied much of the drylands, but is now very restricted following European pastoral settlement. However, Noble *et al.* (2007) found that bettong warrens occupied about 0.6% of a study area in semi-arid NW NSW. There, infiltration rates were about 4.7 times higher (at 81 mm h^{-1}) in the bottom of the bettong warren than on the soils upslope, and the soil bulk density was also lower. The warrens are thought to have contributed to patchiness in these landscapes, and to the trapping of water, seeds, and organic matter (Ludwig *et al.* 1997). In turn, it is likely that the greater plant growth would have contributed a reinforcement of this effect owing to greater production of litter forming obstacles to overland flow. In a similar way, Eldridge (1993) showed that the entrances to ant burrows of the species *Aphaenogaster barbigula* (a common funnel ant whose colonies have large entrances) in the semiarid lands, occupied up to 0.9% of a study landscape south of Cobar in NSW. Infiltration of water ponded over nest entrance sites infiltrated more rapidly (23 ± 1.8 mm min^{-1}) than at non-entrance soils (6 ± 1.4 mm min^{-1}), an effect presumed to arise from the greater abundance of biopores. Although typically surrounded by a torus of excavated materials, the nest entrances were able to intercept overland flow because they could readily be eroded during sheetflow. At the Cobar site, as in the case of the bettong mounds, the biota also played a role in the long-term overturning of the soil materials, and perhaps in the textural differentiation of surface soils. It also increased the degree of surface heterogeneity of the landscape, and the likelihood of water and other resources being conserved. In this way, it can be reasoned that erosion by sheetflow would be diminished in areas occupied by burrowing organisms of the kind discussed here. Termites and many other faunal groups contribute similar kinds of ecosystem modifications (e.g. Whitford *et al.* 1992). Interestingly, Eldridge (1993) noted litter barriers and other evidence of overland flow, even though the soils exhibited very high infiltrability (>300 mm h^{-1} even for nest-free soils) in his study area. Here the explanation may relate to long-duration, low-intensity rains, perhaps associated with a decline in infiltrability to values lower than those found by Eldridge in his 30 min tests. The kinds of hydrological patchiness and resource-trapping mechanisms outlined here from faunal activity increase the level of spatial variability beyond that arising from the plant-related 'fertile islands' referred to previously, and this emphasizes that dryland ecosystems display many surface features capable of modifying conditions for the development and spatial integration of overland flow, and hence, of hillslope sediment transport. Similar roles for fauna have been described from other global drylands. For instance, James *et al.* (2008) reported enhanced soil infiltrability near the nests of two species of ants in the Chihuahuan Desert.

Is Australian dryland ecohydrology distinctive?

The foregoing discussion has sought to highlight the role of biota in fragmenting dryland landscapes into patches of various kinds, and ranging in size from <1 m^2 to >10^3 m^2. This patchiness is fundamental to most of the Australian drylands, owing to their relatively abundant vegetation cover and associated soil fauna, and indeed to drylands generally. The hydrological and geomorphological significance of this patchiness is that runoff is spatio-temporally discontinuous in runoff events except perhaps those exceeding in depth, duration, or rain rate, a threshold beyond which the retention capacity of the patch structure is exceeded. The threshold is itself likely to vary both spatially and temporally, in relation for instance to the antecedent seasonal conditions and the resulting density of plant cover, litter abundance, soil moisture, ant population, etc. The 'resource-holding' function of the patch system correspondingly also limits soil detachment and transport. In the light of these facts, the suggestion often made to the effect that drylands can be characterized by the widespread occurrence of erosive Hortonian overland flow (capable, for instance, of delivering high sediment loads to ephemeral channels, via sheet or rill erosion) is clearly not applicable over large tracts of the

Australian drylands. Instead, compartmentalized runoff–runon systems dominate, and the landscape is strongly conserving of water, organic matter, and other resources. Are the Australian drylands distinctively patchy in their hydrological properties, in comparison with other global drylands? This is a difficult question to answer. However, there are factors that suggest that patchiness may indeed be both more marked and more functionally important in the Australian drylands. The Australian dryland ant and termite fauna, for instance, is large and taxonomically diverse, and it has been suggested that this has probably resulted in larger modifications to soil properties in the Australian drylands than elsewhere (Stafford Smith & Morton 1990). In addition, the topographic gradient is very low over large areas of inland Australia, and this suggests that even relatively subtle surface features, such as animal scratchings, low ant mounds or minor accumulations of litter, may have the capacity to influence the hydraulic behaviour and sediment transport capacity of overland flow.

Does erosion reach its maximum intensity in Australian semi-arid landscapes?

There are important consequences for the widespread role of biota in fragmenting landscapes and in limiting the occurrence of occasions when integrated hillslope overland flow would deliver water and sediments to stream channels. In a drying climate, some plants would be lost, and the patch structuring of the landscape would be diminished in its efficacy. Dunkerley & Brown (1995) argued that this would have been the case in the drier conditions of the late glacial, and that integrated hillslope flow under those conditions accounts for the downslope sorting of the now inactive stone cover on hillslopes at Fowlers Gap. This provides an example of a conclusion that might appear unreasonable if the role of patchiness was not appreciated: that downslope transport of stones can be possible under dry conditions but perhaps not when the climate becomes wetter. That is, although the Langbein–Schumm 'rule' (Langbein & Schumm 1958) asserts that erosion is most efficient when effective rainfall (having deducted evapotranspiration losses from the open-field rainfall) lies in the 300 mm a^{-1} range, this may not generally be so. The changes in hillslope transport of stones by surface runoff envisaged by Dunkerley & Brown (1995) suggest that a modification of the Langbein–Schumm conception is required for low-gradient landscapes with vegetation patches or bands. The efficiency of vegetation bands would be diminished under a drier climate, and occasional large rain events would then be able to drive integrated hillslope runoff, increasing the erosion rate in comparison with that for a wetter regime. Therefore, in these landscapes, it is possible to speculate that erosion would be most intense under arid conditions, and not under the contemporary semi-arid climate. This indicates a lack of applicability of the argument of Langbein & Schumm (1958) to at least major parts of the Australian drylands, and in a similar way, variants on this 'rule' have been proposed for other global drylands (e.g. Walling & Kleo 1979).

Stream channel processes

This review has so far considered some aspects of the rainfall climate of the Australian drylands, and the ways in which biota influences the partitioning of rainfall into soil moisture, surface ponding, and overland flow. The key role of landscape patchiness generated by biota has been emphasized, across spatial scales that range from that of a single dryland plant or ant nest, to entire hillslopes where the vegetation takes the form of contour-aligned groves. In events where rainfall depth or rate in combination exceed the retention capacity of the dryland hillslopes, overland flow becomes temporarily integrated, and flow is shed downslope to the stream channel network. The behaviour of streamflows is unlikely to be significantly affected by faunal activity. However, the influence of vegetation remains large, and we now turn to consider especially the role of trees in dryland channel processes, and the ways in which this role might alter in response to changes of climate. Trees grow in and along many Australian dryland streams, and contribute large volumes of woody debris. Both the trees and the debris contribute to channel roughness, and the estimates reported by Graeme & Dunkerley (1993) are among the highest known for dryland streams. The woody vegetation along Australian dryland streams contrasts with the much smaller and more flexible herbs and shrubs described for Mediterranean dryland streams (e.g. Sandercock *et al.* 2007), and this extensive woody material may well be distinctive of the Australian drylands, with ramifications for the role of trees and debris in stream form and process that contrast with those of more humid areas (e.g. see Gurnell & Petts 2006).

Influence of trees on dryland stream processes

It is common in the Australian drylands for trees to occur principally along ephemeral stream channels, although their growth is not limited to the riparian zone, but rather may extend across broad stream beds, banks, and adjoining floodplain forming channel-associated vegetation (CAV). In these locations the trees can access a larger soil water store than is found in the upland runoff source areas, because percolation into the bed and banks occurs during channel flow (contributing to transmission losses). The channel and its margins therefore experience greater recharge than would arise from rainfall alone, especially around the margins of pools and waterholes, where seepage may continue for days or weeks following stream flow events. Some of the soil and groundwater accessed by trees may be saline, but the locations along channels, perhaps especially the upper banks (where sandy levees may occur) and bed may have improved water quality resulting from flushing by percolating floodwaters (Payne *et al.* 2006). If this is generally true, then water salinity variation related to flood dilution of salts may be more important to the localization of CAV than access to stored flood water. Among the locations for CAV, the bed and banks are perhaps most influential on channel form and process. The influence of trees arises from root reinforcement of the banks, and from the obstruction of the channel by trees and by woody debris that is swept up by flood flows to form large debris barriers. Turbulence around these structures results in localized scour of the bed, and the deposition of lee-side drapes of sand and mud, which can be colonized by plants, eventually to become stable bars influencing the overall channel form (e.g. Tooth & Nanson 2000). However, even isolated trees at locations distributed over the bed, and not necessarily connected with bar formation, can exert a considerable influence on flow speeds and channel roughness (e.g. Graeme & Dunkerley 1993).

As noted above, large areas of the Australian drylands are essentially arheic, and lack large-sale organized fluvial drainage networks. Especially in Western Australia, there are aggraded valley systems dating from the early Cenozoic before the onset of continental aridity (Zheng *et al.* 2002; Hou *et al.* 2008). Elsewhere, rocky uplands result in rainfall higher than in the flat desert floor, owing to orographic uplift, and runoff from the uplands feeds flanking ephemeral drainage systems. These ephemeral streams intermittently receive runoff water that has been generated within the upland systems of plants and soils previously outlined, to yield flow events that represent a subset of the rain events over the catchments (those capable of delivering sufficient runoff to stream channels to overcome upland transmission losses, and pass flows downstream to the distal reaches). For reaches of stream channel lying within the rocky uplands, flow volumes increase with catchment area, but upon leaving the uplands and entering alluvial plains, more rapid transmission losses into the stream bed and underlying aquifers arise during the downstream passage of floodwaters and may exceed any small tributary inflows. In the case of widespread rain, the bed materials may be wet prior to the arrival of the floodwater, but with localized convective storms, the bed may be completely dry and a flood bore may arrive from a considerable distance upstream, traversing a dry stream bed. Owing to transmission losses, runoff depth from dryland catchments tends to decline with increasing catchment area, a trend opposite to that observed in perennial stream environments, and sediment yield follows the same pattern of decline (e.g. Griffiths *et al.* 2006).

As a result of transmission losses and the limited spatial extent of convective rainstorms, ephemeral streamflows are usually spatially discontinuous, as well as being short-lived. For Nahal Eshtemoa in Israel, a well-researched dryland stream, flow occurs for only about 2% of the time (or about 7 days per year) (Reid *et al.* 1998). For Fowlers Creek in western NSW, Cordery *et al.* (1983) reported that flows had been sufficiently large to reach the terminal lake basin (Lake Bancannia) only about once per decade, on average, and these large flow events are likely to have lasted only a few days. Thus, large flows may be exceptionally rare, and there are too few studies of such events for their role in determining channel form to have been resolved. Especially for the more common sub-bankfull flows from localized convective rainfall, the floodwaters must traverse reaches where the bed sediments are dry and highly permeable, resulting in rapid transmission losses. Channel sands and gravels are of limited depth, however, and may overlie contrasting sediments of much lower permeability. This limits the time through which bed seepage can operate at a high rate (e.g. Lekach *et al.* 1998). It is also necessary to recall that modern rainfall and stream gauging data, where they are available in the drylands, may be insufficient for resolving the flow regime of streams, so that palaeoflood hydrology investigations may provide a longer time-window of supplementary data (e.g. Benito & Thorndycraft 2005). Palaeoflood hydrology brings its own set of uncertainties, including completeness of the record and challenges of dating, but especially the problem of estimating channel roughness and hydraulic conditions (e.g. Baker 2008). Studies using contemporary hydrological data have shown that for ordinary flows, a reasonable description of the trend in flow loss is a fixed fraction of the arriving volume lost for each kilometre of travel. For small streams in arid western NSW, figures of 5–15% per km were derived (Dunkerley 1992); for a stream in arid Washington State, USA, Waichler & Wigmosta (2004) derived the value of 19% per km. As a result of transmission loss, flood flows may be totally lost within 10 km of the upland margin, even for streams with catchments of $10-10^2$ km^2 within the rocky uplands. It is often considered that channel form, such as width/depth ratio, and the permeability of the bed sediments, are the primary determinants of transmission loss. However, other factors certainly exert a significant influence, including the clogging of channel margin sediments by inwashed silts and clays (Dunkerley 2008e), and the obstruction of the channel by trees and woody debris barriers associated with them (Dunkerley 2008d). High suspended sediment loads have been reported from a number of Australian dryland streams, exceeding 10 g l^{-1}, and it has been argued that the behaviour of suspended sediment concentrations is consistent with the idea that the stream-bed sediments act to strain sediment particles from the infiltrating water (e.g. Dunkerley & Brown 1999b). If this is a common process in dryland streams, then the rate of transmission loss is dependent not only on channel form and the primary textural characteristics of the channel margin sediments, such as porosity and permeability, but also on the nature of clogging sediments delivered from the catchment. Following the accumulation of hillslope mantles of windblown dust during arid glacial phases, as described for the Flinders Ranges (Williams *et al.* 2001), these materials may be delivered to streams by overland flow and there clog channel margin sediments. Reduced rates of transmission loss would then provide a reinforcement of the greater channel competence for sediment transport arising from higher post-glacial rainfalls.

There are wider ramifications arising from changes to transmission losses along dryland streams, however, that further complicate links between channel processes and changes in external climate. In the Australian drylands, channel-associated vegetation, notably of the river red gum (*E. camaldulensis*), modifies bed and bank stability, as well as channel roughness and flow speeds. All else being equal, a greater abundance of trees in the channel will increase roughness and reduce sediment transport capacity. Roughness arises from the obstacle drag on the trunks of standing trees, and also from the fallen woody detritus of branches and twigs that is swept up by flood waters to form extensive barriers of debris lodged against trees on the bed and banks. Tree growth and survival are dependent on the recharge of groundwater during transmission loss, and if this process is throttled by the clogging of channel margin sediments, then it is likely that the abundance of trees, and of fallen woody debris, would be reduced.

Barriers of woody debris build up against living or dead trees, on both bed and banks (Fig. 9). In a study of woody debris barriers along Fowlers

Creek, Dunkerley (2008f) found that there was little evidence of tree fall contributing to the bulk of the material forming debris barriers. Rather, the size of wood pieces indicated that most were likely to be branches, with the debris barriers reduced in permeability by infilling twigs, leaves, sand and mud. In some cases, densely packed woody debris barriers obstructed virtually the whole channel width, impounding substantial volumes of water upstream. Inspection of the dry channel showed that widespread mud deposition coated bed and banks in these impounded reaches where flow speeds would have been low, so that the impoundments may contribute less to transmission loss by seepage than would occur in the absence of mud deposition.

Fig. 9. (**a**, **b**) Two examples of the forms of woody debris accumulation in ephemeral dryland streams. Both examples are from the ephemeral Fowlers Creek, western NSW, Australia. Both debris barriers accumulated against the upstream side of river red gum trees growing on the bed. Both contain a mixture of large framework logs (derived from fallen limbs), and the porosity of the barriers is reduced by accumulations of twigs, leaves, sand, and mud. The large crescentic scour hole on the upstream side of the barrier in (a) should be noted. Direction of view in both photographs is downstream.

The woody debris barriers seem to exhibit degrees of persistence. Some contain logs that appear not to have moved for many years, whereas other are freshly built. Their significance for flow roughness (Graeme & Dunkerley 1993) and for channel processes suggests a link to the episodic nature of dryland rainfall and runoff events. Long dry periods give rise to the buildup on the bed of debris released by the channel-associated trees. The first flow following a dry period competent to move the fallen branches and other woody debris may then sweep this material up, augmenting old debris barriers or initiating new ones. However, smaller flows not capable of moving the woody debris might have higher transmission losses arising from shallow flow over a bed littered with fallen debris. There is virtually nothing known about the dynamic changes in channel roughness that might arise during major flood flows, as debris barriers are built or swept away. Channel form clearly reveals that major debris barriers can deflect the flow against one bank, resulting in undercutting and lateral migration of the channel. This is most obvious where tree growth on the bed is relatively dense, providing abundant obstacles to permit the lodgement of debris barriers. Thus, whereas riparian trees probably serve to increase bank stability and reduce rates of channel migration, trees on the stream bed may promote local channel instability. This is further accentuated by the formation of scour holes on the upstream side of large debris barriers, and lee-side accumulations of sand and cobbles. These drapes or lee-side bars tend to be colonized by red gum saplings, and may become stabilized, dividing the original channel into two or more sub-channels. The sub-channels in turn may have different bed elevations, so that the higher one may be active only in larger flows. Channel junctions where there is an offset in bed elevation result from this effect, and form an additional sink for flow energy that is genetically linked to the channel-associated vegetation some distance upstream. A similar role for plants was described from the Marshall River in the NT (Tooth & Nanson 2000). There, a teatree (*Melaleuca glomerate*) occupies a similar role, colonizing the bed and then, via lee-side sediment deposition, promoting the growth and stabilization of bars, which result in the development of an anabranching channel form with many more sub-channels than are seen in the channels of western NSW. Thus, contrasting ephemeral channel forms result in different parts of the drylands, indicating diverse roles for CAV. In the case of the Marshall River, there is no role comparable with that of the long-lasting woody debris barriers seen in western NSW. In western NSW, the majority of trees are riparian, but the woody detritus that they release onto the bed (and floodplain, during overbank

flows) becomes critically important to the formation of debris barriers, and thus to the speed and competence of flood flows. Effects on flow conditions must result in changes in the capacity for sediment transport. It may be that where marked roughness is generated by the CAV and debris barriers, coarse bedload is deposited owing to the reduction in flow competence. An association between coarse bed material and bed obstruction by trees and woody debris is a feature of many Australian dryland streams, but the aggregate effect of CAV on sediment transport rates remains to be explored.

It is possible that the abundance of woody debris barriers in the dryland streams of western NSW relates to the very strong ENSO-driven drought and flood cycle there. Multi-year droughts may be associated with the accumulation of detritus on the bed, and the formation of large and numerous debris barriers in early floods of the succeeding La Niña wet phase. Drought conditions may furthermore trigger increased amounts of branch and leaf drop, contributing further to the availability of detritus on the bed. Whether there are systematic shifts in channel roughness related to such ENSO-linked changes in CAV along dryland streams remains to be investigated.

Owing to the low and unreliable rainfall of much of the Australian drylands, trees commonly grow along lowland stream channels, in locations where runoff from rocky uplands is delivered and where channel conditions allow seepage into the bed and banks during flood events. For the establishment of trees, channel stability is needed. Trees are typically lacking along the banks of incised channels such as the arroyos of western NSW, and die when avulsion or channel migration diverts flood water along a new path. Shaw & Cooper (2008) have highlighted the links between catchment properties and the CAV along ephemeral streams, and erected a classification of stream types based on the strength of this linkage.

In summary, we can observe that along Australian dryland streams, as on the catchment hillslopes, vegetation exerts a significant influence on fluvial form and process, notably including an influence on flow competence and on the rate of transmission loss, and hence on sediment transport processes. This influence arises from at least four fundamental sets of mechanisms, as follows.

(1) Plants influence channel form (e.g. cross-sectional shape) via root stabilization of banks, and the stabilization of in-channel bars or islands upon which river red gums, for example, grow. It has been suggested that *E. coolabah*, a common channel-associated tree of the arid interior, is found often on bank crests and similar positions as these are sites of lower average soil water salinity (Costelloe *et al.* 2008). Although no relevant data

exist, it seems likely that this growth habit and the root systems of these trees confer additional stability on the streambanks. The nature of the channel margin and the area of bed and bank submerged by flood flows exerts a first-order control on the rate of transmission loss by percolation into the bed and banks.

(2) Plants growing in the channel increase the channel roughness, reduce flow speed, and increase the time through which the channel margins are submerged and able to conduct seepage away from the channel (Graeme & Dunkerley 1993). Graeme & Dunkerley (1993) tallied the degree of obstruction to flow provided by river red gums and woody debris barriers lodged against them in several ephemeral streams in western NSW. They found that trees and debris barriers contributed up to 50% of the channel roughness, even where the bed was composed of cobbles and the channel form was irregular.

(3) During dry weather, the riparian and in-channel trees drop abundant leaves, twigs and branches to the surface of the bed. These materials are subsequently swept up during the passage of floodwaters, and lodge against trees growing on the bed to form large barriers of woody debris. In extreme cases, these can extend across the channel, effectively obstructing free flow until the barrier is overtopped.

(4) By altering flow speeds, and locally promoting backwater effects where debris barriers impound the flow, channel-associated vegetation may drive the deposition of muds over the channel perimeter, provided that the upstream catchment provides sediment of the appropriate texture. The muds may be carried into the pore spaces of the bed and banks, resulting in the clogging of pore spaces and a lowering of the permeability of the channel margin sediments (Dahan *et al.* 2007). Reaches of ephemeral channels where the bed and bank sediments are clogged would contribute less to transmission loss, and tend to promote the continued passage of the floodwaters (Mudd 2006; Dunkerley 2008*e*). High-magnitude flow events may be capable of stripping some mud and clogged sediments, increasing their relative importance to groundwater recharge via percolation through stream-bed sediments beyond that related to their greater flow duration (Zammouri & Feki 2005).

Changes to these processes that might follow from a climatic change are not simple to envisage. Consequences in the channels depend upon changes in runoff and sediment load delivered from the uplands in the catchment area, and also upon the changes in channel-associated vegetation. It is conceivable that under drier conditions, increased loads of coarse sediment would be delivered from the more integrated hillslope runoff

upstream (see discussion above). In the channels, such sediment would provide a more permeable bed, although the depth of this material would be limited. At the same time, a general reduction in the frequency of flood events might well reduce the abundance of channel-associated vegetation, or a shift from trees to smaller shrubs. Either outcome would tend to reduce the abundance and scale of woody debris barriers, and exert less restriction on the passage of flood flows. Hence, a drying of climate would probably have opposing effects on transmission loss, the coarser sediments promoting loss, but the reduced channel obstruction allowing faster passage of the floodwaters with less time for seepage losses. To this set of factors must be added any change in the rainfall climate, such as an increased inter-storm period, which would tend to leave the stream bed sediments drier, and hence enhance transmission losses. However, this could be partially compensated for if the rain events were more intense. This is because, as Dunkerley & Brown (1999*b*) pointed out, there is evidence that transmission losses are minimized at flows near bankfull, being larger for both small and overbank flows. Finally, consideration must be given to the possibility that the kinds of trees growing along ephemeral channels may have changed in response to Quaternary climatic fluctuations. For example, Smith (2009) reported that charcoal derived from red gum wood only became abundant in the Puritjarra rock shelter after *c.* 15 ka BP. If red gum growth along ephemeral streams was indeed reduced during the preceding glacial maximum, then channel roughness may also have been significantly reduced, and sediment transport capacity increased, making palaeohydrological analyses more challenging. Potential significant changes in the role of wood in channel processes of just this kind have been highlighted by Francis *et al.* (2008) for the case of European rivers, and this work provided further insights into the potential for trees and woody debris to mediate the effects of climate change on rivers. Evidently, there is much to be learned about the role of vegetation in dryland channel processes, and especially about longer-term changes in CAV that might have accompanied Quaternary climate change. Figure 10 presents in schematic form some of the interactions between climate, catchment sediment yield and CAV that influence the rate of transmission loss, and hence the downstream decline in sediment transport capacity in streams of the Australian drylands.

Discussion

This review has sought to highlight the widespread influences exerted by biota on ecohydrological and geomorphological processes in the Australian

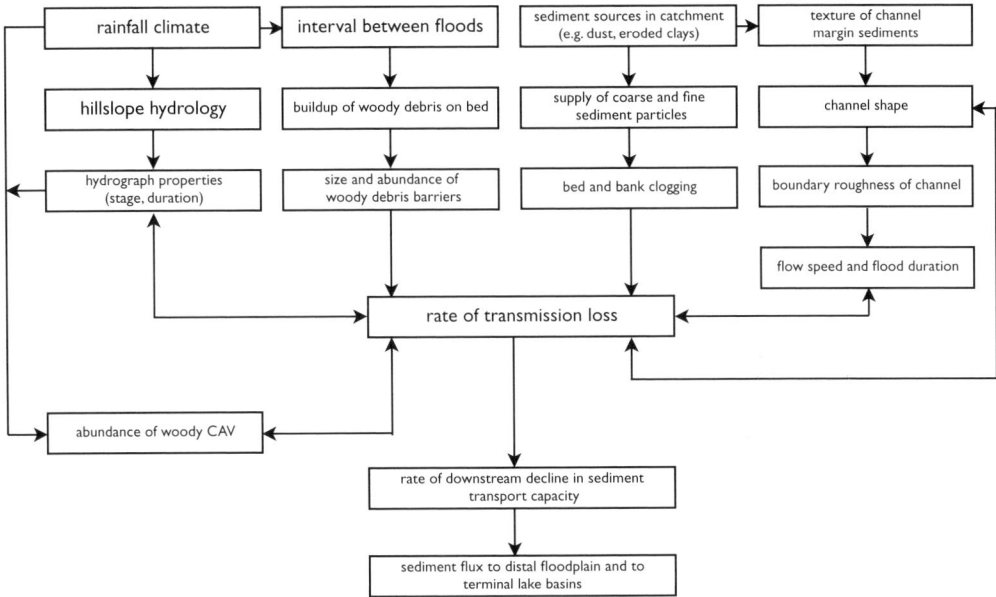

Fig. 10. Diagrammatic representation of some of the linkages between climatic and catchment processes that modify the rate of transmission loss in some Australian dryland streams. The figure refers to streams having extensive CAV, and not to incised (arroyo) channels.

drylands. The coverage has necessarily been geographically selective, because of the paucity of detailed studies over the very large area of these drylands.

One principal reason for preparing this review highlighting selected landscape roles of biota, especially in ecohydrology and geomorphology, is that the landforms and sediments inherited from the Quaternary and Holocene frequently form the basis for interpretations of palaeoclimate, or at least, inferences about changes in rainfall, temperature, or effective moisture. Some examples of this are considered below. In this work, there appears often to be a presumption that the external hydroclimate is imposed on the landscape, and that differences in hydroclimate (and particularly rainfall) during past phases of landscape development can be linked directly to evidence of degradation or aggradation. As repeatedly demonstrated above, however, the land surface mediates the effects of climate change. There are multiple ways in which land surface changes can oppose or reinforce changes in the external climate. It is arguable that a number of these interactions are more significant in the Australian drylands than in the humid zone, owing to some of the characteristics reviewed above, such as the low topographic gradients, the presence of internal drainage basins from which fine materials can be deflated and undergo transport on continental scales, with re-deposition

far downwind, and the nature of hillslope- and channel-associated vegetation. The place of ecosystem processes in mediating the effects of climate change on runoff, erosion and related processes has been highlighted in other environments. The work of Thomas and coworkers (Thomas 2000, 2008; Thomas et al. 2007), primarily in the humid tropics including NE Australia, provides examples of interpretations of this kind. However, there appears to have been no corresponding analysis for the Australian drylands.

The manifold pathways by which climate change may result in altered ecohydrological processes, and in positive and negative feedbacks, suggest that episodes of hillslope erosion or stability, or of channel scour or aggradation, may arise from many sets of precursor events. The actual sequence of landscape processes leading to channel aggradation at one location may not be the same as the causes of aggradation at a different location or during a different climatic fluctuation. Therefore, issues of equifinality in the sedimentary and geomorphological record probably warrant greater attention than they are often given. The most obvious climatic trigger for a landscape change may not be the actual cause of the responses seen in the landscape. Equifinality becomes an issue of concern in many investigations of landscape change, given the complex mesh of processes and developmental pathways. Studies of hydrological response (Ebel & Loague

2006) and of processes of soil erosion (Brazier *et al.* 2000) have highlighted the challenge posed by equifinality. It is also worth observing that the processes involved in many kinds of landscape change, such as aggradation or incision, are by no means fully resolved, even for well-studied cases. The conversion of non-incised dryland streams to incised arroyo channels provides one instance of this. Another example is the case of strath terrace formation, and the nature and causes of the planation and stream incision involved. Strath terrace formation is considered to relate to increased runoff and sediment supply (linked via landscape intermediary processes to increased precipitation). It has been argued that stream-bed aggradation limits vertical stream incision and instead favours lateral planation and strath formation (e.g. see Fuller *et al.* 2009).

A few instances of arguments on the landscape response to climate change can be drawn from other global drylands, including the drylands of SW USA and the Negev desert, where there are informative studies of soil surface conditions, soil salinity, biological soil crusts, and runoff and infiltration behaviour, often made across the rainfall gradient from arid or hyperarid to Mediterranean sub-humid. For example, Yair & Kossovsky (2002) studied runoff from plots and small catchments at a Sede Boqer site in the arid Negev (average annual rainfall 93 mm) and at the Lehavim watershed to the north (average annual rainfall 280 mm). In an extreme storm delivering >105 mm at the Lehavim site and nearly 46 mm at Sede Boqer, no catchment runoff was generated at the former, whereas the arid site yielded streamflow amounting to 37% of rainfall. Differences in the land surface, including areas of rock and loess soil, drove better runoff connectivity, and hence yield, at the arid site. Yair & Kossovsky (2002) concluded that within drylands, it is possible that runoff behaviour is more strongly linked to surface properties than to storm or annual rainfall climate. Thus, they speculated, a climate drying could in fact result in an increase in runoff, owing to loss in plant cover, changes in the soil surface, and better flow path connectivity. Such a conclusion clearly has direct relevance to attempts to infer rainfall change from the sedimentary record, where, for instance, evidence of incision of valley floors might reflect a decrease in rainfall, rather than the seemingly more intuitive expectation that an increase in rainfall would be required. Landscape responses would depend upon the relative rates of adjustment of climate and surface conditions (soil depth, stone cover) and there is the likelihood that sedimentary records of such adjustments would reflect also the mechanisms producing complex response chains (Schumm 1977).

A second example relates to the potential for differential landscape outcomes of a climate change, as a result of regional differences in dryland landscape scale and land surface character. Yair & Raz-Yassif (2004), for example, demonstrated that on longer hillsides in the rocky Negev desert, runoff connectivity was reduced owing to the greater probability of runon absorption in loess soil patches. The effect emerged in catchment-scale flow behaviour partly because of the typically short duration of dryland rain events in this area. The brief rain events ensured that the production of surface runoff would end before integrated overland flowpaths could be established on longer hillsides, owing to the travel time involved. One possible consequence of dust accretion over upland landscapes in the Australian drylands (as documented from the Barrier Ranges in western NSW by Chartres 1982) is therefore a decline in the occurrence of integrated hillslope overland flow. A comparable effect might not be detectable in more extensive, low-gradient landscapes such as those away from the uplands. Importantly, Yair & Raz-Yassif (2004) pointed out that the same decline in occurrence of integrated hillslope runoff production could arise from changes in rainfall behaviour, such as event duration and rain rate, independently of changes in annual or seasonal rainfall total. Once again, these arguments make it clear that the links between landscape behaviour in drylands and external rainfall climate are neither direct nor always as simple as conceptual models might suggest. Moreover, they suggest mechanisms by which a given change in external rainfall climate, such as a 20% decline in summer rainfall amount, might drive different consequences for streamflow and sediment transport in small rocky upland basins or larger lowland systems. If this is so, then apparent conflicts in the apparent history of erosion and aggradation may be explicable only partially in terms of changes in imposed climate. Supplementary information concerning factors that might enhance or reduce hydrological connectivity on hillslopes is clearly also required, such as the history of dust accession to the hillslopes concerned. The same sedimentary signature of a given climate change should therefore not be expected in contrasting landscape contexts such as basins of differing relief and size or those with different environmental histories. For example, basins located differently with respect to the trajectories of dust transport and deposition through an arid phase might be expected to exhibit contrasting responses to a subsequent shift to more humid conditions.

To the arguments just outlined can be added effects arising from changes in plant cover (and perhaps in associated soil disturbance by fossorial fauna) associated with a change in soil moisture status. In the case of biological soil crusts, once again the effects of a climate change may not be in the expected direction. Ram & Yair (2007) examined crusts along a rainfall gradient from 86 mm to 160 mm annual rainfall at a site near the Egypt–Israel border. They found that soil moisture available to support vascular plants was actually less at the higher rainfall sites than at the more arid ones, owing to the abstraction of water onto thicker biological crusts there. At the drier sites, there was less abstraction, and overland runoff could also arise and locally be concentrated downslope to form relatively moist microsites of advantage to vascular plants. These results are consistent with those of Grishkan et al. (2006), who also studied soil crust organisms along a rainfall gradient in Israel. They found that there was only a weak influence of rainfall on the spatial variation in biological crust properties, which instead were largely governed by micro-environmental characteristics such as levels of soil moisture and organic matter. Here, it is important to recall the influence that biological crusts exert on runoff and erosion, discussed above. Faunal excavations exert a somewhat different influence on overland flow, serving as loci for depression storage. The filling of depression stores can contribute to a delay in the integration of overland flow on a hillslope, and to a delay in the commencement of channel flow (or, in small rain events, its complete inhibition). There are also links to rain event characteristics on sub-daily time scales here. Aryal et al. (2007) showed that rain intermittency modulates the conversion of ponded water to overland flow because intervals when rain ceases allow depression stores to partially or completely empty, delaying the recommencement of runoff when rain resumes and reducing the overall runoff yield from a rainfall event. Here then is a mechanism by which the hydrological effects of goanna and echidna foraging pits on the generation of surface roughness will vary with the properties of the rain events. The loss of flowpath connectedness would be most marked in short, or intermittent, low rain-rate events, and this potentially provides a positive feedback loop. Greater infiltration of soil water from ephemeral ponding would better support vascular plant growth, and this in turn would foster further absorption of rain and runon water, as described above. In the case of Quaternary climatic fluctuations, it is necessary to envisage that many of these influences on hydrological response (and on the sediment transport capacity of overland flows) would act simultaneously, with changes in rainfall climate leading to shifts in crust type and abundance, and habitat changes driving shifts in the intensity of faunal activity.

Interpretation of the sedimentary record in terms of past climates is a common task. The late Quaternary history of the Neales River, a major tributary in the Lake Eyre Basin, was discussed by Croke et al. (1996). A diversity of sedimentary sequences in different parts of the stratigraphic record led them to conclude that other controls, including local baselevel changes related to fluctuating lake levels and shoreline positions, might have been responsible. The suggestion arising from the earlier parts of this paper is that local changes in ecosystems, driving shifts in runoff ratios, transmission loss rates, etc., need also to be considered as drivers of sedimentological changes in the record. In the same way, Maroulis et al. (2007) in discussing the sedimentary evidence of fluvial–aeolian events along Cooper Creek offered the interpretation that abundant sand supply for construction of source-bordering dunes implies seasonally much wetter conditions that those of the modern climate. However, there are other possibilities, including moderately higher rainfall in combination with changed runoff ratios in the upstream catchments, or changes in transmission loss and the delivery of water to distal reaches of the stream network. Although undoubtedly wholesale shifts in the external climatic environment are primary drivers of landscape change, the internal readjustments and feedback processes cannot be overlooked. Indeed, interpretations of the sedimentary record in terms of changes to wetter or drier conditions often appear to take little or no account of the principles of geomorphological and hydrological thresholds, nor of the potential for complex response trains to impose time-transgressive changes over significant periods following a change in the external climate.

However, the significance of potential ecosystem changes in attempts to infer past hydrological and climatic conditions have been highlighted in some published studies. For example, Hesse et al. (2004) pointed out that there are disagreements between various climate proxies (speleothem growth chronology, sedimentology of fluvial systems, lakes and lunettes) over when and by how much Quaternary climate in the drylands phases differed from the modern climate. Such regional differences in evidence or its interpretation may suggest real differences in past rainfall climate. However, as Hesse et al. (2004) pointed out, effective moisture supply is influenced by multiple factors, including external climate (rainfall, temperature), the threshold rainfall required to generate runoff, the nature of the vegetation cover, etc., and sophisticated system modelling would be needed to isolate and identify any possible change in rainfall as a single causative factor. It can be added that for the development and application of models of this kind, parameterization would be a severe challenge, as would model validation. Other possible changes in ecosystem form and function were highlighted by Hesse *et al.* (2004), in light of the lower concentration of atmospheric carbon dioxide at the last glacial maximum (LGM), combined with reduced temperatures, especially in the continental interior. Thus, evidence for dune activity and dust deflation might indicate reduced plant cover arising from metabolic effects, rather than being entirely attributable to rainfall decline. But again, in light of the arguments raised in this paper, an appeal to a change in rainfall alone is certainly likely to miss important landscape responses. Rain is received in widely spaced events, and the nature of those events (depth, rain rate, intermittency, duration) and the length of intervening dry periods exert major controls on effective rainfall, soil moisture, biota, the production of overland flow, and the magnitude and spatial extent of streamflows. This in turn would modulate the erosion and transportation of sediments and the nature of the sediments laid down. Unsurprisingly, given the nature of the alluvial record, inferences about past intrusions of monsoon rainfall into the drylands, or of the broad changes in climate, have not operated at time scales finer than, for instance, discussions of changed rainfall total and seaonality (e.g. Bowler *et al.* 2001). Rare flow events can be of great significance to hillslope and channel form in drylands, but as Briant *et al.* (2008) have shown from a temperate study area, the dating and interpretation of field evidence for major flood events is itself faced with severe methodological difficulties.

Taking another view, Nanson *et al.* (2008) have offered the suggestion that most active fluvial conditions, with larger fluvial flows than today, occurred not in last interglacial (MIS stage 5e) times but later, when sea levels had fallen sufficiently to restrict ocean currents through Torres Strait and the Indonesian archipelago. They speculated that this may have diverted some water from the western Pacific warm pool southward along the Australian coast, supplying moisture despite the likelihood that the northern Australian monsoon was weaker by that time. However, two points arise here. If the rainfall were higher than now, then it is reasonable to argue that vegetation cover would have been more extensive. Therefore, runoff ratios may well have been lower owing to interception loss and the evapotranspiration (ET) flux from the plants. Additionally, according to the Langbein–Schumm 'rule' (e.g. Wilson 1973), increased plant cover beyond the semi-arid climate is associated with a declining erosion rate owing to the protective role of plants. It follows that any inference about the amount of rainfall arising from such a displaced warm pool would tend to be an underestimate unless the effects of declining runoff ratios and higher ET losses were allowed for. Once again, key steps in the chain of reasoning

from the sedimentary record to inferences about climate cannot be overlooked: the mediation of all change via ecohydrological functions within the landscape must be incorporated.

The example presented above of the possible development of widespread vegetation banding as moisture supply in the drylands increased in the early Holocene makes the same point about the lack of straightforward links between rainfall and surface processes. As conditions became less severely arid, runoff and hillslope erosion are inferred to have declined owing to the efficient resource trapping that occurs in patterned vegetation, not risen. Thus, the onset of moister conditions ought to be reflected in reduced rates of sediment accumulation in the terminal lake basins within the landscape. It can be hypothesized that such changes in surface process and resulting mediation of climate change effects are likely to be most critical in the drylands. Drylands are widely cited as being sensitive recorders of climate change, but the record is evidently not a straightforward one to interpret. Nor, of course, were climatic conditions stable through the course of the Holocene, with astronomically driven changes in solar radiation driving altered monsoon behaviour, among other climatic variations.

A final instance of the complex links between climate change, biota and landscape response can be drawn from the drylands of the SW USA. There, Wells *et al.* (1987) explored the age and development of piedmont deposits in the Soda Mountains area, near the California–Nevada border. The late glacial in this area was a high-lake phase, and the Holocene drier, but with high-magnitude monsoonal rain events. In contrast to the behaviour hypothesized for the Australian drylands, in which plant cover became more extensive in the Holocene, in the Soda Mountains area, woodland retreated. Wells *et al.* (1987) reasoned that deflation of salts and dusts from deposits of the former pluvial lake Mojave during the Holocene influenced the development of soil carbonates at shallow depth, so reducing the sustained surface infiltrability. In turn, and in combination with reduced plant cover, this promoted greater runoff, and the dissection of the piedmont, despite the drier conditions. Much of this change was time-transgressive, and presumably lagged the driving climate change by millennia owing to the time needed for soil carbonate layers to be set down. Significantly, Wells *et al.* (1987) pointed out that the ecohydrological changes related to soil profile evolution would be hard to differentiate from those that might relate to climate change, and this is probably especially true as the runoff changes seem to relate to some combination of rain event properties, soil infiltrability and vegetation cover. Finally, they pointed out that colluvial aprons at the hillslope base may have absorbed runon water from the hillslopes above, limiting the volume of runoff passing down to the piedmont slopes. This again illustrates a local link between landscape process and runoff that is not directly related to a climate change, but is perhaps most correctly viewed as part of a complex response cascade. In other words, in circumstances such as those described by Wells *et al.* (1987), we see alluvial responses that relate to slow-onset pedogenic and plant cover changes that exhibit both spatial variability and substantial time lags following the climatic change that is reasoned to have been the ultimate trigger for ensuing events. (Similar scenarios have been envisaged elsewhere; e.g. see Bull & Schick 1979.) Major time lags in ecosystem responses to external triggers such as climate change are problematic for attempts to link cause and effect. This becomes an even more significant challenge in light of the fact that pedological evolution can be expected to drive changes in the nature of the vegetation cover. Given that prolonged climatic stability is unusual, we need to enquire how frequently and for what time periods landscape equilibrium is achieved, perhaps in contrast to persistently lagged responses, at least in some of the slowly evolving aspects such as soil depth and hydraulic properties (e.g. see Bracken & Wainwright 2008). We can finally observe that even if major bursts of incision, sheet erosion, etc. occur during extreme rain events, these are preconditioned by the long run of less exceptional years. This is because the presence of erodible soil materials on the hillslopes, required to feed sediment into colluvial aprons, owes its existence to the sustained ecohydrological and weathering processes occurring in the period between extreme events.

Conclusions

This overview of selected aspects of Australian ecohydrology and geomorphology has highlighted several principles.

(1) Although continental aridity determines the broad patterns of ecosystem type and hence the framework for landscape ecohydrology and the modifications of geomorphological processes linked to it, in detail there is considerable independence from strict climatic control. Locally, the availability of water is modified by the effects of isolated plants, groves of plants, faunal activity, and channel processes such as the redistribution of upland runoff during transmission loss, as modified by the channel-associated vegetation. Many of the processes affected by biota modify the soil moisture store, which is the immediate control on plant growth, and which exhibits less intermittency than

open-field rainfall. Some of the influences of biota are considered to have been long-term aspects of the Australian drylands, and have influenced weathering, soil development, and other fundamental aspects of the landscape.

Thus, it is reasonable to suggest that over a considerable part of the drylands, ecosystem structures and processes moderate the impacts of aridity, and allow additional biomass to be supported. Runon flows of various kinds in the drylands add to direct, on-site rainfall, such that there are many locations where the total supply of water greatly exceeds the amount that would be expected if rainfall were the sole source of moisture. For example, resource-conserving mosaic vegetation is very widespread in Australia in comparison with other major global drylands, and plant groves within these landscapes may receive runon flows that are almost as large as direct rainfall. Moreover, the runoff ratio for these landscapes is exceedingly small. Very probably, this serves to reduce erosion rates below the already low rates that would be anticipated in a generally low-relief, arid landscape. However, some fluxes (notably the loss of material from the continent in aeolian transport) are not well known. An additional probable role of the vegetation mosaics, yet to be fully explored, is in conferring resilience on the plant communities, and facilitating recovery from drought (e.g. Dunkerley 1997*b*). This arises from the enduring differentiation in key soil properties, and the subtle surface microtopography that promotes surface ponding adjacent to groves even when plant cover is greatly reduced (Dunkerley & Brown 1999*a*). Despite the evident resilience of patterned vegetation communities, concerns have been raised about the possible ecosystem disturbance arising from changes in floristics and plant physiology related to increasing concentrations of atmospheric carbon dioxide, and associated shifts in temperature and rainfall (e.g. Kefi *et al.* 2007, 2008; Michaelides *et al.* 2009). In the light of our incomplete understanding of dryland ecohydrology and its links to geomorphological processes, these impacts remain uncertain. Nevertheless, it is clear that for many drylands, meteorological rainfall and changes in rainfall are of less direct importance than changes in soil moisture, especially plant-available moisture. Therefore, a deeper understanding of the spatiotemporal controls on soil moisture is vital both for managing drylands used for primary production, and for attempts to model the ecosystem effects of climate change (Tietjen *et al.* 2009).

(2) Dryland biota acts in a number of ways to increase the partitioning of water into the soil moisture store. In so doing, the biota simultaneously reduces the fraction of incident rain that is partitioned into surface runoff. This has immediate

consequences for the rate of soil loss from hillslopes. The set of mechanisms at work in this
effect is spatio-temporally complex, partly owing
to the magnitude–frequency spectrum of incident
rain events. The modification of runoff by biota is
reasoned to have a diminished significance in high-
magnitude rainfall events, but could greatly attenu-
ate runoff in smaller events. Exploration of
the landscape response across the magnitude–
frequency spectrum of rain and runoff events is
needed for improved understanding of Australian
dryland landscapes. Spatial complexity arises from
the diversity of landscape contexts, from local hillslope scale to very large catchments. Differing sensitivity to environmental change across these
spatio-temporal scales suggests that geomorphological outcomes and the alluvial records will vary
in space and time, even for a single notional
change in the external climatic environment. Likewise, the effects of contemporary climatic change
will not be uniform across spatial scales.

(3) Biological responses to external climate
change (augmented by linked pedogenic and other
landscape changes) have the potential to obscure
the timing, direction, and magnitude of that
change in the alluvial archives. The alluvial record
of an external climate change can be expected to
vary regionally and to involve both threshold
effects and complex response in the sense of
Schumm (1977, 1979). It was noted above that an
increased rainfall might result in more developed
microphytic crusts and a decrease in overland
flow, for example. The presumed Holocene development of vegetation banding likewise probably
caused reductions in overland flow despite a
change to wetter climate. In both of these cases,
the magnitude and direction of the change in overland flow depend upon the magnitude of the external
climate change, and on the local site characteristics.
It is not straightforward to generalize regarding
the outcomes for landscape processes, but it is
clear from some of the studies cited here that
landscape characteristics can overshadow changes
in external climate. Consequently, landscape
responses might suggest climate changes unlike
those actually experienced, if incautiously interpreted. Examples include increased erosion as a
response to climate drying (e.g. related to soil carbonate accumulation at shallow depth) or reduced
erosion following increased moisture availability
(e.g. related to the development of contour-aligned
vegetation banding).

(4) There are aspects of the ecohydrology and
geomorphology of the Australian drylands that are
distinct from those of other drylands. Over large
areas, the desert pavements of resistant gibbers in
the Australian drylands are old (e.g. Fujioka et al.
2005), and have not developed key characteristics

in the Holocene, in contrast to events in, for
example, the western USA. However, in some
locations, such as western NSW, the arrival of
dusts during phases of the Quaternary undoubtedly
drove cumulic pedogenesis and continued evolution
of stone mantles. Elsewhere, accessions of dust
derived from the continental drylands have certainly
affected soil properties including soil depth and
texture, and soil modified the hydrological effects
of soil mantles. The role of fire through the late Quaternary has not been elucidated. However, in contrast to the significant erosional aftermath of fire in
steep and dissected terrain (e.g. Frechette &
Meyer 2009) the low topographic gradients over
much of the Australian drylands argue against the
occurrence of extreme post-fire erosion. Nevertheless, fire may have modified the distributions of
particular vegetation communities, and incremental
change in soil properties may have been significant.
Both may have had impacts on the partitioning of
rainfall and on the generation of overland flow.
The low topographic gradients over much of the
Australian drylands appear to have facilitated the
extensive development of banded vegetation communities. The very high runoff–runon trap efficiency of these landscapes incorporates the effects
of extensive faunal activity, including a very
diverse myrmecofauna. Large areas operate under
very strong ENSO-linked rainfall variability, and
taken together, these suggest a set of ecosystem
properties that are distinct from those of other drylands. The extensive CAV along the dryland
streams provides a notable instance of this, with
tree growth and woody debris accumulation on the
bed and banks providing a major component of
channel roughness, with some streams exhibiting
roughness coefficients that are among the highest
known from dryland streams anywhere.

(5) Even in Australian dryland landscapes whose
framework features are ancient, Quaternary and
modern events continued to affect landscape processes. For instance, bodies of alluvium in western
NSW that are rich in fluvially reworked dust slake
readily, and this has been implicated in the pro-
pensity of some streams there to undergo incision
to the unstable, arroyo form, with the consequent
loss of the CAV further driving channel instability.
These changes in channel form and process may
have been triggered by European land-use practices,
but may simply be part of a continuing complex
response cascade relating to increased frequency
of runoff events in the Holocene, and progressive
decline in sediment delivery from the stony
uplands owing to the exhaustion of supply. In the
former case, intervention by land managers would
clearly be supportable, but in the latter case, a
rationale for intervention is less apparent. A sustained development of our understanding of these

and other aspects of dryland behaviour will facilitate informed land stewardship as global climate changes continue to affect these landscapes and their functionally important biota.

This paper benefited from thoughtful reviews by J. Wainwright and D. Eldridge, and I extend my sincere thanks to them for their attempts to improve my writing and argument.

References

ALLEY, N. F. 1998. Cainozoic stratigraphy, palaeoenvironments and geological evolution of the Lake Eyre basin. *Palaeogeography, Palaeoclimatology, Palaeoecology*, **144**, 239–263.

ANGEL, J. R., PALECKI, M. A. & HOLLINGER, S. E. 2005. Storm precipitation in the United States. Part II: soil erosion characteristics. *Journal of Applied Meteorology*, **44**, 947–959.

ARNAU-ROSALÉN, E., CALVO-CASES, A., BOIX-FAYOS, C., LAVEE, H. & SARAH, P. 2008. Analysis of soil surface component patterns affecting runoff generation. an example of methods applied to Mediterranean hillslopes in Alicante (Spain). *Geomorphology*, **101**, 595–606.

ARYAL, R. K., FURUMAI, H., NAKAJIMA, F. & JINADASA, H. K. P. K. 2007. The role of inter-event time definition and recovery of initial/depression loss for the accuracy in quantitative simulations of highway runoff. *Urban Water Journal*, **4**, 53–58.

BAKER, V. R. 2006. Palaeoflood hydrology in a global context. *Catena*, **66**, 161–168.

BAKER, V. R. 2008. Paleoflood hydrology: origin, progress, prospects. *Geomorphology*, **101**, 1–13.

BARBIER, N., COUTERON, P., LEJOLY, J., DEBLAUWE, V. & LEJEUNE, O. 2006. Self-organised vegetation patterning as a fingerprint of climate and human impact on semi-arid ecosystems. *Journal of Ecology*, **94**, 537–547.

BARTLEY, R., ROTH, C. H. *ET AL*. 2006. Runoff and erosion from Australia's tropical semi-arid rangelands: influence of ground cover for differing space and time scales. *Hydrological Processes*, **20**, 3317–3333.

BAUDENA, M., BONI, G., FERRARIS, L., VON HARDENBERG, J. & PROVENZALE, A. 2007. Vegetation response to rainfall intermittency in drylands: results from a simple ecohydrological box model. *Advances in Water Resources*, **30**, 1320–1328.

BEDFORD, D. R. & SMALL, E. R. 2008. Spatial patterns of ecohydrologic properties on a hillslope–alluvial fan transect, Central New Mexico. *Catena*, **73**, 34–48.

BELNAP, J. & LANGE, O. L. (eds) 2001. *Biological Soil Crusts: Structure, Function, and Management*. Springer, Berlin.

BENITO, G. & THORNDYCRAFT, V. R. 2005. Palaeoflood hydrology and its role in applied hydrological sciences. *Journal of Hydrology*, **313**, 3–15.

BHARK, E. W. & SMALL, E. E. 2003. Association between plant canopies and the spatial patterns of infiltration in shrubland and grassland of the Chihuahuan Desert, New Mexico. *Ecosystems*, **6**, 185–196.

BLACKBURN, W. H. 1975. Factors influencing infiltration and sediment production of semiarid rangelands in Nevada. *Water Resources Research*, **11**, 929–937.

BOER, M. & PUIGDEFÁBREGAS, J. 2005. Effects of spatially structured vegetation patterns on hillslope erosion in a semiarid Mediterranean environment: a simulation study. *Earth Surface Processes and Landforms*, **30**, 149–167.

BOWKER, M. A., BELNAP, J., CHAUDHARY, V. B. & JOHNSON, N. C. 2008. Revisiting classic water erosion models in drylands: the strong impact of biological soil crusts. *Soil Biology and Biochemistry*, **40**, 2309–2316.

BOWLER, J. M., WYRWOLL, K.-H. & LU, Y. 2001. Variations of the northwest Australian summer monsoon over the last 300,000 years: the paleohydrological record of the Gregory (Mulan) Lakes system. *Quaternary International*, **83–85**, 63–80.

BOWMAN, D. M. J. S., BOGGS, G. S., PRIOR, L. D. & KRULL, E. S. 2007. Dynamics of *Acacia aneura– Triodia* boundaries using carbon (^{14}C and ∂^{13}C) and nitrogen (∂^{15}N) signatures in soil organic matter in central Australia. *Holocene*, **17**, 311–318.

BRACKEN, L. J. & WAINWRIGHT, J. 2008. Equilibrium in the balance? Implications for landscape evolution from dryland environments. *In*: GALLAGHER, K., JONES, S. J. & WAINWRIGHT, J. (eds) *Landscape Evolution: Denudation, Climate and Tectonics over Different Time and Space Scales*. Geological Society, London, Special Publications, **296**, 29–46.

BRAZIER, R. E., BEVEN, K. J., FREER, J. & ROWAN, J. S. 2000. Equifinality and uncertainty in physically based soil erosion models: application of the gLUE methodology to wEPP—the water erosion prediction project—for sites in the UK and USA. *Earth Surface Processes and Landforms*, **25**, 825–845.

BRIANT, R. M., GIBBARD, P. L., BOREHAM, S., COOPE, G. R. & PREECE, R. C. 2008. Limits to resolving catastrophic events in the Quaternary fluvial record: a case study from the Nene valley, Northamptonshire, UK. *In*: GALLAGHER, K., JONES, S. J. & WAINWRIGHT, J. (eds) *Landscape Evolution: Denudation, Climate and Tectonics over Different Time and Space Scales*. Geological Society, London, Special Publications, **296**, 79–104.

BULL, W. B. 1991. *Geomorphic Responses to Climatic Change*. Oxford University Press, New York.

BULL, W. B. & SCHICK, A. P. 1979. Impact of climatic change on an arid watershed: Nahal Yael, Southern Israel. *Quaternary Research*, **11**, 153–171.

CALDWELL, T., YOUNG, M., ZHU, J. & McDONALD, E. 2008. Gradation of soil hydraulic properties from canopy to interspace on a Mojave Desert soil chronosequence. *Geophysical Research Abstracts*, **10**, EGU2008-A-09980.

CERDA, A. & DOERR, S. 2008. The effect of ash and needle cover on surface runoff and erosion in the immediate post-fire period. *Catena*, **74**, 256–263.

CHARTRES, C. J. 1982. The pedogenesis of desert loam soils in the barrier range, western New South Wales. I. Soil parent materials. *Australian Journal of Soil Research*, **20**, 269–281.

COLLINS, D. B. G., BRAS, R. L. & TUCKER, G. E. 2004. Modeling the effects of vegetation–erosion coupling on landscape evolution. *Journal of Geophysical Research*, **109**, doi:10.1029/2003JF000028.

CORDERY, I., PILGRIM, D. H. & DORAN, D. G. 1983. Some hydrological characteristics of arid western New South Wales. *In*: *Hydrology and Water Resources Symposium 1983: Preprints of Papers. Institution of Engineers, Australia. National Conference Publication*, **83/13**, 287–292.

COSTELLOE, J. F., PAYNE, E., WOODROW, I. E., IRVINE, E. C., WESTERN, A. W. & LEANEY, F. W. 2008. Water sources accessed by arid zone riparian trees in highly saline environments, Australia. *Oecologia*, **156**, 43–52.

CROKE, J. 1997. Australia. *In*: THOMAS, D. S. G. (ed.) *Arid Zone Geomorphology*. 2nd edn. Wiley, Chichester, 563–573.

CROKE, J., MAGEE, J. & PRICE, D. 1996. Major episodes of Quaternary activity in the lower Neales river, northwest of Lake Eyre, Central Australia. *Palaeogeography, Palaeoclimatology, Palaeoecology*, **124**, 1–15.

DAHAN, O., SHANI, Y., ENZEL, Y., YECHIELI, Y. & YAKIREVICH, A. 2007. Direct measurements of floodwater infiltration into shallow alluvial aquifers. *Journal of Hydrology*, **344**, 157–170.

DAVID, B., ROBERTS, R. G. ET AL. 2007. Sediment mixing at Nonda Rock: investigations of stratigraphic integrity at an early archaeological site in Northern Australia and implications for the human colonization of the continent. *Journal of Quaternary Science*, **22**, 449–479.

DUNKERLEY, D. L. 1992. Channel geometry, bed material, and inferred flow conditions in ephemeral stream systems, Barrier Range, Western N.S.W., Australia. *Hydrological Processes*, **6**, 417–433.

DUNKERLEY, D. L. 1997a. Banded vegetation: development under uniform rainfall from a simple cellular automaton model. *Plant Ecology*, **129**, 103–111.

DUNKERLEY, D. L. 1997b. Banded vegetation: survival under drought and grazing pressure from a simple cellular automaton model. *Journal of Arid Environments*, **35**, 419–428.

DUNKERLEY, D. L. 2000. Hydrologic effects of dryland shrubs: defining the spatial extent of modified soil water uptake rates at an Australian desert site. *Journal of Arid Environments*, **45**, 159–172.

DUNKERLEY, D. L. 2002. Infiltration rates and soil moisture in a groved mulga community near Alice Springs, arid central Australia: evidence for complex internal rainwater redistribution in a runoff–runon landscape. *Journal of Arid Environments*, **51**, 199–219.

DUNKERLEY, D. L. 2008a. Intra-storm evaporation as a component of canopy interception loss in dryland shrubs: observations from Fowlers Gap, Australia. *Hydrological Processes*, **22**, 1985–1995.

DUNKERLEY, D. L. 2008b. Identifying individual rain events from pluviograph records: a review with analysis of data from an Australian dryland site. *Hydrological Processes*, **22**, 5024–5036.

DUNKERLEY, D. L. 2008c. Rain event properties in nature and in rainfall simulation experiments: a comparative review with recommendations for increasingly systematic study and reporting. *Hydrological Processes*, **22**, 4415–4435.

DUNKERLEY, D. L. 2008d. Flow chutes in Fowlers Creek, arid western New South Wales, Australia: evidence for diversity in the influence of trees on ephemeral channel form and process. *Geomorphology*, **102**, 232–241.

DUNKERLEY, D. L. 2008e. Bank permeability in an Australian ephemeral dry-land stream: variation with stage resulting from mud deposition and sediment clogging. *Earth Surface Processes and Landforms*, **33**, 226–243.

DUNKERLEY, D. L. 2008f. Woody debris: effects on channel processes and form in an Australian desert stream. Paper presented at the Geological Society of America Conference, Houston, 5–9 October 2008. Summary available at: http://gsa.confex.com/gsa/2008AM/finalprogram/abstract_149420.htm.

DUNKERLEY, D. L. & BOOTH, T. L. 1999. Plant canopy interception and its significance in a banded landscape, arid western New South Wales, Australia. *Water Resources Research*, **35**, 1581–1586.

DUNKERLEY, D. L. & BROWN, K. J. 1995. Runoff and runon areas in patterned chenopod shrubland, arid western New South Wales, Australia: characteristics and origin. *Journal of Arid Environments*, **30**, 41–55.

DUNKERLEY, D. L. & BROWN, K. J. 1999a. Banded vegetation near Broken Hill, Australia: significance of soil surface roughness and soil physical properties. *Catena*, **37**, 75–88.

DUNKERLEY, D. L. & BROWN, K. J. 1999b. Flow behaviour, suspended sediment transport and transmission losses in a small (sub bankfull) flow event in an Australian desert stream. *Hydrological Processes*, **13**, 1577–1588.

EBEL, B. A. & LOAGUE, K. 2006. Physics-based hydrologic-response simulation: seeing through the fog of equifinality. *Hydrological Processes*, **20**, 2887–2900.

ELDRIDGE, D. J. 1993. Effect of ants on sandy soils in semi-arid eastern Australia: local distribution of nest entrances and their effect on infiltration of water. *Australian Journal of Soil Research*, **31**, 509–518.

ELDRIDGE, D. J. & GREENE, R. S. B. 1994. Microbiotic soil crusts: a review of their role in soil and ecological processes in the rangelands of Australia. *Australian Journal of Soil Research*, **32**, 389–415.

ELDRIDGE, D. J. & MENSINGA, A. 2007. Foraging pits of the short-beaked echidna (*Tachyglossus aculeatus*) as small-scale patches in a semi-arid Australian box woodland. *Soil Biology and Biochemistry*, **39**, 1055–1065.

ELDRIDGE, D. J. & RATH, D. 2002. Hip holes: kangaroo (*Macropus* spp.) resting sites modify the physical and chemical environment of woodland soils. *Austral Ecology*, **27**, 527–536.

ELDRIDGE, D. J. & ROSENTRETER, R. 2004. Shrub mounds enhance water flow in a shrub-steppe community in southwestern Idaho, U.S.A. *In*: HILD, A. L., SHAW, N. L., MEYER, S., BOOTH, D. T. & MCARTHUR, E. D. (Compilers) *Seed and Soil Dynamics in Shrubland Ecosystems. USDA Forest Service Proceedings*, **RMRS-P-31**, 77–83.

ELY, L. L. 1997. Response of extreme floods in the southwestern United States to climatic variations in the late Holocene. *Geomorphology*, **19**, 175–201.

ENZEL, Y., ELY, L. L., HOUSE, P. K. & BAKER, V. R. 1993. Paleoflood evidence for a natural upper bound to flood magnitudes in the Colorado river basin. *Water Resources Research*, **29**, 2287–2297.

FINLAYSON, B. L. & MCMAHON, T. A. 1988. Australia v the world: a comparative analysis of streamflow characteristics. *In*: WARNER, R. F. (ed.) *Fluvial Geomorphology of Australia*. Academic Press, San Diego, CA, 17–40.

FRANCIS, R. A., PETTS, G. E. & GURNELL, A. M. 2008. Wood as a driver of past landscape change along river corridors. *Earth Surface Processes and Landforms*, **33**, 1622–1626.

FRECHETTE, J. D. & MEYER, G. A. 2009. Holocene fire-related alluvial-fan deposition and climate in ponderosa pine and mixed-conifer forests, Sacramento Mountains, New Mexico, USA. *Holocene*, **19**, 639–651.

FUJIOKA, T., CHAPPELL, J., HONDA, M., TATSEVICH, I., FIFIELD, K. & FABEL, D. 2005. Global cooling initiated stony deserts in central Australia 2–4 Ma, dated by cosmogenic ^{21}Ne–^{10}Be. *Geology*, **33**, 993–996.

FULLER, T. K., PERG, L. A., WILLENBRING, J. K. & LEPPER, K. 2009. Field evidence for climate-driven changes in sediment supply leading to strath terrace formation. *Geology*, **37**, 467–470.

GEDDES, N. & DUNKERLEY, D. L. 1999. The influence of organic litter on the erosive effects of raindrops and of gravity drops released from desert shrubs. *Catena*, **36**, 303–313.

GIFFORD, G. F. 1978. Rangeland hydrology in Australia – a brief review. *Australian Rangeland Journal*, **1**, 150–166.

GILAD, E., SHACHAK, M. & MERON, E. 2007. Dynamics and spatial organization of plant communities in water-limited systems. *Theoretical Population Biology*, **72**, 214–230.

GOBERNA, M., PASCUAL, J. A., GARCIA, C. & SÁNCHEZ, J. 2007. Do plant clumps constitute microbial hotspots in semiarid mediterranean patchy landscapes? *Soil Biology and Biochemistry*, **39**, 1047–1054.

GOUDIE, A. S. 2002. Australia. *In*: GOUDIE, A. S. (ed.) *Great Warm Deserts of the World*. Oxford University Press, Oxford, 319–359.

GRAEME, D. & DUNKERLEY, D. L. 1993. Hydraulic resistance by the river red gum, *Eucalyptus camaldulensis*, in ephemeral desert streams. *Australian Geographical Studies*, **31**, 141–154.

GRIFFITHS, P. G., HEREFORD, R. & WEBB, R. H. 2006. Sediment yield and runoff frequency of small drainage basins in the Mojave Desert, U.S.A. *Geomorphology*, **74**, 232–244.

GRISHKAN, I., ZAADY, E. & NEVO, E. 2006. Soil crust microfungi along a southward rainfall gradient in desert ecosystems. *European Journal of Soil Biology*, **42**, 33–42.

GURNELL, A. & PETTS, G. 2006. Trees as riparian engineers: the Tagliamento river, Italy. *Earth Surface Processes and Landforms*, **31**, 1558–1574.

HARRINGTON, G. A., COOK, P. G. & HERCZEG, A. L. 2002. Spatial and temporal variability of ground water recharge in Central Australia: a tracer approach. *Ground Water*, **40**, 518–528.

HARRINGTON, G. A., HERCZEG, A. L. & COOK, P. G. 1999. *Groundwater Sustainability and Water Quality in the Ti-Tree Basin, Central Australia*. CSIRO Land and Water, Technical Report, **53/99**.

HESSE, P. P., MAGEE, J. W. & VAN DER KAARS, S. 2004. Late Quaternary climates of the Australian arid zone: a review. *Quaternary International*, **118–119**, 87–102.

HISCOCK, P. 2008. *Archaeology of Ancient Australia*. Routledge, London.

HOLMGREN, M., STAPP, P. ET AL. 2006. A synthesis of ENSO effects on drylands in Australia, North America and South America. *Advances in Geosciences*, **6**, 69–72.

HOU, B., FRAKES, L. A., SANDIFORD, M., WORRALL, L., KEELING, J. & ALLEY, N. F. 2008. Cenozoic Eucla Basin and associated palaeovalleys, southern Australia—climatic and tectonic influences on landscape evolution, sedimentation and heavy mineral accumulation. *Sedimentary Geology*, **203**, 112–130.

HOWES, D. A. & ABRAHAMS, A. D. 2003. Modeling runoff and runon in a desert shrubland ecosystem, Jornada Basin, New Mexico. *Geomorphology*, **53**, 45–73.

HUI, W. J., COOK, B. I., RAVI, S., FUENTES, J. D. & D'ODORICO, P. 2008. Dust–rainfall feedbacks in the West African Sahel. *Water Resources Research*, **44**, doi: 10.1029/2008WR006885.

ISTANBULLUOGLU, E. & BRAS, R. L. 2006. On the dynamics of soil moisture, vegetation, and erosion: implications of climate variability and change. *Water Resources Research*, **42**, doi: 10.1029/2005WR004113.

JACOBY, Y., GRODEK, T., ENZEL, Y., PORAT, N., MCDONALD, E. V. & DAHAN, O. 2008. Late Holocene upper bounds of flood magnitudes and twentieth century large floods in the ungauged, hyperarid alluvial Nahal Arava, Israel. *Geomorphology*, **95**, 274–294.

JAMES, A. I. & ELDRIDGE, D. J. 2007. Reintroduction of fossorial native mammals and potential impacts on ecosystem processes in an Australian desert landscape. *Biological Conservation*, **138**, 351–359.

JAMES, A. I., ELDRIDGE, D. J., KOEN, T. B. & WHITFORD, W. G. 2008. Landscape position moderates how ant nests affect hydrology and soil chemistry across a Chihuahuan desert watershed. *Landscape Ecology*, **23**, 961–975.

JANSEN, J. & BRIERLEY, G. J. 2004. Pool-fills: a window to palaeoflood history and response in bedrock-confined rivers. *Sedimentology*, **51**, 910–925.

JOHNSON, B. J., MILLER, G. H., FOGEL, M. K., MAGEE, J. W., GAGAN, M. K. & CHIVAS, A. R. 1999. 65 000 years of vegetation change in central Australia and the Australian summer monsoon. *Science*, **284**, 1150–1152.

KEFI, S., ALADOS, C. L., PUEYO, Y., PAPANASTASIS, V. P., EL AICH, A. & DE RUITER, P. C. 2007. Spatial vegetation patterns and imminent desertification in Mediterranean arid ecosystems. *Nature*, **449**, 213–218.

KEFI, S., RIETKERK, M. & KATUL, G. G. 2008. Vegetation pattern shift as a result of rising atmospheric CO_2 in arid ecosystems. *Theoretical Population Biology*, **74**, 332–344.

KIRKBY, M. J., BRACKEN, L. J. & SHANNON, J. 2005. The influence of rainfall distribution and morphological factors on runoff delivery from dryland catchments in SE Spain. *Catena*, **62**, 136–156.

KÖCHY, M. 2008. Effects of simulated daily precipitation patterns on annual plant populations depend on life stage and climatic region. *BMC Ecology*, **8**, doi: 10.1186/1472-6785-8-4.

KWARTENG, A. Y., DORVLO, A. S. & VIJAYA KUMAR, G. T. 2008. Analysis of a 27-year rainfall data (1977–2003) in the Sultanate of Oman. *International Journal of Climatology*, doi: 10.1002/joc.1727.

LANGBEIN, W. B. & SCHUMM, S. A. 1958. Yield of sediment in relation to mean annual precipitation. *Transactions of the American Geophysical Union*, **39**, 1076–1084.

LEFEVER, R. & LEJEUNE, O. 1997. On the origin of tiger bush. *Bulletin of Mathematical Biology*, **59**, 263–294.

LEKACH, J., AMIT, R., GRODEK, T. & SCHICK, A. P. 1998. Fluvio-pedogenic processes in an ephemeral stream channel, Nahal Yael, Southern Negev, Israel. *Geomorphology*, **23**, 353–369.

LEWIS, M. F. & WALKER, G. R. 2002. Assessing the potential for significant and episodic recharge in southwestern Australia using rainfall data. *Hydrogeology Journal*, **10**, 229–237.

LIU, Q.-X., JIN, Z. & LI, B.-L. 2008. Numerical investigation of spatial patterns in a vegetation model with feedback function. *Journal of Theoretical Biology*, **254**, 350–360.

LUDWIG, J. A. & TONGWAY, D. J. 1995. Spatial organization of landscapes and its function in semi-arid woodlands, Australia. *Landscape Ecology*, **10**, 51–63.

LUDWIG, J. A., BARTLEY, R., HAWDON, A. A., ABBOTT, B. N. & MCJANNET, D. 2007. Patch configuration non-linearly affects sediment loss across scales in a grazed catchment in North-East Australia. *Ecosystems*, **10**, 839–845.

LUDWIG, J. A., TONGWAY, D., FREUDENBERGER, D., NOBLE, J. & HODGKINSON, K. 1997. *Landscape Ecology, Function and Management: Principles from Australia's Rangelands*. CSIRO, Melbourne.

LYFORD, F. P. & QASHU, H. K. 1969. Infiltration rates as affected by desert vegetation. *Water Resources Research*, **5**, 1373–1376.

MAROULIS, J. C., NANSON, G. C., PRICE, D. M. & PIETSCH, T. 2007. Aeolian–fluvial interaction and climate change: source-bordering dune development over the past ~100 ka on Cooper Creek, central Australia. *Quaternary Science Reviews*, **26**, 386–404.

MARTIN, H. A. 2006. Cenozoic climatic change and the development of arid vegetation in Australia. *Journal of Arid Environments*, **66**, 533–563.

MARX, S. K., MCGOWAN, H. A. & KAMBER, B. S. 2009. Long-range dust transport from eastern Australia: a proxy for Holocene aridity and ENSO-type climate variability. *Earth and Planetary Science Letters*, **282**, 167–177.

MAYOR, A. G., BAUTISTA, S., SMALL, E. E., DIXON, M. & BELLOT, J. 2008. Measurement of the connectivity of runoff source areas as determined by vegetation pattern and topography: a tool for assessing potential water and soil losses in drylands. *Water Resources Research*, **44**, doi: 10.1029/2007WR006367.

MCDONALD, A. K., KINUCAN, R. J. & LOOMIS, L. E. 2009. Ecohydrological interactions within banded vegetation in the northeastern Chihuahuan Desert, USA. *Ecohydrology*, **2**, 66–71.

MCGOWAN, H. & CLARK, A. 2008. Identification of dust transport pathways from Lake Eyre, Australia using Hysplit. *Atmospheric Environment*, **42**, 6915–6925.

MCMAHON, T. A. 1982. World hydrology: does Australia fit? *In: Proceedings of the Hydrology and Water Resources Symposium, Melbourne*. Institution of Engineers, Australia, Canberra, 1–7.

MCMAHON, T. A., MURPHY, R. E., PEEL, M. C., COSTELLOE, J. F. & CHIEW, F. H. S. 2008. Understanding the surface hydrology of the Lake Eyre basin: Part 1—rainfall. *Journal of Arid Environments*, **72**, 1853–1868.

MICHAELIDES, K., LISTER, D., WAINWRIGHT, J. & PARSONS, A. J. 2009. Vegetation controls on small-scale runoff and erosion dynamics in a degrading dryland environment. *Hydrological Processes*, **23**, 1617–1630.

MILLER, G., MANGAN, J., POLLARD, D., THOMPSON, S., FELZER, B. & MAGEE, J. 2005. Sensitivity of the Australian monsoon to insolation and vegetation: implications for human impact on continental moisture balance. *Geology*, **33**, 65–68.

MITCHELL, P. B. & HUMPHREYS, G. S. 1987. Litter dams and microterraces formed on hillslopes subject to rainwash in the Sydney Basin, Australia. *Geoderma*, **39**, 331–357.

MONGER, H. C. & BESTELMEYER, B. T. 2006. The soil – geomorphic template and biotic change in arid and semi-arid ecosystems. *Journal of Arid Environments*, **65**, 207–218.

MORGAN, K. H. 1999. Rainfall recharge to arid–semi-arid phreatic aquifers, Yilgarn Craton, Western Australia. *In: Water 99: Joint Congress; 25th Hydrology & Water Resources Symposium, 2nd International Conference on Water Resources & Environment Research; Handbook and Proceedings*. Institution of Engineers, Australia, Barton, ACT, 1204–1211.

MUDD, S. M. 2006. Investigation of the hydrodynamics of flash floods in ephemeral channels: scaling analysis and simulation using a shock-capturing flow model incorporating the effects of transmission losses. *Journal of Hydrology*, **324**, 65–79.

MUELLER, E. N., WAINWRIGHT, J. & PARSONS, A. J. 2007. Impact of connectivity on the modeling of overland flow within semiarid shrubland environments. *Water Resources Research*, **43**, doi: 10.1029/2006WR005006.

MURPHY, B. F. & TIMBAL, B. 2007. A review of recent climate variability and climate change in southeastern Australia. *International Journal of Climatology*, **24**, 703–721.

MURRAY, A. B., KNAAPEN, M. A. F., TAL, M. & KIRWAN, M. L. 2008. Biomorphodynamics: physical–biological feedbacks that shape landscapes. *Water Resources Research*, **44**, doi: 10.1029/2007WR006410.

NANSON, G. C., PRICE, D. M. *ET AL.* 2008. Alluvial evidence for major climate and flow regime changes during the middle and late Quaternary in eastern central Australia. *Geomorphology*, **101**, 109–129.

NEARING, M. A., PRUSKI, F. F. & O'NEAL, M. R. 2004. Expected climate change impacts on soil erosion rates: a review. *Journal of Soil and Water Conservation*, **59**, 43–50.

NEAVE, M. & ABRAHAMS, A. D. 2001. Impact of small mammal disturbances on sediment yield from grassland and shrubland ecosystems in the Chihuahuan Desert. *Catena*, **44**, 285–303.

NICHOLLS, N. 1991. The El Niño/Southern Oscillation and Australian vegetation. *Vegetatio*, **91**, 23–36.

NICHOLLS, N. & KARIKO, A. 1993. East Australian rainfall events: interannual variations, trends, and relationships with the Southern Oscillation. *Journal of Climate*, **6**, 1141–1152.

NOBLE, J. C., MÜLLER, W. J., DETLING, J. K. & PFITZNER, G. H. 2007. Landscape ecology of the burrowing bettong: warren distribution and patch dynamics in semiarid eastern Australia. *Austral Ecology*, **32**, 326–337.

O'CONNELL, J. F. & ALLEN, J. 2004. Dating the colonization of Sahul (Pleistocene Australia–New Guinea): a review of recent research. *Journal of Archaeological Science*, **31**, 835–853.

O'CONNELL, J. F. & ALLEN, J. 2007. Pre-LGM Sahul (Pleistocene Australia–New Guinea) and the archaeology of early modern humans. *In*: MELLARS, P., BOYLE, K., BAR-YOSEF, O. & STRINGER, C. (eds) *Rethinking the Human Revolution*. McDonald Institute for Archaeological Research, Cambridge, 395–410.

OSTERKAMP, W. R. & FRIEDMAN, J. M. 2000. The disparity between extreme rainfall events and rare floods—with emphasis on the semi-arid American West. *Hydrological Processes*, **14**, 2817–2829.

OSTERKAMP, W. R. & LANE, L. J. 2003. Groundwater recharge estimates in arid areas using channel morphology and a simulation model. *In*: ALSHARHAN, A. S. & WOOD, W. W. (eds) *Water Resources Perspectives: Evaluation, Management and Policy*. Elsevier, Amsterdam, 281–286.

PAYNE, E. G. I., COSTELLOE, J. F., WOODROW, I. E., IRVINE, E. C., WESTERN, A. W. & HERCZEG, A. L. 2006. Riparian tree water use by Eucalyptus coolabah in the Lake Eyre basin. *In*: *Proceedings of the 30th Hydrology and Water Resources Symposium, 4–7 December 2006, Launceston, Tasmania*. Engineers Australia, Canberra, 692–697.

PEEL, M. C., FINLAYSON, B. L. & MCMAHON, T. A. 2007. Updated world map of the Köppen–Geiger climate classification. *Hydrology and Earth System Sciences*, **11**, 1633–1644.

PERKINS, S. R. & MCDANIEL, K. C. 2005. Infiltration and sediment rates following creosotebush control with Tebuthiuron. *Rangeland Ecology and Management*, **58**, 605–613.

PICKUP, G. 1998. Desertification and climate change—the Australian perspective. *Climate Research*, **11**, 51–63.

POWER, S., TSEITKIN, F., MEHTA, V. B., LAVERY, V. B., TOROK, S. & HOLBROOK, N. 1999. Decadal climate variability in Australia during the twentieth century. *International Journal of Climatology*, **19**, 169–184.

PUIGDEFÁBREGAS, J., DEL BARRIO, G., BOER, M. M., GUTIÉRREZ, L. & SOLÉ, A. 1998. Differential response of hillslope and channel elements to rainfall events in a semi-arid area. *Geomorphology*, **23**, 337–351.

PUIGDEFÁBREGAS, J., SOLE, A., GUTIEREZ, L., DEL BARRIO, G. & BOER, M. 1999. Scales and processes of water and sediment redistribution in drylands: results from the Rambla Honda field site in southeast Spain. *Earth-Science Reviews*, **48**, 39–70.

RAKICH, C. S., HOLBROOK, N. J. & TRIMBAL, B. 2008. A pressure gradient metric capturing planetary-scale influences on eastern Australian rainfall. *Geophysical Research Letters*, **35**, doi: 10.1029/2007GL032970.

RAM, A. & YAIR, A. 2007. Negative and positive effects of topsoil biological crusts on water availability along a rainfall gradient in a sandy arid area. *Catena*, **70**, 437–442.

READ, J. L., KOVAC, K.-J. & FATCHEN, T. J. 2005. 'Biohyets': a method for displaying the extent and severity of environmental impacts. *Journal of Environmental Management*, **77**, 157–164.

REANEY, S. M., BRACKEN, L. J. & KIRKBY, M. J. 2007. Use of the connectivity runoff model (CRUM) to investigate the influence of storm characteristics on runoff generation and connectivity in semi-arid areas. *Hydrological Processes*, **21**, 894–906.

REID, I., LARONNE, J. B. & POWELL, D. M. 1998. Flashflood and bedload dynamics of desert gravel-bed streams. *Hydrological Processes*, **12**, 543–557.

RIETKERK, M., DEKKER, S. C., DE RUITER, P. C. & VAN DE KOPPEL, J. 2004. Self-organised patchiness and catastrophic shifts in ecosystems. *Science*, **305**, 1926–1929.

RING, P. J. 1988. Long term changes in storm runoff response in the Macintyre Valley, NSW. *In*: *Proceedings of the Hydrology and Water Resources Symposium, Canberra, 1–3 February 1988*. Institution of Engineers, Canberra.

ROTSTAYN, L. D., COLLIER, M. A. *ET AL*. 2009. Improved simulation of Australian climate and ENSO-related rainfall variability in a global climate model with an interactive aerosol treatment. *International Journal of Climatology*, doi: 10.1002/joc.1952.

SALA, O. E. & LAUENROTH, W. K. 1982. Small rainfall events: an ecological role in semiarid regions. *Oecologia*, **53**, 301–304.

SANDERCOCK, P. J., HOOKE, J. M. & MANT, J. M. 2007. Vegetation in dryland river channels and its interaction with fluvial processes. *Progress in Physical Geography*, **31**, 107–129.

SCHLESINGER, W. H., RAIKES, J. A., HARTLEY, A. E. & CROSS, A. F. 1996. On the spatial pattern of soil nutrients in desert ecosystems. *Ecology*, **77**, 364–374.

SCHUMM, S. A. 1977. *The Fluvial System*. Wiley, New York.

SCHUMM, S. A. 1979. Geomorphic thresholds: the concept and its application. *Transactions of the Institute of British Geographers*, **4**, 485–515.

SHAMIR, E., WANG, J. & GEORGAKAKOS, P. 2007. Probabilistic streamflow generation model for data sparse arid watersheds. *Journal of the American Water Resources Association*, **43**, 1142–1154.

SHAW, J. R. & COOPER, D. J. 2008. Linkages among watershed, stream reaches, and riparian vegetation in dryland ephemeral stream networks. *Journal of Hydrology*, **350**, 68–82.

SHERRATT, J. A. & LORD, G. J. 2007. Nonlinear dynamics and pattern bifurcations in a model for vegetation stripes in semi-arid environments. *Theoretical Population Biology*, **71**, 1–11.

SLATYER, R. O. 1965. Measurements of precipitation interception by an arid zone plant community (*Acacia aneura* F. Muell). *In*: *Methodology of Plant Eco-physiology*. Proceedings of Montpellier Symposium. UNESCO Arid Zone Research, **25**, 181–191.

SMITH, M. A. 2009. Late Quaternary landscapes in central Australia: sedimentary history and palaeoecology of Puritjarra rock shelter. *Journal of Quaternary Science*, doi: 10.1002/jqs.1249.

SMITH, M. A., VELLEN, L. & PASK, J. 1995. Vegetation history from archaeological charcoals in central Australia: the late Quaternary record from Puritjarra rock shelter. *Vegetation History and Archaeobotany*, **4**, 171–177.

STAFFORD SMITH, D. M. & MORTON, S. R. 1990. A framework for the ecology of arid Australia. *Journal of Arid Environments*, **18**, 255–278.

THOMAS, M. F. 2000. Late Quaternary environmental changes and the alluvial record in humid tropical environments. *Quaternary International*, **72**, 23–36.

THOMAS, M. F. 2008. Understanding the impacts of late Quaternary climate change in tropical and sub-tropical regions. *Geomorphology*, **101**, 146–158.

THOMAS, M. F., NOTT, J., MURRAY, A. S. & PRICE, D. M. 2007. Fluvial response to late Quaternary climate change in NE Queensland, Australia. *Palaeogeography, Palaeoclimatology, Palaeoecology*, **251**, 119–136.

THOMPSON, D. B., WALKER, L. R., LANDAU, F. H. & STAK, L. R. 2005. The influence of elevation, shrub species, and biological soil crust on fertile islands in the Mojave Desert, USA. *Journal of Arid Environments*, **61**, 609–629.

TIETJEN, B., ZEHE, E. & JELTSCH, F. 2009. Simulating plant water availability in dry lands under climate change: a generic model of two soil layers. *Water Resources Research*, **45**, doi: 10.1029/2007WR006589.

TOOTH, S. & NANSON, G. C. 2000. The role of vegetation in the formation of anabranching channels in an ephemeral river, Northern Plains, arid central Australia. *Hydrological Processes*, **14**, 3099–3117.

TUCKER, G. E. & SLINGERLAND, R. 1997. Drainage basin responses to climate change. *Water Resources Research*, **33**, 2031–2047.

VETAAS, O. R. 1992. Micro-site effects of trees and shrubs in dry Savannas. *Journal of Vegetation Science*, **3**, 337–344.

WAICHLER, S. R. & WIGMOSTA, M. S. 2004. Application of hydrograph shape and channel infiltration models to an arid watershed. *Journal of Hydrologic Engineering*, **9**, 433–439.

WALLING, D. E. & KLEO, A. H. A. 1979. Sediment yields of rivers in areas of low precipitation: a global review. *In*: The Hydrology of Areas of Low Precipitation. Proceedings of the Canberra Symposium. International Association of Hydrological Sciences Publication, **128**, 479–493.

WALKER, P. H., CHARTRES, C. J. & HUTKA, J. 1988. The effect of aeolian accessions on soil development on granitic rocks in south-eastern Australia. I. Soil morphology and particle-size distributions. *Australian Journal of Soil Research*, **26**, 1–16.

WEBB, S. 2008. Megafauna demography and late Quaternary climatic change in Australia: a predisposition to extinction. *Boreas*, doi: 10.1111/j.1502-3885.2008.00026.x.

WELLS, S. G., MCFADDEN, L. D. & DOHRENWEND, J. C. 1987. Influence of late Quaternary climatic changes on geomorphic and pedogenic processes on a desert piedmont, eastern Mojave Desert, California. *Quaternary Research*, **27**, 130–146.

WEZEL, A., RAJOT, J.-L. & HERBRIG, C. 2000. Influence of shrubs on soil characteristics and their function in Sahelian agro-ecosystems in semi-arid Niger. *Journal of Arid Environments*, **44**, 383–398.

WHITFORD, W. G. 1998. Contribution of pits dug by goannas (*Varanus gouldii*) to the dynamics of banded mulga landscapes in eastern Australia. *Journal of Arid Environments*, **40**, 453–457.

WHITFORD, W. G., LUDWIG, J. A. & NOBLE, J. C. 1992. The importance of subterranean termites in semi-arid ecosystems in south-eastern Australia. *Journal of Arid Environments*, **22**, 87–91.

WILLIAMS, M., PRESCOTT, J. R., CHAPPELL, J., ADAMSON, D., COCK, B., WALKER, K. & GELL, P. 2001. The enigma of a late Pleistocene wetland in the Flinders Ranges, South Australia. *Quaternary International*, **83–85**, 129–144.

WILSON, L. 1973. Variations in mean annual sediment yield as a function of mean annual precipitation. *American Journal of Science*, **273**, 335–349.

WISCHUSEN, J. D. H. 1994. Sustainability of a hard rock aquifer at Kintore, Gibson Desert, central Australia. *In*: Water Down Under 94: Groundwater/Surface Hydrology, Preprints of Papers. Institution of Engineers, National Conference Publication, **94/10**, 343–349.

XIE, Z., LIU, Y., HU, C., CHEN, L. & LI, D. 2007. Relationships between the biomass of algal crusts in fields and their compressive strength. *Soil Biology and Biochemistry*, **39**, 567–572.

YAIR, A. & KOSSOVSKY, A. 2002. Climate and surface properties: hydrological response of small arid and semi-arid watersheds. *Geomorphology*, **42**, 43–57.

YAIR, A. & RAZ-YASSIF, N. 2004. Hydrological processes in a small arid catchment: scale effects of rainfall and slope length. *Geomorphology*, **61**, 155–169.

ZAMMOURI, M. & FEKI, H. 2005. Managing releases from small upland reservoirs for downstream recharge in semi-arid basins (northeast of Tunisia). *Journal of Hydrology*, **314**, 125–138.

ZHENG, H., POWELL, C. McA. & ZHAO, H. 2002. Eolian and lacustrine evidence of Quaternary palaeoenvironmental changes in southwestern Australia. *Global and Planetary Change*, **35**, 75–92.

History of Australian aridity: chronology in the evolution of arid landscapes

TOSHIYUKI FUJIOKA[1,2]* & JOHN CHAPPELL[3]

[1]*Department of Nuclear Physics, Research School of Physics and Engineering, Australian National University, Canberra, ACT, 0200, Australia*

[2]*Present address: Institute for Environmental Research, Australian Nuclear Science and Technology Organisation, Lucas Heights, NSW, 2234, Australia*

[3]*Research School of Earth Sciences, Australian National University, Canberra, ACT, 0200, Australia*

*Corresponding author (e-mail: toshiyuki.fujioka@ansto.gov.au)

Abstract: Australian climate and vegetation, known from marine and lacustrine sediments and fossils, varied dramatically throughout the Cenozoic Era, with several warm reversals superimposed on overall drying and cooling. A suite of landforms, including stony deserts, dunefields and playa lakes, formed in response to the advancing aridity but their age generally remained uncertain until fairly recently, owing to a lack of suitable dating methods. Within the last 5 years, the chronology of Late Quaternary fluctuations of lakes, dunes and dust-mantles has been established by luminescence dating methods, and mid-Pleistocene onset of playa conditions in a few closed basins has been estimated using palaeomagnetic reversal chronology. Only recently has it been shown, by cosmogenic isotope dating, that major tracts of arid landforms including the Simpson Desert dunefield, and stony deserts of the Lake Eyre Basin, were formed in early Pleistocene and late Pliocene times, respectively. These landscapes represent a stepwise response to progressive climatic drying and, speculatively, were accompanied by biological adaptations. Recent molecular DNA studies indicate that Australia's arid-adapted species evolved from mesic-adapted ancestors during the Pliocene or earlier, but whether speciation rapidly accompanied the development of stony deserts and other arid geomorphological provinces awaits further studies of arid landscape chronology.

Australia lay at the edge of Gondwana until Jurassic times, its southern margin adjacent to Antarctica and its western margin bordering Greater India. Around 160 Ma ago, tectonic rifting stretched the continental crust between Australia and Antarctica, and, around 96 Ma ago, oceanic crust started to develop between them (White 1994). Rifting and sea-floor spreading continued at a slow rate of $c.$ 4 mm a^{-1} until the seaway was about 500 km wide, 45 Ma ago. Since then, the seaway has widened by another 2600 km at about 6–7 cm a^{-1}, as the island continent of Australia drifted northwards (Veevers *et al.* 1991). The climatic consequences have been substantial. The Circumpolar Current was established, isolating Antarctica and blocking meridional heat transfer, leading to growth of the Antarctic ice sheet. Australia drifted into the dry climatic zone of subtropical anticyclones, which may have contracted northwards and probably intensified as the southern icecap grew, and the meridional temperature gradient steepened.

In this paper, we review our current knowledge of the history of the development of Australian desert landscapes, and then focus on recent studies of the chronology of Australian arid landforms.

Following an overview of global cooling during the Cenozoic, Australian evidence of Cenozoic climates is reviewed before examining the rather sketchy chronology of Australian arid landforms and their relationships to climatic changes. We suggest that cosmogenic isotope dating is potentially the best method for improving this chronology, which is desirable to forge the links between climatic change, landscape evolution and evolution of Australia's arid-adapted biota, at the million-year time scale.

Cenozoic climate: overview

Marine oxygen isotope data are major archives of global climatic history, and records from benthic (bottom-dwelling) foraminifera, compiled from more than 40 Deep Sea Drilling Project (DSDP) and Ocean Drilling Program (ODP) sites including the Atlantic, Indian and Pacific Oceans, reflect the combined evolution of deep ocean temperature and continental ice volume through the Cenozoic Era (Zachos *et al.* 2001). The record exhibits several phases of warming and cooling, linked to episodes of ice-sheet growth and decay (Fig. 1).

From: Bishop, P. & Pillans, B. (eds) *Australian Landscapes*. Geological Society, London, Special Publications, **346**, 121–139. DOI: 10.1144/SP346.8   0305-8719/10/$15.00 © The Geological Society of London 2010.

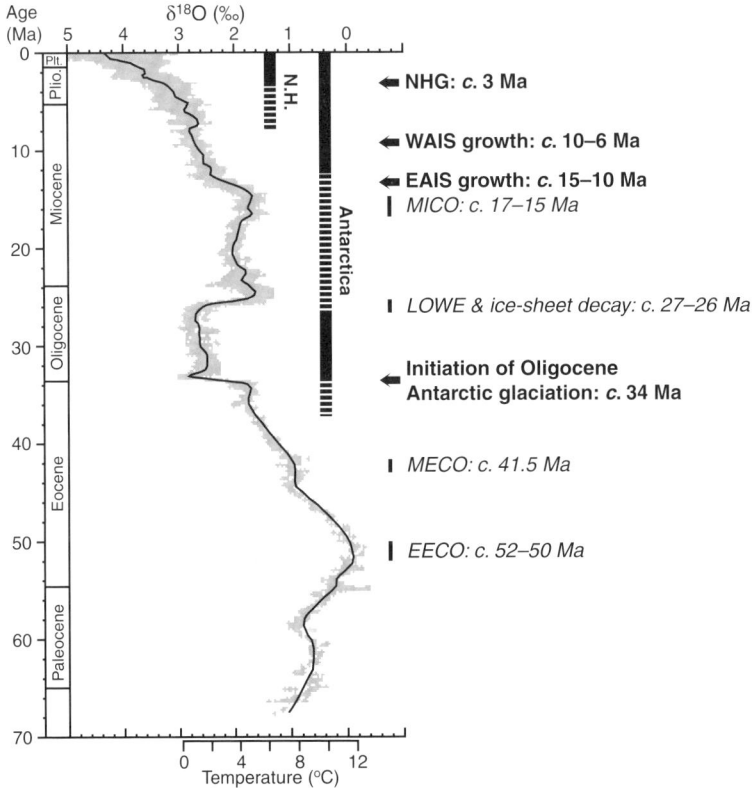

Fig. 1. Global deep-sea oxygen isotope record, based on data compiled from over 40 DSDP and ODP sites, with major climatic events including ice-sheet evolution during the Cenozoic (after Zachos *et al.* 2001). Smoothed-mean benthic $\delta^{18}O$ values are shown. The temperature scale at the bottom shows the deep-sea temperature for an ice-sheet free ocean, and thus is applicable only to the time prior to the initiation of Antarctic glaciation (*c.* 34 Ma). Vertical thick bars show rough qualitative representation of ice volume in Antarctica and Northern Hemisphere (N.H.), with the dashed bar representing partial or ephemeral ice volume ($\leq$50% of present) and the full bar, full-scale and permanent ($>$50% of present). Texts with arrows denote cooling event; texts with short bars denote warming events (after Zachos *et al.* 2001; McGowran *et al.* 2004). NHG, Northern Hemisphere Glaciation; WAIS, West Antarctic Ice Sheet; EAIS, East Antarctic Ice Sheet; MICO, Miocene Climatic Optimum; LOWE, Late Oligocene Warming Event; MECO, Middle Eocene Climatic Optimum; EECO, Early Eocene Climatic Optimum.

The first warming, which commenced in the mid-Paleocene and peaked at the Early Eocene Climatic Optimum (EECO: 52–50 Ma), was followed by Eocene cooling, when deep-sea temperature fell by *c.* 7 °C from *c.* 12 °C to 4.5 °C, but was interrupted by a modest warming at *c.* 42 Ma ago, known as the Middle Eocene Climatic Optimum (MECO: Bohaty & Zachos 2003) (Fig. 1). Until this time, only small or negligible polar ice sheets existed (Zachos *et al.* 2001). An abrupt increase of $\delta^{18}O$ by 1‰ in the earliest Oligocene (*c.* 34 Ma) reflects both a decrease of deep-sea water temperature and an increase of ice volume; roughly half of the signal (*c.* 0.6‰) is thought to reflect a substantial increase of Antarctic ice (Zachos *et al.* 1994), owing to high-latitude cooling following the establishment of the

Circumpolar Current and perhaps boosted by a reduction of atmospheric CO_2 (Zachos & Kump 2005). Relatively high $\delta^{18}O$ ($>$2.5‰) throughout the most of the Oligocene suggests a permanent Antarctic ice sheet with a mass of *c.* 50% of that of the current ice sheet, and bottom-ocean temperatures of *c.* 4 °C (Zachos *et al.* 1993).

Warming in the later Oligocene culminated in the Late Oligocene Warming Event (LOWE: 27–26 Ma), and was succeeded by the Miocene Climatic Optimum (MICO: 17–15 Ma) (Fig. 1). During this period, the Antarctic ice sheets may have nearly collapsed, and bottom-water temperature may have remained relatively high (Miller *et al.* 1991), although other researchers have argued for relatively high Antarctic ice volume during the

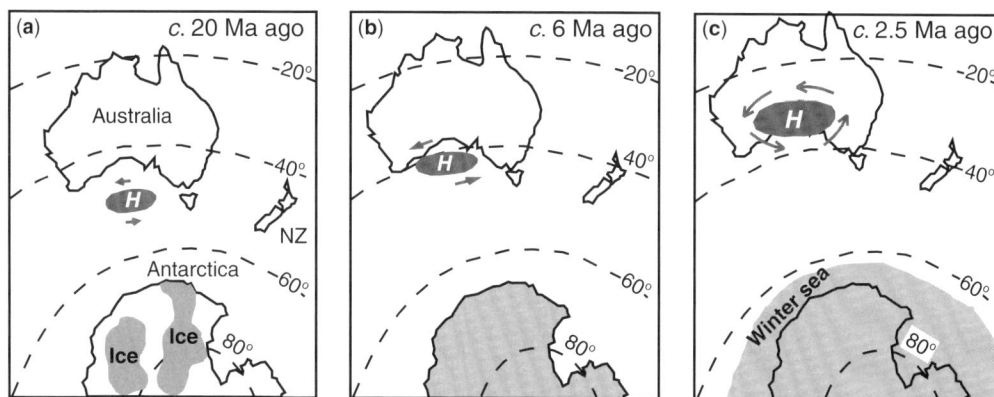

Fig. 2. Conceptual factors influencing climatic changes in the Australia–Antarctic region in the last 20 Ma, according to Bowler (1982). In Bowler's view, the position of the Sub-Tropical High Pressure system (STHP; shown as '*H*' in figure) is affected by development of ice sheet in Antarctica. (**a**) At 20 Ma ago, Australia resided around 10° south of its current position; an ice sheet partly covered Antarctica, and the STHP cell may have been around 50°S latitude south of Australia, which had a wet-tropical climate. (**b**) At 6 Ma ago, while Australia drifted northwards, Antarctic ice sheets became fully developed and the STHP cell moved northwards, perhaps more rapidly than the drifting continent, providing seasonal aridity with summer rain in southern Australia. (**c**) At 2.5 Ma ago, Australia approached to its current latitude, and the STHP cell became positioned at the centre of the continent, *c.* 30°S, causing an arid climate across a broad area of the continent. Substantial glaciation started in the Northern Hemisphere.

Early Miocene (Pekar & DeConto 2006). Rapid cooling of deep water following the MICO event apparently accompanied major regrowth of the East Antarctic Ice Sheet (EAIS) by *c.* 10 Ma (Flower & Kennett 1995), and increasing $\delta^{18}O$ through the later Miocene until the earlier Pliocene (*c.* 6 Ma) indicates additional cooling and may reflect small-scale ice-sheet expansion in West Antarctica (Kennett & Barker 1990) and the Arctic (Thiede & Vorren 1994). The cooling trend continued through the Pliocene, perhaps interrupted by subtle warming in the Early Pliocene (Poore & Sloan 1996). Finally, Quaternary glacial cycles commenced about 3 Ma ago, reflecting the development of Northern Hemisphere continental glaciation (Maslin *et al.* 1998).

Australia's climate during the Cenozoic has been subject to two major factors. One is the development of Antarctic ice and punctuated cooling of the deep ocean, indicated by marine oxygen isotope records; the other is northward drift of Australia itself (Fig. 2). During the passage from a forested continent in the lower Cenozoic to its present semi-arid condition, many secondary changes of climate and landscape processes have occurred.

Cenozoic climate and development of aridity in Australia

Australia today is the driest inhabited continent, but it was wetter in the past. Plant fossils show that rainforest covered much of central Australia until 25–30 Ma ago (Hill 1994; White 1994). Northward drift of the continent, development of circumpolar circulation in the Southern Ocean, and growth of the Antarctic ice sheet led to continent-wide drying as Australia entered the realm of subtropical high-pressure cells, but reversals during Cenozoic global cooling and marine incursions into Australia's interior basins led to departures from the drying trend. This section gives an overview of Australian climate during the Cenozoic, inferred from sedimentological and palaeontological signatures within the inland and the southern marginal basins, and evidence of increased aridity during the Pliocene and Pleistocene is described in the following section. Key field locations mentioned in the following text are shown in Figure 3.

Types of climatic indicators

Most evidence of Cenozoic Australian climate comes from sedimentological and palaeontological studies of the internal basins and southern marginal basins (Fig. 4), and includes palynofloras and plant fossils, marine fossils such as molluscs and foraminifera, coastal and offshore sediments, and inland lake deposits. Palaeotemperatures have been reconstructed from the Perth, Bremer, Eucla, Pirie, St Vincent, Murray, Otway, Bass and Gippsland marginal basins (Benbow *et al.* 1995; McGowran *et al.* 2004), and vegetation has been reconstructed from plant-fossil and pollen evidence from inland sites, including the Lake Eyre, Torrens and Billa

Fig. 3. Locations mentioned in the text. Filled circles denote major cities and towns; open outlines denote lakes and rivers. Areas of stony and sandy deserts and playas are also shown, with the boundary of arid zone; blank areas within the arid zone include upland and riverine deserts, and shield, karst and clay plains (after Mabbutt 1988).

Kalina Basins (Benbow *et al.* 1995; Callen *et al.* 1995; Alley 1998) (Fig. 4). Data from inland basins are usually confined to relatively short records of lacustrine phases. Within these, changes from non-saline to saline evaporitic facies reflect shifts in the balance between precipitation and evaporation.

Cenozoic Australian climate

The general character of Cenozoic climate in southern Australia, inferred from evidence within the inland and the southern marginal basins, is summarized in Figure 5 (Benbow *et al.* 1995; Simon-Coinçon *et al.* 1996; Alley 1998; McGowran *et al.* 2004; Martin 2006) and is reviewed below.

Paleocene (65.0–54.8 Ma). The marine oxygen isotope record indicates that globally the Paleocene was a warm period (Fig. 1). Australia lay close to Antarctica and was moving northwards relatively

slowly (*c.* 4 mm a^{-1}) (Veevers *et al.* 1991). Its climate was relatively wet, and plant fossils and palynofloras show widespread rainforest and swamp vegetation, although the NW may have been relatively dry (Fig. 5; Martin 2006). The south–central part of the continent was dominated by cool-temperate coniferous rainforests (Alley & Clarke 1992), whereas in central Australia the lower part of the Eyre Formation (Late Paleocene) contains variable angiosperm and gymnosperm assemblages with predominantly tropical to subtropical affinities (Sluiter 1991). This apparent contrast of tropical and cool-temperate plants in South Australia may imply that central Australia was significantly warmer than the southern margin (Alley 1998), subject to uncertainty about ecological niches of early Tertiary taxa (Benbow *et al.* 1995). A phase of red, hematitic weathering occurred in southeastern Australia at *c.* 60 Ma (Smith *et al.* 2009). Sediments in the southern marginal basins

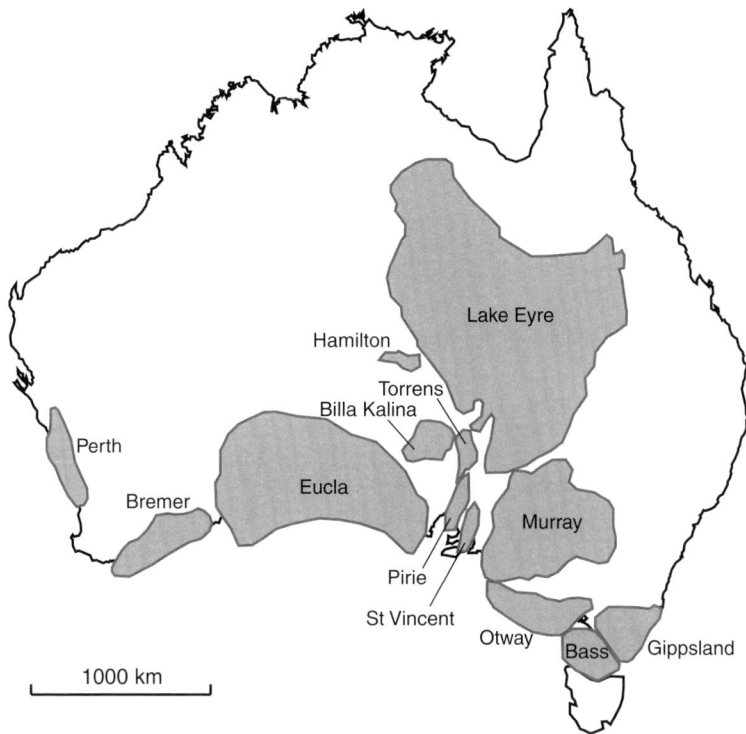

Fig. 4. The southern marginal and inland Cenozoic basins (after Alley 1998; McGowran *et al.* 2004).

are siliciclastic rocks, organic-rich but carbonate-poor (Fig. 5, Sequence I), and include both non-marine and marginal marine facies; marine transgressions are marked by calcareous benthic or planktonic foraminifera and some molluscs (McGowran *et al.* 2004).

Eocene (54.8–33.7 Ma). Globally, the early Eocene was the warmest period in the Cenozoic (Fig. 1). Paleocene forests persisted in Australia through the Early Eocene, with coniferous cool-temperate rainforests dominating the southern margin, whereas tropical rainforests dominated the central region (Fig. 5; Martin 2006). Global cooling in the Middle–Late Eocene, probably associated with initial growth of the Antarctic ice sheets (Fig. 1), was punctuated in Australia's southern basins by a distinct hiatus during the late Early to early Middle Eocene, when the rate of separation between Australia and Antarctica increased to 6–7 cm a^{-1}, around 45 Ma ago (Fig. 5; McGowran *et al.* 2004). Temperatures fell in the Middle to Late Eocene but widespread rainforest indicates continued high humidity (Martin 2006). However, rainforest in central Australia became restricted to valley bottoms and open woodland appeared in the hinterland (Alley 1998), suggesting the onset of monsoonal conditions and latitudinal climatic zonation in South Australia (Benbow *et al.* 1995). In southern marginal basins, marine limestone began to accumulate for the first time since the Palaeozoic, and persisted into the Oligocene (Fig. 5, Sequence II; McGowran *et al.* 2004).

Oligocene (33.7–23.8 Ma). The Eocene–Oligocene boundary is marked both by major changes in oceanic biota (the 'Terminal Eocene Event': Prothero 1994) and by abrupt cooling and growth of the Antarctic ice sheet (Fig. 1) associated with stepwise opening of the Southern Ocean: the Tasman Gateway at *c.* 34 Ma, followed by the Drake Passage around *c.* 30 Ma (Livermore *et al.* 2005). The Australian climate in the Oligocene was generally cool and wet, becoming warm and wet in later Oligocene to the Early Miocene (Fig. 5). Rainforest became less extensive, and the vegetation of southeastern Australia (and, less certainly, Western Australia) suggests seasonal rainfall (Martin 2006). No deposition during this period is recognized in the Lake Eyre Basin, which is inferred to have been a marshy region of low relief, characterized by widespread development of pedogenic silcrete, which is expressed today as the Cordillo Surface (Fig. 5; Wopfner 1974, 1978).

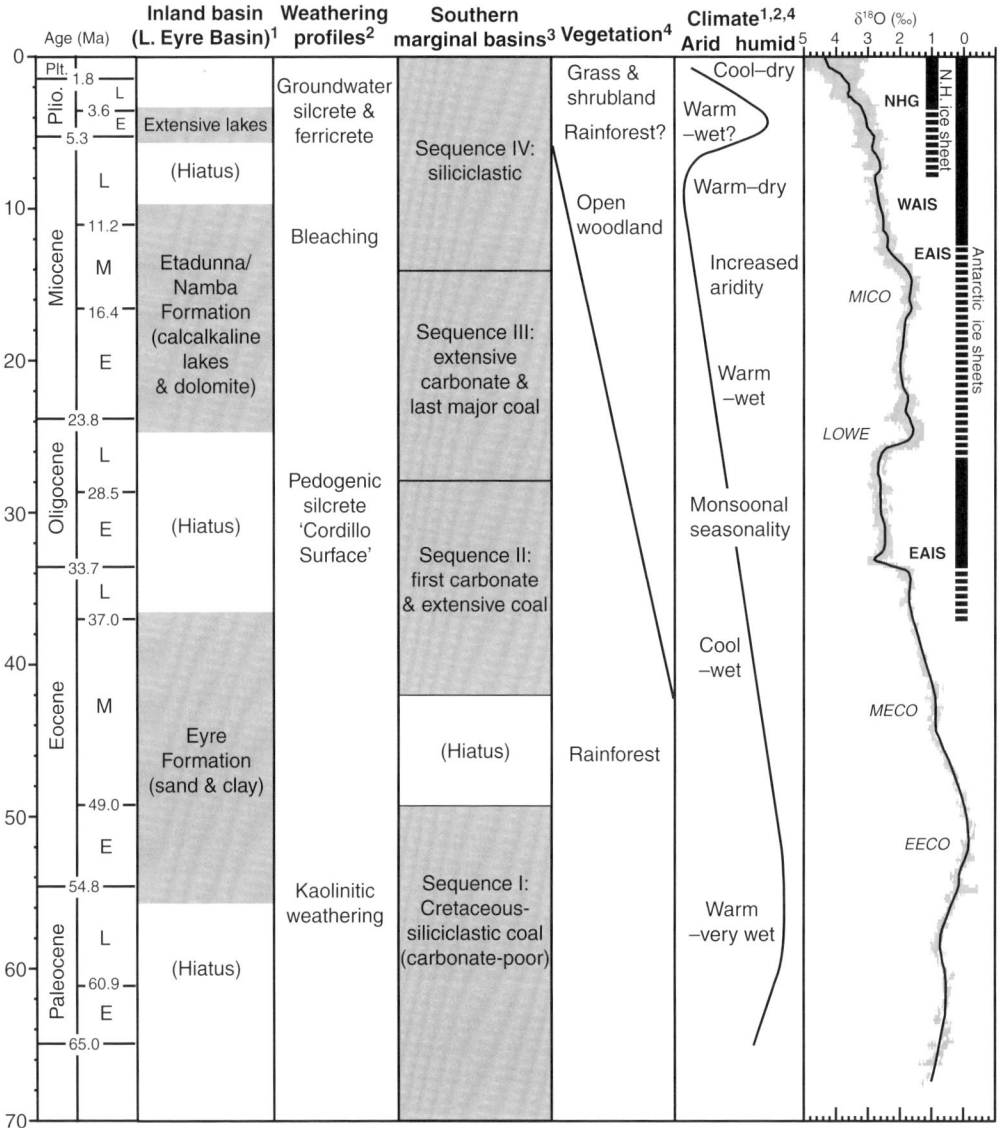

Fig. 5. Cenozoic Australian climate inferred from sedimentological and palaeontological evidence in inland and southern marginal basins and from weathering regimes. Time scale is from the revised geomagnetic polarity time scale (Berggren *et al.* 1995). Deep-sea oxygen isotope record is adopted from Figure 1. References: 1, Alley (1998); 2, Simon-Coinçon *et al.* (1996); 3, McGowran *et al.* (2004); 4, Benbow *et al.* (1995); Martin (2006).

Miocene (23.8–5.3 Ma). Globally warm and humid conditions returned in the Late Oligocene, probably accompanied by retreat of Antarctic ice (Fig. 1). The Nullarbor Limestone was deposited across central southern Australia; its diverse foraminiferal fauna shows that the south Australian shelf sea was warmer in the Early–Middle Miocene than at any other time in the Tertiary (Fig. 5, Sequence III; McGowran *et al.* 2004). Although the inland

climate was significantly wetter than today, the Early–Middle Miocene marks the first step towards aridity: palynoflora suggests increasing aridity over the continent, particularly in central Australia, where rainforest became restricted to small pockets and open woodland was widespread (Benbow *et al.* 1995; Alley 1998). The Lake Eyre Basin contained extensive, shallow, alkaline lakes with extensive dolomite deposition, indicating high temperatures and relatively dry conditions (Krieg *et al.* 1990).

The peak of the Early–Middle Miocene global warm period (MICO, *c.* 15 Ma), was followed by sudden cooling in the Late Miocene, when marine oxygen isotope records indicate redevelopment of the Antarctic ice sheet (Fig. 1). Anticyclonic circulation over Australia probably intensified, as a result, and relative aridity is suggested by a lack of widespread Late Miocene inland deposits (Benbow *et al.* 1995). An episode of red weathering in SE Australia is dated to *c.* 15 Ma (Smith *et al.* 2009). Rainforest was considerably reduced and open woodland became prominent over the continent (Fig. 5; Martin 2006). Widespread bleached weathering profiles developed in shale in central Australia, implying restricted flow of acidic waters with only partial loss of the more soluble components including K, SO_4 and Si (Simon-Coinçon *et al.* 1996). The timing of the shift was, however, regionally variable: cessation of deposition of carbonaceous sediments in the Murray Basin indicates diminishing precipitation by the end of the Middle Miocene (Martin 1977), whereas wet conditions with rainforest persisted in the SE into the Late Miocene (Kemp 1978). The immaturity of karst development and preservation of palaeodrainage in the Nullarbor Limestone suggest aridity in the Eucla Basin since the Miocene (Benbow 1990).

Pliocene (5.3–1.8 Ma). By the Pliocene, the Australian climate was semi-arid and the vegetation was predominantly open woodland with arid shrublands, grasslands and very rare rainforest (Fig. 5; Martin 2006). In the southern basins, carbonate deposits became less and the facies became more siliciclastic (Fig. 5, Sequence IV: McGowran *et al.* 2004).

Within the Late Miocene to Pliocene drying trend an early Pliocene warm, wet reversal has been inferred from extensive fluvial and lacustrine sedimentation in the Eucla and Lake Eyre Basins (Benbow *et al.* 1995). Thick, partly calcareous mudstone on northern Eyre Peninsula and clastic fluvial sediments in the Hamilton Basin, NW of Lake Eyre, may correlate with the upper part of the vertebrate-bearing Etadunna and Namba Formations in the Lake Eyre Basin that may extend into the Early Pliocene; if so, warm, relatively wet conditions are suggested (Martin 1990; Benbow *et al.* 1995). Relatively high water tables have been inferred from ferricrete and groundwater silcrete, thought to be of Pliocene age, in bleached regolith west of Lake Eyre (Simon-Coinçon *et al.* 1996). To the south, Pliocene fluvio-lacustrine evaporites, deposited in palaeovalleys of the Eucla Basin, indicate reduced, intermittent flow along palaeovalleys and suggest significant drying and/or warming (Hou *et al.* 2008).

Towards the end of the Pliocene, the climatic zones of modern Australia become evident in plant-fossil and palynological data, although the climate appears to have been wetter than today (Martin 2006). The tempo of climatic change then quickened as Quaternary glacial cycles commenced, accompanying expansion of Northern Hemisphere ice caps *c.* 3 Ma ago.

Pleistocene aridity and cyclic climatic fluctuations

Quaternary glacial–interglacial cycles in the northern continents were manifest in Australia as alternating cool-arid and warm, less arid intervals. Biologically, the series of repeated climatic changes probably favoured more adaptable taxa, and may have caused many local extinctions (Martin 2006). In Australia's central region, Quaternary vegetation records are poor, whereas fossil vertebrates have a limited but better record (Hope 1982), showing that grazers such as kangaroos appeared during the Early Pliocene as savannah woodland replaced the forests. In the Late Pliocene to Early Pleistocene, fauna resembled those of the arid zone today and included the hare wallaby and rodents. By the Late Pleistocene, lungfish, flamingos and crocodiles were disappearing from the central region and the Murray Basin, as formerly large water bodies vanished (Hope 1982). The Late Quaternary (the last 130 ka) is widely represented in the Lake Eyre Basin by lacustrine, fluvial and aeolian sediments, and is a guide to previous Quaternary cycles in Australia. From previous reviews (Bowler 1982; Hesse *et al.* 2004) and recent research papers, we give a brief overview of Australia's Late Quaternary records, to provide the context for recent studies that reveal new aspects of Pleistocene aridification.

Lacustrine and other sediments in the Lake Eyre Basin contain evidence of alternating high-lake and playa phases, particularly well preserved for the last 130 ka, where high-lake episodes during the last interglacial and early stages of the last glacial cycle gave way to playa conditions and intense deflation of Lake Eyre itself at about the peak of the last ice age. Sedimentary sequences near the lake register earlier episodes of intense deflation and aridity, presumably during previous glacial periods (Magee *et al.* 1995, 2004; Alley 1998; Croke *et al.* 1998; Magee & Miller 1998). Further south, speleothem records from Tasmania indicate that the Holocene and last interglacial were wetter than the preceding glacial periods (Colhoun *et al.* 2010), whereas speleothem deposits dated by U-series in the Naracoorte Caves show a record of episodic activity through the Late Quaternary: periods without speleothem growth indicate major dry periods that apparently coincide with interglacial or late-glacial episodes of the last 300 ka, around 0–20 ka, 115–155 ka and 220–270 ka. Earlier no-growth episodes are also recorded but with lower chronological precision (Ayliffe *et al.* 1998). All such results are relevant in debates about megafaunal extinction in the region, as it is clear that most of the species that went extinct during the last arid cycle had survived multiple, earlier arid periods (Prideaux *et al.* 2007).

Sites in eastern Australia include Lake George, near Canberra, where pollen assemblages show a significant shift from temperate rainforest taxa in the Late Pliocene to open-canopied sclerophyll assemblages in the Pleistocene (Singh *et al.* 1981). In eastern central Queensland, the sedimentary record from Lake Buchanan shows four high-lake episodes in the last *c.* 800 ka, apparently separated by long dry periods, and there are traces of earlier episodes of dry playa alternating with lacustrine conditions, possibly extending beyond 1.6 Ma (Chivas *et al.* 1986). At Fraser Island in south–central coastal Queensland, rainforest vegetation that existed around 600 ka ago gave way to sclerophyllous, drier rainforest around 350 ka ago (Longmore & Heijnis 1999). Fossils from Mt Etna, east Queensland, show that rainforest fauna were replaced by xeric-adapted animals sometime between 280 and 205 ka ago (Hocknull *et al.* 2007). High-quality pollen records from Lynch's Crater in NE Queensland show cyclic alternation of rainforest and sclerophyll woodland that correspond to the Northern Hemisphere interglacials and glacials of the last 250 ka (the last glacial period also shows a substantial increase of detrital charcoal, suggesting burning by humans: Kershaw 1986), whereas in far northern Australia, climatic variability is closely related to monsoon intensity: valley deposits of Magela Creek in Northern Territory show evidence of reduced erosion since *c.* 300 ka ago, suggesting weakening of the Australian monsoon (Nanson *et al.* 1993). To the west, wet periods are represented by greatly enlarged Pleistocene lakes in the semi-arid north, particularly at Gregory Lakes and also at Lake Woods in Northern Territory (Bowler *et al.* 2001), and Pleistocene high-lake phases are recognized at playas in the goldfields region of southern Western Australia (Zheng *et al.* 1998).

Overall, various arid landforms including stony and sandy deserts together with playa lakes developed as aridity intensified. However, the ages of Australia's arid landforms remained rather indeterminate until the relatively recent application of luminescence and exposure age dating methods appeared, which is reviewed in the following sections.

Chronology of Australian arid landforms

Over 75% of Australia is today classified as climatically arid or semi-arid, and its surface is covered by various types of desert landforms including dune fields, stony deserts and playas (Fig. 3; Bowler 1976; Mabbutt 1988). Although alkaline lake deposits (reviewed above) indicate an early phase of Australian aridity in the Early–Middle Miocene, chronological data from the arid landforms themselves are largely from late Quaternary deposits. One difficulty of dating Australia's arid landforms has been the limited time range of the dating methods used so far, such as radiocarbon and luminescence dating. Radiocarbon has a limited time range (<50 ka) and, in arid deposits, suffers from poor preservation of carbonaceous material. Luminescence dating (thermoluminescence or TL, optically stimulated luminescence or OSL, and infrared stimulated luminescence or IRSL) can be used to determine the age of sedimentary deposits, with quartz or feldspar grains as the favoured material (Aitken 1985, 1998). Depending on the amounts of radioactive elements (e.g. K, U and Th) in the grains, the time range of luminescence dating can extend beyond 100 ka, and may reach as old as 700 ka if K, U and Th contents are particularly low (Huntley & Prescott 2001). Other methods with a greater time range have been applied in a limited number of cases, including palaeomagnetic dating (McElhinny & McFadden 2000), oxygen isotope chronology (Lisiecki & Raymo 2005) and cosmogenic isotope dating (Gosse & Phillips 2001).

The interpretation of geochronological data from aeolian deposits is often not straightforward, as the relationship between the dates obtained and the deposition of aeolian materials, such as dune deposits, can be uncertain and also may differ from one dating method to another. These uncertainties are outlined for each chronological method and palaeoclimate proxy referred to below.

Chronology of playas

Playa lakes are a mark of intensifying aridity in Australia, and the time when continuous lake sedimentation gave way to ephemeral (playa) conditions has been estimated from lake sediment cores at a few sites, using palaeomagnetic reversal chronology. Palaeomagnetic dating uses signatures of magnetic reversals recorded in continuously deposited sediments and correlates these with the international standard geomagnetic reversal chronology (Berggren *et al.* 1995; Cande & Kent 1995). The timing of changes of sedimentological facies is interpolated between reversal horizons, with assumptions about sedimentation rates. The interpretation of palaeomagnetic signals in aeolian sediments may be affected by loss by deflation; however, the method has proved valuable in Australia's dry lake sequences. The earliest preserved transition from permanent lake to playa lake so far identified is at Lake Amadeus playa in central Australia, which contains an upper sequence of aeolian and saline gypseous sandy-clay deposits that represent playa conditions (Winmatti Beds), overlying thick, fairly uniform fluvial–lacustrine clays (Uluru Clay). The Brunhes–Matuyama (B–M) palaeomagnetic boundary (0.78 Ma) occurs in the Winmatti Beds, and the transition from Uluru Clay to playa sediments lies immediately above a short, magnetically normal episode within the Matuyama Chron, most probably correlated with the Jaramillo event (1.07–0.99 Ma) or possibly with the Olduvai event (1.95–1.77 Ma) (Chen & Barton 1991). Hence, arid playa conditions at Lake Amadeus began at, or before, 900 ka ago.

A somewhat different result was found at Lake Lewis playa in central Australia, about 250 km north of Lake Amadeus. A core from the playa, composed of thick uniform brown lacustrine clay deposits (Anmatyerre Clay) overlain by relatively thin gypsum-rich clay deposits (Tilmouth Beds), indicates a long period of perennial, relatively non-saline lacustrine sedimentation followed by a period of saline, fluctuating playa conditions. The B–M boundary occurs in the Anmatyerre Clay, c. 9 m below the interface between the two sequences, implying that the transition from non-saline to saline conditions at Lake Lewis occurred within the last 780 ka; probably c. 300–400 ka ago according to likely sedimentation rates (English 2001). Although the age interpretation is affected by uncertain sedimentation rates, the result from Lake Lewis is substantially younger than the age of a comparable transition at Lake Amadeus. This age difference is discussed further below but, in summary, is due to a substantial difference of hydrological conditions at the two lakes.

Palaeomagnetic reversal methods were used to estimate the timing of a similar shift from perennial to ephemeral lacustrine conditions in southeastern Australia, where palaeolake Bungunnia extended across a broad area of the western Murray Basin between 2.5 and c. 0.7 Ma (Stephenson 1986; Bowler et al. 2006). A drill core from Lake Tyrrell, a remnant playa within the area of the palaeolake, shows thick uniform lacustrine clay sediments (Blanchetown Clay) disconformably overlain by relatively thin gypseous saline deposits (Tyrrell Beds). Palaeomagnetic analyses found that the Tyrrell Beds have normal polarity whereas the upper part of the Blanchetown Clay shows reversed polarity; the two formations have been assigned to the Brunhes and Matuyama Chrons, respectively. Although the precise position of the

B–M transition could not be determined because of the stratigraphic hiatus, the transition from fluvial–lacustrine to saline arid conditions at palaeolake Bungunnia is estimated to have occurred 400–700 ka ago (An et al. 1986). In a similar study at Lake Lefroy playa in Western Australia, the B–M transition was identified 1 m below the top of clayey fluvial–lacustrine sediments of the Revenge Formation, which is overlain by gypseous clayey sand (the Roysalt Formation). Thus, arid playa conditions are thought to have commenced relatively early in the Brunhes Chron, probably around 500 ka (Zheng et al. 1998).

Collectively, the results suggest that by about 500 ka many previously perennial, non-saline lakes had become saline playas. In many cases, including Lakes Eyre, Lewis and Gregory, subsequent reversals to perennial lake conditions occurred more than once (Bowler et al. 2001; English 2001; Magee et al. 2004). However, as shown by the difference between Lakes Amadeus and Lewis (discussed at more length below), the timing of the perennial–playa transition for any given lake depends on its catchment hydrology (Bowler 1981), and these results must be interpreted in that light.

Chronology of aeolian sediments and forms

Widespread aeolian sediments including sand dunes and aeolian dust mantles are characteristic of arid regions. Sand deserts, including sand plains and dunefields, cover c. 40% of the Australian continent (Mabbutt 1988), and, where dunefields are dominated by parallel dune-ridges, their orientations represent a continental-scale anticlockwise whorl (Fig. 3; Jennings 1968; Bowler 1976; Wasson et al. 1988). The pattern broadly reflects dominant winds associated with today's winter anticyclones, but may more precisely reflect their position during the Late Pleistocene (Brookfield 1970). Australian aeolian dust not only is distributed across the continent but travels offshore to the NW and SE, occasionally as far as New Zealand (McTainsh 1989). Aeolian dust is often referred to as 'parna' in Australia and can be a significant constituent of local soils in SE Australia (Butler 1956; Chartres et al. 1988).

Dunefield chronology. Little was certain about the age of Australia's desert dunes, prior to the 1980s. Correlations had been made with radiocarbon-dated fluvial and/or lacustrine sediments (Wopfner & Twidale 1967), but it awaited the advent of new dating methods, notably luminescence dating, before the situation improved. Luminescence measurements of the time interval since a sediment was last exposed to sunlight have provided ages of Late Quaternary episodes of dune formation and fluvial activity at a number of central Australian sites (Gardner et al. 1987; Nanson et al. 1988, 1991, 1992a, b, 1995; Readhead 1988; Chen et al. 1990, 1995; English et al. 2001; Twidale et al. 2001; Lomax et al. 2003; Hollands et al. 2006; Sheard et al. 2006; Fitzsimmons et al. 2007; Fujioka et al. 2009). Previously, luminescence ages from dunes such as those summarized in Figure 6 were interpreted as showing that dunes were active during Late Pleistocene glacial stages (MIS 2 and 4, and probably MIS 6 and 8), although the uncertainties are relatively large for earlier episodes (e.g. Nanson et al. 1992b). More recent studies have indicated that dunes were also active during the last interglacial and the Holocene (Fig. 6). Moreover, as datasets of luminescence ages from dunes have grown, in studies from Australia and elsewhere, it has become apparent that they represent continua rather than sharply defined peaks. The lack of definitive episodes of dune activity in the luminescence records may partly reflect the lack of precision in the dating method, but also suggest that dunefields are frequently reactivated, at least locally, through interplay of factors such as windiness, drought and vegetation cover (Bateman et al. 2003; Stone & Thomas 2008).

Luminescence dating, however, may not be the best method for determining the age of inception of Australian dunefields. In favourable sediments, quartz-based luminescence dating may measure ages exceeding c. 700 ka (Huntley & Prescott 2001), but the practical limit of age determinations from most Australian desert dunes is usually 200 ka or less. Moreover, this method is unlikely to give the age of dunefield initiation owing to reworking of the sand, which bleaches luminescence signals from earlier depositional episodes. Palaeosols are sometimes observed within internal dune structure (Readhead 1988, 1991; Lomax et al. 2003; Fitzsimmons et al. 2007; Fujioka et al. 2009), but the luminescence ages from dune sand below a palaeosol horizon will indicate only the maximum age of the palaeosol formation, and not necessarily the timing of initial deposition of the sand itself. The luminescence results summarized in Figure 6 therefore are more likely to represent dune stabilization ages (Hesse et al. 2004) or accumulation phases separated by palaeosol formations, rather than the timing of dunefield inception. To some extent the results suggest that dune activity may have started earlier in the centre of Australia than at the periphery (Nanson et al. 1992b), but we reiterate that the age of dunefield initiation at most sites probably lies well beyond the range of luminescence dating.

Cosmogenic burial dating provides a different approach to the problem of dating the timing of dunefield initiation. This method rests on differences in decay rates of a pair of cosmogenic radioisotopes to determine the duration of burial of mineral grains that accumulated the isotopes before burial. The burial-age calculation is relatively straightforward when grains are deeply buried, whereas the age models become complex when burial depth is shallow (less than c. 2 m). Using ^{10}Be and ^{26}Al, burial ages from 300 ka to as

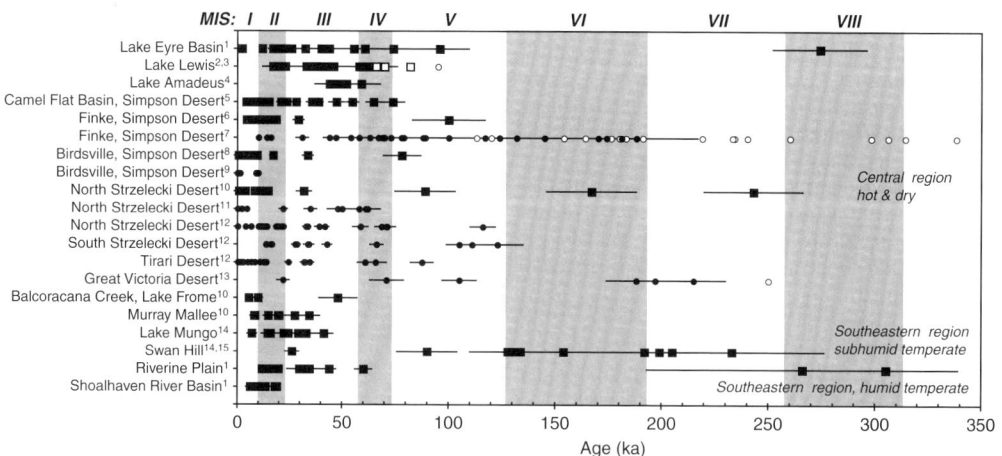

Fig. 6. Summary of luminescence ages for Australian dunes. TL and OSL ages are shown as squares and circles, respectively, with open symbols for saturated ages. One-sigma uncertainties are shown. The Marine Isotope Stage (MIS) is shown by shaded bands as glacial periods. References: 1, Nanson et al. (1991, 1992b); 2, Chen et al. (1995); 3, English et al. (2001); 4, Chen et al. (1990); 5, Hollands et al. (2006); 6, Nanson et al. (1995); 7, Fujioka (2007); Fujioka et al. (2009); 8, Nanson et al. (1992a); 9, Twidale et al. (2001); 10, Gardner et al. (1987); 11, Lomax et al. (2003); 12, Fitzsimmons et al. (2007); 13, Sheard et al. (2006); 14, Readhead (1988); 15, Readhead (1991).

much as 5 Ma may be determined (Granger & Muzikar 2001). The method has recently been applied to drill-core samples from the west Simpson Desert dunefield near Finke in central Australia: in summary, four stratigraphic dune-sand units were identified, separated by three palaeosol horizons, marked by soft hematite nodules. Burial ages of the core samples from the lowest dune-sand unit, determined by paired ^{10}Be–^{26}Al analysis, indicated that the first dunes formed in this region c. 1 Ma ago (Fujioka 2007; Fujioka et al. 2009).

Cosmogenic burial dating using the ^{10}Be–^{26}Al pair is, however, not practical for dating younger deposits with ages less than 300 ka, owing to the unavoidable uncertainty of the method (c. $\pm$ 100 ka) (Granger & Muzikar 2001), but has the advantage over luminescence dating in that isotope accumulation is not reset by subsequent reworking events. Instead, the cosmogenic isotope content in sand grains reflects their entire exposure and burial history, which may include multiple episodes of deposition, deflation and re-exposure at shallow depth (Fujioka et al. 2009). Age models based on cosmogenic data from aeolian deposits may therefore be relatively complex. Nevertheless, the longer age range of the method and lack of resetting are complementary to luminescence dating methods, and the combined use of the two would be a powerful approach for future dune studies.

Aeolian dust. Aeolian dust as an indicator of aridity has been prominent in studies of Australia's environmental history. Australian aeolian dust is relatively rich in clay (>30%), and Australian peridesert loess, known as parna, is present in soils and buried soils of the southeastern highlands (Butler 1956). Its origin lies in the Lake Eyre and Murray–Darling basins, where fine, clay-rich sediments are mobilized by wind in dry periods (Chartres et al. 1988; McTainsh 1989).

The chronology of past episodes of aeolian dust deposition remained obscure until aeolian dust was identified in offshore marine sediment cores and tied to the standard marine oxygen isotope stratigraphy (Hesse 1994). A record extending beyond 600 ka based on six marine sediment cores from the Tasman Sea shows that the first significant episode of dust transport from the Murray–Darling region was in glacial stage MIS 10 (c. 350 ka), and larger fluxes occurred in subsequent glacial stages (MIS 8, 6 and 2; Hesse 1994). A subsequent study of sediment cores from near the outer margin of the NW Australian continental shelf shows that strong dust flux from the northwestern zone occurred during the last two major glacial stages (MIS 6 and 2; Hesse & McTainsh 2003). Together, these records indicate that aeolian sedimentary processes commenced in the Late Pleistocene in the SE

and NW, and were strongest during glacial stages. Notably, the first significant dust pulse from southeastern Australia, as seen in Tasman Sea sediments, appears around 350 ka, which is later than deposition of an aeolian mantle north of Broken Hill in western NSW, which began to accumulate at least 500 ka ago and probably closer to 1 Ma ago, according to cosmogenic burial dating of a stone pavement beneath more than 1 m of indurated dust (Fisher 2003).

Age measurements from stony deserts

Stony deserts are durable landscape features in inland Australia and cover c. 15% of the continent. Typically, stony desert pavement comprises a single layer of siliceous pebble- to cobble-sized clasts, known in Australia as 'gibbers'. Gibber pavements can occur on parent rock, usually silcrete, but more widely they overlie alluvium or float on stone-free clayey aeolian silt (Mabbutt 1977, 1988). Their age was largely a matter of guesswork until recent cosmogenic exposure age dating using cosmogenic ^{10}Be and ^{21}Ne showed that gibber pavements west of Lake Eyre formed 2–4 Ma ago. Results from fan surfaces west of Oodnadatta indicate that dissection of fan-head slopes and gullies effectively ceased around 2 Ma ago, and results from silcrete mesa surfaces suggest that a previous soil-cover had been lost by 1 Ma (Fujioka et al. 2005; Fujioka 2007).

Change of weathering regime

Landforms and deposits such as playas, dunes and stony deserts are among direct indicators of aridity, and changes of pedogenesis in sub-humid areas also indicate climatic change. Weathered sediments at coastal sections south of Adelaide and on Kangaroo Island show evidence of a mid-Pleistocene shift from oxidic to calcitic weathering, in sections typically composed of weathered gravelly sands (Ochre Cove Formation), sharply overlain by grey–green clay, possibly aeolian (Ngaltinga Clay), together with younger deposits. The Ochre Cove sediments show strong mottled oxidic red-weathering, whereas calcareous weathering prevails throughout the Ngaltinga Clay and younger deposits above the Ochre Cove Formation. Chemical remanent magnetization in hematite, a post-depositional weathering product, shows that the B–M boundary lies within the upper part of the Ochre Cove Formation (Pillans & Bourman 2001). The authors infer that the change in weathering regime followed a significant decline of regional rainfall, estimated from the position of the B–M boundary and likely sedimentation rates for the Ochre Cove Formation to have occurred c. 500–600 ka ago.

Discussion: the onset of Australian aridity

The chronological aspects of studies relating to Australia's arid landscapes, reviewed above, are summarized in Figure 7, together with time ranges of the dating methods used. Reported ages of Australian arid landforms range from Holocene to mid-Pleistocene, with a few early Pleistocene–Pliocene ages, and they vary locally, regionally, and between landform types. There also are differences between results obtained with different dating methods. These factors are now discussed.

Bias from dating methods

Radiocarbon was the principal method for dating Australian aeolian deposits until about 25 years ago (Bowler 1976; Wasson 1983), and has since been overtaken by luminescence dating, which expanded the datable age range to *c.* 200 ka (e.g. Gardner *et al.* 1987; Readhead 1988; Nanson *et al.*

1991, 1992*b*). However, luminescence dating has a weakness in that dune reactivation re-exposes sand grains to sunlight, which resets the luminescence signal and induces considerable uncertainty into the dating of dune-building episodes prior to the latest activation. Bioturbation, common in the upper 1–2 m at a dune, has a similar effect (Hesse *et al.* 2004). In contrast, cosmogenic nuclides are not reset by bioturbation, although burial depth assumptions can be affected. Palaeosol formation could preserve an earlier accumulation record, but luminescence ages from the layers below a palaeosol can only be interpreted as the maximum age of the palaeosol formation and not the timing of the first deposition of sands. Moreover, inception of aeolian sediment activation and transport lies beyond the most common range of luminescence dating: dust from southeastern Australia has accumulated in the Tasman Sea since at least 350 ka (Hesse 1994); sediment cores from Lake Amadeus suggest that sand has been mobile in central

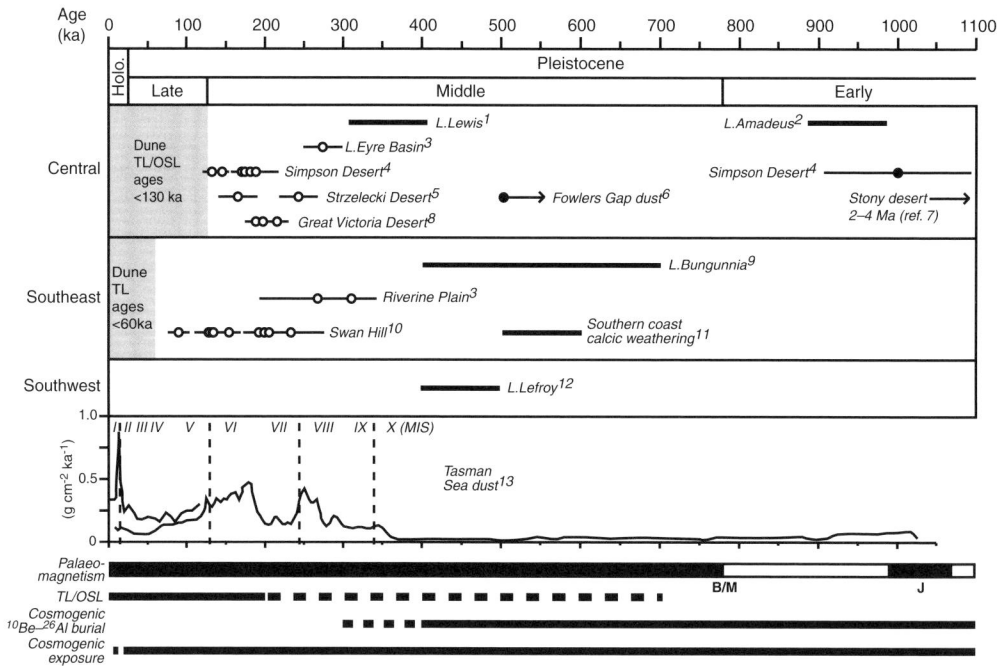

Fig. 7. Summary of previous chronological studies of Australian arid landforms (top) and relevant dating methods and their time scales (bottom). Chronological data are regionally categorized by the sampling locations. Thick bars represent the timing of the transition within playa sequences from continuous lacustrine to gypseous playa sediments and the transition of weathering regime from oxide to carbonate in South Australia coastal sections. Shaded areas represent clusters of luminescence ages of linear dunes and lunettes summarized in Figure 6. Some older dune ages are separately shown as open circles with the standard error. Filled circle represents cosmogenic [10]Be–[26]Al burial ages of dust deposition at Fowlers Gap, western NSW, and of basal dune horizons in the west Simpson Desert. Profiles of the Tasman Sea dust flux are shown in the middle panel, with dashed lines indicating the termination of glacial periods of MIS 10, 8, 6 and 2. References: 1, English (2001); 2, Chen & Barton (1991); 3, Nanson *et al.* (1991, 1992*b*); 4, Fujioka *et al.* (2009); 5, Gardner *et al.* (1987); 6, Fisher (2003); 7, Fujioka *et al.* (2005); 8, Sheard *et al.* (2006); 9, An *et al.* (1986); 10, Readhead (1988, 1991); 11, Pillans & Bourman (2001); 12, Zheng *et al.* (1998); 13, Hesse (1994).

Australia for at least 900 ka (Chen & Barton 1991); and cosmogenic $^{10}Be-^{26}Al$ burial dating shows that dune activity commenced in the west Simpson Desert around 1 Ma ago (Fig. 7; Fujioka *et al.* 2009).

Climatic thresholds of arid landform development

Climatic changes may appear to occur at different times at different places, because the response of dunes, lakes or rivers to a given climatic shift can vary with local characteristics of a particular landform. Moreover, whereas playas may form and dunes may mobilize following a climatic shift to aridity, they may not return quickly to their prior condition if the climatic shift is reversed. For example, the threshold for dune stabilization may require rainfall to be higher than that at which dunes become unstable, owing to greater porosity of mobile sand. Before discussing the onset of aridity across Australia, these concepts of thresholds and lags are briefly reviewed.

Lake types and their hydrological threshold. The condition of a lake (overflow, closed perennial, or ephemeral) depends on its water balance, governed by climatic factors (precipitation; evaporation) together with catchment size and runoff. From a simple water balance, a relationship is readily derived between the ratio of basin catchment area to lake area (A_c/A_l), on one hand, and a function F_c that incorporates precipitation (P), evaporation (E) and runoff coefficient (f), on the other, where $F_c = (E - P)/fP + 1$ (Bowler 1981). (It should be noted that when pan evaporation data are used to estimate E, the data should be multiplied by a pan constant, normally taken as 0.8). Figure 8 shows a graph of selected Australian lakes plotted in terms of (A_c/A_l) and F_c classification (Bowler 1981). The graph separates three classes of lake: permanent, ephemeral and dry. Five large ephemeral lakes (Eyre, Lewis, Gregory, Woods and Buchanan) lie in the central class; all five would become permanent with only a relatively small change of either E, P or f, suggesting a hydrological threshold, separating generally permanent and near-permanent lakes from dry and mostly dry lakes (Fig. 8).

The climatic shift needed for a dry lake to become a permanent lake depends on its distance from the hydrological threshold. For example, playa lakes Frome, Torrens, Amadeus, Gairdner and Tyrrell have relatively small, ineffective catchments and lie far from the threshold, whereas Lake Lewis, a

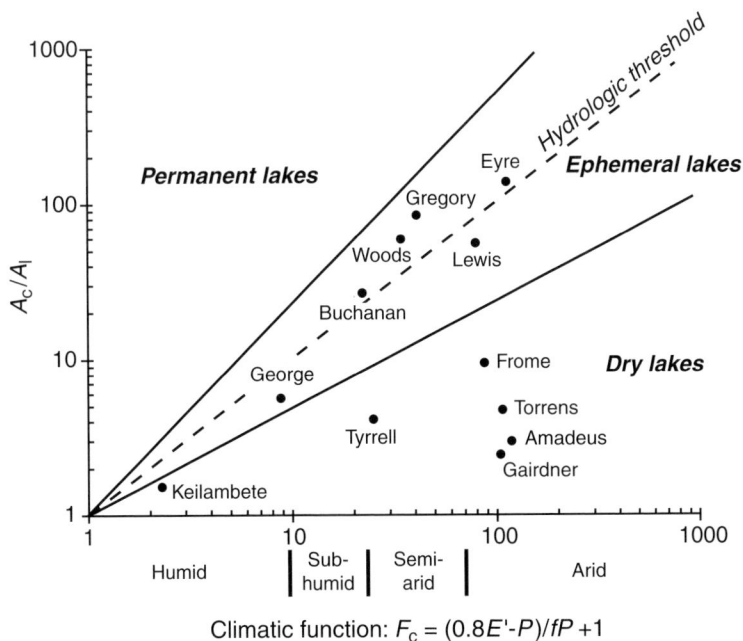

Climatic function: $F_c = (0.8E'-P)/fP + 1$

Fig. 8. Classification of Australian inland lakes and the hydrological threshold (after Bowler 1981). Vertical axis shows catchment area to lake area ratio (A_c/A_l), and horizontal axis shows climatic function (F_c), where E', P and f denote pan-evaporation and precipitation rates per year (mm a^{-1}) and runoff coefficient. For the runoff coefficient, a lower and more realistic value of $f = 0.1$ was used for all lakes shown in the figure. Data for A_c, A_l, E and P, for all the lakes in this figure, are from Bowler (1981, 1986), except for Lake Lewis, from English (2001).

frequently flooded playa more similar to Gregory Lakes than to Lake Amadeus, has a relatively efficient catchment and lies close to the threshold (Fig. 8). Hence, Lake Lewis, which first became a playa around 300–400 ka (English 2001), withstood the Pleistocene trend towards aridity for substantially longer than Lake Amadeus, which became a playa at least 900 ka ago (Chen & Barton 1991).

Thresholds of dune activation and stony desert formation. Australian dunefields are generally well vegetated and are fairly stable under the present climatic regime. The threshold of dune instability depends on wind regime, constrained by vegetation cover on dunes. Both factors can be approximated in simple climatic terms: for example, the percentage of days per year with sand-moving winds (D) can be used as an index of potential sand mobility (M), whereas the ratio of actual to potential evapotranspiration (E_a/E_p) is a surrogate for vegetation cover (Ash & Wasson 1983) (lacking measurements of E_a, annual rainfall may be a fair approximation; Chappell 1983). Using observations of inland and coastal dunes, Ash & Wasson (1983) outlined a dune-mobilization threshold in terms of M and E_a/E_p (Fig. 9), and concluded that Australia's arid

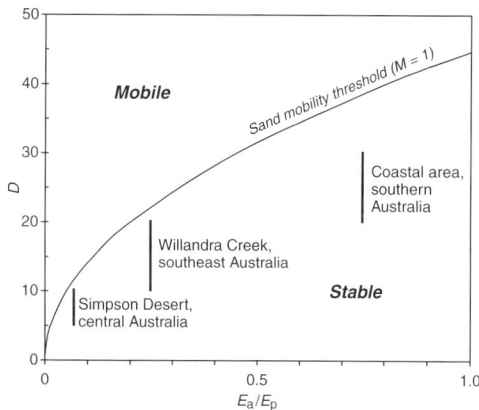

Fig. 9. Threshold of sand mobility (after Ash & Wasson 1983). Sand mobility index M is defined as $M = 0.0005 \times D^2/(E_a/E_p)$, where D, E_a and E_p denote the percentages of days per year with winds sufficiently strong for sand transport (c. 8 m s^{-1}), and actual and potential evapotranspiration, respectively. Dune sands become subject to entrainment when $M > 1$ (upper region, above the sand mobility threshold curve). Australian dunes are generally stable, under current wind regimes and vegetation cover (lower region, below the sand mobility threshold curve). The ratio E_a/E_p for the Simpson Desert, Willandra Creek region and humid coastal area is about 0.07, 0.25 and 0.75 respectively (Ash & Wasson 1983), and wind frequency data are from *Climatic Atlas of Australia* (Bureau of Meteorology 1979).

dunefields could be mobilized by a moderate increase in the frequency of sand-shifting winds, or by a more substantial decrease of E_a/E_p (or P/E). However, it must be noted that, having become destabilized, the threshold for dune revegetation and stabilization is likely to require higher P/E than at the destabilization threshold.

Wind and vegetation are also implicated in the formation of stony deserts, which comprise a stony surface lag deposit (gibber) that remains after deflation of silt and clay from gravelly sediment. Landscape hydrology changes when gibber develops at the expense of prior soil cover; net runoff from gibber can be high (Callen & Benbow 1995) but is subject to considerable mesoscale variation and is much lower in the small vegetated groves that characteristically pattern Australia's gibber deserts (Dunkerley 2002). Although unquantified, the difference in P/E terms between the thresholds of stony desert formation and recovery is probably very much greater than for activation versus stabilization of dunefields. Once formed, the stony monolayer persists at the ground surface, and, unlike other arid landforms such as playa lakes and dunefields, described above, stony deserts appear to have persisted through Pleistocene wetter episodes, when many playas became permanent lakes and dunefields stabilized.

Regional age variations of arid landforms

Dry salt lakes occur throughout Australia except for the sub-humid eastern belt and tropical far north; dunefields, thought to have been active at the last glacial maximum, cover over 40% of the continent (Mabbutt 1988). Luminescence dates from dunes, although limited by their tendency to measure the age of the most recent mobilization or the timing of palaeosol formation rather than that of dunefield inception, suggest that arid landforms became active first in central Australia and that the arid zone later expanded (Nanson *et al.* 1992b).

The ages of arid-zone features, shown by recent cosmogenic ages from dunes and stony deserts together with age estimates of the permanent lake to playa transition, are substantially greater than suggested by luminescence dating; however, the impression remains that arid landforms of the centre developed earlier than those near the periphery. As reviewed earlier and summarized in Figure 7, similar ages of around 1 Ma are reported for basal horizons in western Simpson Desert dunes (Fujioka *et al.* 2009) and for the onset of playa conditions at Lake Amadeus (Chen & Barton 1991). Both are about double the age of onset of playa conditions towards the periphery, as judged from Lakes Bungunnia in Victoria (400–700 ka: An *et al.* 1986) and Lefroy in Western Australia (Zheng *et al.*

1998). Although such comparisons are speculative because catchment conditions for all these lakes almost certainly were different before they became playas, the timing of the transition to playa condition at Lakes Bungunnia and Lefroy coincides with the change from red oxidic to calcitic weathering south of Adelaide, c. 500–600 ka ago (Pillans & Bourman 2001). Finally, it remained until c. 350 ka before aeolian processes in the SE became sufficiently intense for the dust flux to contribute significantly to sedimentation in the Tasman Sea (Hesse 1994).

Concluding remarks

Global cooling in the late Cenozoic was associated with the expansion of Antarctic and the other large continental ice sheets, and with the lowering of global sea levels. Fluctuation of global climate during the Cenozoic certainly affected Australian climate, which can be seen in the changes of biological and geological records in the marine and inland basins, notably in southern marginal basins, where the development of the Southern Ocean directly influenced Australian climate. In the late Cenozoic, intensifying cooling was accompanied by the expansion of aridity in Australia, culminating in repeated arid episodes during Quaternary glacial cycles. Major causes of increased aridity during a glacial episode include intensified atmospheric circulation, reduced atmospheric moisture and increased continentality owing to lowering sea levels. Decreased water availability, owing to reduced precipitation and, if wind activity was enhanced, perhaps to increased evaporation, increased the importance of aeolian processes such as deflation, formation of sand dunes and aeolian silt transport.

Until recently, there have been few age measurements of Australian arid landforms, but dating studies are now providing some constraints. Palaeomagnetic dating has shown that playa lakes began to appear in central Australia at least 900 ka ago, and towards the periphery c. 500 ka ago (An et al. 1986; Chen & Barton 1991; Zheng et al. 1998). Palaeomagnetic methods also give an age of c. 500 ka for the transition from oxidic weathering soils to calcareous weathering in southern South Australia (Pillans & Bourman 2001). Luminescence dating has been widely applied to dunes and fluvial sediments and shows that dune activity in Australia commenced at least 300 ka ago (Gardner et al. 1987; Nanson et al. 1988, 1991, 1992a, b, 1995; Readhead 1988; Chen et al. 1990, 1995; English et al. 2001; Twidale et al. 2001; Lomax et al. 2003; Hollands et al. 2006; Sheard et al. 2006; Fitzsimmons et al. 2007; Fujioka et al. 2009). Dust deposits linked to marine oxygen isotope stratigraphy in Tasman Sea sediments indicate that aeolian mobilization and transport of sediment from southeastern Australia commenced at least 350 ka ago, implying that relatively large amounts of dust were available by that time (Hesse 1994; Hesse & McTainsh 2003). Cosmogenic isotope dating has revealed that dunefields in the western part of the Simpson Deserts began to form 1 Ma ago and stony deserts in central Australia date to 2–4 Ma ago (Fujioka et al. 2005, 2009).

Our knowledge of the chronology and environmental conditions associated with the development of Australian arid landforms nevertheless remains insufficient to reconstruct their history fully. In particular, a lack of chronological data from western regions of the continent is apparent and we have little idea about the evolutionary history of vast western dunefields such as the Great Sandy, Great Victoria and Gibson Deserts. Dated playas and dunes indicate that the degree of aridity in central Australia increased about 1 Ma ago, whereas the ages of similar landforms are younger elsewhere. However, with the exception of cosmogenic ages of 2–4 Ma for stony desert pavements west of Lake Eyre, none of these results extends back into the pre-Quaternary, when the main features of Australia's bedrock landforms were shaped and the arid-adapted biota largely evolved.

Recent phylogenetic studies based on DNA analysis suggest that the Australian arid-zone species began to supplant their mesic-adapted predecessors during the Pliocene or earlier (Pepper et al. 2006; Crisp & Cook 2007; Byrne et al. 2008; Hugall et al. 2008), implying the presence of large tracts of stony or sandy deserts that isolated communities and induced speciation (Pepper et al. 2006, 2008; Byrne et al. 2008; Shoo et al. 2008). Although adaptations such as hybridization and parthenogenesis (Kearney et al. 2006; Byrne et al. 2008) may have occurred within the time range of the dunes, gibbers and playas that have been dated so far, the broader picture lies deeper in time. Only when the ages of Australia's desert landforms have been better determined may the co-evolution of landscape and biota become clear. Of the dating methods used so far, cosmogenic exposure-age and burial dating appear to offer the best prospect for achieving this.

The authors thank M. Williams and D. Thomas for their constructive reviews, and the handling-editor P. Bishop for his helpful comments.

References

AITKEN, M. J. 1985. *Thermoluminescence Dating.* Academic Press, London.

AITKEN, M. J. 1998. *An Introduction to Optical Dating.* Oxford University Press, Oxford.

ALLEY, N. F. 1998. Cainozoic stratigraphy, palaeoenvironments and geological evolution of the Lake Eyre Basin.

Palaeogeography, Palaeoclimatology, Palaeoecology, **144**, 239–263.

ALLEY, N. F. & CLARKE, J. D. A. 1992. Stratigraphy and palynology of Mesozoic sediments from the Great Australian Bight area, southern Australia. *BMR Journal of Australian Geology and Geophysics*, **13**, 113–129.

AN, Z., BOWLER, J. M., OPDYKE, N. D., MACUMBER, P. G. & FIRMAN, J. B. 1986. Palaeomagnetic stratigraphy of Lake Bungunnia: Plio-Pleistocene precursor of aridity in the Murray Basin, southeastern Australia. *Palaeogeography, Palaeoclimatology, Palaeoecology*, **54**, 219–239.

ASH, J. E. & WASSON, R. J. 1983. Vegetation and sand mobility in the Australian desert dunefield. *Zeitschrift für Geomorphologie Supplementband*, **45**, 7–25.

AYLIFFE, L. K., MARIANELLI, P. C., MORIARTY, K. C., WELLS, R. T., MCCULLOCH, M. T., MORTIMER, G. E. & HELLSTROM, J. C. 1998. 500 ka precipitation record from southeastern Australia: Evidence for interglacial relative aridity. *Geology*, **26**, 147–150.

BATEMAN, M. D., THOMAS, D. S. G. & SINGHVI, A. K. 2003. Extending the aridity record of the Southwest Kalahari: current problems and future perspectives. *Quaternary International*, **111**, 37–49.

BENBOW, M. C. 1990. Tertiary coastal dunes of the Eucla Basin, Australia. *Geomorphology*, **3**, 9–29.

BENBOW, M. C., ALLEY, N. F., CALLEN, R. A. & GREENWOOD, D. R. 1995. Geological history and palaeoclimate. *In*: DREXEL, J. F. & PREISS, W. V. (eds) *The Geology of South Australia, Vol. 2: The Phanerozoic*. Geological Survey of South Australia Bulletin, **54**, 208–217.

BERGGREN, W. A., KENT, D. V., SWISHER, C. C. I. & AUBRY, M.-P. 1995. A revised Cenozoic geochronology and chronostratigraphy. *In*: BERGGREN, W. A., KENT, D. V., AUBRY, M.-P. & HARDENBOL, J. (eds) *Geochronology Time Scales and Global Stratigraphic Correlation*. SEPM Special Publication, **54**, 129–212.

BOHATY, S. M. & ZACHOS, J. C. 2003. Significant Southern Ocean warming event in the late middle Eocene. *Geology*, **31**, 1017–1020.

BOWLER, J. M. 1976. Aridity in Australia: Age, origins and expression in aeolian landforms and sediments. *Earth-Science Reviews*, **12**, 279–310.

BOWLER, J. M. 1981. Australian salt lakes—A palaeohydrologic approach. *Hydrobiologia*, **82**, 431–444.

BOWLER, J. M. 1982. Aridity in the Tertiary and Quaternary of Australia. *In*: BARKER, W. R. & GREENSLADE, P. J. M. (eds) *Evolution of the Flora and Fauna of Arid Australia*. Peacock Publications, Adelaide, 35–46.

BOWLER, J. M. 1986. Spatial variability and hydrologic evolution of Australian lake basins: Analogue for Pleistocene hydrologic change and evaporite formation. *Palaeogeography, Palaeoclimatology, Palaeoecology*, **54**, 21–41.

BOWLER, J. M., WYRWOLL, K.-H. & LU, Y. 2001. Variations of the northwest Australian summer monsoon over the last 300 000 years: The paleohydrological record of the Gregory (Mulan) Lakes System. *Quaternary International*, **83–85**, 63–80.

BOWLER, J. M., KOTSONIS, A. & LAWRENCE, C. R. 2006. Environmental evolution of the Mallee region, western

Murray Basin. *Proceedings of the Royal Society of Victoria*, **118**, 161–210.

BROOKFIELD, M. 1970. Dune trends and wind regime in Central Australia. *Zeitschrift für Geomorphologie Supplementband*, **10**, 121–153.

BUREAU OF METEOROLOGY 1979. *Climatic Atlas of Australia*. Map set 8, Wind roses. Department of Science and the Environment, Australian Government Publishing Service, Canberra.

BUTLER, B. E. 1956. Parna: an aeolian clay. *Australian Journal of Science*, **18**, 145–151.

BYRNE, M., YEATES, D. K. ET AL. 2008. Birth of a biome: insights into the assembly and maintenance of the Australian arid zone biota. *Molecular Ecology*, **17**, 4398–4417.

CALLEN, R. A. & BENBOW, M. C. 1995. The deserts—playas, dunefields and watercourses. *In*: DREXEL, J. F. & PREISS, W. V. (eds) *The Geology of South Australia, Vol. 2: The Phanerozoic*. Geological Survey of South Australia Bulletin, **54**, 244–251.

CALLEN, R. A., ALLEY, N. F. & GREENWOOD, D. R. 1995. Lake Eyre Basin. *In*: DREXEL, J. F. & PREISS, W. V. (eds) *The Geology of South Australia, Vol. 2: The Phanerozoic*. Geological Survey of South Australia Bulletin, **54**, 188–194.

CANDE, S. C. & KENT, D. V. 1995. Revised calibration of the geomagnetic polarity timescale for the Late Cretaceous and Cenozoic. *Journal of Geophysical Research*, **100**, 6093–6095.

CHAPPELL, J. 1983. Thresholds and lags in geomorphologic changes. *Australian Geographer*, **15**, 357–366.

CHARTRES, C. J., CHIVAS, A. R. & WALKER, P. H. 1988. The effect of aeolian accessions on soil development on granitic rocks in south-eastern Australia. II. Oxygen-isotope, mineralogical and geochemical evidence for aeolian deposition. *Australian Journal of Soil Research*, **26**, 17–31.

CHEN, X. Y. & BARTON, C. E. 1991. Onset of aridity and dune-building in central Australia: sedimentological and magnetostratigraphic evidence from Lake Amadeus. *Palaeogeography, Palaeoclimatology, Palaeoecology*, **84**, 55–73.

CHEN, X. Y., PRESCOTT, J. R. & HUTTON, J. T. 1990. Thermoluminescence dating on gypseous dunes of Lake Amadeus, central Australia. *Australian Journal of Earth Sciences*, **37**, 93–101.

CHEN, X. Y., CHAPPELL, J. & MURRAY, A. S. 1995. High (ground)water levels and dune development in central Australia: TL dates from gypsum and quartz dunes around Lake Lewis (Napperby), Northern Territory. *Geomorphology*, **11**, 311–322.

CHIVAS, A. R., DE DECKKER, P., NIND, M., THIRIET, D. & WATSON, G. 1986. The Pleistocene palaeoenvironmental record of Lake Buchanan: an atypical Australian playa. *Palaeogeography, Palaeoclimatology, Palaeoecology*, **54**, 131–152.

COLHOUN, E. A., KIERNAN, K., BARROWS, T. T. & GOEDE, A. 2010. Advances in Quaternary studies in Tasmania. *In*: BISHOP, P. & PILLANS, B. (eds) *Australian Landscapes*. Geological Society, London, Special Publications, **346**, 165–183.

CRISP, M. D. & COOK, L. G. 2007. A congruent molecular signature of vicariance across multiple plant lineages.

Molecular Phylogenetics and Evolution, **43**, 1106–1117.

CROKE, J. C., MAGEE, J. W. & PRICE, D. M. 1998. Stratigraphy and sedimentology of the lower Neales River, West Lake Eyre, Central Australia: from Palaeocene to Holocene. *Palaeogeography, Palaeoclimatology, Palaeoecology*, **144**, 331–350.

DUNKERLEY, D. 2002. Systematic variation of soil infiltration rates within and between the components of the vegetation mosaic in an Australian desert landscape. *Hydrological Processes*, **16**, 119–131.

ENGLISH, P. 2001. *Cainozoic evolution and hydrogeology of Lake Lewis Basin, central Australia*. PhD thesis, Australian National University, Canberra.

ENGLISH, P., SPOONER, N. A., CHAPPELL, J., QUESTIAUX, D. G. & HILL, N. G. 2001. Lake Lewis basin, central Australia: environmental evolution and OSL chronology. *Quaternary International*, **83–85**, 81–101.

FISHER, A. G. 2003. *The Geomorphic Evolution of Australian Stony Deserts: An Application of* In-situ-*produced Cosmogenic Radionuclide Analysis at the Fowlers Gap Arid Zone Research Station*. PhD thesis, University of New South Wales, Sydney.

FITZSIMMONS, K. E., RHODES, E. J., MAGEE, J. W. & BARROWS, T. T. 2007. The timing of linear dune activity in the Strzelecki and Tirari Deserts, Australia. *Quaternary Science Reviews*, **26**, 2598–2616.

FLOWER, B. P. & KENNETT, J. P. 1995. Middle Miocene deepwater paleoceanography in the southwest Pacific: Relations with East Antarctic Ice Sheet development. *Paleoceanography*, **10**, 1095–1112.

FUJIOKA, T. 2007. *Development of* In situ *Cosmogenic ^{21}Ne Exposure Dating, and Dating of Australian Arid Landforms by Combined Stable and Radioactive* In situ *Cosmogenic Nuclides*. PhD thesis, Australian National University, Canberra.

FUJIOKA, T., CHAPPELL, J., HONDA, M., YATSEVICH, I., FIFIELD, L. K. & FABEL, D. 2005. Global cooling initiated stony deserts in central Australia 2–4 Ma, dated by cosmogenic ^{21}Ne–^{10}Be. *Geology*, **33**, 993–996.

FUJIOKA, T., CHAPPELL, J., FIFIELD, L. K. & RHODES, E. J. 2009. Australian desert dune fields initiated with Pliocene–Pleistocene global climatic shift. *Geology*, **37**, 51–54.

GARDNER, G. J., MORTLOCK, A. J., PRICE, D. M., READHEAD, M. L. & WASSON, R. J. 1987. Thermoluminescence and radiocarbon dating of Australian desert dunes. *Australian Journal of Earth Sciences*, **34**, 343–357.

GOSSE, J. C. & PHILLIPS, F. M. 2001. Terrestrial *in situ* cosmogenic nuclides: theory and application. *Quaternary Science Reviews*, **20**, 1475–1560.

GRANGER, D. E. & MUZIKAR, P. F. 2001. Dating sediment burial with *in situ*-produced cosmogenic nuclides: theory, techniques, and limitations. *Earth and Planetary Science Letters*, **188**, 269–281.

HESSE, P. P. 1994. The record of continental dust from Australia in Tasman Sea sediments. *Quaternary Science Reviews*, **13**, 257–272.

HESSE, P. P. & MCTAINSH, G. H. 2003. Australian dust deposits: modern processes and the Quaternary record. *Quaternary Science Reviews*, **22**, 2007–2035.

HESSE, P. P., MAGEE, J. W. & VAN DER KAARS, S. 2004. Late Quaternary climates of the Australian arid zone: a review. *Quaternary International*, **118–119**, 87–102.

HILL, R. S. 1994. *The History of Australian Vegetation and Flora: Cretaceous to Recent*. Cambridge University Press, Cambridge.

HOCKNULL, S. A., ZHAO, J.-X., FENG, Y.-X. & WEBB, G. E. 2007. Responses of Quaternary rainforest vertebrates to climate change in Australia. *Earth and Planetary Science Letters*, **264**, 317–331.

HOLLANDS, C. B., NANSON, G. C., JONES, B. G., BRISTOW, C. S., PRICE, D. M. & PIETSCH, T. J. 2006. Aeolian–fluvial interaction: evidence for Late Quaternary channel change and wind-rift linear dune formation in the northwestern Simpson Desert, Australia. *Quaternary Science Reviews*, **25**, 142–162.

HOPE, J. H. 1982. Late Cainozoic vertebrate faunas and the development of aridity in Australia. *In*: BARKER, W. R. & GREENSLADE, P. J. M. (eds) *Evolution of the Flora and Fauna of Arid Australia*. Peacock Publications, Adelaide, 85–100.

HOU, B., FRAKES, L. A., SANDIFORD, M., WORRALL, L., KEELING, J. & ALLEY, N. F. 2008. Cenozoic Eucla Basin and associated palaeovalleys, southern Australia—Climatic and tectonic influences on landscape evolution, sedimentation and heavy mineral accumulation. *Sedimentary Geology*, **203**, 112–130.

HUGALL, A. F., FOSTER, R., HUTCHINSON, M. N. & LEE, M. S. Y. 2008. Phylogeny of Australasian agamid lizards based on nuclear and mitochondrial genes: implications for morphological evolution and biogeography. *Biological Journal of the Linnean Society*, **93**, 343–358.

HUNTLEY, D. J. & PRESCOTT, J. R. 2001. Improved methodology and new thermoluminescence ages for the dune sequence in south–east South Australia. *Quaternary Science Reviews*, **20**, 687–699.

JENNINGS, J. N. 1968. A revised map of the desert dunes of Australia. *Australian Geographer*, **10**, 408–409.

KEARNEY, M., BLACKET, M. J., STRASBURG, J. L. & MORITZ, C. 2006. Waves of parthenogenesis in the desert: evidence for parallel loss of sex in a grasshopper and a gecko from Australia. *Molecular Ecology*, **15**, 1743–1748.

KEMP, E. M. 1978. Tertiary climatic evolution and vegetation history in the Southeast Indian Ocean region. *Palaeogeography, Palaeoclimatology, Palaeoecology*, **24**, 169–208.

KENNETT, J. P. & BARKER, P. F. 1990. Latest Cretaceous to Cenozoic climate and oceanographic developments in the Weddell Sea, Antarctica: an ocean-drilling perspective. *In*: BARKER, P. F., KENNETT, J. P. ET AL. (eds) *Proceedings of the Ocean Drilling Program, Scientific Results*, **113**, 937–960.

KERSHAW, A. P. 1986. Climate change and Aboriginal burning in north-east Australia during the last two glacial/interglacial cycles. *Nature*, **322**, 47–49.

KRIEG, G. W., CALLEN, R. A., GRAVESTOCK, D. I. & GATEHOUSE, C. G. 1990. Geology. *In*: TYLER, M. J., TWIDALE, C. R., DAVIES, M. & WELLS, C. B. (eds) *Natural History of the Northeast Deserts*. Occasional Publications of the Royal Society of South Australia, **6**, 1–26.

LISIECKI, L. E. & RAYMO, M. E. 2005. A Pliocene–Pleistocene stack of 57 globally distributed benthic δ^{18}O records. *Paleoceanography*, **20**, PA1003, doi: 10.1029/2004PA001071.

LIVERMORE, R., NAKIVELL, A., EAGLES, G. & MORRIS, P. 2005. Paleogene opening of Drake Passage. *Earth and Planetary Science Letters*, **236**, 459–470.

LOMAX, J., HILGERS, A., WOPFNER, H., GRÜN, R., TWIDALE, C. R. & RADTKE, U. 2003. The onset of dune formation in the Strzelecki Desert, South Australia. *Quaternary Science Reviews*, **22**, 1067–1076.

LONGMORE, M. E. & HEIJNIS, H. 1999. Aridity in Australia: Pleistocene records of palaeohydrological and palaeoecological change from the perched lake sediments of Fraser Island, Queensland, Australia. *Quaternary International*, **57/58**, 35–47.

MABBUTT, J. A. 1977. *An Introduction to Systematic Geomorphology, 2: Desert Landforms*. Australian National University Press, Canberra.

MABBUTT, J. A. 1988. Australian desert landscapes. *Geo-Journal*, **16.4**, 355–369.

MAGEE, J. W. & MILLER, G. H. 1998. Lake Eyre palaeohydrology from 60 ka to the present: beach ridges and glacial maximum aridity. *Palaeogeography, Palaeoclimatology, Palaeoecology*, **144**, 307–329.

MAGEE, J. W., BOWLER, J. M., MILLER, G. H. & WILLIAMS, D. L. G. 1995. Stratigraphy, sedimentology, chronology and palaeohydrology of Quaternary lacustrine deposits at Madigan Gulf, Lake Eyre, South Australia. *Palaeogeography, Palaeoclimatology, Palaeoecology*, **113**, 3–42.

MAGEE, J. W., MILLER, G. H., SPOONER, N. A. & QUESTIAUX, D. 2004. Continuous 150 k.y. monsoon record from Lake Eyre, Australia: Insolation-forcing implications and unexpected Holocene failure. *Geology*, **32**, 885–888.

MARTIN, H. A. 1977. The Tertiary stratigraphic palynology of the Murray Basin in New South Wales. I. The Hay–Balranald Wakool districts. *Royal Society of New South Wales, Journal and Proceedings*, **110**, 41–47.

MARTIN, H. A. 1990. The palynology of the Namba Formation in the Wooltana-1 bore, Callabonna Basin (Lake Frome), South Australia, and its relevance to Miocene grasslands in central Australia. *Alcheringa*, **14**, 247–255.

MARTIN, H. A. 2006. Cenozoic climatic change and the development of the arid vegetation in Australia. *Journal of Arid Environments*, **66**, 533–563.

MASLIN, M. A., LI, X. S., LOUTRE, M.-F. & BERGER, A. 1998. The contribution of orbital forcing to the progressive intensification of northern hemisphere glaciation. *Quaternary Science Reviews*, **17**, 411–426.

McELHINNY, M. W. & McFADDEN, P. L. 2000. *Paleomagnetism: Continents and Oceans*. Academic Press, London.

McGOWRAN, B., HOLDGATE, G. R., LI, Q. & GALLAGHER, S. J. 2004. Cenozoic stratigraphic succession in southeastern Australia. *Australian Journal of Earth Sciences*, **51**, 459–496.

McTAINSH, G. H. 1989. Quaternary aeolian dust processes and sediments in the Australian region. *Quaternary Science Reviews*, **8**, 235–253.

MILLER, K. G., WRIGHT, J. D. & FAIRBANKS, R. G. 1991. Unlocking the ice house: Oligocene–Miocene oxygen isotopes, eustasy, and margin erosion. *Journal of Geophysical Research*, **96**, 6829–6848.

NANSON, G. C., YOUNG, R. W., PRICE, D. M. & RUST, B. R. 1988. Stratigraphy, sedimentology and Late Quaternary chronology of the Channel Country of western Queensland. *In*: WARNER, R. F. (ed.) *Fluvial Geomorphology of Australia*. Academic Press, London, 151–175.

NANSON, G. C., PRICE, D. M., SHORT, S. A., PAGE, K. J. & NOTT, J. F. 1991. Major episodes of climate change in Australia over the last 300 000 years. *In*: GILLESPIE, R. (ed.) *Quaternary Dating Workshop 1990*. Department of Biogeography and Geomorphology, Australian National University, Canberra, 45–50.

NANSON, G. C., CHEN, X. Y. & PRICE, D. M. 1992a. Lateral migration, thermoluminescence chronology and colour variation of longitudinal dunes near Birdsville in the Simpson Desert, Central Australia. *Earth Surface Processes and Landforms*, **17**, 807–819.

NANSON, G. C., PRICE, D. M. & SHORT, S. A. 1992b. Wetting and drying of Australia over the past 300 ka. *Geology*, **20**, 791–794.

NANSON, G. C., EAST, T. J. & ROBERTS, R. G. 1993. Quaternary stratigraphy, geochronology and evolution of the Magela Creek catchment in the monsoon tropics of northern Australia. *Sedimentary Geology*, **83**, 277–302.

NANSON, G. C., CHEN, X. Y. & PRICE, D. M. 1995. Aeolian and fluvial evidence of changing climate and wind patterns during the past 100 ka in the western Simpson Desert, Australia. *Palaeogeography, Palaeoclimatology, Palaeoecology*, **113**, 87–102.

PEKAR, S. F. & DECONTO, R. M. 2006. High-resolution ice-volume estimates for the early Miocene: Evidence for a dynamic ice sheet in Antarctica. *Palaeogeography, Palaeoclimatology, Palaeoecology*, **231**, 101–109.

PEPPER, M., DOUGHTY, P. & KEOGH, J. S. 2006. Molecular phylogeny and phylogeography of the Australian *Diplodactylus stenodactylus* (Gekkota: Reptilia) species-group based on mitochondrial and nuclear genes reveals an ancient split between Pilbara and non-Pilbara *D. stenodactylus*. *Molecular Phylogenetics and Evolution*, **41**, 539–555.

PEPPER, M., DOUGHTY, P., ARCULUS, R. & KEOGH, J. S. 2008. Landforms predict phylogenetic structure on one of the world's most ancient surfaces. *BMC Evolutionary Biology*, **8**, 152.

PILLANS, B. & BOURMAN, R. 2001. Mid Pleistocene arid shift in southern Australia, dated by magnetostratigraphy. *Australian Journal of Soil Research*, **39**, 89–98.

POORE, R. Z. & SLOAN, L. C. 1996. Introduction: climates and climate variability of the Pliocene. *Marine Micropaleontology*, **27**, 1–2.

PRIDEAUX, G. J., ROBERTS, R. G., MEGIRIAN, D., WESTAWAY, K. E., HELLSTROM, J. C. & OLLEY, J. M. 2007. Mammalian responses to Pleistocene climate change in southeastern Australia. *Geology*, **35**, 33–36.

PROTHERO, D. R. 1994. *The Eocene–Oligocene transition: Paradise Lost*. Columbia University Press, New York.

READHEAD, M. L. 1988. Thermoluminescence dating study of quartz in aeolian sediments from southeastern Australia. *Quaternary Science Reviews*, **7**, 257–264.

READHEAD, M. L. 1991. Thermoluminescence dating of sediments from Lake Mungo and Nyah West. *In*: GILLESPIE, R. (ed.) *Quaternary Dating Workshop*

1990. Department of Biogeography and Geomorphology, Australian National University, Canberra, 35–37.

SHEARD, M. J., LINTERN, M. J., PRESCOTT, J. R. & HUNTLEY, D. J. 2006. Great Victoria Desert: new dates for South Australia's ?oldest desert dune system. *MESA Journal*, **42**, 15–26.

SHOO, L. P., ROSE, R., DOUGHTY, P., AUSTIN, J. J. & MELVILLE, J. 2008. Diversification patterns of pebble-mimic dragons are consistent with historical disruption of important habitat corridors in arid Australia. *Molecular Phylogenetics and Evolution*, **48**, 528–542.

SIMON-COINÇON, R., MILNES, A. R., THIRY, M. & WRIGHT, M. J. 1996. Evolution of landscapes in northern South Australia in relation to the distribution and formation of silcretes. *Journal of the Geological Society, London*, **153**, 467–480.

SINGH, G., OPDYKE, N. D. & BOWLER, J. M. 1981. Late Cainozoic stratigraphy, paleomagnetic chronology and vegetational history from Lake George, N.S.W. *Journal of the Geological Society of Australia*, **28**, 435–452.

SLUITER, I. R. K. 1991. Early Tertiary vegetation and climates, Lake Eyre region, northeastern South Australia. *In*: WILLIAMS, M. A. J., DE DECKKER, P. & KERSHAW, A. P. (eds) *The Cainozoic in Australia: a Re-appraisal of the Evidence*. Geological Society of Australia Special Publication, **18**, 99–166.

SMITH, M. L., PILLANS, B. J. & MCQUEEN, K. G. 2009. Paleomagnetic evidence for periods of intense oxidative weathering, McKinnons Mine, Cobar, New South Wales. *Australian Journal of Earth Sciences*, **56**, 201–212.

STEPHENSON, A. E. 1986. Lake Bungunnia—A Plio-Pleistocene megalake in southern Australia. *Palaeogeography, Palaeoclimatology, Palaeoecology*, **57**, 137–156.

STONE, A. E. C. & THOMAS, D. S. G. 2008. Linear dune accumulation chronologies from the southwest Kalahari, Namibia: challenges of reconstructing late Quaternary palaeoenvironments from aeolian landforms. *Quaternary Science Reviews*, **27**, 1667–1681.

THIEDE, J. & VORREN, T. O. 1994. The Arctic Ocean and its geologic record: Research history and perspectives. *Marine Geology*, **119**, 179–184.

TWIDALE, C. R., PRESCOTT, J. R., BOURNE, J. A. & WILLIAMS, F. M. 2001. Age of desert dunes near Birdsville, southwest Queensland. *Quaternary Science Reviews*, **20**, 1355–1364.

VEEVERS, J. J., POWELL, C. M. & ROOTS, S. R. 1991. Review of seafloor spreading around Australia. I. Synthesis of the patterns of spreading. *Australian Journal of Earth Sciences*, **38**, 373–389.

WASSON, R. J. 1983. The Cainozoic history of the Strzelecki and Simpson dunefields (Australia), and the origin of the desert dunes. *Zeitschrift für Geomorphologie Supplementband*, **45**, 85–115.

WASSON, R. J., FITCHETT, K., MACKEY, B. & HYDE, R. 1988. Large-scale patterns of dune type, spacing and orientation in the Australian continental dunefield. *Australian Geographer*, **19**, 89–104.

WHITE, M. E. 1994. *After the Greening, the Browning of Australia*. Kangaroo Press, Kenthurst, NSW.

WOPFNER, H. 1974. Post-Eocene history and stratigraphy of northeastern South Australia. *Transactions of the Royal Society of South Australia*, **98**, 1–12.

WOPFNER, H. 1978. Silcretes of northern South Australia and adjacent regions. *In*: LANGFORD-SMITH, T. (ed.) *Silcrete in Australia*. Department of Geography, University of New England, Armidale, NSW, 93–142.

WOPFNER, H. & TWIDALE, C. R. 1967. Geomorphological history of the Lake Eyre Basin. *In*: JENNINGS, J. N. & MABBUTT, J. A. (eds) *Landform Studies from Australia and New Guinea*. Australian National University Press, Canberra, 118–143.

ZACHOS, J. C. & KUMP, L. R. 2005. Carbon cycle feedbacks and the initiation of Antarctic glaciation in the earliest Oligocene. *Global and Planetary Change*, **47**, 51–66.

ZACHOS, J. C., LOHMANN, K. C., WALKER, J. C. G. & WISE, S. W. 1993. Abrupt climate change and transient climates during the Paleogene: a marine perspective. *Journal of Geology*, **101**, 191–213.

ZACHOS, J. C., STOTT, L. D. & LOHMANN, K. C. 1994. Evolution of early Cenozoic marine temperatures. *Paleoceanography*, **9**, 353–387.

ZACHOS, J., PAGANI, M., SLOAN, L., THOMAS, E. & BILLUPS, K. 2001. Trends, rhythms, and aberrations in global climate 65 Ma to present. *Science*, **292**, 686–693.

ZHENG, H., WYRWOLL, K.-H., LI, Z. & POWELL, C. M. 1998. Onset of aridity in southern Western Australia – a preliminary palaeomagnetic appraisal. *Global and Planetary Change*, **18**, 175–187.

The Australian desert dunefields: formation and evolution in an old, flat, dry continent

PAUL P. HESSE

Department of Environment and Geography, Macquarie University, Sydney, NSW 2109, Australia (e-mail: Paul.Hesse@mq.edu.au)

Abstract: A new map, the first based on interpretation of satellite imagery, reveals both the complexity of Australia's dunefields and their relationships with topography, climate and substrate. Of the five main sand seas, the Mallee, Strzelecki and Simpson in eastern Australia cover Quaternary sedimentary basins whereas the Great Victoria and Great Sandy dunefields in the west are formed by reworking of valley and piedmont sediments in a non-basinal landscape of low-relief ridge and valley topography. These dunefields cover large areas of the arid zone and semi-arid zone and small areas of dunes in sub-humid areas around the margins of the continent reflecting past expansion of arid climates during glacial stages of the last several glacial cycles. Several areas of low relief stand out as being largely dune-free: the limestone Nullarbor Plain, clay plains of the Georgina Basin and floodplains of rivers in the Carpentaria, Lake Eyre and Murray–Darling drainage basins where sand is rare or not transported by diminished Late Quaternary rivers. The Yilgarn Block of southwestern Australia is also surprisingly free of dunes, possibly as a result of long, deep weathering. Everywhere the history of climate change is evident in dune morphology and distribution, including large areas where the sand dune orientations are markedly divergent from modern sand moving wind directions.

The Australian desert dunefields are at once generally well understood and poorly known in detail. The dunefields were the least receptive part of the Australian continent to European exploration and settlement and remain comparatively undisturbed today. Although there is a recent resurgence in dating of dunes in the Simpson and Strzelecki Deserts (e.g. Fitzsimmons *et al.* 2007), large parts of the western dunefields remain almost unknown in terms of their aeolian geomorphology and chronology.

Scientific investigation of the dunefields lagged that of most other Australian landscapes. Sturt reached and crossed parts of the Strzelecki Desert dunefield in 1845 and saw in them no water and little of interest (Sturt 1849). Giles (1889) was turned back by dunefields of the Gibson Desert in 1873 and 1874 before crossing the continent through the better vegetated dunes of the Great Victoria Desert in 1875, and Gregory (1906) recommended against the attempt to extend agriculture in the 'dead heart' of the Simpson Desert in 1901, by which time the first wave of cattle grazing was already failing. The isolation continued into the twentieth century such that the first rough map of the continental dunefield was sketched by Madigan only in 1936, based on early aerial surveys and arduous ground surveys (Madigan 1936). The first accurate map of the continent-wide dunefield was not published until 1968 by Jennings (1968), although the published map scale was around 1:22 500 000 and derived from the first

national topographic mapping series at 1:250 000 scale. Nevertheless, King (1960) and Wopfner & Twidale (1967) had already published high-quality maps of the South Australian dunefields based on aerial photography. Subsequent original maps have all been based either on aerial photography at a regional scale (Bowler & Magee 1978; Wasson *et al.* 1988) or on the continental 1:250 000 topographic map series, which, in turn, was derived from the aerial photography (Wasson *et al.* 1988).

What were clear to the first explorers, but not fully expounded until Mabbutt (1986), were the fundamental factors governing the distribution of the dunefields: topography and climate. The dunefields occupy the lowlands and basins of arid Australia (Wopfner & Twidale 1967), separated by dune-free landscapes varying from steep ranges to stark, gently elevated gibber plains. European exploration and settlement followed the uplands, where water was available in gorges and rocky waterholes, and avoided the dunefields, where creeks and rivers died out into the sand. This reflects the pattern of aboriginal occupation over interstadial arrival and dispersal, glacial refuge and Holocene re-expansion (Smith 1989). Dunefields were also identifiable as the characteristic landform of the arid inland, rather than the humid fringes of the island continent where rainfall has a concentric pattern reaching its lowest mean value near Lake Eyre (Gentilli 1986).

Nevertheless, some characteristics of the dunefields have escaped much evaluation, at least at a

From: Bishop, P. & Pillans, B. (eds) *Australian Landscapes*. Geological Society, London, Special Publications,
346, 141–164.   DOI: 10.1144/SP346.9   0305-8719/10/$15.00 © The Geological Society of London 2010.

continental scale, possibly because a comprehensive continental-scale analysis of the context and formative factors has not been attempted. Why, for example, are some lowland areas of the arid zone (or within the range of known dunes) so devoid of dunes? Although there has been a vigorous debate about the mode of formation of the dunes, the distance sand has been transported in the dunes and even the provenance of the quartz, there is little synthesis and there are many unanswered questions. These issues are bound together but each is highly contentious. King (1960) and, more recently, Hollands *et al.* (2006) proposed wind rifting of a sandy substrate as the process by which the extensive longitudinal dunes of the Australian dunefields were formed. In support of their argument Hollands *et al.* (2006) gave evidence pointing to the lack of dune extension and in favour or vertical growth. However, a diametrically opposed model, of dune extension, has been proposed by others (Wopfner & Twidale 1967; Twidale 1972) based on the increasing organization of dunes away from deflation basins, observed growth of downwind noses and (unpublished) seismic surveys of internal dune structure. Geochemical fingerprinting of quartz and zircon sand grains was used by Pell *et al.* (1999, 2001), and later Reid & Hou (2006), to argue against inter-basin transport of sand in the dunefields and, therefore, a limited role for downwind sand transport. Both these studies identified the provenance of sand grains, pointing both to aeolian as well as fluvial transport pathways in a complex, multi-stage, evolution of the modern dunefields.

In addition, recognition of dunes for a long time was biased against the stable, well-vegetated landforms that characterize the Australian dunefields. Sturt (1849, p. 247) was convinced that the parallel sand ridges were far too regular and extensive to have been formed by wind and that they were water-formed, which happened to agree with his hypothesis of a vanished inland sea. He did not name the comparatively well-vegetated Simpson or Strzelecki dunefields he discovered or refer to them as deserts, naming only the dune-free gibber plain between them the Stony Desert. Other aspects of the dunes encouraged a dismissive attitude. Gregory (1906, p. 65) wrote of the dunes near Lake Eyre that:

> the supposed sand-hills were not real dunes, but ridges of loam, with banks of sand along the summits. Instead of the hills being composed of heaps of sand from thirty to fifty feet in thickness, the sand was only some three to five feet thick on the crest, with thinner layers on each flank of the ridges.

A survey of the Barkly Tablelands (Stewart 1954) called low, vegetated dune ridges 'Tertiary plain with aeolian limestone', presumably because of

their substantial pedogenesis and low relief. The map of Wasson *et al.* (1988) was remarkable for its recognition of all dunes, no matter their apparent activity, including those in the more humid fringes of the continent. The same map, unfortunately, did not categorize the basic type or morphology of up to 20% of dunes, leaving them as 'irregular' or 'confused' (Wasson 1986), and adds to some confusion about the extent of the dunefield, especially in the SW of Western Australia, because of the methods used.

A further factor identified in the formation of the Australian dunefields is climate change; in other words, the variable extent of arid areas during the glacial cycles. At least going back to King (1960) the prior greater activity and initiation of dune activity has been accepted, even though the exact age has not. The clear evidence of longitudinal dunes submerged beneath the sea in northwestern Australia (Jennings 1975), the sometimes dense vegetation on the dunes, noted as early as 1875 by Giles, who 'tunnelled' his way through the mallee eucalypts of the Great Victoria Desert (Giles 1889), and the restricted evidence of contemporary sand movement (Wopfner & Twidale 1967) all point to prior greater dune activity. The emergence of luminescence dating methods (Gardner *et al.* 1987; Readhead 1988) helped settle some resistance to accepting the Pleistocene age of the dunes (see Wopfner & Twidale 1988; Twidale *et al.* 2001) and suggested a strong association between dune formation and glacial stages (Hesse *et al.* 2004) of the last few glacial cycles. Recent intensive dating of dunes in the Strzelecki Desert (Fitzsimmons *et al.* 2007) has shown a more finely structured age relationship, with less clear climatic interpretation, and luminescence dates have pushed back the age of earliest formation of the dunefields to the mid-Pleistocene (Nanson *et al.* 1992; Sheard *et al.* 2006). Cosmogenic nuclide burial ages on Simpson Desert dunes (Fujioka *et al.* 2009) indicate basal ages over 1 Ma.

This paper takes a continent-wide view of the Australian dunefields and the principal factors determining their distribution and character: topography, climate and lithology. In addition, the evolution of the dunefield is considered in relation to the well-known history of Quaternary climate changes but also the evolution of both topography and lithological controls over this period. This new assessment is based on the first continent-scale mapping of the dunefield based on satellite imagery.

Methods

Dunes and dunefields were mapped from three-band (green, near-IR, mid-IR) Landsat TM images

(GeoCover; available from https://zulu.ssc.nasa.gov/mrsid/). Mosaics of georeferenced and enhanced images cover 5° of latitude and 6° longitude, corresponding to Universal Transverse Mercator (UTM) zones. Initial mapping was undertaken using the available 'c. 1990' imagery with pixel size 28.5 m. TIFF images were imported to Illustrator graphics software for interpretation and mapping, sometimes with images coarsened to 57 m pixel size to limit file size. A second stage of more detailed mapping was undertaken after refinement of an empirical morphological mapping scheme using 'c. 2000' imagery with pixel size 14.25 m, sometimes coarsened to 28.5 m. Actual on-screen scale of the imagery while mapping was between 1:50 000 and 1:100 000. A total of 24 images were mapped in whole or part and then combined to produce a continental-scale map. The map excludes the largely dune-free far north (Kimberley and Arnhem Land) and east coast and highlands, including Tasmania. Small areas of non-coastal dunes have been described in these areas (Bowden 1983; Nott & Price 1991; Thom et al. 1994; Hesse et al. 2003), and more are probably yet undescribed, but they add little to the total picture described here.

The availability of free imagery came at the cost of control over spectral bands and image enhancement. This was sometimes a problem where dunes were low, indistinct and/or thickly vegetated. These conditions occurred most commonly in the dunefield margins of southeastern and southwestern Australia. The well-described palaeo-channel source-bordering dunes of the Riverine Plain (Butler et al. 1973), for example, were largely indistinguishable because of thick vegetation. Disturbed and cropped vegetation also worked to conceal the presence of low dunes, whereas native vegetation mostly helped to accentuate them. An exception is in the northern dunefields where fire scars in spinifex (Triodia spp.) produced complex patterns completely independent of the dune morphology. A peculiarity of the imagery is that it is mostly taken from particularly wet years and therefore waterways, lakes and pans are particularly clear in many images but vegetation is sometimes unusually lush (Hesse & Simpson 2006).

To overcome the limitations of scale (resolution) and 2D aerial view, the original 14.25 m pixel 'c. 2000' Landsat TM images were consulted regularly and SPOT imagery in Google Earth, where available, was consulted for its much higher resolution. In limited areas of the Mallee and northwestern NSW dunefields 1:50 000 aerial photographs were available. Only on the SPOT imagery and aerial photographs were details of superimposed crest features (small transverse or barchanoid dunes) visible. In addition, mapping was compared with a range of papers, topographical maps, geological maps, soil surveys and geomorphological maps where available. During the term of this project ground-truthing was conducted opportunistically in the Mallee, northwestern NSW, Strzelecki and Simpson dunefields (Hesse & Simpson 2006).

Mapped attributes included extent (i.e. presence or absence) of dunes and dunefields, dune orientation and dune morphology. Dunefields and sand seas (i.e. groups of dunes without breaks greater than dune wavelength) (Cooke et al. 1993) and single dunes outside the limits of the dunefields were mapped. Orientation (i.e. relative to formative wind direction) was determined from the direction of slip faces, where present, but these are rare in the Australian dunefield and at this scale. Most commonly the orientation of Y-junctions was taken as indicative of orientation of longitudinal dunes, following accepted practice and understanding (Wasson et al. 1988; Werner 1995). This assumption obviates the debate over the terminology of longitudinal or linear dunes (Cooke et al. 1993; Livingstone et al. 2007). Basic dune morphological types were mapped, initially following the classification of McKee (1979) and the hierarchical structure of Cooke et al. (1993) as well as local terminology (Wasson et al. 1988). However, all of these were inadequate for the task of describing the morphological variations found and a refinement of these schemes (essentially, several layers of additional morphological characteristics) was developed but will be described in detail in a subsequent paper. The additional properties described include the relative frequency of dune junctions and terminations (Mabbutt & Wooding 1983; Werner 1995; Bullard et al. 1996), variations on crest morphology and complexity (Wasson et al. 1988) and relative dune width and distinctness. In the absence of a reliable technique, the activity of the dunefields was not mapped. In general terms, areas of bare sand are uncommon but the level of sand movement on less vegetated dunes is highly dependent on interannual and decadal rainfall patterns (Hesse & Simpson 2006).

Results

Continental distribution and extent of the dunefields

The main areas of dunes lie in a diagonal band across the continent from the Mallee dunefield in the SE to the Great Sandy Desert in the NW (Fig. 1). These two dunefields with the Strzelecki Desert, Simpson Desert and Great Victoria Desert dunefields constitute the largest distinct sand seas (using the terminology of Cooke et al. 1993).

In addition there are many smaller dunefields surrounding these sand seas and isolated dunes over an even broader area.

Not all the named Australian deserts describe dunefields (e.g. Tanami (largely dune-free), Gibson (complex mosaic of dunes and ranges)) and not all dunefields have common or widely used names. For example, the southeastwards extension of the Great Victoria Desert dunefield into the Eyre Peninsula all the way to Spencer Gulf is not normally included in the Great Victoria Desert but has no acknowledged name and is large enough to constitute a sand sea in its own terms. Twidale (2008) has used the name Crocker Dunefield to describe both this area as well as dunes on Yorke Peninsula, Kangaroo Island, the Adelaide plain and the Mallee, which are all independent dunefields (now and in the past) in the sense of Cooke *et al.* (1993), and unhelpfully changed a widely accepted dunefield name in the process. In this study the dunefield of the Eyre Peninsula is referred to as the Eyre Dunefield, the large dunefields in the coastal hinterland of Western Australia are termed the Gascoyne Dunefield and the large dunefields of the Northern Territory are called the Wiso Dunefield and Barkly Dunefield, following Jennings & Mabbutt (1986). Those workers also gave names to other smaller dunefields of central and northern Australia but most have gained no acceptance, and therefore the dunefields remain effectively nameless, and the Jennings & Mabbutt names have not been followed here. There is renewed interest in the Jennings & Mabbutt physiographic regions as a layer within the Australian Soil Resources Information System (Pain 2008), which may popularize some names. Unfortunately, an alternative scheme, biogeographical regions (bioregions), which is widely used in conservation management in Australia, uses both different boundaries and different names despite being, at least partly, based on geomorphology (Morgan & Terrey 1992).

Compared with the best previous continental-scale map, that of Wasson *et al.* (1988), the new

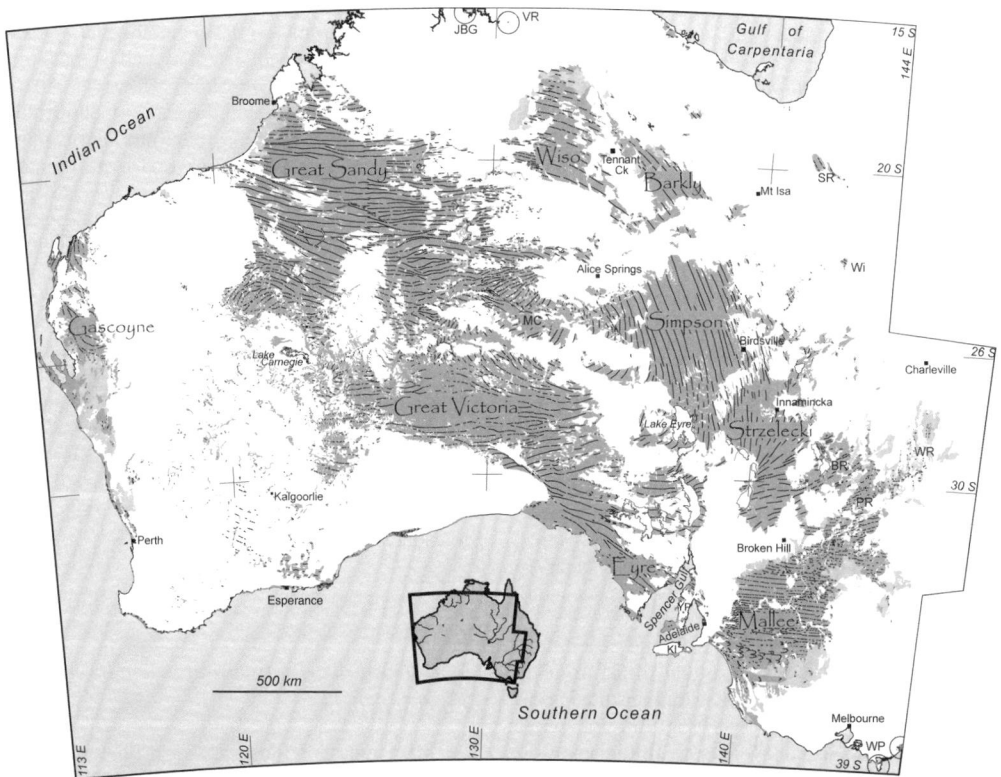

Fig. 1. The Australian dunefields mapped from Landsat TM imagery. Light grey areas are sand plains; medium grey areas are dunefields. Representative longitudinal and parabolic dune orientations are shown by black lines. The names of the principal dunefields used in this paper are shown. Locations and dunefields referred to in the text: JBG, Joseph Bonaparte Gulf; VR, Victoria River; SR, Saxby River; Wi, Winton; MC, Mount Connor; BR, Bulloo River; PR, Paroo River; WR, Warrego River; YP, Yorke Peninsula; KI, Kangaroo Island; WP, Wilsons Promontory.

map shows the same gross distribution of dunes in all areas except, glaringly, the SW of Western Australia. The Wasson *et al.* map was created in a fundamentally different way, using different sources. In their map the presence of dunes is shown only by arrows indicating the direction of dune elongation but is based on the presence or absence of dunes in a 0.25° × 0.25° grid. Although most of the interior of Western Australia was mapped from 1:250 000 topographic maps, the SW of WA was mapped using 1:250 000 photo mosaics or larger-scale air photographs in limited areas (Wasson *et al.* 1988). In addition to longitudinal sand dunes, Wasson *et al.* also mapped lunette dunes associated with the thousands of small salt lakes now found along the remnant palaeo-drainage. Satellite imagery shows a scattering of small longitudinal and parabolic dunes and many small lunettes over the area (Fig. 1), commonly associated with the remnant palaeo-drainage lines. The type of dune, different mode of representation and different sources must be borne in mind when interpreting both maps. On the eastern fringe, some small areas of dunes on Wilson's Promontory at the southernmost point of the continent (Hill & Bowler 1995), larger areas of dunes and sandsheet in the Paroo and Warrego fans of New South Wales and Queensland (Hesse 2006, 2007), and small dunefields south of Winton and on the Saxby River in western Queensland have been mapped that were not mapped by Wasson *et al.* On the northern fringe, there are small fields of longitudinal dunes in the southern shore of the Gulf of Carpentaria, as well as in the valley of the Victoria River and Joseph Bonaparte Gulf in the NW.

The continental dunefields therefore extend from 15°S to 39°S (and further in Tasmania) and to both south and north coasts as well as to the west coast and east coast (Thom *et al.* 1994) of the continent. Nevertheless, the main dunefields fall within narrower limits than implied by those extremes and there are large areas near the centre of the continent in which no dunes are found even though they occur in surrounding areas. Wasson *et al.* (1988) emphasized the extent of the dunefields, covering '40 per cent of 15 minute quadrates'. However, as noted above, their mapping technique is inclined to give a result larger than the true area of the dunes. Although not quantified here, the area of the Australian dunefields is likely to be somewhat less than 40% of the continental area but they still form a very extensive landscape.

The influence of climate on the distribution and formation of the dunefields

The major dunefields lie within the arid and semi-arid climate zones, although some smaller dunefields occur in areas which are at present sub-humid or humid and densely vegetated (Hesse *et al.* 2003; Fig. 2). The broadly concentric pattern of rainfall distribution over the Australian continent is the product of summer monsoonal rainfall over the tropical north decreasing towards the south and winter westerly frontal rainfall in the south decreasing to the north. As is well recognized, Australia as a whole has low rainfall and extensive arid areas but does not at present experience hyper-aridity. The driest part of the continent, at the northern end of Lake Eyre, still averages over 100 mm of rain each year and is largely vegetated to a degree that is sufficient to stabilize dune crests. Vascular plant cover on dune crests of the eastern Simpson Desert and Strzelecki Desert ranges from 20 to 80% for 'average' conditions but can be as low as 12% for the less common case where small bare patches of sand with slip faces have formed (Hesse & Simpson 2006). In dry years ephemeral forbs die and bare sand is exposed but in most cases it requires some substantial disturbance, such as grazing pressure, to sufficiently reduce perennial plants and cyanobacterial crusts to a level such that substantial dune activation occurs. However, over much of the Mallee and Great Victoria dunefields there is a native vegetation cover of sparse small trees and shrubs that are relatively insensitive to inter-annual rainfall variability and there is no contemporary dune activity (Ash & Wasson 1983). In the Great Sandy, western Great Victoria, Gascoyne and Wiso dunefields fire-promoting spinifex (*Triodia* spp.) grass forms perennial clumps in combination with varying densities of shrubs and small trees but dune activity is again very low.

In the centre of the continent, rainfall in either season is possible but infrequent and extremely variable (van Etten 2009). The median annual (July–June) rainfall at Birdsville (summer dominated) is 133 mm but has varied from 14 mm in 1924 to 659 mm in 1917 (Bureau of Meteorology data). There is a strong multi-decadal cycle in climatic records (Fig. 3b) as well as inter-annual variability related to El Niño–Southern Oscillation (ENSO). At Birdsville, on the eastern margin of the Simpson Desert, there is a noticeable annual cycle of greatest sand drift potential (calculated after Kalma *et al.* 1988) in the spring (September–November) and weakest in autumn (April–June) as the resultant drift direction alternates from SW in winter, through southerly in spring, to SE in summer and back again. The strong spring winds are also associated with the highest frequency of dust storms (McTainsh *et al.* 1998; Ekstrom *et al.* 2004; Fig. 3a). Since 1960, when wind recording began at Birdsville, there have been several quasi-decadal cycles of stronger winds (greater drift

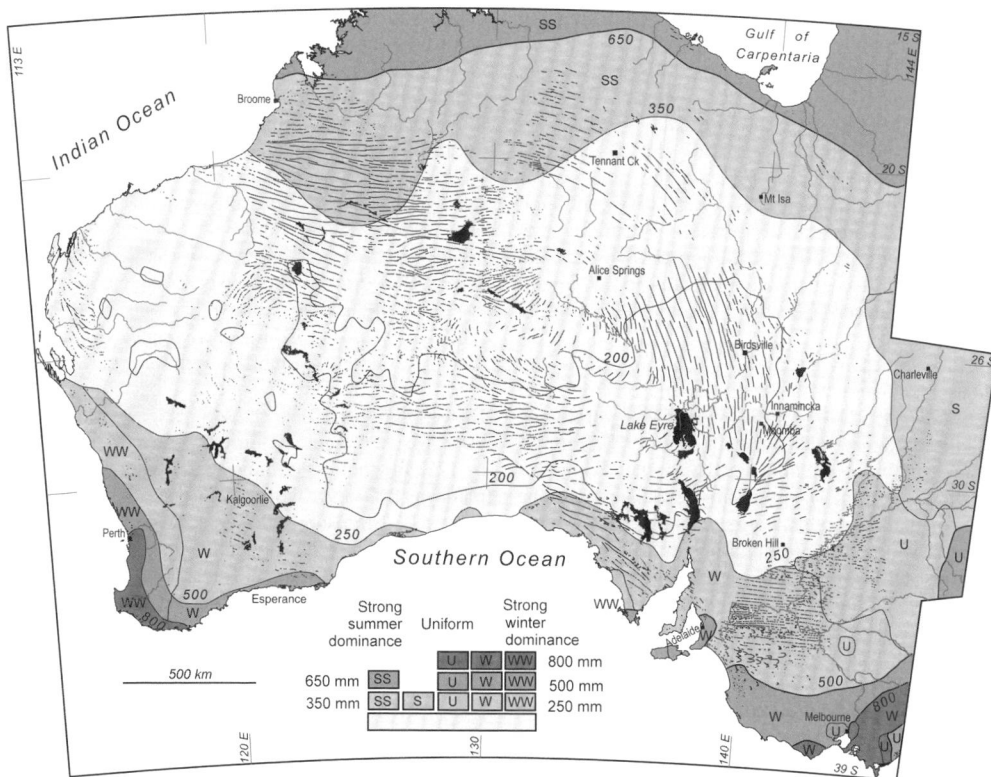

Fig. 2. Median annual rainfall and seasonality based on 100 year period from 1900 to 1999. Source: Bureau of Meteorology (Australia). Dunefields are represented by representative longitudinal and parabolic dune crests (black lines). Major salt lakes or playas are shown as black.

potential) with peaks in 1964, 1969, 1984 and 1997–1999 (Fig. 3d). These are broadly synchronous with positive excursions of the Interdecadal Pacific Oscillation (Fig. 3c), which modulates the impact of ENSO in Australia, more frequent dust storms and periods of lower rainfall (Fig. 3b). These intervals of stronger than average winds are also associated with a shift in the monthly resultant sand drift direction to the east, relative to the mean state. Conversely, intervals of weaker than average winds are associated with wind shifts to the west and higher than average rainfall. Therefore, under the current climate regime the major drought cycles are also the times of greatest windiness and sand drift potential. This pattern of circulation changes adds to the complexity of understanding dune orientations and sand transport through analysis of average, or short-term, wind data (Brookfield 1970).

The anticlockwise whorl of longitudinal dune orientations has long been recognized and its general similarity to the anticlockwise winds of the sub-tropical high-pressure system have been noted. Nevertheless, dune orientation is expected

to reflect the long-term resultant drift direction at any point and the subtropical ridge has a pronounced seasonal cycle of movement north and south over the southern half of the continent such that the dunefield does not represent the synoptic wind flow of any single period or season. The new mapping reveals more completely the pattern of dunes at the centre of the whorl at c. 26°S. There is a distinct double centre to the pattern of dunes; one centred over Lake Carnegie (26°S, 123°E) and the other NE of Mt Connor (25.5°S, 132.5°E). To the west, a complex centre of east–west convergence and north–south divergence of dune orientations occurs near the West Australian coast south of the Gascoyne River along the same latitude, and to the east there is a similar node centred over the Cooper Creek floodplain south of Windorah. Outside this central axis the dunes are oriented with great regularity and very little evidence of local intersection or overprinting.

Although Madigan (1936) wrote that the modern wind roses matched dune orientations 'exactly' (p. 208) later workers noted either variable

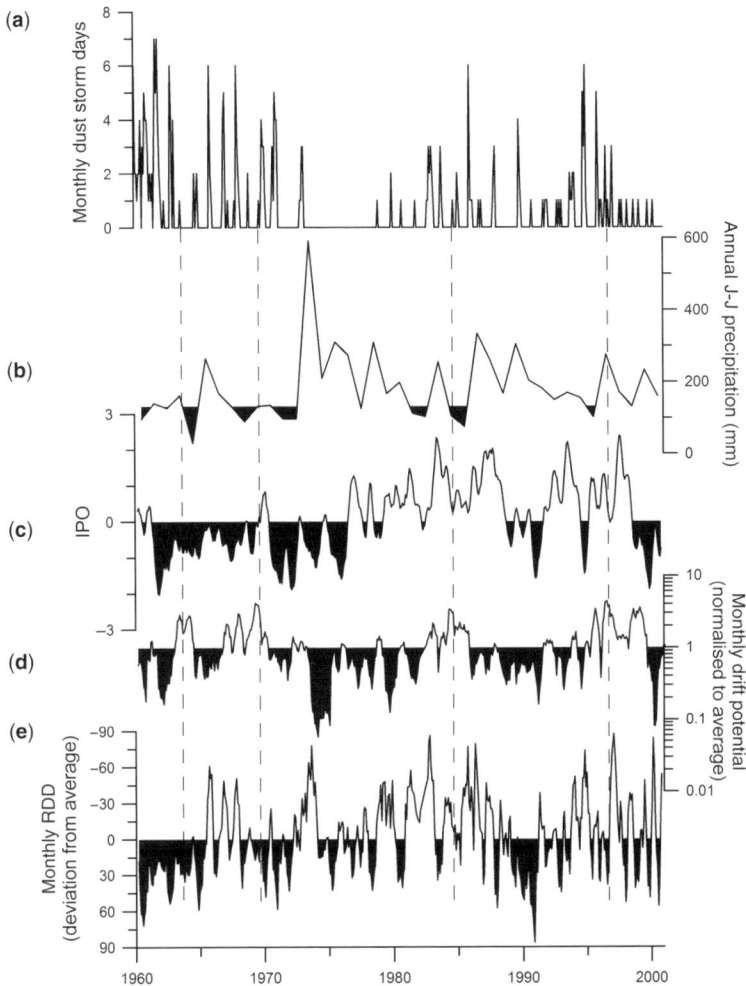

Fig. 3. Climate series for Birdsville, eastern Simpson Desert (data from Bureau of Meteorology, Australia). (**a**) Monthly frequency of dust storm days; (**b**) annual July–June precipitation; (**c**) inter-decadal Pacific Oscillation index (5 month moving average); (**d**) monthly sand drift potential, calculated following Kalma *et al.* (1988), normalized to the average for each month (5 month moving average); (**e**) monthly resultant sand drift direction (calculated following Kalma *et al.* 1988) as a deviation from the average for each month (5 month moving average). Negative values represent eastward deviation; positive values represent westward deviation.

deviations (Brookfield 1970) or consistent patterns of deviation of varying degree (Sprigg 1982; Nanson *et al.* 1995), possibly varying systematically across the continent. Comparison of the new dune map with the annual resultant drift direction (RDD) determined by Kalma *et al.* (1988) (Fig. 4) shows that there are regions in which there is close agreement and areas of divergence and even opposed wind and dune directions. There is reasonably close agreement (within 15°) between RDD and dune orientation in large parts of the Mallee, Strzelecki, Simpson, Barkly and Wiso dunefields.

However, dunes over most of the Great Victoria, Eyre, southwestern Mallee, Great Sandy and Gascoyne are oblique (up to 75° divergent) and there are areas of the Gascoyne, northern Great Sandy and along the axis of the dune whorl in central Australia where dunes are opposed or even reversed with respect to the modern RDD. Clearly, the divergences cannot be extrapolated from one region to another and require sophisticated explanations beyond simple latitudinal translation of the circulation pattern (see Harrison 1993; Hesse 1994).

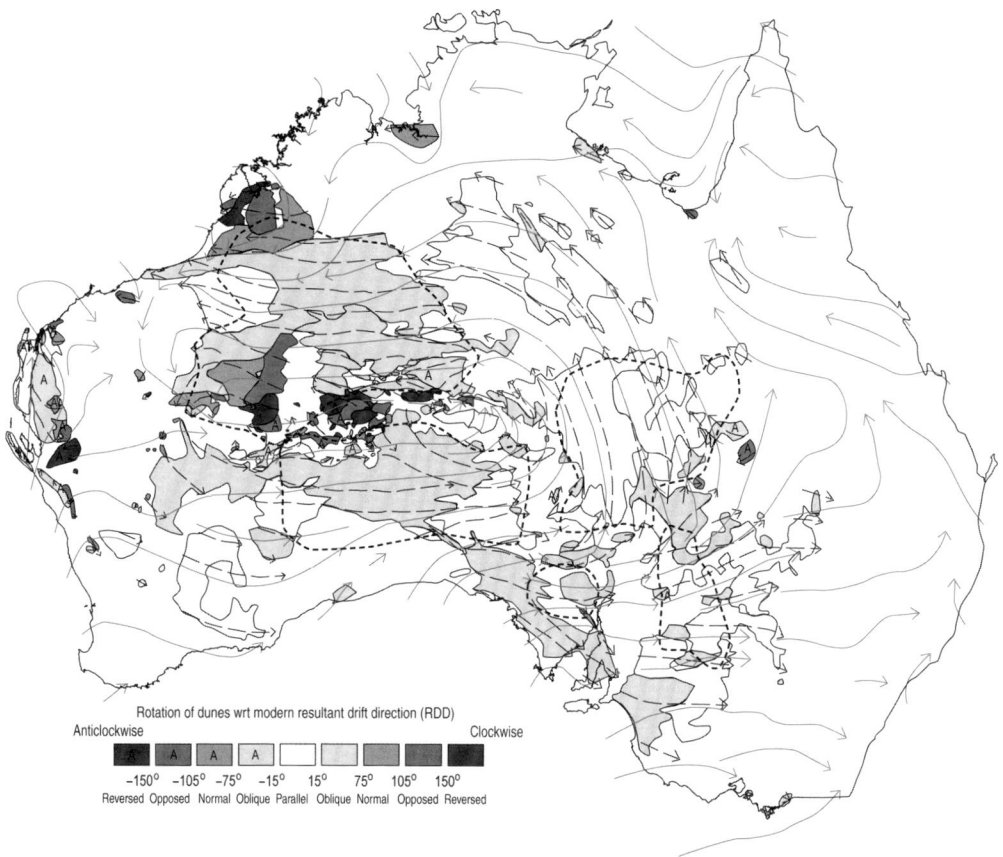

Fig. 4. Comparison of mapped dune orientations (shown here as long-dashed arrows) with the flowlines of annual resultant sand drift direction (RDD) determined by Kalma *et al.* (1988). Shading shows deviation of dunes with respect to the modern RDD in either clockwise or anticlockwise direction and degree of deviation. Short-dashed lines indicate areas of no meteorological records used in the analysis of modern RDD. In these areas the RDD flowlines have been slightly modified from Kalma *et al.* (1988) to give a more conservative reconstruction.

Wasson *et al.* (1988) illustrated the systematic geographical distribution of different dune types across the continent. Here all dunes could be classified into several morphological categories to give a more complete analysis. Single-crested longitudinal dunes (Figs 5b & 6a) are the most common type in Australia and dominate in all dunefields except the Great Sandy, where multi-crested longitudinal dunes dominate (Figs 5c & 6b). These groupings include several variant morphologies with distinct distributions, including forms that are transitional to network dunes. Transverse dunes (excluding shoreline features) are rare in the Australian dunefield and are not included here (Fig. 6h). Around and between the dual centres of the whorl, network dunes (Fig. 6f) are common, in areas where there is greater curvature in the dune streamline pattern; that is, where dunes change orientation over relatively short distances (Fig. 5b). The network dunes are really short, linked linear segments varying from no detectable preferred orientation to dunes with a dominant set of longitudinal dunes but with connecting segments of different orientations. At the centre of the whorl, particularly in the west, there are areas of isolated mounds (Figs 5a & 6g): mostly small dunes separated by sandless plains but often with a regular arrangement. These are in some ways similar to star dunes, reflecting a highly variable wind regime, but are formed in areas of very low sand availability. Within the longitudinal dunefields there are areas of parabolic dunes (Figs 5d & 6e), especially towards the margins of the dunefield in areas where longitudinal streamlines are parallel and show little curvature, in the Mallee, Eyre and Wiso dunefields and in patches of the extreme SW Great Victoria dunefield and

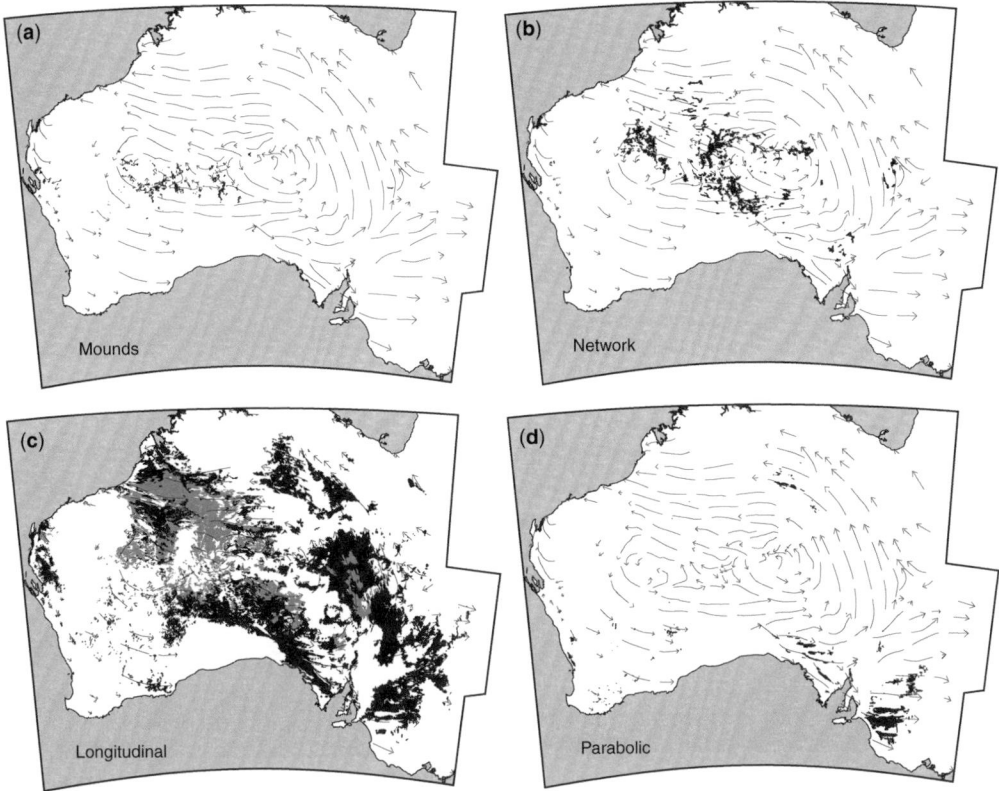

Fig. 5. Distribution of major dune types in the Australian dunefields. Arrows indicate the orientation of longitudinal dunes. (**a**) Mounds; (**b**) network; (**c**) longitudinal (black, single crest; grey, multi-crest); (**d**) parabolic.

through the SW. In general, the distribution of dunes at this broad scale conforms with expectations of dune morphological response to wind direction (Wasson & Hyde 1983; Werner 1995).

Dune preservation is determined by the balance between constructive aeolian processes and destructive erosion (e.g. owing to water). One instance where this may occur is on steeper and rockier ground where runoff is more effective (see below). A second instance is climatically driven, particularly at the margins of the dunefield where the climate today is more humid. This is particularly noticeable in the north, where dunes in the Wiso and Barkly dunefields are nearly all of suppressed expression without clear crests. In the northern Simpson Desert, between the Hay and Plenty Rivers, there is a sharp transition from sharp-crested dunes in the south to very broad dunes in the north (Fig. 6c). In these areas it appears that the modern climate may be contributing to destruction of dunes by runoff. Rills are reasonably common on the lower flanks of dunes in the Simpson and Stzelecki dunefields.

The influence of topography on the distribution and formation of the dunefields

Topography has a strong influence on the location of dunefields and dunes at all scales: it limits the supply of sand and erodible sediment in low-lying areas, influences the erosivity of runoff, and induces acceleration or deceleration of wind. At the broadest continental scale topography has an immediately obvious relationship with the supply of erodible sand. There are rare instances of small dunefields in upland areas associated with localized sand sources (Nott & Price 1991; Hesse *et al.* 2003) but most dunefields occur in valleys, piedmonts, coastal plains or lowland basins from sedimentary accumulations.

In the eastern half of the continent there is a clear topographic distinction between erosional uplands and sedimentary basins (Fig. 7). With a few exceptions dunefields occur in the lowland basins and are absent from the Great Dividing Range of the east coast and the extensive low hills of the inland, notably in Queensland and the Flinders, Mount

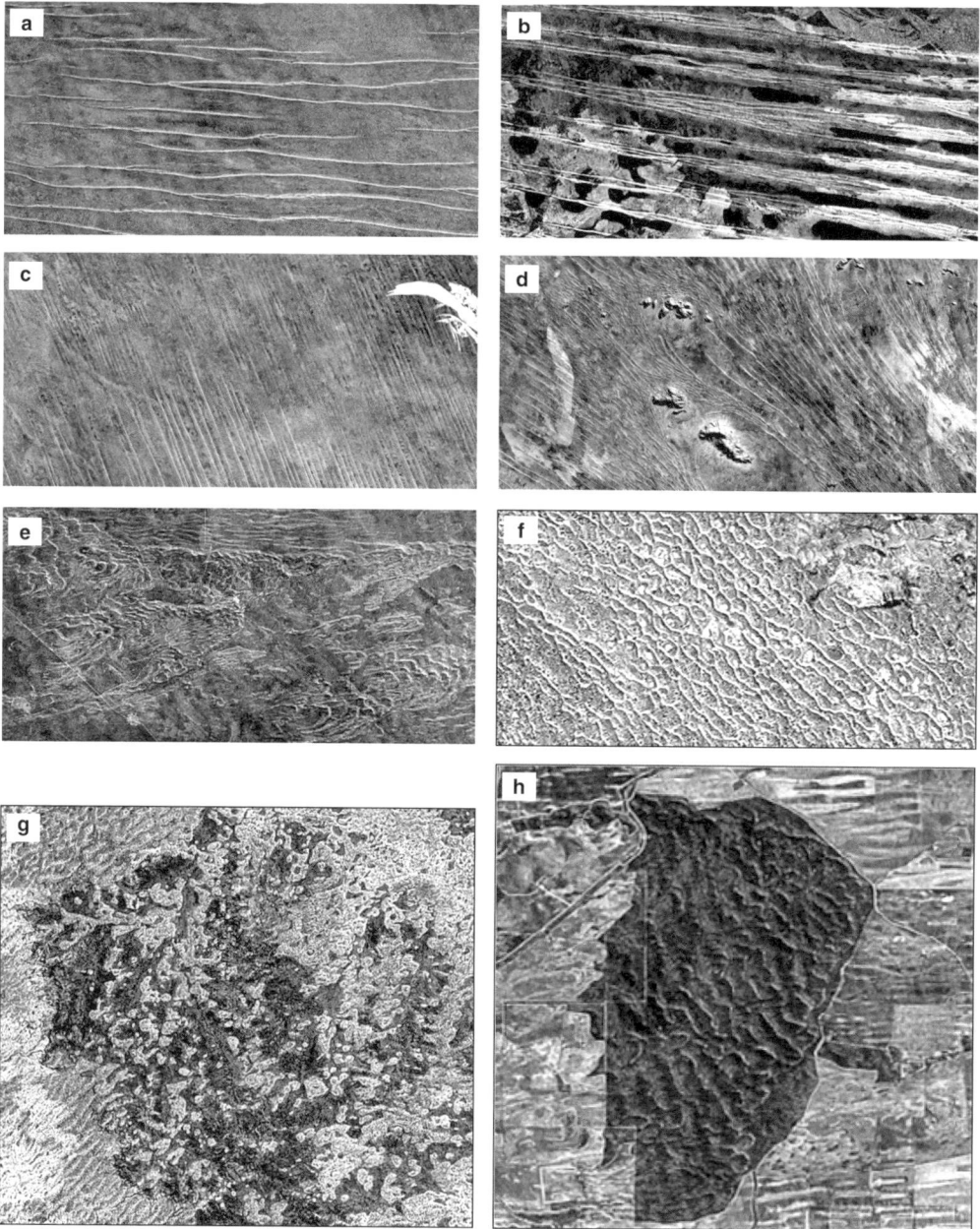

Fig. 6. Landsat images of major dune morphological types. (**a**) Straight single-crest longitudinal dunes in the SE Great Sandy Desert; (**b**) straight multi-crest longitudinal dunes in the SE Great Sandy Desert; (**c**) the transition from sharp straight single-crest longitudinal dunes (bottom) to broad crested dunes (top) in the far north of the Simpson Desert; (**d**) topographic steering of longitudinal dunes around inselbergs in the central northern Great Victoria Desert; (**e**) parabolic dunes (centre) in the Mallee dunefield with neighbouring short straight longitudinal dunes (top); (**f**) network dunes (with preferred orientation SE–NW) in the NW Simpson Desert; (**g**) isolated mounds (centre) in the central north Great Victoria Desert surrounded by longitudinal dunes (SW) and network dunes (NE); (**h**) the only clear example of free transverse ridges in Australia from the southern Mallee Dunefield (dark, wooded area) with parabolic dunes (east and west) and short straight longitudinal dunes (north and south). All images oriented with north to top. Images (a)–(e) *c.* 10 km east–west field of view; images (f)–(h) *c.* 5 km east–west field of view.

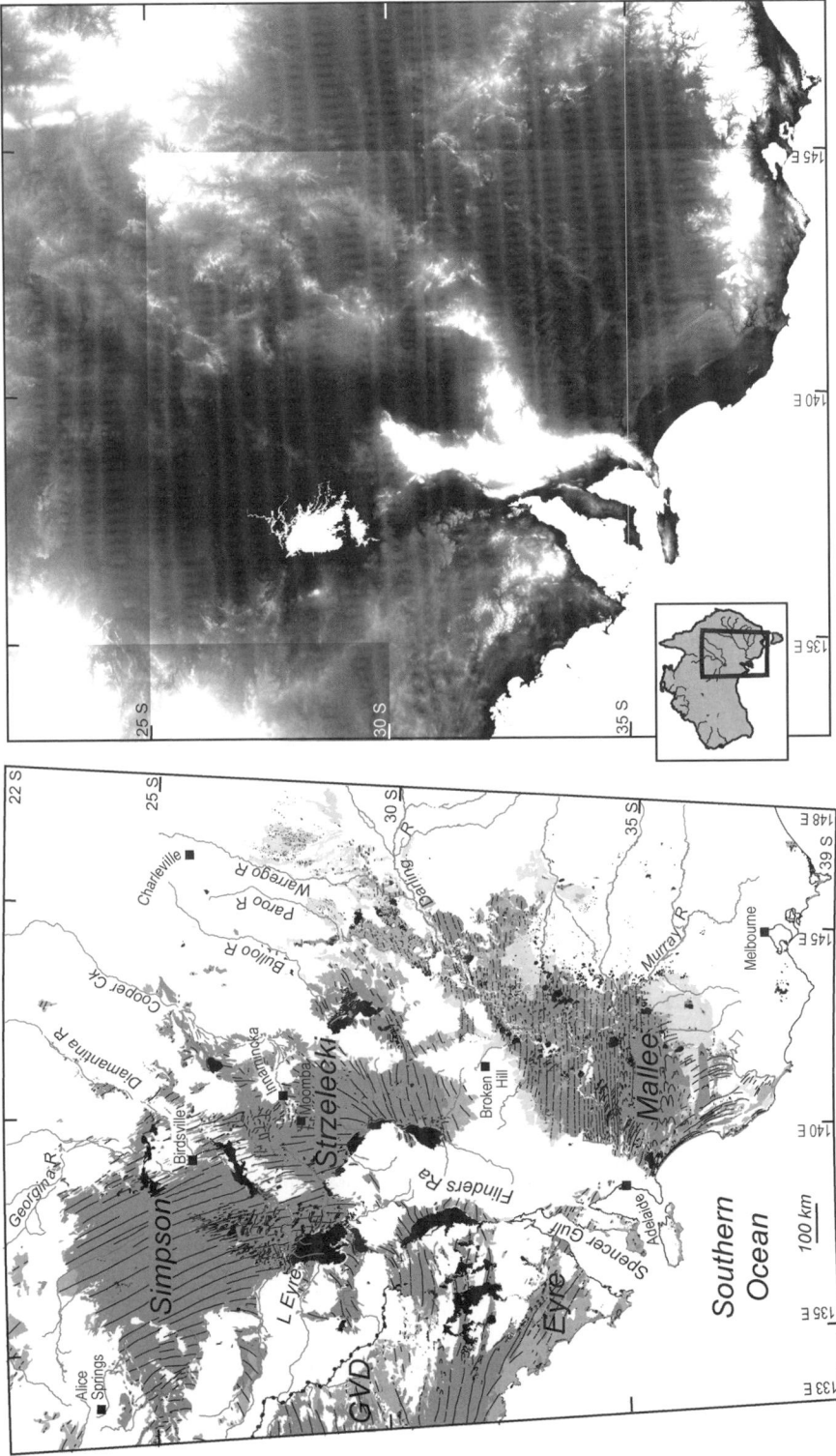

Fig. 7. Dunefields of eastern Australia compared with DEM topography (Global Map Australia 1M, 2004, Geoscience Australia). Sand plains shown as light grey, indistinct dunefields as mid-grey and major dunefields as dark grey. Salt lakes, playas and pans shown as black. Escarpments shown as thick black lines with circles. Prominent relict strandlines in Mallee dunefield shown by thick black lines. GVD, Great Victoria Desert.

Lofty and Barrier Ranges of South Australia and western NSW. The Mallee Dunefield occupies the western portion of the Cenozoic Murray Basin (Lawrence 1966) (Fig. 7) and extends westwards to cover a portion of the Late Mesozoic–Cenozoic Otway–Gambier Basins (Drexel & Preiss 1995) near the coast. There is often no clear distinction between coastal dunes and 'continental' dunes and the source of sand in the continental dunes appears to be at least partly the Pliocene and Pleistocene coastal strandlines of the Murray and Gambier basins (Bowler et al. 2007). The Cenozoic Lake Eyre Basin (Drexel & Preiss 1995) underlies both the Simpson and Strzelecki dunefields, which are separated by the low structural ridge of the Gason Dome on which Sturt's Stony Desert is found (Fig. 7). There has been debate about the lacustrine (Loffler & Sullivan 1987) or fluvial (Wasson 1983) sediments from which the dunes are derived, but there is agreement that they are formed by winnowing of non-marine basin sediments. Two extensive Neogene basins are notable exceptions to this pattern: the Carpentaria Basin of northern Australia and the Darling Riverine Plain of the SE. The Carpentaria Basin is a Mesozoic–Cenozoic basin with Neogene sequences but only insignificant dunefields. The absence of dunes may be due both to the abundance of clay and scarcity of sand from weathering of basalt in the Great Dividing Range headwaters and the presence of cemented hardpan soils throughout much of the Gulf area (Nanson et al. 1991) presenting an unerodible surface. The Darling Riverine Plain is a shallow late Cenozoic basin with an extensive clay plain covering its eastern portion (Martin 1997). Dunes are restricted to rare small transverse dunes adjacent to sandy palaeochannels of the main rivers (Watkins 1993) but most of the plain is unerodible clay, a very similar situation to the eastern Murray Basin (Butler et al. 1973). However, the floodplains of the western tributaries of the Darling River, the Warrego and Paroo Rivers, have extensive low sand dunes and sand sheets (Fig. 7). Although they occupy a more arid climate zone the controlling factor is probably the supply of sand from Cretaceous sediments forming the extensive low ridges in their catchments. In all these cases the basins are repositories of large volumes of sediment that have been reworked into sand dunes in extensive, continuous dunefields.

In the western half of the continent, the older cratonic crust has less dramatic relief and lacks Neogene basins. The extensive Great Victoria and Great Sandy dunefields occur on landscapes of subtle but distinct ridge and valley topography (Fig. 8). The topography and extensive palaeodrainage network (Fig. 8) date to the late Mesozoic to Early Cenozoic (van de Graaff et al. 1977; Clarke

1994). Only channels in the wetter (e.g. Sturt Creek) and steeper parts of the network remain as active fluvial systems today. It is thought that this relict landscape was the product of advancing aridity through the Neogene (Martin 1990) leading to loss of effective fluvial transport. Subsequent evolution towards an arid landscape saw the development of saline groundwater windows, playas and aeolian landforms by at least 1 Ma ago (Chen & Barton 1991). The Great Victoria and Great Sandy dunefields are extensive but less continuous than the dunefields in the east, with frequent exposure of the low ridges as bare areas, hills and ranges between the dunes (Fig. 8). Although the topography is often low enough that sand dunes can climb over and cover the ridges between valleys, it is clear from the topographic dependence that the valleys are the source of the sediment and that aeolian processes have not only reshaped the valley floors but also blown sand out to cover much of the ridges between (Fig. 8). In this respect they are very different from the eastern dunefields or even most dunefields of Africa or Asia of comparable extent, which occupy geological and topographic basins (Breed & Grow 1979).

The central ranges between the Great Victoria Desert and the Barkly Tableland provide another distinct landscape exerting strong topographic control on dunefield distribution. Prominent dune-free ranges in Archaean–Proterozoic cratonic rocks up to several hundred metres high are separated by broad lowlands formed in Proterozoic–Palaeozoic basin sequences. The Musgrave Ranges of the Musgrave Block, and McDonnell and northern ranges of the Arunta Block alternate with the Amadeus lowland and Burt Plain lowland (between Lake Lewis and Lake Mackay). Further north the Wiso plain lies between the Arunta Block and the Ashburton–Davenport Ranges, and the Barkly Plains extend beyond these. In these lowlands lie extensive dunefields (Fig. 8), although none have names in common use. All of these lowlands have Quaternary sediment accumulations but they are generally thin accumulations of lacustrine, playa, fluvial piedmont and aeolian sediments (Chen & Barton 1991; English et al. 2001). The aeolian landforms are commonly found in association with lakes and playas (gypseous and carbonate-rich) and fluvial sediments (quartz-rich) as source-bordering small dunes and dunefields. The larger longitudinal dunefields have a less obvious specific source but appear to be derived from winnowing and saltation of sand from the broader non-lacustrine piedmont sediments.

Ranges and uplands form some of the boundaries of the Great Sandy and Great Victoria dunefields where the dunefields occupy the lowest parts of the landscape. Elsewhere the topographic

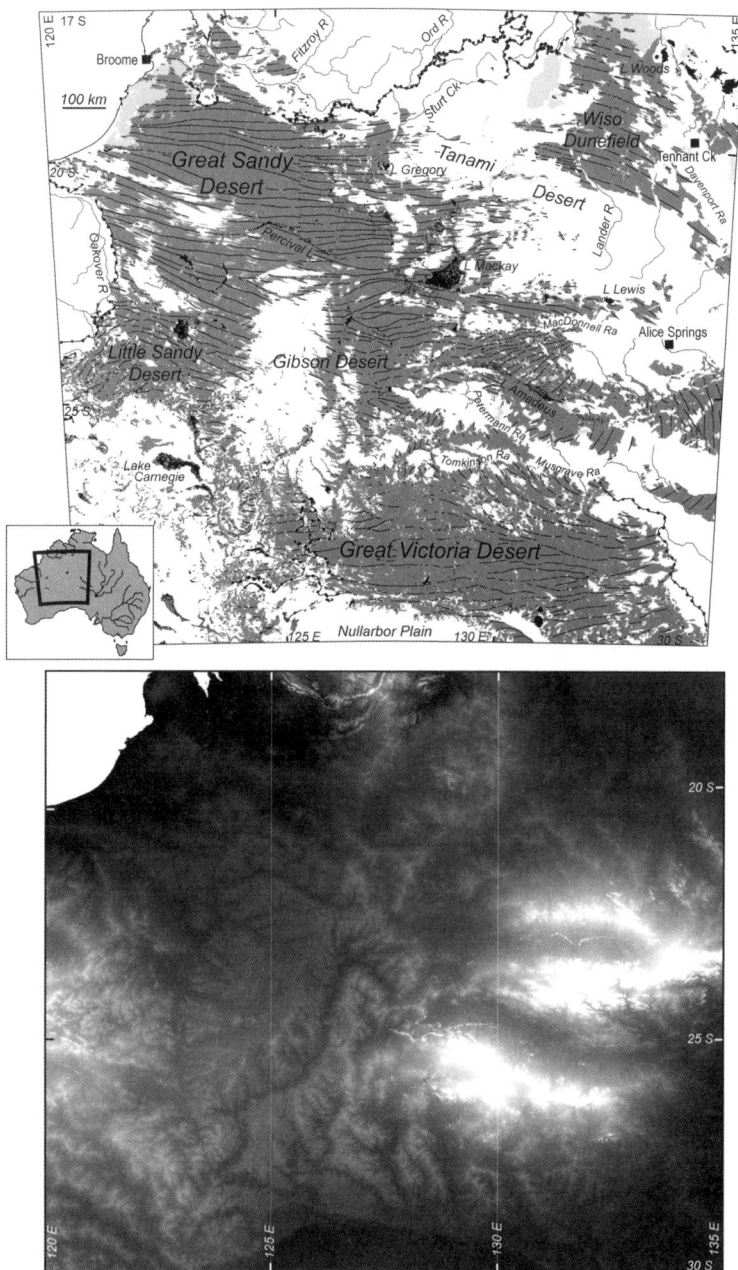

Fig. 8. Dunefields of western and central Australia compared with DEM topography (Global Map Australia 1M, 2004, Geoscience Australia). Sand plains shown as light grey and major dunefields as dark grey. Salt lakes, playas and pans shown as black. Escarpments shown as lines with circles.

relationship is reversed and the dunefields sit above escarpments (Fig. 8). The escarpment marking the western boundary of the Lake Eyre catchment is well known. Dunes from the Great Victoria Desert on the western (upwind) plateau sit above a dune-free landscape of higher relief to the east. Likewise, an escarpment forms the western boundary of the Great Sandy Desert above the Oakover River at the edge of the Pilbara Ranges and a very long escarpment in places forms the northern boundary of the Great Sandy dunefield above the Fitzroy River. In all three cases the escarpments are not particularly high (no more than 100– 200 m) but mark sudden changes in topography, geomorphological processes and surface cover. In all cases the escarpments mark a boundary between landscapes with coherent fluvial drainage networks below and landscapes without above. Conversely, the landscapes below the escarpments are dune-free, or nearly so, and the landscapes above are dominated by sand dunes. These escarpments illustrate the relationship between topography, runoff and dune formation. Flow separation in the wakes of the escarpments should provide preferential areas for sand deposition but, instead of sand accumulation, the steeper escarpment slopes are bare of sand. Whereas the absence of dunes from hills and ranges can be interpreted as evidence of wind speed-up not allowing dune formation, the escarpments illustrate that runoff efficiency on steeper ground is at least as important, and maybe more important, in limiting sand accumulation and either dune formation or preservation.

In the southwestern portion of the continent another relict valley and ridge landscape is preserved, similar to the Great Sandy and Great Victoria Deserts, but without extensive dunefields (Fig. 9). This difference is not due to climate, as most of the Yilgarn is as dry today as the Great Sandy dunefield, or drier than it (Fig. 2). The valleys and their sediments are likely to have been formed at around the same time in both areas (van de Graaff *et al.* 1977; Clarke 1994) but they are formed in contrasting geological provinces. The landscape of the SW is formed on the Archaean cratonic crust of the Yilgarn Block, characterized by granite and greenstone (ultramafic) lithologies, whereas the Great Sandy and Great Victoria deserts are largely underlain by Proterozoic to Mesozoic basin sediments. Dunes in the Yilgarn are widespread but small. They occur mostly along the valley floors associated with lakes and playas but also, rarely, as small areas on the interfluve plateaux, particularly east and SE of Meekatharra (Fig. 9). Sand accumulation and dune formation are thus restricted to the flattest parts of the landscape on plateaux and valley floors (Zheng *et al.* 2003). This may be because the valleys of the eastern Yilgarn are sufficiently deep and slope processes active enough to prevent sand accumulation or preservation of aeolian landforms or because of a scarcity of erodible sand (see below). The apparent representation of a large dunefield in this region by Wasson *et al.* (1988) is due to the presence of these small dunes in nearly every $0.25° \times 0.25°$ cell of their grid.

The most extensive dunefields of the SW occur outside the structural Yilgarn Block (Fig. 10). Those

north of Esperance, however, are still linked to the palaeodrainage system and are apparently formed by aeolian reworking of sediment accumulated in these extremely flat, low-relief valleys (Fig. 9). Along the west coast there are extensive dune and shoreline barrier systems on the Swan Coastal Plain (Bastian 1996). These merge eastward with extensive sandplains on uplifted parts of the Palaeozoic–Mesozoic Perth Basin (Fink *et al.* 2000). The sand plain becomes more continuous northward but dunes are limited mostly to either the coastal plain or topographic depressions near the scarp of the Darling Fault, forming the western boundary of the Yilgarn Block. From Shark Bay northwards the extensive Gascoyne Dunefield has formed in the coastal plain and linked valleys of the Wooramel, Gascoyne and other rivers draining the Pilbara region. The rugged inland valleys of the Pilbara have very little accumulated sediment and very few aeolian landforms. The Gascoyne Dunefield ranges from 'continental' longitudinal dunes inland to coastal aeolianites and intermediate complexes where the two are genetically linked and difficult to separate.

The greater relief and steep slopes of the central ranges present aerodynamic obstacles in clear contrast to the subdued landscape of the Great Victoria and Great Sandy dunefields. Topographic steering of winds and sand transport is evident in the complex patterns of dune elongation around many ranges (Hollands *et al.* 2006), particularly at the western end of the Musgrave Block. Nevertheless, the subdued topography of the Great Victoria and Great Sandy Dunefields is sufficient to create its own effects on dune orientation with noticeable deflections of dune orientation going into and out of the palaeodrainage depressions (Fig. 8). Both these landscapes contrast with the unwaveringly consistent orientation of the longitudinal dunes in the basin-filling Simpson, Strzelecki and Mallee dunefields in which there are no topographic obstacles.

The influence of substrate lithology on the distribution and formation of the dunefields

Topography and climate together explain the distribution of most of the Australian dunefields. Nevertheless, there are some regions of the arid and semi-arid zones of low relief that are conspicuously dune free. Some of these, the Carpentaria Basin and Upper Darling Riverine Plain, have been mentioned above and explained in terms of unerodible hardpan soils and sand-free clay soils. However, these characteristics of unerodible and sand-poor substrates are shared by many other areas (Fig. 10).

The Nullarbor Plain is the most conspicuous of these: an extremely flat, arid plain with no dunes,

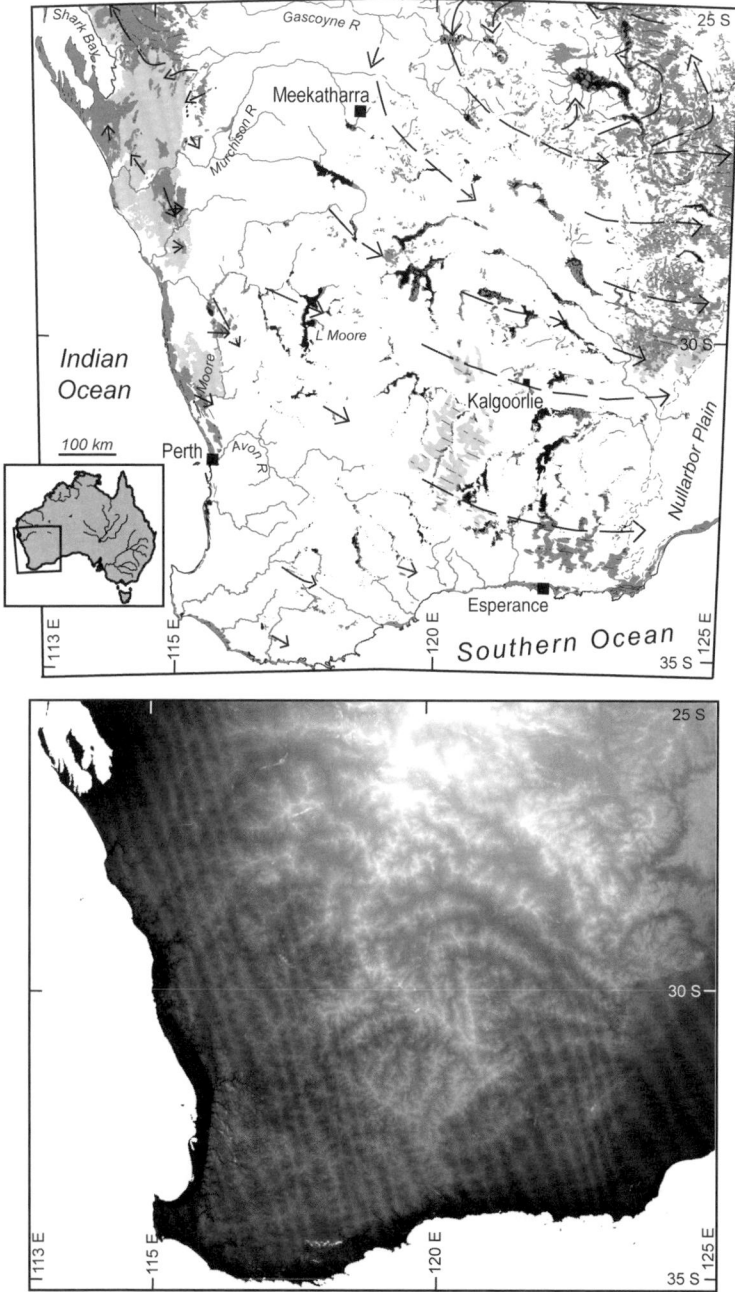

Fig. 9. Dunefields of southwestern Australia compared with DEM topography (Global Map Australia 1M, 2004, Geoscience Australia). Sand plains shown as light grey, indistinct dunefields as mid-grey and major dunefields as dark grey. Salt lakes, playas and pans shown as black. Dune orientation indicated by large long-dashed arrows.

except along its northern and eastern margins. The plain is formed by undeformed raised Eocene–Miocene limestone of the Eucla Basin (Drexel & Preiss 1995). Weathering does not produce sand and

there is no regressive clastic shoreline series (as for the Murray Basin) although there is an Eocene high-stand clastic shoreline (Ooldea Ridge) (Benbow 1990) from which Pleistocene longitudinal dunes

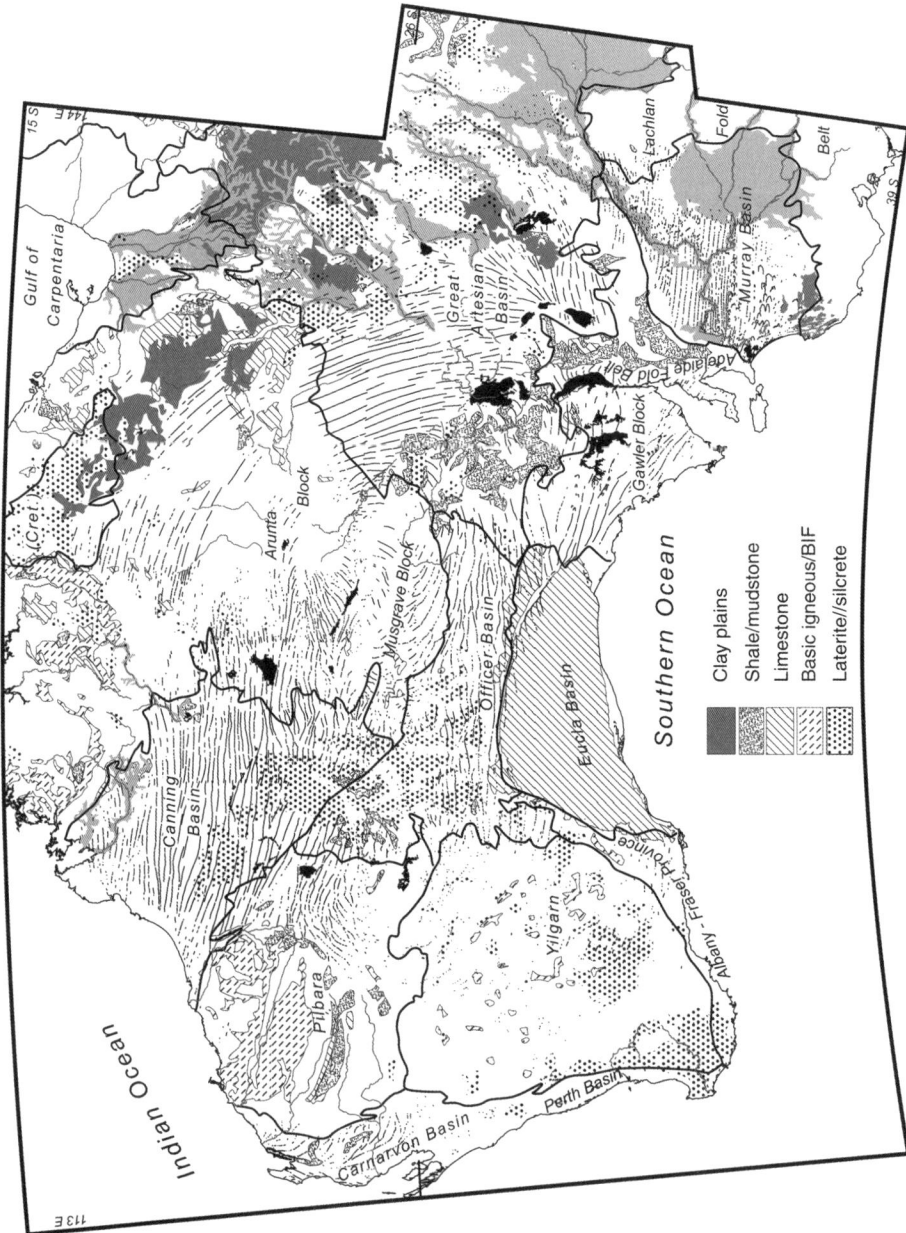

Fig. 10. Geological provinces and selected quartz-poor and weathering-resistant surface lithologies (AUSLIG 1988).

have been formed by reworking (Sheard *et al.* 2006). The absence of a regressive shoreline sequence can in part be explained by sediment starvation. The Eocene highstand flooded the palaeodrainage system of the Yilgarn Block and since that time the climate was too dry to allow re-establishment of efficient fluvial systems in the valleys to transport sand to the Eucla Basin (Clarke 1994). Some confirmation of this comes from the late Eocene shoreline, where zircons are derived from the Musgrave Block but not the Yilgarn Block or Albany–Fraser Province in the west (Reid & Hou 2006). Other large limestone terrains of northern Australia (Fig. 10) are also dune-free.

Pleistocene dunes of the western margin of the Great Victoria Desert do contain zircons from the adjacent Yilgarn Block and more distant Albany–Fraser Province, but dunes from the central or eastern Great Victoria Desert do not (Pell *et al.* 1999). Coastal dunes from near Perth are apparently derived from the relatively small Leeuwin Block and the Albany–Fraser Province but not the extensive Yilgarn Block from which rivers drain to the coastal plain (Sircombe & Freeman 1999). The scarcity of Yilgarn-derived sand in the dunefields marginal to the Yilgarn Block could be explained by the argument of inefficient transport of sediment through the drainage sytem because of the relatively arid climate and low runoff. However, the Yilgarn is dominantly composed of granite and it is surprising that retention of granite-derived weathering products within the Yilgarn has not led to more extensive dunefields. It may be that the extensive deep weathering and cementation of secondary colluvial deposits (Anand & Paine 2002) has limited sand availability for aeolian reworking, except under conditions associated with saline valley floors.

Resistant weathering mantles of silcrete and ferricrete (laterite) are widespread over large areas of Australia (Fig. 10). These mantles may incorporate quartz sand from the host material but relatively little is released by breakdown of the duricrust. Laterites are common as caps on plateaux in the Yilgarn and also within the Officer and Canning Basins, forming some of the largest dune-free areas within the Great Victoria and Great Sandy dunefields. Silcrete (often with some iron) is a common occurrence in Cretaceous sediments of the Great Artesian Basin (also known as the Great Australian Basin and Eromanga Basin), forming both caps on high plateaux and low-relief surfaces beneath the margins of the dunefields and rendering large parts of the catchment of Lake Eyre a poor source of sand. In addition, gravel (gibber) surfaces are areas of high sand transport rates because of the long saltation lengths (Bagnold 1941).

The Great Artesian Basin also contains extensive, thick shale and mudstone units (as well as sandstone), further limiting sand supply to the Simpson and Strzelecki dunefields. Deep clay soils on the low slopes of the interfluves feed clay to extensive floodplains of the 'Channel Country', the wide, multi-channel floodplains of the Georgina, Diamantina and Cooper systems (Nanson *et al.* 1988). The upper catchment of the Georgina River and the Barkly Tablelands also contain extensive dune-free clay plains formed by weathering of shales and mudstones of Cambrian Georgina Basin sediments.

Whereas the eastern dunefields occur in basin depocentres and their sand is derived from reworking of sand-rich Neogene coastal, shoreline and fluvial sediments derived from lithologically diverse catchments, the western dunefields have less clear pathways of formation. The sand of the Great Victoria, Great Sandy, Wiso and Barkly dunefields appears to be largely reworked from shallow valley and piedmont accumulations. Whereas the Yilgarn appears not to have been a good source of sand for surrounding low-lying regions, the Central Australian craton, parts of the Pilbara block and the Neoproterozoic–Mesozoic sediments of the Canning and Officer Basins seem to have yielded large volumes of sand.

The low supply of sand generally and low storage in many landscapes, particularly in the west, has contributed to the distinctive nature of the Australian dunefield with its dominance of longitudinal dune forms. Longitudinal dunes are associated with low sand supply under moderately variable wind conditions (Wasson & Hyde 1983) and can be simulated in models under these conditions (Werner 1995). The general absence of transverse dunes is the corollary of this, requiring unimodal winds and also high sand supply, which is found only in small areas typically associated with lake basins. Perhaps less well appreciated until now is the occurrence of mound and network dune morphologies under conditions of low supply and highly variable winds (Fig. 5), rather than star dunes, which require very large amounts of sand (Wasson & Hyde 1983; Werner 1995).

Discussion

Extension or rifting? Models of dune formation and sand transport

The formation of longitudinal dunes, which dominate the Australian continental dunefields, remains relatively poorly understood (Livingstone *et al.* 2007). Although there has recently been renewed debate about the potential for large-scale downwind sand transport on longitudinal dunes (Hollands et al. 2006; Twidale 2008) some dunefields show evidence that supports dune extension over at least tens of kilometres distance. Pell et al. (1999, 2000, 2001) found evidence that sand in the Mallee, Simpson–Strzelecki and Great Victoria dunefields was derived within their respective basins and not transported between basins. Within these constraints, however, it is known that downwind dune extension does occur. The idea that dunes in the Simpson Desert originated from deflation of playa lakes north of Lake Eyre has some support because of the rapid increase in dune organization and apparent decrease in equivalent sand thickness away from these basins (Wopfner & Twidale 1967; Mabbutt & Wooding 1983) and the colour and clay content of dunes downwind of playas (Wasson 1983). However, not only is the volume of the lake basins insufficient to account for the entire dunefield, in large parts of the Simpson and in nearly all of the other dunefields there is simply an absence of deflation basins at the

upwind margin of the dunefields. This mechanism is not a general model for development of the Simpson longitudinal dunefield or other dunefields in Australia. However, lake basins and their shorelines are a common local source of sand, which can be deflated and blown in dunes over distances of at least tens of kilometres; for example, from the Willandra lakes of the northeastern Mallee dunefield (Bowler & Magee 1978; Fig. 11).

A greater distance of dune extension may be required to explain several other dunefields. The low gibber-covered ridge that separates the Strzelecki and Simpson dunefields provides a compelling case for downwind longitudinal dune extension over distances up to around 100 km (Fig. 7). The ridge (Sturt's Stony Desert) has no local sources of sand but there are several prominent longitudinal dunes that are continuous from the Strzelecki dunefield upwind, across the ridge to Goyder's lagoon on the northern side. This is one of the clearer examples of extension over these distances, although there are many instances of dune

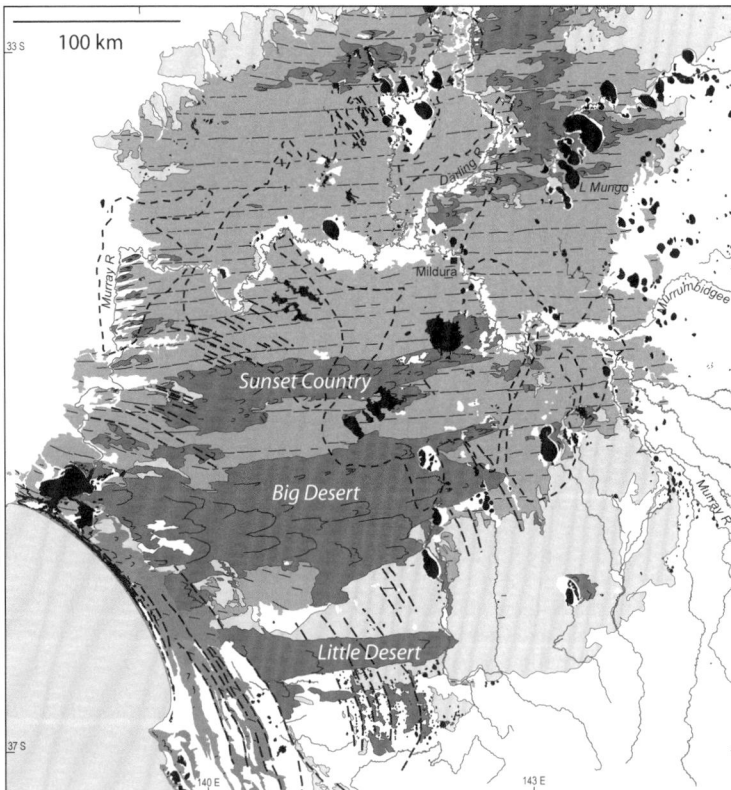

Fig. 11. The Mallee dunefield and the former extent of the Late Pliocene Lake Bungunnia (Bowler et al. 2007), shown by the short-dashed line. Relict strandlines shown by heavy long-dashed lines and representative longitudinal and parabolic dune crests by thin black lines. Sand plain areas shown as light grey, longitudinal dunefields as mid-grey and parabolic dunefields as dark grey. Salt lakes, playas and pans shown as black.

extension from small valleys and basins, over otherwise dune-free slopes and low ridges, in all areas but particularly in Central Australia. Nevertheless, it is equally clear that sand is not transported in great quantities over great distances and is mostly constrained within the topographic limits of each dunefield. Large or steep ranges are effective barriers to sand movement (Hollands et al. 2006) and the efficiency of runoff on these surfaces may also inhibit the formation and preservation of dunes over such barriers where sand transport rates are relatively low.

A more equivocal case of dune extension is provided by the Mallee dunefield, where short longitudinal dunes and parabolic dunes extend for tens of kilometres to over 100 km over the former bed of the late Pliocene–early Quaternary Lake Bungunnia (Bowler et al. 2007; Fig. 11). The Lake Bungunnia sediments are largely sand-free clays and silts (McLaren et al. 2009), which lie between and over the irregular surface of Miocene–Pliocene strandlines. To form the longitudinal dunes that nearly everywhere cover the lacustrine sediments requires either aeolian transport of sand from the lake shorelines or emergent strandlines, or the (complete) reworking of a previously unrecognized sand unit deposited over the lacustrine sediment. Although no such stratum has previously been recognized it is a plausible response to late Pliocene draining of the lake and re-establishment of the fluvial system across the former lake floor. However, the short longitudinal dunes appear to have accreted vertically and to be non-migratory (Bowler & Magee 1978), arguing against dune extension from lake shorelines or strandlines. Nevertheless, adjacent to these dunes are parabolic dunefields elongated in the same direction as the longitudinal dunes and extending for up to 250 km downwind. Although there are some arguments against large-scale extension of these dunes (Blackburn et al. 1965, 1967) the concordant orientations of the dunes and cross-cutting orientation of the strandlines or other substrate structures render extension the only plausible explanation for their downwind elongation.

Dunefield response to evolving climate, topography and substrates

There is no doubt that during intervals in the late Quaternary dunefields have been more active than they are currently (Hesse et al. 2004; Fitzsimmons et al. 2007) and that they formed in areas of the humid continental fringes in which conditions are no longer suitable (Fig. 2). The arid zone has expanded repeatedly in response to Milankovitch cyclical forcing with associated increase in aeolian sand and dust transport (Hesse 1994). These arid

glacial stages were accompanied by changes in the circulation patterns, which affected the distribution of raised dust (Hesse 1994) as well as the orientation of the sand dunes (Fig. 4). There is growing evidence that the longitudinal dunes over large areas were formed by 200 ka (Nanson et al. 1992; Sheard et al. 2006; Fujioka et al. 2009) or even the mid-Pleistocene (Fujioka et al. 2009). Although data are still sparse, some areas, such as the Strzelecki, appear to have formed later and have greatest ages from the penultimate glacial cycle (Lomax et al. 2003; Fitzsimmons et al. 2007).

Stratigraphic and geomorphological evidence from these studies indicates that the dunes have grown since their initial formation but have not been reoriented by changing wind patterns since. In rare cases younger dunes have formed from fresh sediment on floodplains in alignment with prevailing winds, which have not reoriented older pre-existing dunes (Nanson et al. 1995; Hollands et al. 2006). Therefore the dunes preserve the resultant sand drift direction at the time of their formation, dominantly during glacial intervals, although the ages are unknown for large areas. There are dramatic divergences between dune orientations and modern resultant sand drift directions in several areas (Fig. 4): regions of onshore wind flow in the Gascoyne, Eyre and western Great Sandy dunefields, dunefields along the centre of the dune whorl at 26°S, and possibly over larger areas of the Great Sandy and Great Victoria dunefield (where wind data were not available to Kalma et al. 1988). Relatively minor shifts in seasonal strength or latitudinal position of wind systems could easily produce large responses in resultant drift direction along the axis of the whorl but the areas of onshore winds suggest changes in strong winds at a regional scale related to altered land–sea temperature contrast, which influences the monsoon in the north and sea-breezes in the south, and/or an effect of shoreline migration with sea-level change. It is important to include these regional patterns of variation in discussions of the response of the circulation system to climate change and in considering studies that have used dune–wind divergence but with contrasting results (Brookfield 1970; Sprigg 1982; Nanson et al. 1995).

The formation of the extensive dunefields in Australia is one manifestation of the transformation of the continent through progressive aridification beginning in the Miocene in response to the reorganization of hemispheric circulation patterns (Byrne et al. 2008). Whereas the age of the oldest 'arid' landforms, such as playas and dunes, varies regionally there has been an apparent progressive decline in river discharges (Nanson et al. 1992; Maroulis et al. 2007) and mega-lake stands (Bowler et al. 2001) and increasing dust flux

(Hesse 1994) and dune building since at least MIS 10–11. In many of the dunefields this transition has seen the breakdown of former rivers and drainage systems and decreased introduction of new sediment into the desert lowlands. The climatic sensitivity of dune movement points to periodic supply limitation as a result of vegetation cover (Hesse & Simpson 2006) but there is also a longer trend of supply limitation as fresh sediment sources are reduced and the desert surface becomes increasingly armoured.

Since mid-MIS 3 (40–50 ka) decline in discharge of Cooper Creek has reduced the transfer of sand to the Strzelecki Desert and seen the formation of extensive clay floodplains (Maroulis et al. 2007). The extensive dunefields within the Strzelecki Desert attest to the contribution of sand from sandstone members and other quartz-rich rocks in the catchment (largely comprising Great Artesian Basin sediments) to Quaternary sediments. However, the post-40 ka clay floodplain now covers older sandy fluvial sediments, from which dunes were derived, and is burying and isolating dunes (Maroulis et al. 2007). Clay pellets now contribute to the dunes on the downwind side of the floodplains (Wasson 1983). The same pathway of river evolution has occurred in the Murray Basin, although the transfer to suspended load channels occurred at the last glacial termination (Page et al. 1996). The sand of the Mallee dunefield is also not being replenished under the modern climate regime. It is ultimately derived from the transfer by rivers of sand from the headwaters to shallow seas in the western depocentre present through the late Miocene and Pliocene, where it was sorted into extensive strandlines in a recessive sequence still evident today (Bowler et al. 2007). These strandlines, extending from the inland limits of the Mallee dunefield to the present coastline, were then reworked into continental dunefields post-mid-Quaternary as aridification intensified (Bowler et al. 2007). A very similar sequence of events has created parts of the Great Victoria and Eyre dunefields by reworking of coastal clastic shorelines. In many of these dunefields supply limitation is compounded by pedogenesis within the dunes, reducing erodibility (Bowler & Magee 1978; Sheard et al. 2006).

Australia has experienced relative tectonic stability in the Cenozoic but has nevertheless seen widespread low-amplitude long-wave flexure (Sandiford 2003) and localized faulting (Quigley et al. 2006) in response to the modern in situ stress field. In some parts of the Australian desert there is some evidence that subtle seismogenic movements have affected topography (Braun et al. 2009) and affected drainage organization. For example, palaeovalleys in the Great Sandy Desert are in places disconnected by warping, creating basins that host aeolian–lacustrine features (Sandiford et al. 2009; Fig. 8). The eastern margin of the Strzelecki Desert is also subject to neotectonic movement (Nanson et al. 2008) affecting rivers and sedimentation. The scale and age of the movements is not well constrained but the contribution to more widespread drainage disruption, leading to formation of closed basins, terminal lakes and associated aeolian deflation, is possible. The Bulloo overflow (a terminal lake and associated dunefields) is a candidate for this scenario. Dunes on the downwind (eastern) side of the Bulloo basin climb to the crest of the gentle interfluve (Fig. 7) and are currently in a state of erosional degradation with a drainage network formed between the dunes. This dunefield has possibly been elevated by tectonic uplift after dune formation and, therefore, within the last 200–300 ka. The nearby playa and dunefield of the Bindegolly–Currawinya corridor, linking the upper Bulloo River to the lower Paroo River, is also possibly a casualty of tectonic uplift in this area.

Conclusions

The new mapping from satellite imagery shows the limited extent but wide distribution of dunes in the SW of the continent and extends the known occurrence of sand dunes in the currently humid margins of the continent compared with the earlier map by Wasson et al. (1988) using different sources and methods. Dunefields cover a large portion of the Australian continent in areas with favourable topography and climate and sand supply. Most of the dunefields today receive little or no new supplies of sand from fluvial sources but rework old coastal, fluvial and lacustrine sediments from relict shorelines, terrestrial basin deposits, piedmonts and valley floors. In the eastern part of the continent the Simpson, Strzelecki and Mallee dunefields occupy the depocentres of large Neogene sedimentary basins. The Great Victoria and Great Sandy dunefields in the western sector have fundamentally different settings in subdued ridge and valley topography with sand derived from deflation of shallow valley-filling sediments and piedmont deposits. Large low-relief areas of the arid zone are dune-free and these are related to low sand supply because of substrate or catchment lithology or sediment starvation. Examples are found in the Nullarbor Plain (limestone), Georgina Basin (shale–mudstone), Carpentaria basin (clay soils and hardpans) and parts of the Darling Riverine Plain and Murray–Murrumbidgee Riverine Plain (clay soils). The Yilgarn Block of southwestern Australia appears to have been a poor source of sand for dune formation in its subdued landscape or surrounding areas during the Cenozoic, possibly as a result of prolonged, deep weathering and duricrust formation. The generally low sand supply has contributed to the characteristic dune morphologies: longitudinal, network and mound.

The repeated expansion of arid conditions in glacial stages of the Quaternary glacial cycles has left dunefields extending from the most arid parts of the continent into the semi-arid zone and isolated patches in present-day humid areas. Growing geochronological evidence points to dune formation beginning in the mid-Pleistocene and accelerating as the climate has become increasingly arid in subsequent glacial cycles. However, the dunes appear to have undergone little modification of orientation after their initial formation, possibly as a result of stabilization by pedogenesis and vegetation, and their orientation preserves the alignment owing to sand drift direction at that time. There are substantial areas of large divergence between sand dune orientations and modern resultant sand drift direction that point to past altered circulation patterns: areas of onshore wind flow in the Great Sandy, Gascoyne and Eyre dunefields, and areas along the central axis of the dune whorl at 26°S. Some combination of latitudinal shift in response to hemispheric temperature gradients, altered land–sea temperature contrast and regional variations in proximity to the coastline as sea level fell may explain this regional pattern of variation. There is no simple, continent-wide displacement of dune orientations, and therefore simple latitudinal shift of circulation patterns, as has been frequently postulated, is not an adequate explanation of the observed patterns.

The trend towards aridity experienced since the Miocene and modulated by the Milankovich cycles has seen progressive landscape change, spread of aeolian landforms, development of playa lakes and loss of fluvial drainage networks. The dunefields exhibit many features reflective of the evolution of the climate as well as related changes in sediment supply.

R. Hyde calculated the sand drift potentials and directions. I thank R. Wasson and G. Wiggs for helpful reviews of the paper.

References

ANAND, R. R. & PAINE, M. 2002. Regolith geology of the Yilgarn Craton, Western Australia: implications for exploration. *Australian Journal of Earth Sciences*, **49**, 3–162.

ASH, J. E. & WASSON, R. J. 1983. Vegetation and sand mobility in the Australian dunefield. *Zeitschrift für Geomorphologie, Neue Folge, Supplementband*, **45**, 7–25.

AUSLIG 1988. *Geology and Minerals. Atlas of Australian Resources, 5*. Australian Surveying and Land Information Group, Canberra.

BAGNOLD, R. A. 1941. *The Physics of Blown Sand and Desert Dunes*. Methuen, London.

BASTIAN, L. V. 1996. Residual soil mineralogy and dune subdivision, Swan Coastal Plain, Western Australia. *Australian Journal of Earth Sciences*, **43**, 31–44.

BENBOW, M. C. 1990. Tertiary coastal dunes of the Eucla Basin, Australia. *Geomorphology*, **3**, 9–29.

BLACKBURN, G., BOND, R. D. & CLARKE, A. R. P. 1965. *Soil Development Associated with Stranded Beach Ridges in South-East South Australia*. CSIRO Soil Publication, **22**.

BLACKBURN, G., BOND, R. D. & CLARKE, A. R. P. 1967. *Soil Development in Relation to Stranded Beach Ridges of County Lowan, Victoria*. CSIRO Soil Publication, **24**.

BOWDEN, A. R. 1983. Relict terrestrial dunes: legacies of a former climate in coastal northeastern Tasmania. *Zeitschrift für Geomorphologie, Neue Folge, Supplementband*, **45**, 153–174.

BOWLER, J. M. & MAGEE, J. W. 1978. Geomorphology of the Mallee region in semi-arid northern Victoria and western New South Wales. *Proceedings of the Royal Society of Victoria*, **90**, 5–21.

BOWLER, J. M., WYRWOLL, K.-H. & LU, Y. 2001. Variations of the northwest Australian monsoon over the last 300 000 years: the paleohydrological record of the Gregory (Mulan) Lakes System. *Quaternary International*, **83–85**, 63–80.

BOWLER, J. M., KOTSONIS, A. & LAWRENCE, C. R. 2007. Environmental evolution of the Mallee region, western Murray Basin. *Proceedings of the Royal Society of Victoria*, **118**, 161–210.

BRAUN, J., BURBIDGE, D. R., GESTO, F. N., SANDIFORD, M., GLEADOW, A. J. W., KOHN, B. P. & CUMMINS, P. R. 2009. Constraints on the current rate of deformation and surface uplift of the Australian continent from a new seismic database and low-*T* thermochronological data. *Australian Journal of Earth Sciences*, **56**, 99–110.

BREED, C. S. & GROW, T. 1979. Morphology and distribution of dunes in sand seas observed by remote sensing. *In*: MCKEE, E. D. (ed.) *A Study of Global Sand Seas*. US Geological Survey, Professional Papers, **1052**, 253–302.

BROOKFIELD, M. 1970. Dune trends and wind regime in Central Australia. *Zeitschrift für Geomorphologie, Supplementband*, **10**, 121–135.

BULLARD, J. E., THOMAS, D. S. G., LIVINGSTONE, I. & WIGGS, G. F. S. 1996. Wind energy variations in the southwestern Kalahari Desert and implications for linear dunefield activity. *Earth Surface Processes and Landforms*, **21**, 263–278.

BUTLER, B. E., BLACKBURN, G., BOWLER, J. M., LAWRENCE, C. R., NEWELL, J. W. & PELS, S. 1973. *A Geomorphic Map of the Riverine Plain of South-Eastern Australia*. Australian National University Press, Canberra.

BYRNE, M., YEATES, D. K. ET AL. 2008. Birth of a biome: insights into the assembly and maintenance of the Australian arid zone biota. *Molecular Ecology*, **17**, 4398–4417.

CHEN, X. Y. & BARTON, C. E. 1991. Onset of aridity and dune-building in central Australia: sedimentological and magnetostratigraphic evidence from Lake Amadeus. *Palaeogeography, Palaeoclimatology, Palaeoecology*, **84**, 55–73.

CLARKE, J. D. A. 1994. Geomorphology of the Kambalda region, Western Australia. *Australian Journal of Earth Sciences*, **41**, 229–239.

COOKE, R., WARREN, A. & GOUDIE, A. 1993. *Desert Geomorphology*. UCL Press, London.

DREXEL, J. F. & PREISS, W. V. (eds) 1995. *The Geology of South Australia, 2, The Phanerozoic*. Geological Survey of South Australia, Adelaide.

EKSTROM, M., MCTAINSH, G. H. & CHAPPELL, A. 2004. Australian dust storms: temporal trends and relationships with synoptic pressure distributions (1960–99). *International Journal of Climatology*, **24**, 1581–1599.

ENGLISH, P., SPOONER, N. A., CHAPPELL, J., QUESTIAUX, D. G. & HILL, N. G. 2001. Lake Lewis basin, central Australia: environmental evolution and OSL chronology. *Quaternary International*, **83–85**, 81–102.

FINK, D., MCKELVEY, B., HANNAN, D. & NEWSOME, D. 2000. Cold rocks, hot sands: *in-situ* cosmogenic applications in Australia at ANTARES. *Nuclear Instruments and Methods in Physics Research B*, **172**, 838–846.

FITZSIMMONS, K. E., RHODES, E. J., MAGEE, J. W. & BARROWS, T. T. 2007. The timing of linear dune activity in the Strzelecki and Tirari Deserts, Australia. *Quaternary Science Reviews*, **26**, 2598–2616.

FUJIOKA, T., CHAPPELL, J., FIFIELD, L. K. & RHODES, E. J. 2009. Australian desert dune fields initiated with Pliocene–Pleistocene global climatic shift. *Geology*, **37**, 51–54.

GARDNER, G. J., MORTLOCK, A. J., PRICE, D. M., READHEAD, M. L. & WASSON, R. J. 1987. Thermoluminescence and radiocarbon dating of Australian desert dunes. *Australian Journal of Earth Sciences*, **34**, 343–357.

GENTILLI, J. 1986. Climate. *In*: JEANS, D. (ed.) *The Natural Environment. A Geography of Australia, 1*. Sydney University Press, Sydney, 14–48.

GILES, E. 1889. *Australia Twice Traversed*. Sampson Low, Marston, Searle and Rivington, London.

GREGORY, J. W. 1906. *The Dead Heart of Australia*. John Murray, London.

HARRISON, S. P. 1993. Late Quaternary lake-level changes and climates of Australia. *Quaternary Science Reviews*, **12**, 211–231.

HESSE, P. P. 1994. The record of continental dust from Australia in Tasman Sea sediments. *Quaternary Science Reviews*, **13**, 257–272.

HESSE, P. P. 2006. *Geomorphology of Nocoleche Nature Reserve*. Access Macquarie Ltd, Macquarie University, for NSW Department of Environment and Conservation, Sydney.

HESSE, P. P. 2007. *Geomorphology of Ledknapper Nature Reserve*. Access Macquarie Limited, Sydney.

HESSE, P. P. & SIMPSON, R. L. 2006. Variable vegetation cover and episodic sand movement on longitudinal desert sand dunes. *Geomorphology*, **81**, 276–291.

HESSE, P. P., HUMPHREYS, G. S. ET AL. 2003. Late Quaternary aeolian dunes on the presently humid Blue Mountains, Eastern Australia. *Quaternary International*, **108**, 13–32.

HESSE, P. P., MAGEE, J. W. & VAN DER KAARS, S. 2004. Late Quaternary climates of the Australian arid zone: a review. *Quaternary International*, **118–119**, 87–102.

HILL, S. M. & BOWLER, J. M. 1995. Linear dunes at Wilson's Promontory and south-east Gippsland, Victoria: relict landforms from periods of past aridity. *Proceedings of the Royal Society of Victoria*, **107**, 73–81.

HOLLANDS, C. B., NANSON, G. C., JONES, B. G., BRISTOW, C. S., PRICE, D. M. & PIETSCH, T. J. 2006. Aeolian–fluvial interaction: evidence for Late Quaternary channel change and wind-rift linear dune formation in the northwestern Simpson Desert, Australia. *Quaternary Science Reviews*, **25**, 142–162.

JENNINGS, J. N. 1968. A revised map of the desert dunes of Australia. *Australian Geographer*, **10**, 408–409.

JENNINGS, J. N. 1975. Desert dunes and estuarine fill in the Fitzroy estuary (north-western Australia). *Catena*, **2**, 215–262.

JENNINGS, J. N. & MABBUTT, J. A. 1986. Physiographic outlines and regions. *In*: JEANS, D. N. (ed.) *Australia: a Geography, 1 The Natural Environment*. Sydney University Press, Sydney, 80–96.

KALMA, J. D., SPEIGHT, J. G. & WASSON, R. J. 1988. Potential wind erosion in Australia: a continental perspective. *Journal of Climatology*, **8**, 411–428.

KING, D. 1960. The sand ridge deserts of South Australia and related aeolian landforms of the Quaternary arid cycles. *Transactions of the Royal Society of South Australia*, **83**, 99–108.

LAWRENCE, C. R. 1966. Cainozoic stratigraphy and structure of the Mallee region, Victoria. *Proceedings of the Royal Society of Victoria*, **79**, 517–553.

LIVINGSTONE, I., WIGGS, G. F. S. & WEAVER, C. M. 2007. Geomorphology of desert sand dunes: a review of recent progress. *Earth-Science Reviews*, **80**, 239–257.

LOFFLER, E. & SULLIVAN, M. E. 1987. The development of the Strzelecki Desert Dunefields, Central Australia. *Erdkunde*, **41**, 42–48.

LOMAX, J., HILGERS, A., WOPFNER, H., GRUN, R., TWIDALE, C. R. & RADTKE, U. 2003. The onset of dune formation in the Strzelecki Desert, South Australia. *Quaternary Science Reviews*, **22**, 1067–1076.

MABBUTT, J. A. 1986. Desert lands. *In*: JEANS, D. (ed.) *Australia: A Geography, 1 The Natural Environment*. Sydney University Press, Sydney, 180–202.

MABBUTT, J. A. & WOODING, R. A. 1983. Analysis of longitudinal dune patterns in the northwestern Simpson Desert, central Australia. *Zeitschrift für Geomorphologie, Neue Folge, Supplementband*, **45**, 51–69.

MADIGAN, C. T. 1936. The Australian sand-ridge deserts. *Geographical Review*, **26**, 205–227.

MAROULIS, J. C., NANSON, G. C., PRICE, D. M. & PIETSCH, T. 2007. Aeolian–fluvial interaction and climate change: source-bordering dune development over the past ~100 ka on Cooper Creek, central Australia. *Quaternary Science Reviews*, **26**, 386–404.

MARTIN, H. A. 1990. Tertiary stratigraphic palynology and palaeoclimate of the inland river systems in New South

Wales. *In*: WILLIAMS, M. A. J., KERSHAW, P. & DE DECKKER, P. (eds) *The Cainozoic in Australia: A Re-appraisal of the Evidence*. Geological Society of Australia, Sydney, 181–194.

MARTIN, H. A. 1997. The stratigraphic palynology of bores along the Darling River, downstream from Bourke, New South Wales. *Proceedings of the Linnean Society of New South Wales*, **119**, 87–102.

McKEE, E. D. 1979. Introduction to a study of global sand seas. *In*: McKEE, E. D. (ed.) *A Study of Global Sand Seas*. US Geological Survey, Professional Papers, **1052**, 1–20.

McLAREN, S., WALLACE, M. W., PILLANS, B. J., GALLAGHER, S. J., MIRANDA, J. A. & WARNE, M. T. 2009. Revised stratigraphy of the Blanchetown Clay, Murray Basin: age constraints on the evolution of paleo Lake Bungunnia. *Australian Journal of Earth Sciences*, **56**, 259–270.

McTAINSH, G. H., LYNCH, A. W. & TEWS, E. K. 1998. Climatic controls upon dust storm occurrence in eastern Australia. *Journal of Arid Environments*, **39**, 457–466.

MORGAN, G. & TERREY, J. 1992. *Nature Conservation in New South Wales*. National Parks Association of NSW, Sydney.

NANSON, G. C., YOUNG, R. W., PRICE, D. M. & RUST, B. R. 1988. Stratigraphy, sedimentology and late Quaternary chronology of the Channel Country of Western Queensland. *In*: WARNER, R. F. (ed.) *Essays in Fluvial Geomorphology*. Academic Press, Sydney, 151–176.

NANSON, G. N., PRICE, D. M., YOUNG, R. W., SHORT, S. A. & JONES, B. G. 1991. Comparative uranium–thorium and thermoluminescence chronologies for weathered alluvial sequences in the seasonally dry tropics of Northern Queensland, Australia. *Quaternary Research*, **35**, 347–366.

NANSON, G. C., PRICE, D. M. & SHORT, S. A. 1992. Wetting and drying of Australia over the past 300 ka. *Geology*, **20**, 791–794.

NANSON, G. C., CHEN, X. Y. & PRICE, D. M. 1995. Aeolian and fluvial evidence of changing climate and wind patterns during the past 100 ka in the western Simpson Desert, Australia. *Palaeogeography, Palaeoclimatology, Palaeoecology*, **113**, 87–102.

NANSON, G. C., PRICE, D. M. *ET AL.* 2008. Alluvial evidence for major climate and flow regime changes during the middle and late Quaternary in eastern central Australia. *Geomorphology*, **101**, 109–129.

NOTT, J. F. & PRICE, D. M. 1991. Late Pleistocene to Early Holocene aeolian activity in the upper and middle Shoalhaven catchment, New South Wales. *Australian Geographer*, **22**, 168–177.

PAGE, K., NANSON, G. & PRICE, D. 1996. Chronology of Murrumbidgee River palaeochannels on the Riverine Plain, southeastern Australia. *Journal of Quaternary Science*, **11**, 311–326.

PAIN, C. 2008. A revised map of Australia's physiographic regions: a hierarchical background for digital soil mapping. *Geophysical Research Abstracts*, **10**, doi: 1607-7962/gra/EGU2008-A-05870.

PELL, S. D., CHIVAS, A. R. & WILLIAMS, I. S. 1999. Great Victoria Desert: development and sand provenance. *Australian Journal of Earth Sciences*, **46**, 289–299.

PELL, S. D., CHIVAS, A. R. & WILLIAMS, I. S. 2000. The Simpson, Strzelecki and Tirari Deserts: development and sand provenance. *Sedimentary Geology*, **130**, 107–130.

PELL, S. D., CHIVAS, A. R. & WILLIAMS, I. S. 2001. The Mallee Dunefield: development and sand provenance. *Journal of Arid Environments*, **48**, 149–170.

QUIGLEY, M., CUPPER, M. L. & SANDIFORD, M. 2006. Quaternary faults of southern Australia: palaeoseismicity, slip rates and origin. *Australian Journal of Earth Sciences*, **53**, 285–301.

READHEAD, M. L. 1988. Thermoluminescence dating study of quartz in aeolian sediments from southeastern Australia. *Quaternary Science Reviews*, **7**, 257–264.

REID, A. J. & HOU, B. 2006. Source of heavy minerals in the Eucla Basin palaeobeach placer province, South Australia: age data from detrital zircons. *MESA Journal*, **42**, 10–14.

SANDIFORD, M. 2003. Geomorphic constraints on the late Neogene tectonics of the Otway Ranges. *Australian Journal of Earth Sciences*, **50**, 69–80.

SANDIFORD, M., QUIGLEY, M., DE BROEKERT, P. & JAKICA, S. 2009. Tectonic framework for the Cenozoic cratonic basins of Australia. *Australian Journal of Earth Sciences*, **56**, S5–S18.

SHEARD, M. J., LINTERN, M. J., PRESCOTT, J. R. & HUNTLEY, D. J. 2006. Great Victoria Desert: new dates for South Australia's ?oldest desert dune system. *MESA Journal*, **42**, 15–26.

SIRCOMBE, K. N. & FREEMAN, M. J. 1999. Provenance of detrital zircons on the Western Australian coastline—implications for the geologic history of the Perth basin and denudation of the Yilgarn craton. *Geology*, **27**, 879–882.

SMITH, M. A. 1989. The case for a resident human population in the Central Australian Ranges during full glacial aridity. *Archaeology in Oceania*, **24**, 93–105.

SPRIGG, R. C. 1982. Alternating wind cycles of the Quaternary era and their influences on aeolian sedimentation in and around the dune deserts of south-eastern Australia. *In*: WASSON, R. J. (ed.) *Quaternary Dust Mantles of China, New Zealand and Australia*. Australian National University, Canberra, 211–240.

STEWART, G. A. 1954. Geomorphology of the Barkly region. *In*: *Survey of the Barkly region, Northern Territory and Queensland, 1947–48*. CSIRO Land Research Series, **3**, 42–58.

STURT, C. 1849. *Narrative of an Expedition into Central Australia*. T and W Boone, London.

THOM, B., HESP, P. & BRYANT, E. 1994. Last glacial 'coastal' dunes in Eastern Australia and implications for landscape stability during the Last Glacial Maximum. *Palaeogeography, Palaeoclimatology, Palaeoecology*, **111**, 229–248.

TWIDALE, C. R. 1972. Evolution of sand dunes in the Simpson Desert, Central Australia. *Transactions of the Institute of British Geographers*, **56**, 77–109.

TWIDALE, C. R. 2008. The study of desert dunes in Australia. *In*: GRAPES, R. H., OLDROYD, D. & GRIGELIS, A. (eds) *History of Geomorphology and Quaternary Geology*. Geological Society, London, Special Publications, **301**, 215–239.

TWIDALE, C. R., PRESCOTT, J. R., BOURNE, J. A. & WILLIAMS, F. M. 2001. Age of desert dunes near Birdsville, southwest Queensland. *Quaternary Science Reviews*, **20**, 1355–1364.

VAN DE GRAAFF, W. J. E., CROWE, R. W. A., BUNTING, J. A. & JACKSON, M. J. 1977. Relict early Cainozoic drainages in arid Western Australia. *Zeitschrift für Geomorphologie, Neue Folge*, **21**, 379–400.

VAN ETTEN, E. J. B. 2009. Inter-annual rainfall variability of arid Australia: greater than elsewhere? *Australian Geographer*, **40**, 109–120.

WASSON, R. J. 1983. The Cainozoic history of the Strzelecki and Simpson dunefields (Australia), and the origin of the desert dunes. *Zeitschrift für Geomorphologie, Neue Folge, Supplementband*, **45**, 85–115.

WASSON, R. J. 1986. Geomorphology and Quaternary history of the Australian continental dunefields. *Geographical Review of Japan, Series B*, **59**, 55–67.

WASSON, R. J. & HYDE, R. 1983. Factors determining desert dune type. *Nature*, **304**, 337–339.

WASSON, R. J., FITCHETT, K., MACKEY, B. & HYDE, R. 1988. Large-scale patterns of dune type, spacing and orientation in the Australian continental dunefield. *Australian Geographer*, **19**, 89–104.

WATKINS, J. J. 1993. Thermoluminescence dating of Quaternary sediments from the Nyngan–Walgett area. *Geological Survey of New South Wales, Quaterly Notes*, **89**, 23–29.

WERNER, B. T. 1995. Eolian dunes: computer simulations and attractor interpretation. *Geology*, **23**, 1107–1110.

WOPFNER, H. & TWIDALE, C. R. 1967. Geomorphological history of the Lake Eyre Basin. *In*: JENNINGS, J. N. & MABBUTT, J. A. (eds) *Landform studies from Australian and New Guinea*. Australian National University Press, Canberra, 119–143.

WOPFNER, H. & TWIDALE, C. R. 1988. Formation and age of desert dunes in the Lake Eyre depocentres in central Australia. *Geologische Rundschau*, **77**, 815–834.

ZHENG, H., POWELL, C. M. & ZHAO, H. 2003. Eolian and lacustrine evidence of late Quaternary palaeoenvironmental changes in southwestern Australia. *Global and Planetary Change*, **35**, 75–92.

Advances in Quaternary studies in Tasmania

ERIC A. COLHOUN[1], KEVIN KIERNAN[2], TIMOTHY T. BARROWS[3]*
& ALBERT GOEDE[2]

[1]*Earth Sciences, University of Newcastle, Callaghan, NSW 2308, Australia*

[2]*School of Geography and Environmental Studies, University of Tasmania, Hobart, 7001, Australia*

[3]*Geography, College of Life and Environmental Sciences, University of Exeter, Exeter, EX4 4RJ, UK*

Corresponding author (e-mail: T.Barrows@exeter.ac.uk)

Abstract: The last 35 years have seen rapid advances in our knowledge of climate change during the Quaternary Period in Tasmania. Extensive mapping and new dating studies, particularly since the advent of exposure dating, have revealed that maximum ice advance occurred 1 Ma ago and later advances were less extensive. Ice advances occurred several times during the last 100 ka, not only during the Last Glacial Maximum. Deglaciation was rapid after 18 ka and complete by 14 ka. Ice strongly affected limestone and produced extensive glaciokarst with deranged surface drainage. Glacial sediment plugged conduits to underground passages partially filled with glacio-fluvial gravels. Periglacial erosion, and human impact since late oxygen isotope stage (OIS) 3, enhanced sediment influxes. New pollen records, particularly from Lake Selina, provide a 125 ka vegetation and climate record representative of the Southern Hemisphere. Finally, stable isotope studies of speleothem growth have revealed wide swings in climate. The climate was warm and moist during OIS 5e and early in OIS 1. Climate was cold and dry during OIS 5d and 4, and prevented speleothem growth during OIS 3 and OIS 2.

Located at 40–43°S, Tasmania is the most southerly part of Australia. It has a cool maritime temperate climate and is separated from the rest of Australia by a shallow continental shelf. Currently Tasmania has no glaciers, but the formation of ice caps and glaciers on several occasions during the Cenozoic contributed to major modification of many mountain areas of western Tasmania, and is an important record of terrestrial climate change. Our understanding of major shifts between glacial and interglacial climatic conditions in the Late Cenozoic, particularly during the Quaternary, has greatly improved during the last 35 years. This paper outlines some of the most significant advances.

Charles Gould, a government surveyor of geology, first recognized in 1859–1860 that Tasmania had been glaciated when he identified moraines in the Cuvier Valley west of Mount Olympus in the Central Highlands (Banks *et al.* 1987). The first model of glaciation was developed by Arndell Neil Lewis, a Hobart lawyer, parliamentarian and military officer, and published posthumously (Lewis 1945). The model presented a three-fold picture of glaciations with an early ice cap covering much of western Tasmania, subsequent extensive valley glaciation, and finally, restricted glaciation mainly by cirque glaciers. Although he could not date the glaciations, Lewis suggested that they might represent the last three European glaciations identified by Penck and Bruckner (1909). The lack of dating led later workers to suggest that Lewis's three glaciations could be accommodated within the Last Glaciation, although it was recognized that evidence for multiple glaciation would eventually be found (Jennings & Banks 1958; Davies 1967, 1974; Derbyshire 1972). This highlighted the need to date the glacial deposits and determine how many glaciations had affected Tasmania in the Cenozoic, particularly during the Quaternary.

Cenozoic glaciations

Glaciation occurred in Tasmania several times during the Cenozoic between the Oligocene and Pliocene. The evidence is localized and not conspicuous in the present landscape, and will not be considered here. It has been presented by Sutherland and Hale (1970), Augustinus and Idnurm (1993) and Macphail *et al.* (1993).

Early Pleistocene glaciations

During the Early Pleistocene ice covered *c.* 7000 km^2 of the Central Plateau and western mountain regions (Fig. 1). Maximum ice thickness was 500–800 m close to, and west of, the Central Highlands,

From: BISHOP, P. & PILLANS, B. (eds) *Australian Landscapes*. Geological Society, London, Special Publications, **346**, 165–183.   DOI: 10.1144/SP346.10   0305-8719/10/$15.00 © The Geological Society of London 2010.

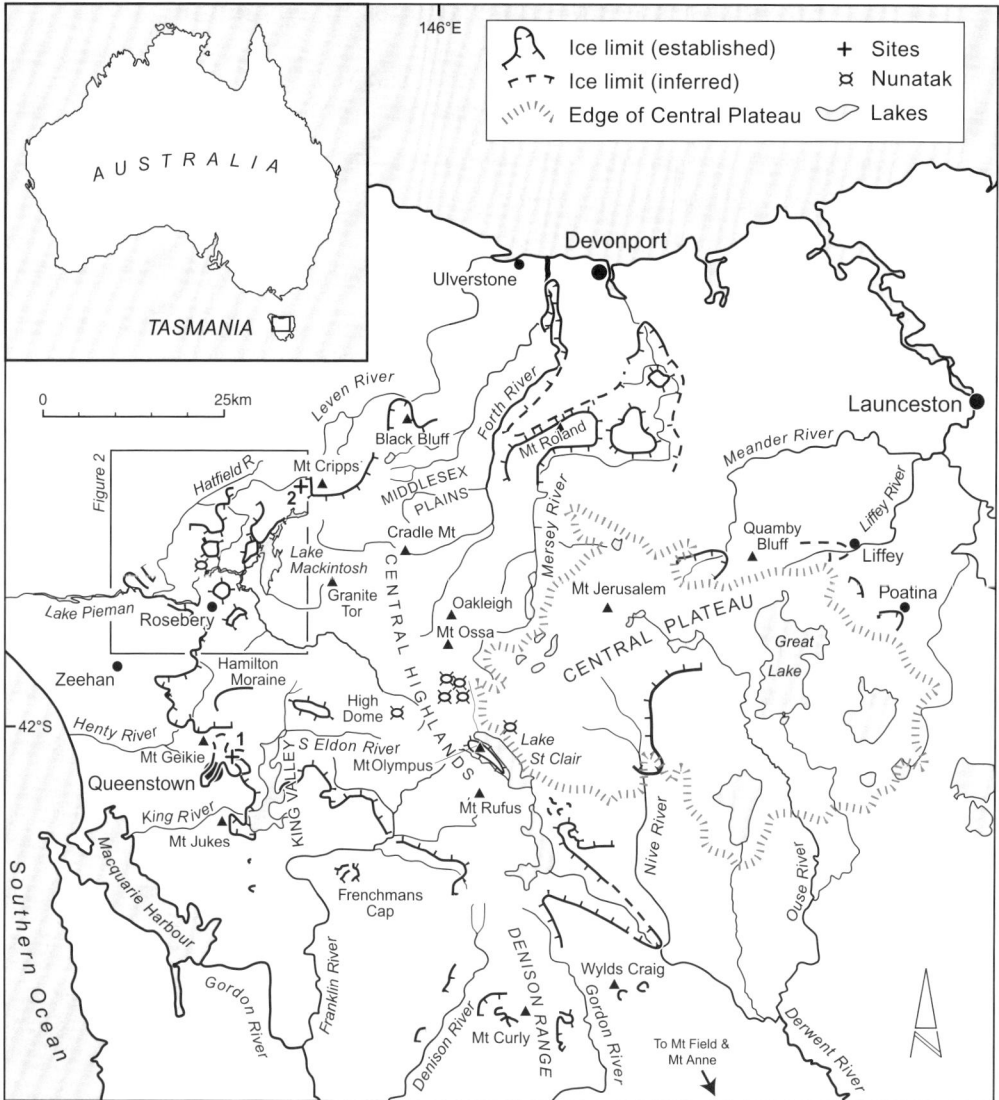

Fig. 1. The maximum extent of Late Cenozoic glaciation in Tasmania (mainly Early Pleistocene Linda Glaciation). The major areas of ice cap cover were between the Central Plateau and West Coast Range north of Queenstown, with outlet glaciers extending along the major river valleys (after Kiernan 1990*a*).

particularly in the Murchison–Mackintosh catchment (Kiernan 1990*a*; Augustinus 1999*a*). The outlet glaciers in the valleys extending west, north and south of the ice cap were about 350–150 m thick, depending on distance from the ice source and size of the glacier system. Cirque and short valley glaciers also formed on several mountain ranges in the SW.

The Linda Glaciation appears to have been multi-phase and the deposits are well exposed adjacent to the central West Coast Range. The deposits

are deeply chemically weathered with the thickness of weathering rinds formed on subsurface dolerite clasts averaging between 57 and 90 mm at different sites, with some clasts having rinds as thick as 300 mm. Using the mean thickness of rinds developed on dolerite erratics within deposits of Last Glacial Maximum (LGM) OIS 2 age and assuming a linear rate of weathering with time it was initially estimated that the deposits were at least 600 ka old (Kiernan 1983*a*; Colhoun 1985). As the weathering rate is unlikely to have been linear but would have proceeded more slowly with rind development the age of the glacial deposits was regarded as likely to be considerably greater.

Moraines and deeply weathered deposits attributed to early glaciation have been differentiated throughout the West Coast Range region. They were first found in the Linda Valley, which is the type site (1 in Fig. 1) (Colhoun 1985). They also occur in the King Valley to the SE (Fitzsimons *et al.* 1992) and in the Bulgobac–Pieman Valley system NW of the range (2 in Fig. 1). In the Bulgobac Valley an outer and older but undated stage of glaciation, the Que, has also been identified (Augustinus & Colhoun 1986).

Determining the age of the Linda Glaciation(s) has had a pivotal role in understanding the glacial history of Tasmania. A wood sample held in The Tasmanian Museum and reported to have been collected from varved clays closely associated with moraines at Linda was dated to $26\,480 \pm 800$ years BP (W-323) (Gill 1956). For nearly 30 years this date was accepted as evidence that nearly all former ice in Tasmania belonged to the Last Glaciation (Würm or Wisconsin). However, another wood sample from glacial deposits in the same area gave an age of $>40\,000$ years BP (R-488) (M. Banks, pers. comm.). Its significance as an infinite date was not recognized and the record was not published.

Realization that it was important to establish if the wood samples from the glacial deposits near the head of the Linda Valley were beyond the range of radiocarbon dating led to further field examination. The wood fragments dated by Gill appeared to have been obtained in Idaho Creek, a tributary of Linda Creek. There, tree stumps in growth position and fallen trunks of *Phyllocladus* occurred within and on a palaeosol surface formed in alluvial fan gravels. The fan gravels abutted conglomerate bedrock and were juxtaposed to Linda glacial deposits, but were clearly alluvial deposits. Radiocarbon dating of the wood, several times, finally showed that it exceeded 48 500 years BP in age (ANU 3413) (Colhoun & Fitzsimons 1990). This indicated that the previously dated wood samples were likely to have been contaminated by modern carbon, and the original date of Gill (1956) thus cannot be accepted as placing the Linda glacial deposits within the Last Glaciation of OIS 2.

An old age for the Linda Glaciation(s) was further suggested because the 1.5 m thick organic palaeosol was rich in pollen. Analysis of the pollen indicated a Late Pliocene age for the palaeosol (Fitzsimons *et al.* 1992; Macphail *et al.* 1995). The age of the Linda Glaciation(s) was thus post-Late Pliocene.

The complexity of early glaciations in Tasmania has yet to be fully resolved. Multiple ice cap and glacier advances have been found to occur more widely than in the West Coast Range region.

Deeply chemically weathered glacial deposits are known north of the Central Plateau in the Mersey and Forth valleys (Hannan & Colhoun 1987; Kiernan 1989*a*, 1990*a*; Kiernan & Hannan 1991), on the southern margin of the Central Plateau in the Derwent and upper Franklin valleys (Kiernan 1989*a*, 1990*b*), and further south in the Florentine, Weld and Huon valleys (Kiernan 1983*a*, 1990*b*; Kiernan *et al.* 2001).

A probable Early Pleistocene age for the ice advances was indicated when it was shown that varved clays in lake deposits associated with the Linda moraines have reversed magnetic polarity (Barbetti & Colhoun 1988). Extensive palaeomagnetic work throughout western Tasmania showed that the deeply weathered deposits have reversed polarity and can be separated on the basis of spatial distribution, stratigraphy and post-depositional modification from the later much less weathered deposits with normal polarity (Pollington *et al.* 1993; Augustinus *et al.* 1995; Fig. 2). Hence, the glacial deposits and multiple ice advances associated with the Linda Glaciation are over 783 ka old and of at least Matuyama Chron age (Spell & Mc Dougall 1992). Being younger than Late Pliocene and older than the Matuyama–Brunhes boundary, it is likely that most deposits and ice advances referred to the Linda Glaciation(s) are of Early Pleistocene age. Similarly, in the Pieman Valley most deposits referred to the Bulgobac Glaciation are probably of approximately the same age.

In the Pieman Valley on the divide between the Huskisson and Marionoak rivers a short sequence of deposits shows glacial silts with reversed magnetization overlying organic-rich silts with normal magnetization. The glacial silts belong to the Bulgobac Glaciation and the organic silts contain pollen of numerous extinct taxa known from the Late Pliocene. The normal magnetization suggests that the organic silts may have been deposited during the Jaramillo event, the phase of normal magnetization preceding the Matuyama–Brunhes reversal. The Jaramillo event has an age of 0.99–1.07 Ma, and if the silts were deposited during that period it suggests that the Bulgobac Glaciation occurred shortly after 1 Ma (Augustinus & Macphail 1997).

Although to date most evidence suggests that the glacial deposits identified as Linda and Bulgobac are of Early Pleistocene age, the moraines and deposits at the head of the Linda Valley are very complex and older ice advances could have occurred in the Pliocene. Similarly, older ice advances such as the Que north of Bulgobac are of unknown age.

Middle Pleistocene glaciations

During the Middle Pleistocene ice accumulated on the western Central Plateau, the Central Highlands,

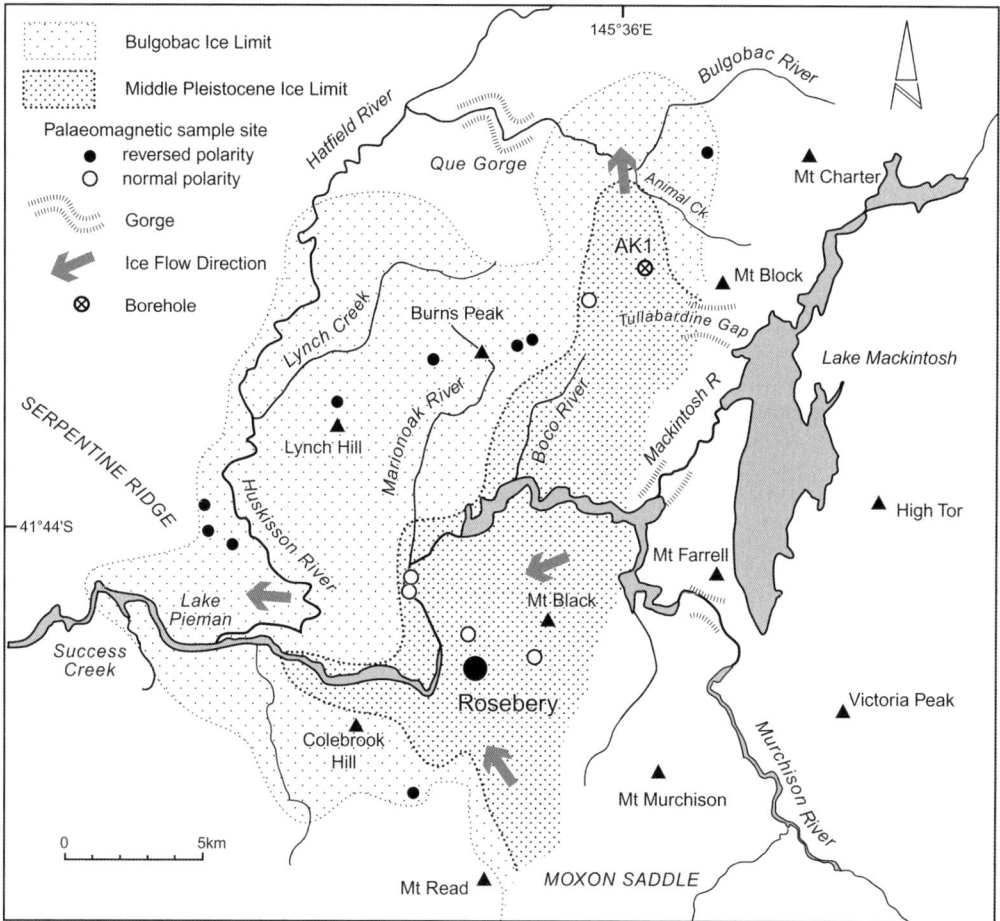

Fig. 2. Separation of Early Pleistocene Bulgobac glacial deposits (= Linda) from Middle Pleistocene glacial deposits between Success Creek and the Bulgobac River in the upper Pieman catchment on the basis of reversed v. normal magnetic polarity (after Augustinus *et al.* 1995).

the West Coast Range, and on several mountain ranges in the SW. The outlet glaciers did not extend as far and were not as thick as during the Early Pleistocene. The moraines and associated ice-contact landforms are better preserved and the deposits are less chemically weathered. The mean thicknesses of weathering rinds on subsurface dolerite erratics at different sites range from 5 to 15 mm. All tested lake deposits associated with Middle Pleistocene tills and moraines in the West Coast Range region are normally magnetized, indicating a Brunhes Chron age of <0.783 Ma.

NW of the West Coast Range ice advances belonging to three distinct glaciations, the Animal Creek, Bobadil and Boco, are represented by deposits in the Pieman catchment. They have been analysed for magnetic polarity, weathering of dolerite and andesite clasts, and iron from ferricretes within

sediment core AK 1 which has been dated by the U/Th method (Augustinus & Colhoun 1986; Augustinus *et al.* 1995, 1997) (Table 1).

The difference in dolerite weathering rind thicknesses between the Boco and Bobadil deposits suggests that the latter are at least twice as old. The very high values of weathering rind thicknesses for the Bulgobac deposits demonstrate that they are much older. The gradual density decrease in the andesite clasts is consistent with increase in relative age from the Boco to Bulgobac glacial deposits, but little is known about how andesite weathering varies with time. The minimum ages of the glacial events have been determined by U/Th assays on iron pans and ferricretes in which iron from the glacial deposits was leached during the subsequent interglacial period and precipitated lower in the sediment profile.

Table 1. *Characteristics of deposits that suggest age differences for the Early and Middle Pleistocene glaciations of the Pieman and Bulgobac valleys*

Glaciation	Magnetic polarity	Dolerite weathering rind thickness (mm)	Andesite density (g cm^{-3})	U/Th age of ferricretes in Core AK 1 (ka)
Boco	Normal	7.1	2.46–2.6	$78 + 23/-20$
Bobadil	Normal	13.1	2.22–2.24	$178 + 20/-18$
Animal Creek	Normal	–	–	$240 + 45/-35$
Bulgobac	Reversed	130–240	2.05–2.25	–

Collectively, the data suggest that the Animal Creek, Bobadil and Boco glaciations represent separate glacial stages of Middle Pleistocene age. The U/Th dates indicate that they occurred during the later rather than earlier part of the Middle Pleistocene and long succeeded the glacial events associated with the Bulgobac Glaciation. The Boco sediments were probably deposited during OIS 6, Bobadil during OIS 8 and Animal Creek during OIS 10, with the weathering intervals when the ferricretes were formed represented by the interglacials of OIS 5, 7 and 9 (Augustinus *et al.* 1994, 1997; Augustinus 1999*a*, *b*).

Middle Pleistocene glacial deposits are also well represented east of the West Coast Range in the middle King Valley. Three separate ice advances occurred during OIS 6, each represented by till, moraine and outwash deposits. Buried deposits of two older ice advances attributed to the Moore Glaciation are separated by organic deposits with an amino-acid racemization age of OIS 10 (Fitzsimons *et al.* 1992).

Extensive Middle Pleistocene ice advances occurred north of the Central Highlands and Central Plateau. From Cradle Mountain National Park, glaciers extended northwards onto Middlesex Plains (Thrush 2008). From the northeastern Central Highlands and northwestern Central Plateau, large glaciers extended down the Forth and Mersey valleys (Hannan & Colhoun 1987; Kiernan & Hannan 1991). SE of the Central Highlands and Central Plateau a large glacier flowed via Lake St Clair into the upper Derwent and Nive valleys, and to the south and west glaciers flowed into the upper Gordon and Franklin valleys (Kiernan 1990*a*, 1991, 1995). In SW Tasmania, Middle Pleistocene age glaciers extended SE from the Mt Anne massif into the Weld Valley. They also extended from Mt Field West southwards into the Humboldt Valley and NW into Lawrence Rivulet (Kiernan 1989*b*, 1990*b*, *c*; Kiernan *et al.* 2001). The ice limits have been determined by mapping of landforms, and deposits have been dated using weathering techniques, palaeomagnetic

signatures, U/Th dating, and cosmogenic isotope exposure age dating methods.

A significant recent finding is that the very large Hamilton Moraine in the West Coast Range, long considered to mark the ice limit of the Last Glaciation (Lewis 1945; Colhoun 1985), is older than OIS 2. This moraine is a complex structure formed during more than one period of ice advance. The distal slope appears to be 300 m high when viewed from the west but the deposits are draped over a buried rock shelf and may be only 100–150 m thick. Exposure age dating has provided six cosmogenic isotope age values for the moraine south of Basin Lake. Three ages were determined at the Australian National University (ANU) by Barrows *et al.* (2002) during a study of the age of the Last Glacial Maximum in Tasmania. Three were determined at the Australian Nuclear Science and Technology Organisation (ANSTO) during a study of exposure ages of glacial deposits in the West Coast Range, details of which are yet to be published. Barrows *et al.* (2002) obtained ^{10}Be ages ranging from 190–350 ka on quartzitic Owen conglomerate boulders using a production rate of 5.02 ± 0.27 atoms g^{-1} a^{-1} at 1013 mbar pressure (sea level) scaled for latitude and altitude, assuming that 3% of ^{10}Be production at sea level is due to muon reactions. The ANSTO data include paired assays for ^{10}Be and ^{26}Al using calculated site production rates of 1465 atoms g^{-1} a^{-1} (samples HM 10 and 11) and 1589 (HM 12) to provide a mean weighted dual isotope age for the sample. The samples determined at ANSTO gave the following ages: OZ 0739 ARC-HM-10, 218.1 ± 25.7 ka; OZ 0741 ARC-HM-11, 214.8 ± 12.1 ka; OZ 0742 ARC-HM-12, 195.3 ± 10.8 ka.

All exposure ages indicate that the Hamilton Moraine was formed long before OIS 2. Four of the ages fall within OIS 7 and one within OIS 8, which, given that the ages are likely to be minimal, suggests that ice last advanced to and constructed a large part of the moraine complex south of Basin Lake during OIS 8. The oldest value of 352 ± 31 ka may not be an anomalous age, as the moraine is known to have a core of strongly weathered deposits exposed on Boulder Hill. This exposure age suggests that moraines were formed SW of Basin Lake during OIS 10 or in an earlier glacial stage and some boulders are derived from older till. The most likely explanation for the complexity and multiple age character of the southern part of the Hamilton Moraine complex is that the glacier that expanded from the cirques east of Mt Geikie and east of the crest of the West Coast Range through the gap in the mountains at Lake Margaret attained approximately the same position on several occasions. The northern part of the Hamilton Moraine extending around Basin Lake onto the slopes of Mt Geikie, which is topographically the most sharply accentuated and potentially the youngest major moraine in the complex, has not been dated.

In the western Arthur Range of SW Tasmania preliminary exposure ages of 68–204 ka have been obtained from large moraines in two valleys.

They are certainly older than OIS 2 in age and some are probably older than OIS 6. These ages confirm that many of the large moraines in Tasmania were not formed during the Last Glacial Maximum and point to greater age and complexity of ice advances in many of the mountain ranges of western Tasmania than has previously been recognized (Kiernan *et al.* 2008).

Late Pleistocene glaciations

During the Last Glacial Maximum (OIS 2) ice cover is estimated at around 1100 km^2 (Fig. 3). The largest areas of 550 km^2 and 230 km^2 occurred on the western Central Plateau and in the Lake St Clair–upper Franklin Valley, but ice also occurred in the West Coast Range and many mountain ranges of the SW. The ice thickness was greatest (250 m) on the western part of the Central Plateau, where precipitation was highest, and decreased northeastward to 100–50 m as precipitation declined. Moraine

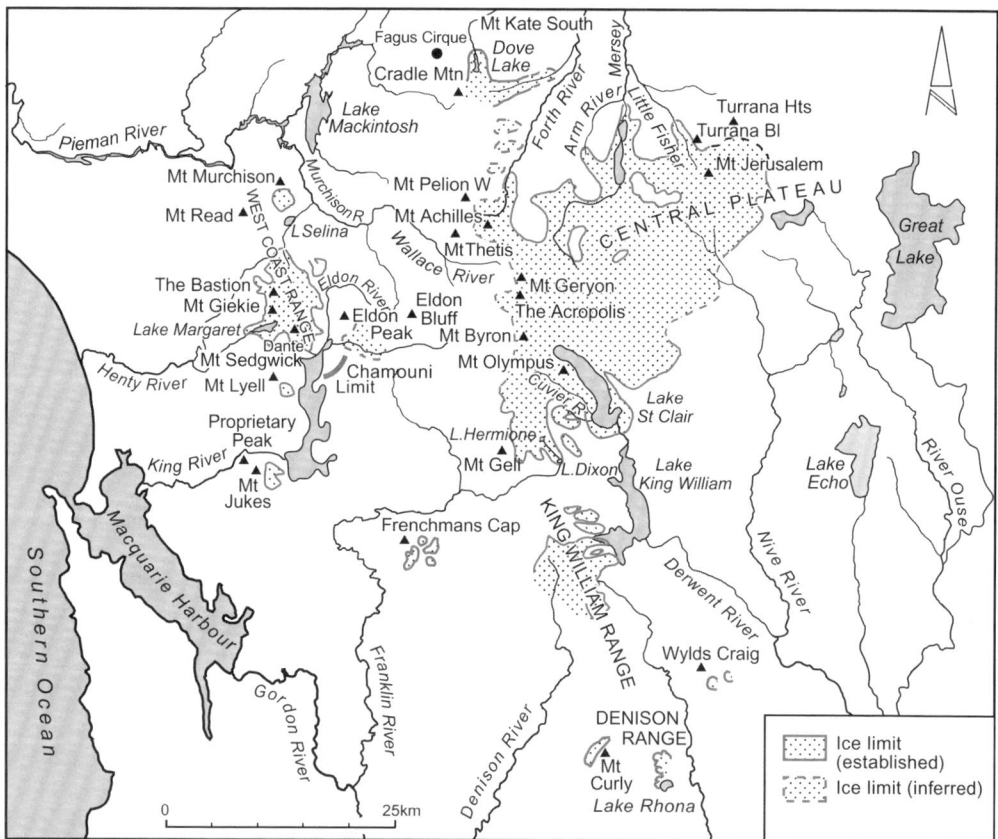

Fig. 3. Major areas of ice occurred on the Central Plateau, Central Highlands and West Coast Range during the Last Glacial Maximum. The Central Plateau ice cap did not connect with the small ice cap on the West Coast Range as it did during the Early Pleistocene.

ridges are much smaller than those formed during OIS 8 and OIS 6 glaciations, and dolerite boulders in the deposits have weathering rinds of about 1 mm thickness restricted to the upper 30 cm of the weathering profile (Kiernan 1991, 1992; Colhoun et al. 1996).

The LGM ice advance has been radiocarbon dated at Dante Rivulet in the King Valley to 18.8 ± 0.5 ka BP and 19.1 ± 0.17 ka BP (22.3–22.5 cal. ka) (Kiernan 1983b; Gibson et al. 1987; Colhoun & Fitzsimons 1996). Many cosmogenic isotope exposure ages from sites across Tasmania show that during OIS 2 ice advances attained their limits between 22 and 17 ka BP with a mean of around 19–20 ka BP (Barrows et al. 2002; Kiernan et al. 2004). After 18–19 ka BP, ice retreated rapidly from all areas during the next 4 ka, with the youngest moraines dated so far having ages of 15.7 ka BP at Fagus Creek cirque in Cradle Valley and between 15.8 and 14.2 ka BP in the Lake Rhona Valley of the Denison Range (McMinn et al. 2008; Thrush 2008). No readvances of ice have been identified after 15–14 ka BP. Thus, currently there is no evidence of glacial conditions that could have resulted from either the Antarctic Cold Reversal or the Younger Dryas phases of cooling.

Cosmogenic isotope exposure ages suggest that in some localities ice advances were locally more extensive during OIS 3 than during OIS 2 (Table 2). Scattered dolerite boulders and degraded moraine extend to 800 m beyond the OIS 2 ice limits south of Lake Ada on the eastern part of the high Central Plateau (Fink et al. 2000). The dolerite boulders have given cosmogenic isotope exposure ages of 24, 31 and 53 ka BP (Table 2), which suggest that a more extensive ice advance occurred during OIS 3. Impressive evidence for an ice advance during OIS 3 to beyond the OIS 2 ice limits occurs in the Broad Valley at Mt Field in south–central Tasmania, where several moraines with short outwash plains span the valley floor. There, large dolerite erratics known as the Griffith

and Taylor boulders occur on the outwash plain distal to the fourth moraine ridge beyond the LGM ice limit, and have been dated to 41 and 44 ka BP (Mackintosh et al. 2006).

Cosmogenic isotope exposure ages of OIS 3 age have also been obtained for glacial boulders, in Dove Gorge at Mt Kate South, and at adjacent Smiths Creek in Cradle Mountain National Park; on cirque moraines of Beatties Tarn and Lake Nicholls at Mt Field East; at Lake Hartz in SE Tasmania; and east of Mt Jukes in the southern West Coast Range (Table 2; Barrows et al. 2002; Thrush 2008). These exposure ages confirm the evidence for pre-OIS 2 ice advances previously obtained by U/Th dating of speleothems associated with glaciofluvial sediments at Welcome Stranger Cave in the Florentine Valley west of Mt Field, and from rock weathering studies (Kiernan et al. 2001) (see the section on karst below).

In addition to OIS 3 glaciation it has also been suggested that an ice advance extended beyond the OIS 2 ice limits in the King Valley during the Last Glaciation (Fig. 3). Glacial deposits underlying the dated LGM deposits at Dante Rivulet have been interpreted as coeval with a broad terrace of till and outwash deposits that extends down-valley and the ice advance is termed the Chamouni Advance. Glacial lake silts associated with this ice advance contain fossil wood and leaf fragments that have been radiocarbon dated to 48.7 ka BP (Fitzsimons et al. 1992). This was the first site in Tasmania where last glacial cycle glacial deposits older than the LGM were identified. An early Last Glaciation age was suggested on the basis of the radiocarbon date being interpreted as infinite and on the limited weathering of the deposits where they occur below the Dante outwash fan. However, the radiocarbon date, although a minimum, does not necessarily unequivocally demonstrate an OIS 4 age for the ice advance, as an early OIS 3 age cannot be excluded. The limited weathering of the deposits also does not necessarily indicate a Last Glaciation

Table 2. *Exposure ages potentially indicating OIS 3 glaciation more extensive than OIS 2 glaciation*

Location	Isotope	Age (ka BP)
Central Plateau[1]	^{36}Cl	24 ± 3, 31 ± 4 and 53 ± 7
Mt Field, Broad Valley[2]	^{36}Cl	44.1 ± 2.2 and 41 ± 2
BeattiesTarn-LakeNicholls[3]	^{36}Cl	33.1 ± 2.4 and 43.1 ± 3
Lake Hartz[3]	^{36}Cl	41.8 ± 3.1 and 48.8 ± 3.1
Mt Kate South, Cradle Park[4]	^{10}Be	43.3 ± 3.0 and 29.9 ± 2.1
Mt Jukes[3]	^{10}Be	42.3 ± 3.7

[1]Fink et al. 2000.
[2]Mackintosh et al. 2006.
[3]Barrows et al. 2002.
[4]Thrush 2008.

age, as the deposits may have been buried for much of the period and the ice advance could have been of OIS 6 age. Thus, the age of the Chamouni ice advance remains uncertain.

Glacial climate

During the last million years, Tasmania's climate alternated between cool to cold humid oceanic and drier conditions. Glaciers were temperate, and climatic snowlines were influenced strongly by westerly precipitation. The localization of glacier systems was strongly influenced by topography, with north–south-trending ridges producing a lee-side effect on snow accumulation and east–west-trending ridges facilitating glacier survival, particularly in SE-facing valleys (Davies 1967).

Davies (1967, 1969) reconstructed the altitude of the climatic snowline for the period of maximum glaciation (then regarded as the Last Glaciation) having regard primarily to precipitation. He showed that the snowline became higher northeastward from about 610 m in the southwestern mountains to 1220 m over the Central Plateau in a distance of 80 km. This trend reflects the pattern of winter snowfall, which is brought mainly by SW winds following the passage of cold fronts. The rise in the snowline across Tasmania is accompanied by a diminished thickness of the icecap on the Central Plateau from about 250 m in the SW to under 100 m in the east, and the presence of an ice-free window in the Walls of Jerusalem area in the north (Hannan & Colhoun 1991).

The rise in snowline eastwards has also been demonstrated by a spatial analysis of the floor altitudes of over 300 cirques. The floor levels of the outermost and lowest cirques represent approximately the mean summer 0 °C temperature during glaciation, which mirrored the snowline trend (Peterson 1968; Peterson & Robinson 1969).

The snowline analysis by Peterson and Robinson also reveals a marked contrast between the cirques of the western mountains, which have over-deepened basins from which glaciers extended a short distance down-valley, and those of the eastern mountains in the SW, where isolated cirques were occupied by discrete glaciers that rarely extended beyond the basins. This pattern indicates that the cirques of the western mountains experienced an oceanic glacial regime with higher annual inputs and losses of snow than did the drier eastern mountains, which had much lower inputs and losses.

The Equilibrium Line Altitudes (ELAs) of the Pleistocene glaciers were determined by air temperature, precipitation, insolation, cloud cover, near-surface wind speed, and humidity. The most important factors were the amounts of winter snow input and summer ablation. Changes in glacial

temperatures compared with present values have been estimated from the relationship between the altitudes of former glacier ELAs and both mean annual and summer (December–February) atmospheric freezing levels calculated from meteorological station data. The Accumulation Area Ratio method (AAR) has been applied as in New Zealand (Porter 1975), where the temperate glacial systems have 65% of their area above the ELA and 35% below it. A lapse rate of 0.65 °C per 100 m, measured between sea level and the summit of Mt Wellington in SE Tasmania (Nunez & Colhoun 1986), has also been used. For a snowline 1000 m lower than at present and using mean annual level of atmospheric freezing, average temperatures were colder by c. 6.5 °C at the LGM compared with present lowland mean temperatures of 11–12 °C (Colhoun 1985). If summer atmospheric freezing levels are used mean temperatures were depressed by about 10 °C and snowlines were about 1500 m lower (Thrush 2008).

The close proximity of the OIS 3 and possible OIS 4 ice limits to that of OIS 2 indicates that the temperature and snowline depression need have been little different from those of OIS 2. Because a small decrease in mean temperature, given sufficient precipitation, can result in considerable ice expansion, temperatures during the ice advances of the Middle and Early Pleistocene glaciations were probably not more than 1–2 °C colder than during OIS 2, and snowlines were only 150–300 m lower (Colhoun 1985). In such mountainous terrain where the altitude of local snowline can vary by as much as 300 m, such estimates are, of course, approximations.

Interactions between glacial and karst processes

Quaternary climate changes had profound effects on processes that operated on the extraglacial landscape. These processes caused changes in the significance of chemical versus physical weathering, slope stability, river flow including underground flow identifiable from cave passages and the large calibre and magnitude of fluvial sediment loads, and in the foci of coastal processes (Davies 1967, 1974). Of particular importance were their impacts on Tasmanian karst regions, which occur throughout the valleys and lowland areas of the west and south, many of which are underlain by dolomites of Precambrian age or by limestone of the Ordovician Gordon Formation. The impact of former glaciations has resulted in formation of the only glaciokarsts in Australia, a type of landscape otherwise recorded from southern temperate latitudes only in the Kahurangi and Fiordland national parks in New Zealand, and from one site in Patagonia

(Maire 2004; Williams 2004). The morphology of the Tasmanian glaciokarsts, and their resulting sub-surface hydrology, displays a clearer relationship to past glaciations than has been demonstrated from their southern counterparts.

The oldest reliable evidence for karst develop-ment in Tasmania is of Middle Devonian age, and occurs at Eugenana (Fig. 4; Banks & Burns 1962). However, most deep underground caves and pas-sages were formed during the late Mesozoic and Cenozoic as river and valley networks were developed on the post-Middle Jurassic dolerite

intrusion surface and superimposed onto the under-lying complex structures of the western Tasmanian Ridge and Valley Province. Although exploration continues to reveal new cave systems, the largest known are over 320 m deep and 20 km long. The large cave systems often have active stream pas-sages with tiers of abandoned passages at higher levels. Such complexes are likely to be of consider-able antiquity.

Some 53 formerly glaciated karst areas have been identified in Tasmania (Kiernan 1996). Many were not glaciated during the Last Glaciation but

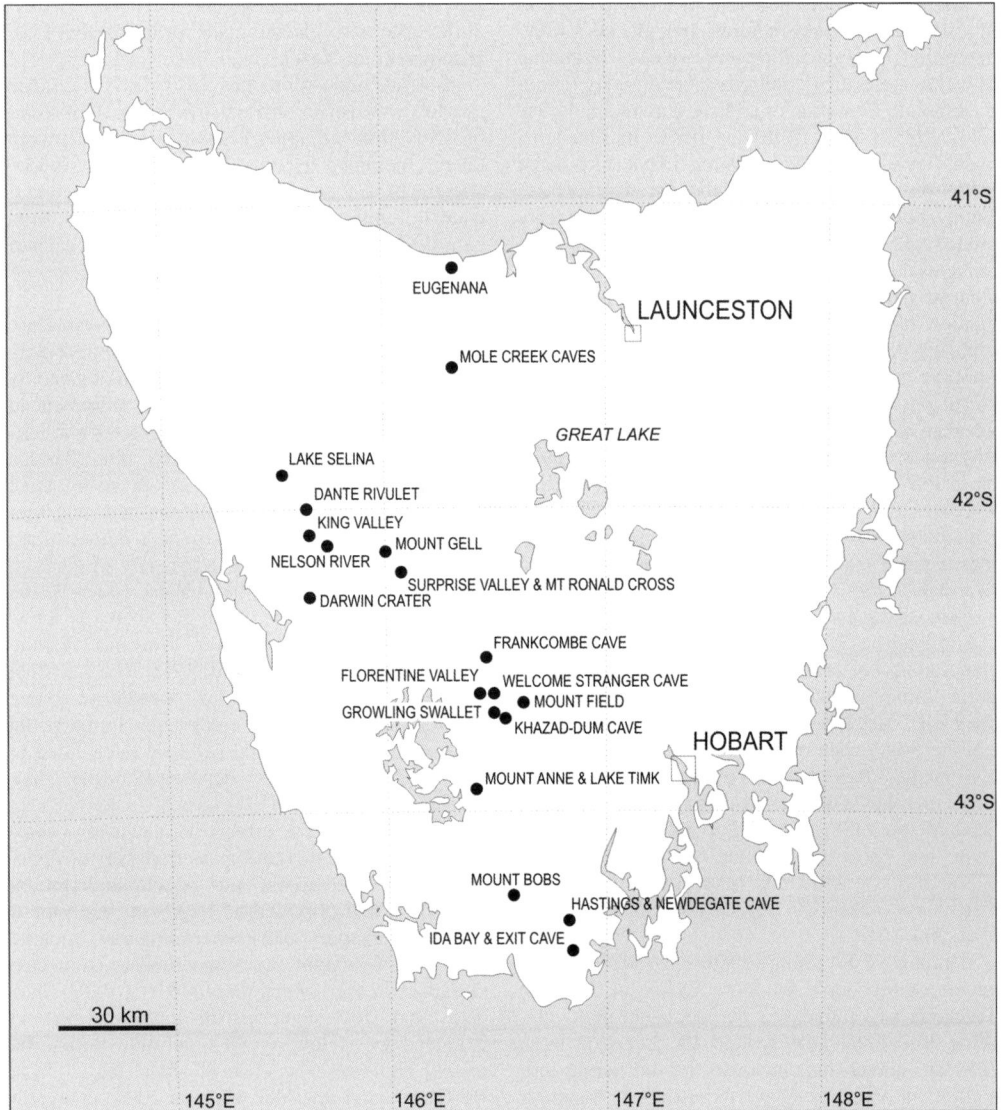

Fig. 4. Location of pollen, karst and speleothem sites referred to in the text.

the impact of surface glacial erosion and sedimentation into the sinks, passages and chambers was extensive during earlier glaciations and gives an essentially glacial aspect to the karstic land surface and shallow underground landforms. Many of the underground systems contain impressive assemblages of large speleothems, which appear to have formed during the later part of the Quaternary. Although studies of the effects of glaciers on these karst areas are in their infancy, there is nevertheless sufficient evidence to show that ice and meltwater had a strong impact above ground, and that meltwater played a major role in cave passage evolution (Jennings & Sweeting 1959; Kiernan 2006).

Glaciers have the potential to erode away previously formed karst features, but glacial valley-deepening, paraglacial pressure-release fracturing of carbonate bedrock and other factors may favour or stimulate formation of a glacio-karstic landscape (Ford 1983; Ford & Williams 1989). In Tasmania, glacial erosion has had varying impacts on karst evolution. At Mt Ronald Cross in the Surprise Valley, opening of joints on the flanks of the deeply eroded valley by glacial unloading seems to have influenced the location of the stream-sink caves. In contrast, in the Timk Valley at Mt Anne similar rock dilation has caused large-scale collapse and has probably destroyed shallow caves (Kiernan 2006). In some cases landforms in Tasmania's glaciated karsts pose the 'chicken or egg' question, as to whether their karst or glacial features developed first. For example, at Lake Timk, which is a large basin that drains underground, it remains unclear whether it was originally formed by ice erosion with karst channels developing subsequently, or was a polje that formed a focus for ice erosion (Kiernan 1990c).

Glacial meltwater carrying a highly abrasive clastic load has been important in gorge and cave elaboration. For example, the 30 m deep limestone gorge through which the Dante Rivulet flows from the West Coast Range was formed partly as a subglacial meltwater channel. The involvement of proglacial meltwater in cave evolution is implied by small moraines constructed during the LGM at the head of the valleys leading to the deep Khazad-Dum and Growling Swallet caves at Mt Field, although it is evident that substantially larger glaciers were present on earlier occasions (Kiernan *et al.* 2001).

Jennings & Sweeting (1959) suggested that the underground course of Mole Creek in northern Tasmania was influenced by proglacial meltwater being forced to flow around the edge of large outwash gravel fans against the bordering hill margins in which caves were developed. A similar situation of diversion of glacial meltwater appears to explain the development of passages towards the downstream end of the large Exit Cave system at Ida Bay (Goede 1969). The underground passages at both Mole Creek and Exit Cave lie outside known Last Glaciation ice limits and the passages were formed during earlier glaciations.

The potential for sediments preserved in fossil proglacial cave passages to shed light on broader landscape evolution is highlighted in the Florentine Valley just west of the Mt Field massif. There meltwater spilled from the Lawrence Glacier over a low divide and invaded Welcome Stranger Cave, which has been partly filled by deposition of at least two generations of gravel. U/Th dating of speleothems stratigraphically associated with the gravels shows that a glacial advance occurred during the last glacial cycle prior to the LGM (Kiernan *et al.* 2001).

Glacial deposition has also had significant impacts on Tasmanian karst. On the western slopes of Mt Gell in the upper Franklin River catchment, lateral moraines deposited across small tributary streams have focused the diversion underground of waters draining from the surrounding slopes. More commonly, glacial deposition has inhibited subsurface drainage or forced temporary reversion of direction to surface drainage courses.

At Lake Sydney in the Mt Bobs area of southern Tasmania, an underground conduit system appears to be too well developed to have formed since the LGM. The conduit is being exhumed from beneath young till, which suggests that it precedes at least the most recent glaciation of the area (Kiernan 1983b, 1989b). The plugging of caves by much older glacial and glaciofluvial sediment and their subsequent flushing out has also been demonstrated in the Nelson River Valley, a tributary of the King Valley east of the southern West Coast Range (Kiernan 1983b). Some of the glaciokarst is shown to be ancient in the lower Lawrence Rivulet Valley adjacent to Mt Field West. The karstified surface is mantled by thick, weathered glacial deposits dating from the earliest glaciation of the area, while the karst conduits have been filled by glaciofluvial sediments deposited during later glaciations.

In addition to the effects of glacial processes, periglacial processes and mass transport of debris have contributed to the surface and underground sediments, and provided an important but hitherto largely untapped palaeoenvironmental archive. Analysis of angular limestone rubbles from cave entrances in the lower Franklin River Valley indicates that frost shattering of limestone bedrock extended to as low as 80 m altitude during the LGM. The cessation of frost shattering *c.* 14 ka BP (Kiernan *et al.* 1983; Kiernan 2008) is broadly coincident with formation of the youngest moraines identified in SW Tasmania (McMinn *et al.* 2008).

Late Quaternary vegetation-indicated climate changes

Many Quaternary lake, bog and swamp sites have been drilled in western Tasmania and pollen from the sediments has been analysed. Most Early and Middle Pleistocene records are from short sediment sequences that give snapshots of the types of vegetation and climate conditions for brief periods. The longest record is from the 62.5 m core obtained from Darwin Crater in western Tasmania, which has a basal age of *c*. 0.7 Ma. Only the top 20 m has been analysed for pollen, which shows that the OIS 9–OIS 10 boundary occurs at 1935 cm depth. Unfortunately, several sand and gravel beds occur within the lake clays above 880 cm. Although a record of vegetation change has been obtained from the site between late OIS 10 and OIS 1, the climatic fluctuations after mid-OIS 7 (880 cm) are not clearly defined (Colhoun & van de Geer 1998).

The best record with clearly defined Late Quaternary climate and vegetation changes for Tasmania has been obtained from Lake Selina at 516 m in the northern West Coast Range. The Selina record is summarized here. It provides details of change for the Last Interglacial–Last Glacial cycle, and the Holocene (*c*. 125–0 ka) (Colhoun *et al.* 1999).

The climatic climax vegetation for the Lake Selina area is cool temperate rainforest in which the two main taxa are *Nothofagus cunninghamii* and *Phyllocladus aspleniifolius*, although the modern vegetation has been much altered by burning. The vegetation changes that occurred during the last 125 ka are shown in the summary pollen diagram (Fig. 5). The pollen diagram is based on two cores: a short 45 cm near-surface core and the main core extending to 397 cm depth (the gap in Fig. 5 separates the cores).

The percentage of pollen for total woody taxa, which include rainforest species, other lowland trees and shrubs, and alpine and subalpine trees and shrubs, is shown by the top of the purple curve, which is regarded as a pollen representation of treeline. Estimates of temperature variations from present for the isotope substages are based on the amplitude of variation in the treeline curve.

Figure 5 shows that the Last Interglacial OIS 5e had more rainforest than the Holocene, suggesting that the climate may have been 1–2 °C warmer and precipitation greater. Although this was almost certainly the case (Petit *et al.* 1999), the Lake Selina pollen record alone does not demonstrate it conclusively, as the differences in vegetation probably partly reflect the impact of Aboriginal burning of vegetation in western Tasmania during the Holocene (Fletcher & Thomas 2007).

Between OIS 5e and OIS 5d there was a very sharp decrease in rainforest at the beginning of the Last Glaciation. The amount of change suggests that mean temperatures decreased by at least 4 °C, as rainforest was restricted to below the pollen site, which lies 600 m below the contemporary climatic rainforest limit. Precipitation may also have decreased, but it is difficult to judge this from the pollen record as it would have required a decrease of over 40–50% or prolonged summer drought to have had an impact on the rainforest. OIS 5b had a similar mean temperature to OIS 5d.

Rainforest and lowland woody vegetation (top of orange curve in Fig. 5) were more abundant during OIS 5c and 5a than in OIS 5d or 5b, but were not as abundant as during the Last Interglacial or Holocene, despite the likelihood that Holocene rainforest had not developed optimally because of Aboriginal burning. A mean temperature of 2–3 °C below present is likely for OIS 5c and 5a.

The coldest and driest part of the Last Glaciation vegetation record includes OIS 4, OIS 3 and OIS 2. The proportion of herbaceous taxa during OIS 4 is nearly equal to that of OIS 2. The magnitude of vegetation change suggests a reduction in mean temperature of at least 5 °C during OIS 4 and over 6 °C during OIS 2 compared with today.

In Tasmania, rainforest changes reflect minimum values of temperature depression because rainforest taxa can occur in shrub-form at treeline and still contribute abundant pollen to the alpine zone. However, the massive reduction of rainforest pollen indicates that little rainforest was present in western Tasmania during much of the period, and rainforest was probably nearly absent (except in protected local refugia) during OIS 2. As the modern treeline in the West Coast Range occurs at 1100 m, the pollen data are compatible with a mean temperature reduction of 6–7 °C during OIS 2.

The degree to which the abundances of grasses and herbs during OIS 4 and 2 are due to reduced temperature, reduced precipitation or human impact is difficult to say. The herbaceous vegetation of OIS 4 must have been a response to climate as Aborigines did not reach Tasmania until around 35 ka BP. The longer and stronger herb peak during OIS 2, although probably primarily influenced by cooling, may have been reinforced by burning as Aborigines occupied the valleys of western Tasmania during the LGM (Kiernan *et al.* 1983).

OIS 3 has a small amount of woody taxa but consists predominantly of herbs, which suggests that the climate was closer to glacial than interglacial. A cooling trend occurs from the beginning of OIS 3 into the LGM, and mean temperature during OIS 3 probably averaged about 4 °C below present. Of particular interest are four fluctuations in the vegetation record that suggest short-period variations of temperature with amplitudes of around

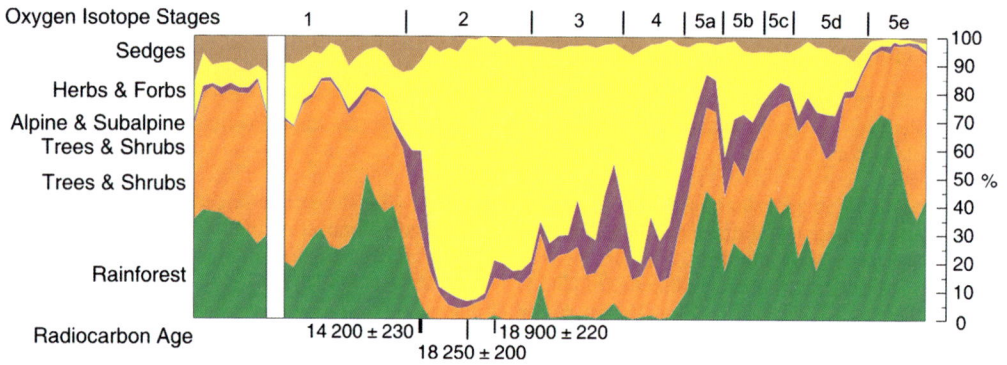

Fig. 5. Summary pollen diagram from Lake Selina. (See Fig. 4 for location.) The major pollen-reconstructed vegetation formations are shown on the left-hand side and percentage counts on the right-hand side. Radiocarbon dates are given below with oxygen isotope stages inferred from the pollen curves above. The top of the purple curve is taken to represent the tree line, which separates alpine from forest and shrub vegetation (complete pollen record for Lake Selina given by Colhoun *et al.* 1999).

1–1.5 °C, but with high values probably around 3 °C and low values 5 °C below present.

Following the strong herb peak of OIS 2 when temperatures were over 6 °C colder and climate was probably considerably drier than at present, rainforest expanded after 14 ka BP in the lowlands and valleys. The initial expansion developed from many local refugia through a pioneer phase of dominant *P. aspleniifolius* to *N. cunninghamii* without evidence for renewed climatic cooling.

Rainforests attained their maximum development during the Early Holocene 'climatic optimum' period of 10–7 ka BP, which was wetter and probably 1–2 °C warmer than today (Macphail 1979). Middle to Late Holocene (6–0 ka BP) records from numerous sites in western Tasmania show increases of *Eucalyptus* and reductions in rainforest pollen. The increase in sclerophyllous vegetation suggests that the climate became cooler and drier, and more variable, and that Aboriginal burning had a greater

impact on Late Holocene vegetation than on early Holocene vegetation.

Climatic significance of the Lake Selina vegetation record

The pollen-inferred treeline curve at Lake Selina (Fig. 5) closely replicates changes in the δD record for the Vostok ice core from central Antarctica (Fig. 6; Petit *et al.* 1999). The parallelism indicates that the regional changes in vegetation at Lake Selina are not solely the product of local climate changes but must represent wider climatic changes in the SW Pacific region and Southern Hemisphere.

The wider significance of the Lake Selina pollen and climatic record is supported by a similar record from Lake Okarito Pakihi at 43°S in southern New Zealand (Vandergoes *et al.* 2005). Barrows *et al.* (2007*b*) have found that there is a close relationship

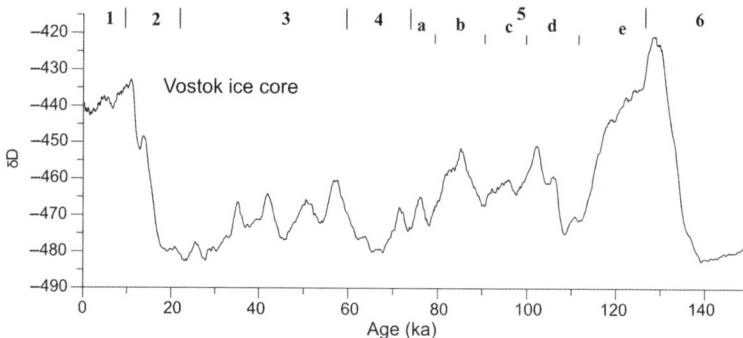

Fig. 6. Deuterium record from Vostok ice core, central Antarctica (after Petit *et al.* 1999). (Note the four climatic fluctuations in OIS 3 and compare them with pollen fluctuations during OIS 3 in Fig. 5.)

between Late Quaternary climate records from Antarctica, the Southern Ocean and temperate latitudes of New Zealand, and the relationship also appears to extend over southern Australia. Of particular importance are the four small-amplitude fluctuations during OIS 3, which, though undated at Lake Selina, appear to correlate with those at Vostok, and to be cyclical fluctuations with a periodicity of around 7–8 ka. Similar fluctuations and inferred short-term climatic events have been recorded by ice-rafted detritus in the southern South Atlantic Ocean (Kanfoush et al. 2000).

Also of note at Lake Selina and in other Tasmanian late-glacial pollen diagrams is rainforest vegetation that does not expand until after the Antarctic Cold Reversal of 13.5–11.7 ka. In addition, rainforest expansion commences in some records during, and in others after, the Younger Dryas period of the North Atlantic region between 12.9 and 11.7 ka but there is no response of vegetation to a phase of cooling at that time.

Late Quaternary environmental change from Tasmanian speleothems

Cave interiors are climatically stable and temperatures rarely vary more than 0.5 °C from mean annual surface temperatures. Information about environmental changes has been obtained from changes in the abundance of speleothem growth and from variations in the rate of growth of single speleothems. Calcite makes up the bulk of speleothems and can be analysed for light stable isotopes ($^{18}O/^{16}O$ and $^{13}C/^{12}C$) and minor elements (Goede & Vogel 1991; Goede 1994). The uranium content has been used to determine accurate and precise age estimates, especially since the introduction of mass spectrometric techniques (thermal ionization mass spectrometry; TIMS). Stalagmites have been widely used for analysis, as they show characteristic growth layers and may grow continuously over long time periods (10^3–10^5 years). Speleothem growth is most abundant and rapid when high temperatures are combined with moist conditions.

In Tasmania, $\delta^{18}O$ variations show a positive relationship with temperature. The temperature effects are indirect because changes are dominated by large changes in the isotopic composition of precipitation over time that are largely, but not entirely, controlled by changes in temperature. Hence quantitative estimates of temperature change are best avoided. Variations in $\delta^{13}C$ are strongly influenced by the bioproductivity on the surface, which in turn depends on the available moisture during the growing season. Magnesium variations are strongly influenced by percolation rates of water and often correlate significantly with $\delta^{13}C$

variations. Strontium variations are influenced by temperature but over longer time periods may be dominated by the amount of atmospheric dust that settles on the surface above the cave (Goede et al. 1998). In inland areas, variations in Br are believed to be related to the frequency and intensity of bushfires. A stalagmite from Lynds Cave at Mole Creek in northern Tasmania (Fig. 4) shows a strong decline in Br between 15 and 12 ka BP (Fig. 7) (Goede 1998). Many cave sites in western Tasmania were abandoned by Aboriginal people during this period, as changing conditions inhibited fires that had been used traditionally to keep the country open, and previously dry caves became wetter and less desirable places in which to live.

The age distribution of Tasmanian stalagmites shows abundant speleothem growth during the Last Interglacial OIS 5e and Holocene OIS 1 (Fig. 8). There was very little growth between 70 and 30 ka BP, reflecting predominantly cool and dry conditions (Xia et al. 2001).

An 840 mm tall stalagmite from Newdegate Cave at Hastings in southeastern Tasmania provides a growth record for the period 155–100 ka BP (Fig. 9). Slow growth at a rate of 18.7 mm ka^{-1} occurred between 155 and 142 ka BP, and the rate decreased to only 5.9 mm ka^{-1} at the end of OIS 6. At 129 ka BP the growth rate increased dramatically to 61.5 mm ka^{-1} before slowing to 16.1 mm ka^{-1} between 122 and 117 ka BP, and then virtually ceasing. Precise TIMS dating shows that the warm and wet Last Interglacial lasted only 7000 years between 129 and 122 ka BP. It probably occurred when the sea attained its highest level and Tasmania had maximum insularity in the westerly wind belt. Cessation of growth by 100 ka BP indicates that the early Last Glacial OIS 5d was very dry and cold (Zhao et al. 2001).

A record for the period 108–95 ka is provided by the basal 600 mm of a 1420 mm stalagmite from

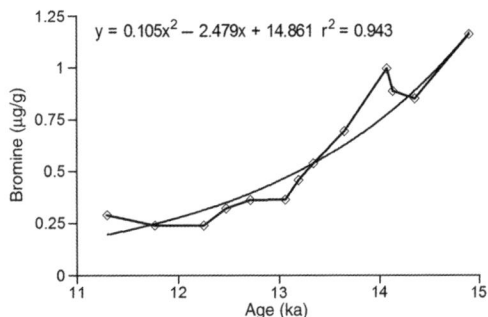

$y = 0.105x^2 - 2.479x + 14.861 \; r^2 = 0.943$

Fig. 7. Lynds Cave stalagmite shows a marked decrease in bromine between 15 and 12 ka BP, which is suggested to represent the impact of aboriginal burning of vegetation.

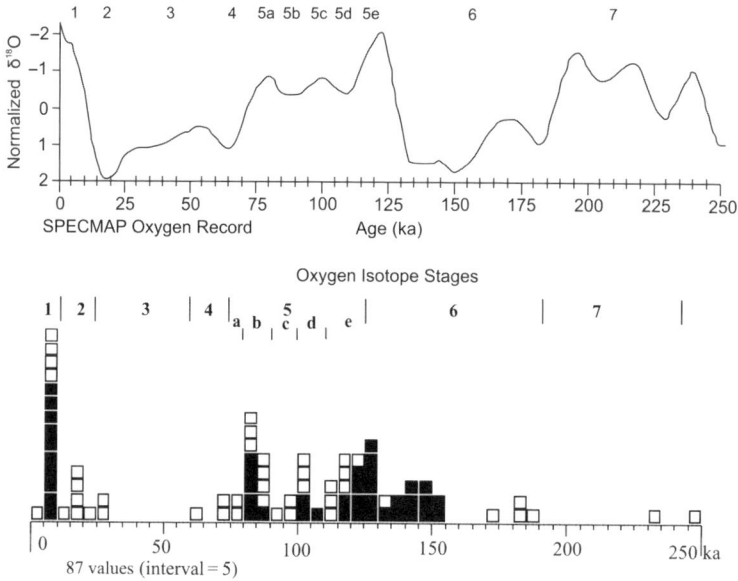

Fig. 8. Frequency distribution of U/Th age determinations for Tasmanian stalagmites between OIS 6 and OIS 1 compared to the SPECMAP oxygen isotope record of the last 250 000 years. Note the relative absence of records during early OIS 6 and between OIS 4 and OIS 2. □, α-counting data of Goede & Harmon (1983); ■, unpublished TIMS data courtesy of Xia & Collerson, University of Queensland.

Little Trimmer Cave (current Mean Annual Temperature (MAT) 9.5 °C) at Mole Creek, northern Tasmania, but the time scale is imprecise because alpha spectrometry was used for dating. The record suggests fully glacial conditions from 108 to 100 ka BP followed by a very significant increase in temperature between 100 and 97 ka BP marking the onset of OIS 5c (Desmarchelier & Goede 1996).

A record for the period 84–56 ka BP is provided by the basal 860 mm of a 907 mm tall stalagmite

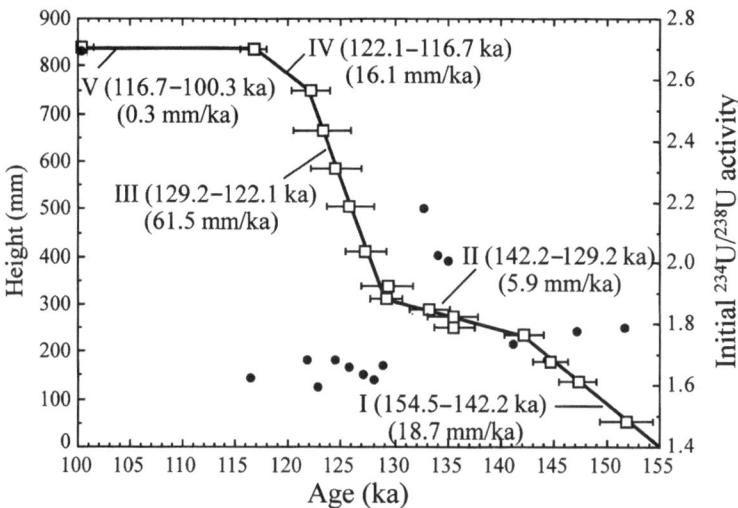

Fig. 9. Growth rates for stalagmite NEW-B in Newdegate Cave between OIS 6 and OIS 5d. The stalagmite grew very slowly during the Penultimate Glaciation, then rapidly during the Last Interglacial between 129 and 122 ka BP, and growth virtually ceased during OIS 5d. Open squares and black dots represent heights and initial $^{230}U/^{238}U$ activity ratios. The best-fit line through the open squares shows a five-stage growth history (after Zhao *et al.* 2001).

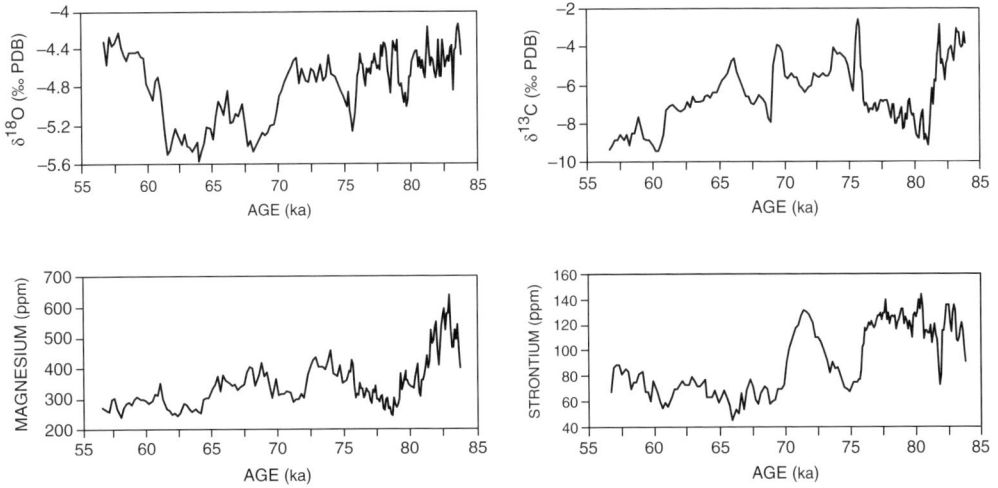

Fig. 10. Stable isotope records of $\delta^{18}O$ and $\delta^{13}C$ compared with the trace element records of magnesium and strontium between 55 and 83 ka BP from Frankcombe Cave stalagmite in the Florentine Valley (after Goede 1998; Goede *et al.* 1998).

from Frankcombe Cave (current MAT 8.3 °C) in the Florentine Valley of central western Tasmania (Fig. 10). It shows that from 84 to 67 ka BP there was a gradual decrease in $\delta^{18}O$ values reflecting decline in temperatures towards the onset of OIS 4. This stage shows two cold peaks separated by a short rise. Temperatures may have been nearly as low as at the LGM. The event was terminated abruptly by a rise at about 60 ka BP marking the beginning of OIS 3 (Goede *et al.* 1998).

The carbon isotope record shows two peaks of bioproductivity at *c.* 80 and 60 ka BP when summer insolation was at a minimum (Goede 1998). The peaks represent greater moisture availability, reflecting the combined effects of changes in precipitation and evaporation. This record correlates significantly with the magnesium record with peaks at 82, 72, 68 and 61 ka BP, and with an overall increase in moisture availability over the period of cooling (Goede 1998).

The strontium record shows two periods of enrichment from Sr-rich dust derived from recent marine sediments, a long period from 84 to 75 ka BP and a short period centred on 70 ka BP (Fig. 10). The most likely dust source is the exposed floor of Bass Strait, which supports observations that sea levels were much lower than today (Goede *et al.* 1998).

There are few speleothem records from OIS 3 or OIS 2 as climatic conditions appear to have been largely too dry for speleothem growth. Not until 15 ka BP, at the end of the deglaciation period, did stalagmites begin to grow again during the warmer and more humid conditions of the late glacial and

Early Holocene. The Holocene experienced a relatively uniform climate, but stable isotope analyses from stalagmites in Lynds Cave at Mole Creek and Frankcombe Cave in the Florentine Valley show the occurrence of small millennial variations (Goede *et al.* 1990; Xia *et al.* 2001; Fig. 11).

The 1127 mm long Lynds Cave stalagmite from Mole Creek (current MAT 9.5 °C) grew between 9.2 and 5.1 ka BP (Fig. 11, right). The record shows variable conditions before 8 ka BP, when an abrupt warming event occurred within 100 years. This 'Climatic Optimum' continued until 6.6 ka BP, with warmer and wetter conditions than at present. After 6.6 ka BP the climate became more variable, cooler and drier until 5.1 ka BP, when the stalagmite ceased to grow (Xia *et al.* 2001).

A 720 mm long stalagmite from Frankcombe Cave in the Florentine Valley (current MAT

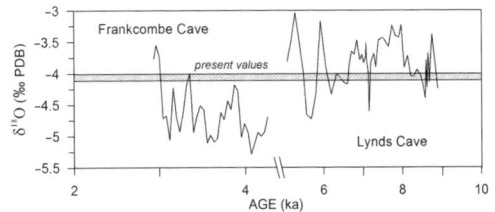

Fig. 11. Holocene $\delta^{18}O$ records from Frankcombe (left) and Lynds Cave (right). It should be noted that $\delta^{18}O$ values indicate temperatures generally above present-day values during the early to middle Holocene and lower than present-day values during the late Holocene (after Goede *et al.* 1990; Xia *et al.* 2001).

8.3 °C) covers part of the Late Holocene record (Fig. 11, left). The oxygen isotope values show that temperatures fluctuated but were significantly lower than at present between 4.2 and 3.0 ka BP. These data clearly indicate Neoglacial conditions but low carbon isotope values show that surface bioproductivity remained high (Goede *et al.* 1990).

Discussion and conclusions

Research work during the last 35 years has shown that Tasmania has had a complex history of Quaternary glaciation during the last million years and associated climate changes. Many ice advances occurred during the Early, Middle and Late Quaternary, and each known ice advance covered a smaller area than the preceding one. Application of new dating techniques has allowed a time-sequenced model of glaciation to be established.

Comparison of the Tasmanian glacial record with mainland Australia and New Zealand is difficult because of the short record for the Mount Kosciuszko region and the limiting effect on the New Zealand record owing to strong tectonic uplift. However, there are some important points of comparison and difference. Although the Mount Kosciuszko region was probably glaciated prior to the Last Glaciation, known geomorphological evidence is confined to the Last Glaciation. Exposure dating in the Blue Lake Valley shows a sequence from a boulder field and degraded moraine BL-I with a minimum age of 59.3 ka BP, through three sets of later moraines BL-II dated to 32 ka BP, BL-III to 19.1 ka BP and BL-IV to 16.8 ka BP (Barrows *et al.* 2001). The sequence shows similarity to the record in Tasmania in that the OIS 4 and OIS 3 ice advances were more extensive than the OIS 2 ice limit. In Blue Lake Valley the OIS 2 ice limit dated to 19.1 ka BP and the youngest ice limit dated to 16.8 ka BP are closely comparable with the age of the LGM ice limit and the youngest ages of glaciers in Tasmania (Barrows *et al.* 2002). It is also important that there is no evidence for formation of ice associated with renewed cold conditions during the Antarctic Cold Reversal or Younger Dryas events.

The oldest glacial events in New Zealand are the Ross Glaciation dated to 2.4–2.6 Ma and Porikan dated to 2.2–2.1 Ma BP (Suggate 1990). It is not known if terrestrial evidence for glaciation is missing as a result of uplift and erosion before OIS 10. There is no comparable record in New Zealand of the Tasmanian glaciations around 1 Ma but both areas were glaciated during OIS 10, 8, 6, 4, 3 and 2. Of particular significance is the establishment in both regions of glaciation during OIS 4 to 2, with a well-recorded series of six ice advances in New Zealand at Aurora Cave near Te Anau in

Fiordland (Williams 1996), and isolated records for OIS 4 and OIS 3 in Tasmania (this paper) making correlation before the LGM difficult.

The LGM of OIS 2 had multiple ice advances in both Tasmania and western New Zealand. An advance in Cradle Mountain is tentatively dated to 29 ka BP in Tasmania whereas the first main Westland advance culminated at 28 ka BP. In Tasmania the main LGM ice advances have a mean age of 20–19 ka BP, comparable with Westland ice advances between 21.5 and 19 ka BP (Suggate & Almond 2005). Following maximum glaciation in Tasmania ice retreat was rapid, as in western New Zealand, and although ice survived in some mountain valleys and cirques until 16–14 ka BP there is no dated evidence for late readvances of ice as recorded at Franz Josef Glacier in New Zealand, where the Waiho Loop Moraine has been exposure dated to 10.48 ka BP, making it of Early Holocene age (Barrows *et al.* 2007*a*).

Tasmania is the only Australian landscape where areas of glacial karst have been formed. Most detectable evidence was probably formed during the Middle and Late Quaternary. Surface and underground drainage patterns have been modified locally in several areas, and cave entrances and passages have been blocked by outwash gravels and till in areas adjacent to or within the boundaries of former ice cover. Establishing connection between glacial events and underground karst landforms and sediments of climatic significance is difficult in Tasmania, and the evidence is disparate and poorly dated. However, the U/Th dates from Welcome Stranger Cave indicating glaciation during either OIS 4 or OIS 3 show the potential of underground study to contribute to surface glacial studies, and to studies of climate change.

The pollen–vegetation sequence from Lake Selina provides the most highly resolved record of Late Quaternary climatic events for southeastern Australia. Its widespread significance for the SW Pacific region is demonstrated by its close similarity to the record from Okarito Pakihi in south Westland New Zealand, to sea surface temperature records and to the Vostok ice core deuterium record (Petit *et al.* 1999; Vandergoes *et al.* 2005; Barrows *et al.* 2007*b*).

Both Lake Selina and Okarito Pakihi show the same pollen–climatic fluctuations for OIS 5a, 5b, 5c, 5d and 5e, with rainforest present during 5a, 5c and 5e, and alpine–subalpine vegetation during 5b and 5d. Both also show herbaceous and subalpine vegetation indicating cold conditions during OIS 4 and OIS 2. In addition, OIS 3 is characterized in each by three fluctuations of cooling during a downward trend towards the LGM. Although not closely dated, the fluctuations are similar to Heinrich events, and of magnitude and duration similar to the millennial-scale events reported from the South Atlantic (Kanfoush *et al.* 2000). The cooling events in the South Atlantic have been suggested to have resulted from either advances of the Antarctic ice cap or, more probably, northward extensions of sea ice that would compress the westerly wind belt. The Lake Selina and Okarito Pakihi pollen records strongly suggest that the climatic fluctuations experienced in the southern middle latitudes during OIS 3 were driven by cooling events at high latitudes in the Southern Hemisphere. There is a slight difference in the development of rainforest at the two sites during Termination I. At Lake Selina rainforest developed after 14 ka BP without a later reversal whereas at Okarito Pahiki there was a slight reversal at the beginning of Termination I between 14.4 and 11 ka BP, the period encompassing both the Antarctic Cold Reversal and Younger Dryas events.

The isotope studies of stalagmites in northern and central Tasmania have produced more refined climatic results than the glacial and pollen studies, particularly during OIS 5, 4 and 1. They have also demonstrated changes in precipitation. Of particular significance is the demonstration that OIS 5e and the early Holocene were the warmest and moistest periods since OIS 6. Both were short periods with maximum durations of around 7 ka. Conditions were drier and colder at all other times, but especially during OIS 5d, 4, 3 and 2, and in the Late Holocene. The marked double peak of drier and colder conditions separated by a slightly milder interval in OIS 4 contrasts with the single cold peak in the Vostok record but it is known only from Frankombe Cave. In addition, the high strontium levels during the early Last Glaciation, indicating low sea levels in Bass Strait, and the falling bromine levels between 15–12 ka BP, when burning of the vegetation by Aborigines was markedly reduced, are currently unique to the Tasmanian environmental record.

The authors acknowledge P. Augustinus, F. Hopf, Q. Xia, J.-X. Zhao and K. Collerson for supplying data and contributing figures. They also thank O. Rey-Lescure of The University of Newcastle, NSW, Australia and S. Rouillard of the University of Exeter for invaluable assistance in drafting the maps and diagrams. Figures 1 and 3 have been modified from figures published by Colhoun (2004) by Elsevier, Amsterdam. We gratefully acknowledge the Australian and New Zealand Geomorphology Group and The Geological Society of London for covering the publication costs for Figure 5.

References

AUGUSTINUS, P. C. 1999*a*. Reconstruction of the Bulgobac Glacial System, Pieman river basin, western Tasmania. *Australian Geographical Studies*, **37**, 24–36.

AUGUSTINUS, P. C. 1999*b*. Dating the Late Cenozoic glacial sequence, Pieman river basin, western Tasmania, Australia. *Quaternary Science Reviews*, **18**, 1335–1350.

AUGUSTINUS, P. C. & COLHOUN, E. A. 1986. Glacial history of the upper Pieman and Boco valleys, western Tasmania. *Australian Journal of Earth Sciences*, **33**, 181–191.

AUGUSTINUS, P. C. & IDNURM, M. 1993. Palaeomagnetism of ferricrete from Vale of Belvoir, western Tasmania: implications for Tasmanian Cenozoic glacial history. *Exploration Geophysics*, **24**, 301–302.

AUGUSTINUS, P. C. & MACPHAIL, M. K. 1997. Early Pleistocene stratigraphy and timing of the Bulgobac glaciation, western Tasmania, Australia. *Palaeogeography, Palaeoclimatology, Palaeoecology*, **128**, 253–267.

AUGUSTINUS, P. C., SHORT, S. A. & COLHOUN, E. A. 1994. Pleistocene stratigraphy of the Boco Plain, western Tasmania. *Australian Journal of Earth Sciences*, **41**, 581–591.

AUGUSTINUS, P. C., POLLINGTON, M. J. & COLHOUN, E. A. 1995. Magnetostratigraphy of the Late Cenozoic glacial sequence, Pieman river basin, western Tasmania. *Australian Journal of Earth Sciences*, **42**, 509–518.

AUGUSTINUS, P. C., SHORT, S. A. & HEIJNIS, H. 1997. Uranium/thorium dating of ferricretes from mid- to late Pleistocene glacial sediments, western Tasmania, Australia. *Journal of Quaternary Science*, **12**, 295–308.

BANKS, M. R. & BURNS, K. L. 1962. Eugenana beds. *Journal of the Geological Society of Australia*, **9**, 185–186.

BANKS, M. R., COLHOUN, E. A. & HANNAN, D. 1987. Early discoverers XXXIV: Early discoveries of the effects of ice action in Australia. *Journal of Glaciology*, **33**, 231–235.

BARBETTI, M. & COLHOUN, E. A. 1988. Reversed magnetisation of glaciolacustrine sediments from western Tasmania. *Search*, **19**, 151–153.

BARROWS, T. T., STONE, J. O., FIFIELD, L. K. & CRESSWELL, R. G. 2001. Late Pleistocene glaciation of the Kosciuszko Massif, Snowy Mountains, Australia. *Quaternary Research*, **55**, 179–189.

BARROWS, T. T., STONE, J. O., FIFIELD, L. K. & CRESSWELL, R. G. 2002. The timing of the last glacial maximum in Australia. *Quaternary Science Reviews*, **21**, 159–173.

BARROWS, T. T., JUGGINS, S., DE DECKKER, P., CALVO, E. & PELEJERO, C. 2007*a*. Long-term sea surface temperature and climate change in the Australian–New Zealand region. *Paleoceanography*, **22**, PA 2215, doi: 10.1029/2006PA001328.

BARROWS, T. T., LEHMAN, S., FIFIELD, L. K. & DE DEKKER, P. 2007*b*. Absence of cooling in New Zealand and the adjacent ocean during the younger Dryas chronozone. *Science*, **318**, 86–89, doi: 10.1126/science.1145873.

COLHOUN, E. A. 1985. The glaciations of the West Coast Range, Tasmania. *Quaternary Research*, **24**, 39–59.

COLHOUN, E. A. 2004. Quaternary glaciations of Tasmania and their ages. *In*: EHLERS, J. & GIBBARD, P. L. (eds) *Quaternary Glaciations—Extent and Chronology Part III: South America, Asia, Africa, Australasia, Antarctica*. Developments in Quaternary Science, **2**, 353–360.

COLHOUN, E. A. & FITZSIMONS, S. J. 1990. Late Cainozoic glaciation in western Tasmania, Australia. *Quaternary Science Reviews*, **9**, 199–216.

COLHOUN, E. A. & FITZSIMONS, S. J. 1996. Additional radiocarbon date from Dante outwash fan, King Valley, and dating of the late Wisconsin glacial maximum in western Tasmania. *Papers and Proceedings of the Royal Society of Tasmania*, **130**, 81–84.

COLHOUN, E. A. & VAN DE GEER, G. 1998. Pollen analysis of 0–20 m at Darwin Crater, Western Tasmania, Australia. *In*: HORIE, S. (ed.) *International Project on Palaeolimnology and Late Cenozoic Climate*. IPPCCE Newsletter, **11**, 68–89.

COLHOUN, E. A., HANNAN, D. & KIERNAN, K. 1996. Late Wisconsin glaciation of Tasmania. *Papers and Proceedings of the Royal Society of Tasmania*, **130**, 33–45.

COLHOUN, E. A., POLA, J. S., BARTON, C. E. & HEIJNIS, H. 1999. Late Pleistocene vegetation and climate history of Lake Selina, western Tasmania. *Quaternary International*, **57–58**, 5–23.

DAVIES, J. L. 1967. Tasmanian landforms and Quaternary climates. *In*: JENNINGS, J. N. & MABBUTT, J. A. (eds) *Landform Studies from Australia and New Guinea*. Australian National University Press, Canberra, 1–25.

DAVIES, J. L. 1969. *Landforms of Cold Climates*. Australian National University Press, Canberra.

DAVIES, J. L. 1974. Geomorphology and Quaternary environments. *In*: WILLIAMS, W. D. (ed.) *Biogeography and Ecology in Tasmania*. W. Junk, The Hague, 17–27.

DERBYSHIRE, E. 1972. Pleistocene glaciation of Tasmania: review and speculations. *Australian Geographical Studies*, **10**, 79–94.

DESMARCHELIER, J. M. & GOEDE, A. 1996. High resolution stable isotope analysis of a Tasmanian speleothem. *Papers and Proceedings of the Royal Society of Tasmania*, **130**, 7–13.

FINK, D., MCKELVEY, B., HANNAN, D. & NEWSOME, D. 2000. Cold rocks, hot sands: *In-situ* cosmogenic applications in Australia at ANTARES. *Nuclear Instruments and Methods in Physics Research*, **B172**, 838–846.

FITZSIMONS, S. J., COLHOUN, E. A., VAN DE GEER, G. & POLLINGTON, M. 1992. The Quaternary geology and glaciation of the King Valley. *Tasmanian Department of Mines, Geological Survey Bulletin*, **68**, 1–57.

FLETCHER, M. S. & THOMAS, I. 2007. Holocene vegetation and climate change from near Lake Pedder, south–west Tasmania, Australia. *Journal of Biogeography*, **34**, 665–677.

FORD, D. & WILLIAMS, P. 1989. *Karst Geomorphology and Hydrology*. Unwin Hyman, London.

FORD, D. C. (ed.) 1983. Castleguard Cave and Karst, Columbia Icefields area, Rocky Mountains area of Canada: a symposium. *Arctic and Alpine Research*, **15**, 425–554.

GIBSON, N., KIERNAN, K. W. & MACPHAIL, M. K. 1987. A fossil bolster plant from the King River, Tasmania. *Papers and Proceedings of the Royal Society of Tasmania*, **121**, 35–42.

GILL, E. D. 1956. Radiocarbon dating for glacial varves in Tasmania. *Australian Journal of Science*, **19**, 80.

GOEDE, A. 1969. Underground stream capture at Ida Bay, Tasmania, and the relevance of cold climatic conditions. *Australian Geographical Studies*, **7**, 41–48.

GOEDE, A. 1994. Continuous early last glacial palaeoenvironmental record from a Tasmanian speleothem based on stable isotope and minor element variations. *Quaternary Science Reviews*, **13**, 283–291.

GOEDE, A. 1998. *Quaternary Studies of Caves and Coasts*. PhD thesis, University of Tasmania, Hobart.

GOEDE, A. & HARMON, R. S. 1983. Radiometric dating of Tasmanian speleothems—evidence of cave evolution and climate change. *Journal of the Geological Society of Australia*, **30**, 89–100.

GOEDE, A. & VOGEL, J. C. 1991. Trace element variations and dating of a Late Pleistocene Tasmanian speleothem. *Palaeogeography, Palaeoecology, Palaeoclimatology*, **88**, 121–131.

GOEDE, A., VEEH, H. H. & AYLIFFE, L. K. 1990. Late Quaternary palaeotemperature records for two Tasmanian speleothems. *Australian Journal of Earth Sciences*, **37**, 267–278.

GOEDE, A., MCCULLOCH, M., MCDERMOTT, F. & HAWKESWORTH, C. 1998. Aeolian contribution to strontium and strontium isotope variations in a Tasmanian speleothem. *Chemical Geology*, **149**, 37–50.

HANNAN, D. G. & COLHOUN, E. A. 1987. Glacial stratigraphy of the Upper Mersey Valley, Tasmania. *Australian 1Geographical Studies*, **25**, 36–46.

HANNAN, D. G. & COLHOUN, E. A. 1991. When were the Walls of Jerusalem last glaciated? *Papers and Proceedings of the Royal Society of Tasmania*, **125**, 1–5.

JENNINGS, J. N. & BANKS, M. R. 1958. The Pleistocene glacial history of Tasmania. *Journal of Glaciology*, **3**, 298–303.

JENNINGS, J. N. & SWEETING, M. M. 1959. Underground breach of a divide at Mole Creek, Tasmania. *Australian Journal of Science*, **21**, 262–263.

KANFOUSH, S. L., HODELL, D. A., CHARLES, C. D., THOMAS, P. G., MORTYN, P. G. & NINNEMANN, U. S. 2000. Millennial-scale instability of the Antarctic ice sheet during the last glaciation. *Science*, **288**, 1815–1818.

KIERNAN, K. 1983*a*. Weathering evidence for an additional glacial stage in Tasmania. *Australian Geographical Studies*, **21**, 197–220.

KIERNAN, K. 1983*b*. Relationship of cave fills to glaciation in the Nelson River Valley, Central Western Tasmania. *Australian Geographer*, **15**, 367–375.

KIERNAN, K. 1989*a*. Multiple glaciation of the Upper Franklin Valley, Western Tasmania Wilderness World Heritage Area. *Australian Geographical Studies*, **27**, 208–233.

KIERNAN, K. 1989*b*. Drainage evolution in a Tasmanian glaciokarst. *Helictite*, **27**, 2–12.

KIERNAN, K. 1990*a*. The extent of late Cainozoic glaciation in the Central Highlands of Tasmania. *Arctic and Alpine Research*, **22**, 341–354.

KIERNAN, K. 1990*b*. The alpine geomorphology of the Mt Anne massif, south-western Tasmania. *Australian Geographer*, **21**, 113–125.

KIERNAN, K. 1990*c*. Bathymetry and origin of Lake Timk, Tasmania. *Helictite*, **28**, 18–21.

KIERNAN, K. 1991. Glacial history of the upper Derwent Valley, Tasmania. *New Zealand Journal of Geology and Geophysics*, **34**, 157–166.

KIERNAN, K. 1992. Glacial geomorphology and the last glaciation at Lake St Clair. *Papers and Proceedings of the Royal Society of Tasmania*, **126**, 47–57.

KIERNAN, K. 1995. A reconnaissance of the geomorphology and glacial history of the upper Gordon River Valley, Tasmania. *Tasforests*, **7**, 51–76.

KIERNAN, K. 1996. *An Atlas of Tasmanian Karst*. Tasmanian Forest Research Council, Hobart.

KIERNAN, K. 2006. Tasmania's cold caves: an island of alpine karst. *In*: GOEDE, A. & BUNTON S. (eds) *Proceedings of the 25th Biennial Conference*. Australian Speleological Federation, Broadway, Hobart, 41–48.

KIERNAN, K. 2008. Extent of the Late Cainozoic periglacial domain in southwest Tasmania, Australia. *Zeitschrift für Geomorphologie*, **52**, 325–348.

KIERNAN, K. & HANNAN, D. 1991. Glaciation of the Upper Forth River Catchment, Tasmania. *Australian Geographical Studies*, **29**, 155–173.

KIERNAN, K., JONES, R. & RANSON, D. 1983. New evidence from Fraser Cave for glacial age man in south–west Tasmania. *Nature*, **301**, 28–32.

KIERNAN, K., LAURITZEN, S. E. & DUHIG, N. 2001. Glaciation and cave sediment aggradation around the margins of the Mt Field Plateau, Tasmania. *Australian Journal of Earth Sciences*, **48**, 251–263.

KIERNAN, K., FIFIELD, L. K. & CHAPPELL, J. 2004. Cosmogenic nuclide ages for last glacial maximum moraine at Schnells Ridge, Southwest Tasmania. *Quaternary Research*, **61**, 335–338.

KIERNAN, K., GRIEG, D. & FINK, D. 2008. Pre-Last Glacial age for morphologically fresh moraines in the western Arthur Range, southwest Tasmania. *In*: COHEN, T. & HOUSHOLD, I. (eds) *Australian and New Zealand Geomorphology Group 13th Conference, Queenstown, Tasmania, Program and Abstracts*. Australia and New Zealand Geomorphology Group, 58.

LEWIS, A. N. 1945. Pleistocene glaciation in Tasmania. *Papers and Proceedings of the Royal Society of Tasmania*, 41–56.

MACKINTOSH, A. N., BARROWS, T. T., COLHOUN, E. A. & FIFIELD, L. K. 2006. Exposure dating and glacial reconstruction at Mt. Field, Tasmania, Australia, identifies MIS 3 and MIS 2 glacial advances and climatic variability. *Journal of Quaternary Science*, **21**, 363–376.

MACPHAIL, M. K. 1979. Vegetation and climates in Southern Tasmania since the last glaciation. *Quaternary Research*, **11**, 306–341.

MACPHAIL, M. K., COLHOUN, E. A., KIERNAN, K. & HANNAN, D. 1993. Glacial climates in the Antarctic region during the late Paleogene: Evidence from northwest Tasmania, Australia. *Geology*, **21**, 145–148.

MACPHAIL, M. K., COLHOUN, E. A. & FITZSIMONS, S. J. 1995. Key periods in the evolution of the Cenozoic vegetation and flora in Western Tasmania: the Late Pliocene. *Australian Journal of Botany*, **43**, 505–526.

MAIRE, R. 2004. Patagonia marble karst, Chile. *In*: GUNN, J. (ed.) *Encyclopedia of Caves and Karst Science*. Fitzroy Dearborn, New York, 572–573.

MCMINN, M. S., KIERNAN, K. & FINK, D. 2008. Cosmogenic nuclide dating in the Denison range, Southwestern Tasmania. *In*: COHEN, T. & HOUSHOLD, I. (eds) *Australian and New Zealand Geomorphology Group 13th Conference, Queenstown, Tasmania, Program and Abstracts*. Australian and New Zealand Geomorphology Group, 63.

NUNEZ, M. & COLHOUN, E. A. 1986. A note on air temperature lapse rates on Mt. Wellington, Tasmania. *Papers and Proceedings of the Royal Society of Tasmania*, **120**, 11–15.

PENCK, A. & BRUCKNER, E. 1909. *Die Alpen im Eiszeitalter*. Tauchnitz, Leipzig.

PETERSON, J. A. 1968. Cirque morphology and Pleistocene ice formation conditions in southeastern Australia. *Australian Geographical Studies*, **6**, 67–83.

PETERSON, J. A. & ROBINSON, G. 1969. Trend surface mapping of cirque floor levels. *Nature*, **222**, 75–76.

PETIT, J. R., JOUZEL, J. ET AL. 1999. Climate and atmospheric history of the past 420 000 years from the Vostok ice core, Antarctica. *Nature*, **399**, 429–436.

POLLINGTON, M. J., COLHOUN, E. A. & BARTON, C. E. 1993. Palaeomagnetic constraints on the ages of Tasmanian glaciations. *Exploration Geophysics*, **24**, 305–310.

PORTER, S. C. 1975. Equilibrium-line altitudes of late Quaternary glaciers in the Southern Alps, New Zealand. *Quaternary Research*, **5**, 27–47.

SPELL, T. L. & MCDOUGALL, I. 1992. Revisions to the age of the Brunhes–Matuyama boundary and the Pleistocene geomagnetic polarity timescale. *Geophysical Research Letters*, **19**, 1181–1184.

SUGGATE, R. P. 1990. Late Pliocene and Quaternary glaciations of New Zealand. *Quaternary Science Reviews*, **9**, 175–197.

SUGGATE, R. P. & ALMOND, P. C. 2005. The Last Glacial Maximum (LGM) in western South Island, New Zealand: implications for the global LGM and MIS 2. *Quaternary Science Reviews*, **24**, 1923–1940.

SUTHERLAND, F. L. & HALE, G. E. A. 1970. Cainozoic Volcanism in and around Great Lake, Central Tasmania. *Papers and Proceedings of the Royal Society of Tasmania*, **104**, 17–36.

THRUSH, M. N. 2008. *The Pleistocene Glaciations of the Cradle Mountain Region, Tasmania*. PhD thesis, University of Newcastle, New South Wales.

VANDERGOES, M., NEWNHAM, R. M. ET AL. 2005. Regional insolation forcing of late Quaternary climate change in the Southern Hemisphere. *Nature*, **436**, 242–245.

WILLIAMS, P. 2004. New Zealand. *In*: GUNN, J. (ed.) *Encyclopedia of Caves and Karst Science*. Fitzroy Dearborn, New York, 540–544.

WILLIAMS, P. W. 1996. A 230 ka record of glacial and interglacial events from Aurora Cave, Fiordland, New Zealand. *New Zealand Journal of Geology and Geophysics*, **39**, 225–241.

XIA, Q., ZHAO, J. & COLLERSON, K. D. 2001. Early–Mid-Holocene climatic variations in Tasmania, Australia: multi-proxy records in a stalagmite from Lynds Cave. *Earth and Planetary Science Letters*, **194**, 177–187.

ZHAO, J., XIA, Q. & COLLERSON, K. D. 2001. Timing and duration of the Last Interglacial inferred from high resolution U-series chronology of stalagmite growth in Southern Hemisphere. *Earth and Planetary Science Letters*, **184**, 635–644.

'Of droughts and flooding rains': an alluvial loess record from central South Australia spanning the last glacial cycle

DAVID HABERLAH[1,2]*, PETER GLASBY[3], MARTIN A. J. WILLIAMS[3], STEVEN M. HILL[1,2], FRANCES WILLIAMS[4], EDWARD J. RHODES[5], VICTOR GOSTIN[1], ANTHONY O'FLAHERTY[6] & GERALDINE E. JACOBSEN[7]

[1]*Geology & Geophysics, School of Earth and Environmental Sciences, University of Adelaide, Adelaide, SA 5005, Australia*

[2]*Cooperative Research Centre for Landscape Environments and Mineral Exploration (CRCLEME), Geoscience Australia, Symonston, ACT 2609, Australia*

[3]*Geographical & Environmental Studies, School of Social Sciences, University of Adelaide, Adelaide, SA 5005, Australia*

[4]*Archaeometry Laboratory, School of Chemistry & Physics, University of Adelaide, Adelaide, SA 5005, Australia*

[5]*Department of Environmental & Geographical Sciences, Manchester Metropolitan University, Chester Street, Manchester M1 5GD, UK*

[6]*School of Land Information Management Systems, TAFE South Australia, O'Halloran Hill Campus, Adelaide, SA5158, Australia*

[7]*Institute for Environmental Research, ANSTO, PMB 1, Menai, NSW 2234, Australia*

Corresponding author (e-mail: david.haberlah@adelaide.edu.au)

Abstract: Deposits of proximal dust-derived alluvium (alluvial loess) within the catchments of the now semi-arid Flinders Ranges in South Australia record regionally synchronous intervals of fluvial entrainment, aggradation and down-cutting spanning the last glacial cycle. Today, these floodplain remnants are deeply entrenched and laterally eroded by ephemeral traction load streams. The north–south aligned ranges are strategically situated within the present-day transitional zone, receiving both topographically enhanced winter rainfall from the SW and convectional downpours from summer monsoonal incursions from the north. We develop a regional chronostratigraphy of depositional and erosional events emphasizing the Last Glacial Maximum (LGM). Based on 124 ages (94 accelerator mass spectrometry radiocarbon and 30 optically stimulated luminescence) from the most significant terrace remnants on both sides of the Ranges, we conclude that the last glacial cycle including the LGM was characterized by major environmental changes. Two pronounced periods of pedogenesis between *c.* 36 and 30 ka were followed by widespread erosion and reworking. A short-lived interval of climatic stability before *c.* 24 ka was followed by conditions in which large amounts of proximal dust (loess) were deposited across the catchments. These loess mantles were rapidly redistributed and episodically transported downstream by floods. The termination of this regime *c.* 18–16 ka was marked by rapid incision.

I love a sunburnt country,

A land of sweeping plains,

Of ragged mountain ranges,

Of droughts and flooding rains.

I love her far horizons,

I love her jewel-sea,

Her beauty and her terror –

The wide brown land for me!

Dorothea Mackellar (1907) (Mackellar 1971)

Thirty years have elapsed since Bowler (1978) demonstrated the importance of Quaternary depositional history in clarifying some of the climatic and tectonic influences responsible for fashioning the Riverine Plain of southeastern Australia, observing

From: BISHOP, P. & PILLANS, B. (eds) *Australian Landscapes*. Geological Society, London, Special Publications, **346**, 185–223. DOI: 10.1144/SP346.11 0305-8719/10/$15.00 © The Geological Society of London 2010.

that although 'denudation chronology may indeed be dead in some parts of the world; its closest relative, depositional chronology, is alive and well in others' (Bowler 1978, p. 72). Over 30 years earlier Crocker (1946) had argued that the presence of calcium carbonate at depth in the red–brown earth soils of southeastern Australia reflected the past influx of calcareous loess derived from continental shelves and beach deposits exposed during times of lower glacial sea levels, a conclusion endorsed by Sprigg (1979) and confirmed by recent strontium isotopic analysis of calcretes in semi-arid South Australia (Dart *et al.* 2007). Our aim in this paper is to integrate these two themes of depositional chronology and loess influx into an analysis of the late Quaternary depositional history of the semi-arid Flinders Ranges of South Australia. Studies of desert margin systems in and outside Australia are highly informative of past climatic influences (Williams *et al.* 1991; Talbot *et al.* 1994) as they not only reflect the interplay between alternating and contrasting morphogenetic systems, but the depositional evidence in question has often been well preserved as a result of low rates of erosion linked to the prevailing aridity.

Our understanding of landscape evolution is only as good as our understanding of the timing, rate and duration of depositional and erosional events. This is particularly the case for landscape features for which no modern analogue exists such as late Pleistocene loess-derived valley-fills and floodplains. This is illustrated by the longstanding controversy over the nature of the Namib Silts (Srivastava *et al.* 2006), which have been reinterpreted over the past decades as lacustrine (Goudie 1972), low-energy alluvial (Vogel 1982; Eitel *et al.* 2005) and high-energy flood deposits (Ward 1987; Smith *et al.* 1993; Heine & Heine 2002; Leopold *et al.* 2006). Similar late Pleistocene fine-grained valley-fills from the Sinai Peninsula (Issar & Eckstein 1969; Issar & Bruins 1983; Rögner *et al.* 2004) and those, largely unnoticed but equally spectacular, from the Flinders Ranges, South Australia, have been reinterpreted in a similar manner (Callen & Reid 1994; Cock *et al.* 1999; Preiss 1999; Williams *et al.* 2001; see Haberlah 2006). The most recent studies from each of these sites independently suggest an aeolian source (Eitel *et al.* 2001; Rögner *et al.* 2004; Williams & Nitschke 2005) and converge on a depositional model of flood deposition with back-flooded areas where slackwater deposits are preserved (Heine & Heine 2002; Haberlah *et al.* 2007). Our present study focuses on the fine-grained aggradational sequence from central South Australia, hereafter referred to as the 'Flinders Silts'. In a collaborative effort, we have established a regional chronostratigraphy based on 14 strategically selected stratigraphic sections from major catchments across the Flinders Ranges (Fig. 1).

The Flinders Ranges is the longest and highest mountain range in South Australia, extending as a series of mainly north–south-trending strike-ridges for over 400 km from the Spencer Gulf inlet of the Indian Ocean deep into the arid heart of the Australian continent. The Ranges consist of a series of tilted, uplifted and dissected weakly metamorphosed Precambrian and Palaeozoic sedimentary rocks (Preiss 1987), rising abruptly from the surrounding plains and reaching 1170 m at St Mary Peak (Ngarri Mudlanha). To the west, north and east, the Flinders Ranges are flanked by large structural basins occupied by the playas (salt flats) Lake Torrens, Lake Callabonna–Lake Blanche and Lake Frome.

The 124 numerical ages, most of which have not been presented before (Table 1), are discussed and interpreted in terms of regional periods of aeolian and fluvial aggradation, and of reworking and downcutting events. Insights are thus provided that potentially resolve questions associated with the less well-dated Namib and Sinai Silts. The catchment-based chronostratigraphic record for the Flinders Silts further assists in interpreting continental-scale records from adjacent arid central Australia (Lake Eyre) and SE Australia (Willandra Lakes) that register intricate responses to both regional and distal climatic events and that are currently being reassessed by independent and more accurate dating methods (e.g. Kemp & Spooner 2007; Nanson *et al.* 2008). The catchments studied here occur across a region trending from SW to NE, roughly parallel to contemporary (Schwerdtfeger & Curran 1996) and palaeo-wind directions, as inferred from dune patterns in the region (Jennings 1968). The regional synchronicity of recorded events is tested and catchment-specific neotectonic as well as sedimentological threshold overprints of climatic signals are ruled out for the recorded period (see also Quigley *et al.* 2007). Although the timing of deposition of the Flinders Silts covers most of the last glacial cycle, the main focus of the dating programme is on the exceptionally well-preserved continuous sequence covering the Last Glacial Maximum (LGM) (21 ± 2 ka: Mix *et al.* 2001); from an early lead-up and throughout the period from c. 24–18 ka into the early Deglacial. The aim of this paper is to describe the present chronostratigraphic record and discuss the timing, nature and scope of events resulting in the formation of regolith in this part of the world. It provides a framework for current and future palaeo-environmental studies drawing on the rich embedded biological, geophysical and geochemical records.

Field area

Despite early work on Australian loess (Crocker 1946; see Haberlah 2007, 2008), the full significance of the contribution of proximal silt-sized

Fig. 1. Overview of field area. Location of stratigraphic sections (red dots); HK refers to sections within the Hookina catchment, BRA to sections within the Brachina catchment, WL to sections within the Wilkawillina catchment, and CAS-1 is situated within the Parachilna catchment. The inset figure indicates the general location of the study area in Australia in terms of relief (Shuttle Radar Topography Mission (SRTM) data) and seasonality (adapted from Gentilli 1986).

dust on landscape evolution in Australia remains poorly understood (Hesse & McTainsh 2003). Recently described dust records offshore from South Australia (Gingele & De Deckker 2005) and downwind from subtropical eastern Australia (Petherick *et al.* 2008a, b, 2009) that span the LGM and beyond, highlight sources of significant amounts of distal fine-grained dust from central and SE Australia. The Flinders Ranges are strategically located in the midst of this late Pleistocene 'dust bowl'. The more than 400 km long sequence of mountain ridges forms a longitudinal topographic barrier with a general elevation above 300 m above sea level and a chain of summits culminating at 1170 m. Hence, the Ranges efficiently harvest both low-lying clouds and proximal dust entrained from adjacent plains (Goossens 1988). As such, the Flinders Silts are a proximal equivalent to distal dust records that reach as far as Antarctica (Revel-Rolland *et al.* 2006). It follows that we look at a different scale of sedimentation rates, with >10 m thick sequences of loess washed into the valleys and plains and choking the narrow gorges. At specific intervals over the last glacial

Table 1. *Flinders Silts age estimates, presented within stratigraphic order of the sections*

Section name and sample code	Depth from top (in cm)	Sample material	^{14}C age estimates (^{14}C a BP ± error)	^{14}C age estimates (a cal. BP, 1 SD) [OSL age estimates (FMM depositional population/CDM)]	^{14}C age estimates (a cal. BP, 2 SD) [OSL age estimates (MAM with 1 SD range)]
HK07-D (elev. 193 m; 31.77733°S, 138.31807°E)					
K1890	85	Q SA		19340 ± 1560	17370 (15140–19540)
K1889	410	Q SA		82450 ± 13480	68790 (60180–77790)
K1888	490	Q SA		–	108460 (93380–118740)
K1887	765	Q SA		84630 ± 6980	78040 (66750–85470)
a1139869					
HK07-L (elev. 162 m; 31.80446°S, 138.26218°E)					
K1892	265	Q SA		30230 ± 1660	15180 (12510–17910)
K1891	325	Q SA		36120 ± 2970	28700 (23150–30610)
HK07-M (elev. 154 m; 31.79532°S, 138.25908°E)					
AdGL-08006	40	Q SA		18040 ± 1620	4400 (2640–5300)
AdGL-08005	637	Q SA		85590 ± 17880	97800 (88800–105730)
AdGL-08004	1075	Q SA		108230 ± 11140	19940 (16870–23430)
BRA-SA (c. 31.337°S, 138.619°E)					
OZE022	405	S	18500 ± 180	22170 ± 310	21550–22790
OZC706	440	C	17300 ± 200	20790 ± 240	20310–21270
OZC705	440	C	18150 ± 350	21860 ± 450	20960–22760
OZC704	440	S	19100 ± 180	22980 ± 210	22560–23400
Beta-84140	550	C	20320 ± 90	24290 ± 180	23930–24650
Beta-84141	600	C	20840 ± 90	24830 ± 80	24670–24990
BRA-SG (elev. 344 m; 31.3388°S, 138.611°E)					
Wk-6548	20	S	14827 ± 87	18100 ± 240	17620–18580
Wk-6554	55	S	15891 ± 85	18970 ± 110	18750–19190
Wk-6555	72	S	16173 ± 89	19340 ± 160	19020–19660
Wk-6552	78	S	16172 ± 93	19340 ± 160	19020–19660
Wk-6550	93	S	16365 ± 94	19620 ± 160	19300–19940
Wk-6549	117	S	16960 ± 100	20390 ± 110	20170–20610
Wk-6553	150	S	16928 ± 94	20360 ± 100	20160–20560
Wk-6556	165	S	17148 ± 96	20600 ± 120	20360–20840
BRA07-SD (elev. 340 m; 31.337339°S, 138.60660°E)					
SSAMS ANU-4107	57	OV	15160 ± 100	18280 ± 230	17820–18740
SSAMS ANU-4109	89	C	16170 ± 120	19340 ± 200	18940–19740
SSAMS ANU-4207	118	OV	17420 ± 120	20920 ± 170	20580–21260
AdGL-96005	c. 160	Q LA		18000 ± 1200	–
SSAMS ANU-4206	200	OV	18410 ± 100	22110 ± 250	21610–22610
SSAMS ANU-4110	214	OV	18520 ± 120	22230 ± 260	21710–22750
SSAMS ANU-4111	249	OV	18610 ± 130	22410 ± 180	22050–22770
SSAMS ANU-4112	271	OV	19360 ± 140	23230 ± 150	22930–23530
AdGL-96007	c. 325	Q LA		22400 ± 2300	–
SSAMS ANU-4113	339	OV	20130 ± 160	24080 ± 240	24560–23600
SSAMS ANU-4114	345	C	18880 ± 140	22680 ± 150	22380–22980
OZJ905	c. 353	S	20460 ± 140	24450 ± 190	24070–24830
SSAMS ANU-4209	370	OV	19670 ± 120	23540 ± 120	23300–23780
SSAMS ANU-4116	468	OV/C	21120 ± 150	25160 ± 240	24680–25640
AdGL-96004	c. 475	Q LA		21600 ± 1200	–
SSAMS ANU-2030	475	C	21890 ± 160	26260 ± 310	25640–26880
SSAMS ANU-2035	501	C	24110 ± 210	28990 ± 390	28210–29770
SSAMS ANU-4117	505	OV	19820 ± 180	23710 ± 200	23310–24110
SSAMS ANU-4205	505	OV	19910 ± 180	23830 ± 250	23330–24330
OZJ909	508	S	21510 ± 310	25660 ± 460	24740–26580
SSAMS ANU-2036	516	C	27630 ± 280	32210 ± 260	31690–32730
SSAMS ANU-2037	516	C	27740 ± 290	32310 ± 290	31730–32890
Beta-96166	c. 565	S	28120 ± 160	32570 ± 260	32050–33090

OSL age estimates (FMM inherited populations)	OSL age estimates (FMM post-depositional populations)	$\delta^{13}C$ (ppm) [H_2O content (%)]	Comments
			Hookina Section D
38920 ± 2700	–	5.1	Below uppermost Bca-horizon
105870 ± 32480	–	7.2	Chromatic band topping 2 pronounced Bca-horizons
119700 ± 12840	–	4.2	Chromatic band between 2 pronounced Bca-horizons
–	–	5.5	Base of lowermost Bca-horizon
			Hookina Section L
–	*11630 ± 1330*	6.5	Upper band of chromatic 'twin marker horizon'
85680 ± 22820	–	5.5	Lower band of chromatic 'twin marker horizon'
			Hookina Section M
52820 ± 8850	8250 ± 0870	6.2	Uppermost Bca-horizon
125140 ± 9620	–	3.5	'Orange Sf band' (reworked dune?)
–	22970 ± 2990	5.2	Onset of Silts, below major cut and fill
			Brachina Section A
		−4.74	(Williams *et al.* 2001)
		−25	(Williams *et al.* 2001)
		−25	(Williams *et al.* 2001)
		0	(Williams *et al.* 2001)
		–	(Williams *et al.* 2001)
		–	(Williams *et al.* 2001)
			Brachina Section G
		−6.6 ± 0.2	(Williams *et al.* 2001)
		−8.1 ± 0.2	(Williams *et al.* 2001)
		−7.6 ± 0.2	(Williams *et al.* 2001)
		−7.4 ± 0.2	(Williams *et al.* 2001)
		−8.3 ± 0.2	(Williams *et al.* 2001)
		−9.1 ± 0.2	(Williams *et al.* 2001)
		−6.8 ± 0.2	(Williams *et al.* 2001)
		−7.7 ± 0.2	(Williams *et al.* 2001)
			Slippery Dip
		−26 ± 3	Topmost undisturbed organic veneer
		−26 ± 3	Topmost undisturbed light band
		−22 ± 3	Base of central light band
		5.3	(Williams *et al.* 2001) based on TSAC/XRS DR
		−20 ± 3	Top of light '5er marker bands'
		−19 ± 3	Base of light '5er marker bands'
		−14 ± 3	Top of continuous 'tufa band'
		−27 ± 3	Top of light '4er marker bands'
		23.7	(Williams *et al.* 2001) based on TSAC/XRS DR
		−25 ± 3	Topping central light band of '3er marker bands'
		−22 ± 3	Central light band of '3er marker bands'
		−7.4	From monolith
		−18 ± 3	Tufa capping lower light band of '3er marker bands'
		−29 ± 3	Top of 'pink band'
		9.5	(Williams *et al.* 2001) based on TSAC/XRS DR
		−28 ± 2	'Pink band'
		−26 ± 3	Base of lowermost light band
		−15 ± 3	lowermost organic veneer
		−21 ± 3	Lowermost organic veneer
		–	Top of 'palaeosol'
		−24 ± 2	Base of 'palaeosol'
		−29 ± 2	Independent run of SSAMS ANU-2036
		−6	(Cock *et al.* 1999) Section C

(Continued)

Table 1. *Continued*

Section name and sample code	Depth from top (in cm)	Sample material	^{14}C age estimates (^{14}C a BP ± error)	^{14}C age estimates (a cal. BP, 1 SD) [OSL age estimates (FMM depositional population/CDM)]	^{14}C age estimates (a cal. BP, 2 SD) [OSL age estimates (MAM with 1 SD range)]
Beta-96679	c. 565	C	29800 ± 180	33560 ± 300	32960–34160
AdGL-96003	c. 600	Q LA		32800 ± 2800	–
AdGL-96006	c. 635	Q LA		24900 ± 1400	–
OZJ904	632	C	29160 ± 380	33570 ± 410	32750–34390
OZJ908	639	S	27990 ± 710	32640 ± 630	31380–33900
OZJ907	667	S	27320 ± 280	31970 ± 200	31570–32370
OZJ906	684	S	27200 ± 500	31900 ± 390	31120–32680
SSAMS ANU-2039	695	C	23600 ± 300	28560 ± 420	27720–29400
SSAMS ANU-1811	695	S	23270 ± 90	28080 ± 80	27920–28240
AdGL-96008	c. 830	Q LA coarse		24000 ± 1600	–
AdGL-96002	c. 830	Q LA		24900 ± 2500	–
Southern End (Section D) (elev. 341 m; c. 31.33935°S, 138.60737°E)					
Beta-96169	c. 120	S	16150 ± 80	19300 ± 140	19020–19580
Beta-96168	c. 145	S	16200 + 60	19380 ± 120	19140–19620
Beta-96167	c. 195	S	17070 ± 70	20500 ± 90	20320–20680
OZC710	255	S	17450 ± 350	20930 ± 390	20150–21710
OZC709	285	S	17650 ± 140	21160 ± 170	20820–21500
Beta-96170	c. 455	S	20650 ± 80	24680 ± 80	24520–24840
BRA-LW (elev. 333 m; c. 31.33593°S, 138.60231°E)					
Wk-6558	1150	C	26430 ± 350	31210 ± 380	30450–31970
Wk-6562	1230	C	25330 ± 310	30240 ± 290	29660–30820
OZC707	1265	C	27100 ± 260	31820 ± 160	31500–32140
Beta-96171	1265	C	27710 ± 220	32250 ± 230	32710–31790
Wk-6561	1275	C	13715 ± 98	16940 ± 70	16800–17080
Wk-6564	1300	C	16380 ± 120	19650 ± 190	19270–20030
BRA-ABC (elev. 316 m; c. 31.3334°S, 138.594°E)					
OZE086	15	T	11650 ± 110	13540 ± 130	13280–13800
Wk-7295	52	OC	15780 ± 140	18910 ± 140	18630–19190
BRA07-AR (elev. 300 m; 31.32897°S, 138.58563°E)					
OZK516	155	S	15550 ± 130	18710 ± 70	18570–18850
OZK002	253	S	16280 ± 130	19500 ± 210	19080–19920
OZK003	338	S	15750 ± 120	18860 ± 110	18640–19080
SSAMS ANU-1812	380	S	14470 ± 60	17720 ± 60	17600–17840
SSAMS ANU-2038	380	C	15210 ± 220	18310 ± 270	17770–18850
OZK517	508	C	16820 ± 180	20210 ± 220	19770–20650
BRA07-G (elev. 270 m; 31.34134°S, 138.56968°E)					
AdGL-08003	375	Q SA		16110 ± 1290	13170 (11660–14590)
AdGL-08002	1330	Q SA		30710 ± 2730	27450 (24120–30860)
ANU BG42	1350	S	24000 ± 240	28900 ± 410	28080–29720
ANU BG27	1400	S	30400 ± 480	34630 ± 400	33830–35430
AdGL-08001	1620	Q SA		32360 ± 1660	23860 (21240–26340)
WL08-UFP (elev. 399 m; 31.27082°S, 138.85221°E)					
Wk23464	20	S	7390 ± 45	8240 ± 60	8120–8360
Wk23465	75	C	13021 ± 52	15640 ± 100	15440–15840
Wk23524	733	S	25084 ± 176	30010 ± 160	29690–30330
Wk23467	780	S	19034 ± 127	22930 ± 180	22570–23290
Wk23525	853	S	31398 ± 367	35300 ± 390	34520–36080

OSL age estimates (FMM inherited populations)	OSL age estimates (FMM post-depositional populations)	δ^{13}C (ppm) [H$_2$O content (%)]	Comments
		−25.8	(Cock *et al.* 1999) Section C
		18.3	(Williams *et al.* 2001) based on TSAC/XRS DR
		17.4	(Williams *et al.* 2001) based on TSAC/XRS DR
		−24.5	Large piece of charcoal (8 cm × 2 cm)
		−	Lower Unit lighter band
		−7.1	Lower Unit darker band
		−	Lower Unit darker band with large calc. rhizocretes
		−27 ± 2	Present base of Lower Unit
		−23 ± 2	Present base of Lower Unit
		19.4	Independent run of AdGL-96002 on coarser fraction
		19.4	(Williams *et al.* 2001) based on TSAC/XRS DR
		−8.2	(Cock *et al.* 1999), *c.* 250 m downstream of BRA07-SD
		−9.1	(Cock *et al.* 1999), *c.* 250 m downstream of BRA07-SD
		−7.9	(Cock *et al.* 1999), *c.* 250 m downstream of BRA07-SD
		0	−
		0	−
		−9.6	(Cock *et al.* 1999), *c.* 250 m downstream of BRA07-SD
			Lubra Waterhole
		−26.0 ± 0.2	Disturbed Silts towards base of 1350 cm section
		−23.1 ± 0.2	Disturbed Silts towards base of 1350 cm section
		−25	Disturbed Silts towards base of 1350 cm section
		−24.9	Disturbed Silts towards base of 1350 cm section
		−24.4 ± 0.2	Disturbed Silts towards base of 1350 cm section
		−23.0 ± 0.2	Disturbed Silts towards base of 1350 cm section
			Brachina ABC Narrows
		0	Topmost tufa cap
		−23.8 ± 0.2	30 cm black organic clay band below tufa (Williams *et al.* 2001)
			Brachina AR
		−9.3	Uppermost gravel–Silts couplet
		−7.6	Gravel–Silts couplet
		−8.5	Gravel–Silts couplet
		−17 ± 1	Lowermost gravel–Silts couplet
		−23 ± 2	Lowermost gravel–Silts couplet
		−24.9	Clast-supported basal gravel (up to *c.* 25 cm)
			Brachina 18 M Section
26740 ± 4500	−	3.7	Towards base of final wedge of local colluvium
38880 ± 4420	−	5.8	Below onset of multiple gravel sheets
		−	(Williams *et al.* 2001)
		−	(Williams *et al.* 2001)
50280 ± 4510	*21650 ± 1960*	4.4	Lowermost exposed band of Silts
			Wilkawillina Upper FP
		−	Uppermost 'red drape'
		−25.5 ± 0.2	Towards base of Silts–capping gravel band
		−	Within 'chromatic band'
		−	Within incipient Bca-horizon below 'chromatic band'
		−7.1 ± 0.2	Pocket of intact shells below 2nd Bca-horizon in Lowermost 'chromatic band'

(Continued)

Table 1. *Continued*

Section name and sample code	Depth from top (in cm)	Sample material	^{14}C age estimates (^{14}C a BP $\pm$ error)	^{14}C age estimates (a cal. BP, 1 SD) [OSL age estimates (FMM depositional population/CDM)]	^{14}C age estimates (a cal. BP, 2 SD) [OSL age estimates (MAM with 1 SD range)]
WL07-FP (elev. 399 m, elev. 382; 31.26960°S, 138.86783°E)					
CLW-95/4	85	Q SG		17070 ± 1620	9810 (9090–10530)
AdGL-07009	85	Q SA		17360 ± 1650	14600 (12630–16430)
SSAMS ANU-1813	240	S	18500 ± 60	22350 ± 110	22130–22570
AdGL-07008	255	Q SA		24900 ± 1350	20810 (18370–23070)
OZK014	268	S	18610 ± 180	22310 ± 310	21690–22930
SSAMS ANU-2033	500	C	24440 ± 200	29290 ± 360	28570–30010
SSAMS ANU-2032	520	C	22800 ± 170	27500 ± 360	26780–28220
OZK012	745	C	32760 ± 430	37120 ± 840	35440–38800
OZK013	750	C	32210 ± 400	36650 ± 930	34790–38510
OZK011	760	S	29300 ± 300	33700 ± 350	33000–34400
AdGL-07007	855	Q SA		38870 ± 2850	23290 (20150–26100)
OZK010	880	C	36630 ± 580	41690 ± 410	40870–42510
AdGL-07006	1320	Q SA		45520 ± 3930	48030 (39520–55780)
AdGL-08007	1467	Q SA		46720 ± 4620	41460 (36960–45630)
WL07-S (elev. 376 m; 31.26887°S, 138.88176°E)					
SSAMS ANU-2001	85	C/OV	29590 ± 370	33890 ± 360	33170–34610
SSAMS ANU-2227	235	OV	21410 ± 120	25440 ± 230	24980–25900
SSAMS ANU-2229	275	OV	24520 ± 190	29410 ± 300	28810–30010
WL07-G (elev. 367 m; 31.27243°S, 138.87921°E)					
CLW-95/5	107	Q SG		17050 ± 1410	13030 (11270–14080)
CLW-95/3	156	Q SG		18330 ± 1380	16650 (15760–17540)
CLW-95/2	310	Q SG		23970 ± 1170	18390 (17000–19730)
OZK019	360	C	19800 ± 160	23670 ± 170	23330–24010
OZK020	370	C	19760 ± 210	23660 ± 220	23220–24100
CLW-95/1	715	Q SG		23970 ± 2000	19070 (16660–21520)
CAS06-1 (elev. 332 m; 31.15334°S, 138.56692°E to 31.15365°S, 138.57011°E)					
AdGL07002	22	Q SA		23160 ± 1100	21680 (19830–23370)
OZJ893	75	C	15080 ± 110	18250 ± 230	17790–18710
OZJ895	225	C	21560 ± 250	25670 ± 400	24870–26470
OZJ897	265	C	22590 ± 260	27310 ± 400	26510–28110
OZJ892	300	C	3940 ± 190	4400 ± 280	3840–4960
OZJ894	382	C	10370 ± 200	12160 ± 350	11460–12860
OZJ896	477	C	12790 ± 180	15310 ± 270	14770–15850
AdGL07001	609	Q SA		29370 ± 2120	27930 (25180–30350)
Other					
OZE026	70	C	300 ± 60	390 ± 70	530–250
OZE025	70	S	3450 ± 70	3730 ± 90	3910–3550
OZC713	–	E	40300 ± 3300	44470 ± 2870	38730–50210
OZJ902	TOP	S	14620 ± 140	17770 ± 90	17590–17950
OZE021	c. 700	S	33250 ± 1050	38170 ± 1800	34570–41770
OZE023	420	C	27900 ± 300	32430 ± 320	31790–33070
OZE024	850	C	28900 ± 500	33330 ± 520	32290–34370

Sample material is indicated: Q, quartz (SA, small aliquots; SG, single grains; LA, large aliquots); C, charcoal; S, carbonate shells of gastropods; (Weninger & Jöris 2008). OSL ages are calculated using the Finite Mixture Model (FMM) of Galbraith (2005), interpreting resultant D_e single population, the Central Dose Model (CDM) is employed (Galbraith 2005). For comparison, results of the Minimum Age Model (MAM)

OSL age estimates (FMM inherited populations)	OSL age estimates (FMM post-depositional populations)	$\delta^{13}C$ (ppm) [H_2O content (%)]	Comments
			Wilkawillina Floodplain
23360 ± 11480	*7400 ± 1.130*	7.2	2–3 cm below 'red drape' within incipient Bca-horizon
24630 ± 2540	–	7.3	2–3 cm below 'red drape' within incipient Bca-horizon
		−6 ± 1	Upper of red 'twin Sf marker bands'
50280 ± 9150	–	6.7	Between red 'twin Sf marker bands'
		−5.7	Lower of red 'twin Sf marker bands'
		−26 ± 2	Onset of Silts above widespread gravel sheet
		−23 ± 2	Onset of Silts above widespread gravel sheet
		−24.7	Massive Silts below multiple gravel bands
		−25.5	Massive Silts below multiple gravel bands
		−2.2	Massive Silts below multiple gravel bands
–	22730 ± 1350	3.8	Massive Silts below gravel drape of lower section
		−23.8	Massive Silts below gravel drape of lower section
72060 ± 5900	–	4.0	Lowermost Bca-horizon
65750 ± 7540	–	2.5	Onset of Silts aggradation below calctrete pans
			Wilkawillina Slackwater
		−25 ± 2	Topmost organic veneer above more 'chromatic Silts'
		−15.6 ± 1.2	Lowermost gravel–Silts couplet
		−14.7 ± 1.3	Lowermost organic veneer
			Wilkawillina Gorge
38080 ± 4730	–	4.4	Distinct light band sandwiched by organic veneers
29370 ± 2200	–	4.9	Termination of massive disturbed Silts below gravel
32640 ± 1800	*13830 ± 1220*	4.0	Light Silts with organic veneers
		−25.8	Onset of light disturbed Silts
		−24.6	Onset of light disturbed Silts
43400 ± 2630	*16010 ± 1840*	7.4	Second brown Silts band from base incorporating Platy gravel and kankar
			Cascades 1
30300 ± 1810	–	4.2	Top without Bca-horizon
		−25	<100 µg! run on ANTARES accelerator
		−25.4	Run on STAR accelerator
		−24	Run on STAR accelerator
		−25	<100 µg! run on ANTARES accelerator
		−25	<100 µg! run on ANTARES
		−25	<100 µg! run on ANTARES
38130 ± 2640	–	4.0	Disturbed Silts at base of section
			Other
		−9.84	Between eroded burial and palaeosol
		−1.65	Between eroded burial and palaeosol
		0	Dune site on Hawker/Parachilna road (palaeosol)
		−8.2	Termination of Silts capped by tufa bench at CAS06-2
		−8.59	From 10 m terrace sampled 1998
		−26.05	Price Creek bank section, 5.8 m terrace
		−25.0	Bunyeroo Creek, 10 m terrace, red–brown clay

OV, veneers of organic detritus; OC, organic clay; T, tufa; E, emu egg shell. All ^{14}C ages are calibrated using the CalPal-2007[Hulu]-calibration dataset populations in terms of 'depositional', 'inherited' and 'post-depositional' age estimates (if relative proportion >25%). Where D_e values fall into a are given (Galbraith 2005). Values in italic type indicate D_e-populations of <25% and are treated as outliers.

cycle, silt- and fine sand-sized dust fallout that mantled the slopes was entrained by runoff and episodically transported downstream. Today, the slopes are largely stripped of dust mantles and ephemeral traction load channels deeply entrench the silts. The aeolian provenance of the Flinders Silts is evident from a number of field and analytical observations, particularly their widespread occurrence regardless of the underlying bedrock lithology (see Williams *et al.* 2001, 2006; Williams & Nitschke 2005). The fine-grained deposits make a significant contribution to local closed catchments within narrow quartzite synclines (e.g. Wilpena Pound) and they interfinger with angular and probably frost-shattered quartzite clasts at the base of the westernmost range front. Pockets of loess are preserved on all major ridges and return geochemical signatures inconsistent with *in situ* weathering (Williams & Nitschke 2005). Finally, aggradation rates far exceed weathering rates within the respective catchments (Williams *et al.* 2001).

Age proxies and models

The chronostratigraphy of the Flinders Silts is based on two independent dating methods: radiocarbon dating and optically stimulated luminescence dating (OSL). The 94 radiocarbon age estimates are based on different types of material, including organic materials (charcoal (39), veneers of charred plant detritus (13) and organic clay (one)) and calcium carbonates (aquatic gastropod carbonate shells (39), tufa (one) and emu egg shell (one)). The 30 OSL age estimates are largely based on small aliquots (i.e. roughly 15 grains (19), and single grains of quartz (five), 180–212 µm in size or, where insufficient, 125–180 µm), and are analysed by various age models (Table 1). A brief discussion of their significance and reliability as proxies to infer the depositional age of host sediments will facilitate the interpretation of the stratigraphic sections.

Flinders Silts radiocarbon dates

Radiocarbon samples were pre-treated, and the carbon was extracted and converted to graphite using standard methods (Hua *et al.* 2001). The graphite was dated by accelerator mass spectrometry (AMS), by measuring the amount of residual ^{14}C in the material and calculating the age in radiocarbon years (years BP) (Tuniz *et al.* 1998). In most cases the precision, expressed as one standard deviation from the best ^{14}C age estimate, is better than 1% and reflects the combined errors and uncertainties from counting statistics, standards, measurements and the natural background. All ages fall within the detection limit of ^{14}C, generally given as eight half-lives (i.e. *c.* 46 000 radiocarbon years; Walker 2005). Atmospheric radiocarbon concentrations are elevated throughout the time range covered, particularly during the LGM and its lead-up. This results in considerable underestimation in terms of calendar years. Recently, calibration using independent numerical dating methods beyond the limits of annual tree-rings (Ferguson *et al.* 1966) and varved marine sediments, which extend to only *c.* 14.7 ka (Reimer *et al.* 2004), now allows the calibration of ^{14}C dates from the LGM and beyond (Fairbanks *et al.* 2005). All ^{14}C ages in this study are calibrated by the currently most reliable and precise high-resolution marine-derived ^{14}C-dated sedimentological and geochemical record from the Cariaco Basin, tied to the high-resolution ^{230}Th-dated Hulu Cave speleothem record (Hughen *et al.* 2006), using the integrated CalPal-2007^{Hulu}-calibration dataset (Weninger & Jöris 2008) as part of the CalPal-2007 calibration and palaeoclimate research software package (Weninger *et al.* 2008). Despite all recent advances, uncertainties in calibrating the older half of the ^{14}C time scale remain (Hoffman *et al.* 2008). As a consequence, the error margin of the calibrated ^{14}C age estimates is presented as two standard deviations, and is expressed as a range between minimum and maximum ages before 1950 (cal. BP; see Table 1). Abrupt and large shifts in ^{14}C concentration beyond 28 ka cal. BP result in prolonged 'age plateaux', reflected in large error ranges.

When all ^{14}C age estimates from the Flinders Silts are plotted as a function of depth below the top of respective stratigraphic sections and sample material, two important conclusions arise. First, the dates obtained from carbonate shells and veneers of organic detritus correlate reasonably well along an approximately linear trend (Pearson coefficient $r = 0.79$, excluding two Holocene ages), suggesting similar deposition rates and concurrent termination of the fine-grained aggradational sequences across a range of geomorphological settings and different catchments (Fig. 2). Second, and in contrast, the charcoal-derived age estimates are scattered with regard to depth ($r = 0.32$, excluding two Holocene ages). Here, dates both older than 30 ka cal. BP and younger than 18 ka cal. BP are derived from similar sample depths ranging from *c.* 25 to 1300 cm below top. Throughout the LGM, many dates obtained from charcoal depart from the stratigraphic age–depth trend based on ^{14}C ages from shells and organic veneers, as well as on luminescence ages.

This observation is supported by single pieces of charcoal sampled from organic veneers that, according to their low δ^{13}C values, represent woody elements of vegetation and return much older ages,

FLINDERS SILTS   AMS Age-Depth Plot

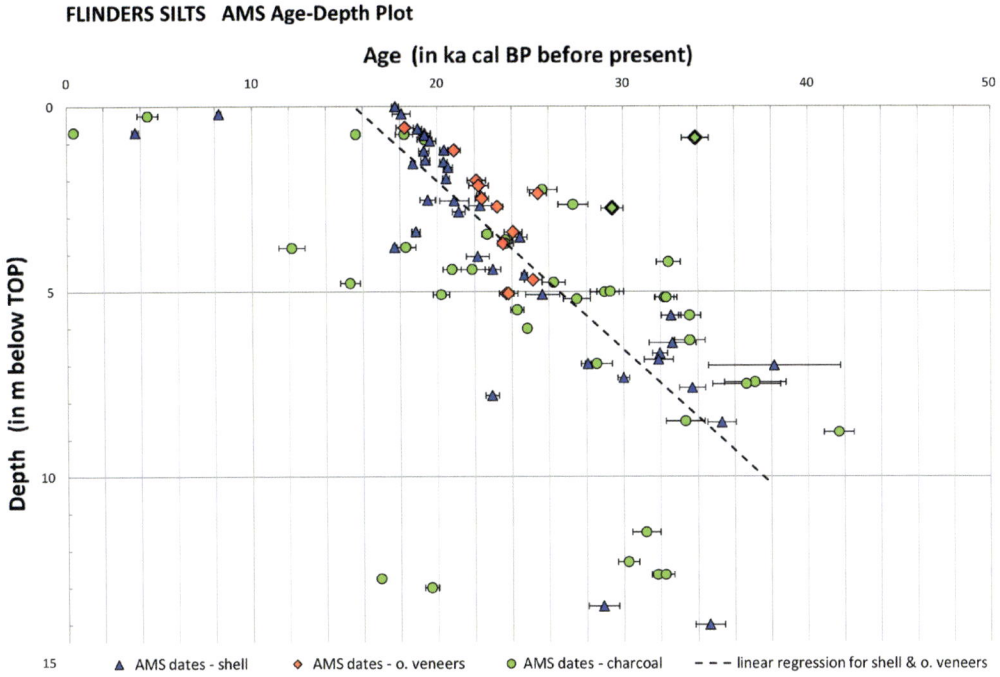

Fig. 2. Flinders Silts AMS radiocarbon data, colour-coded by sample material. The trend line (linear regression) for AMS-dated veneers of organic detritus and carbonate shells (excluding two Holocene samples) is indicated.

presenting outliers in the respective chronostratigraphic sequences (e.g. bold green diamonds in Fig. 2 and Table 1). It is likely that this reflects reworking of charcoal caused by erosion and subsequent redeposition (Blong & Gillespie 1978). In conclusion, ^{14}C ages derived from charcoal in a depositional environment of alluvial loess need to be treated with caution and interpreted as maximum ages. Although charcoal fragments represent a poor proxy for dating depositional events, collectively they do reflect intervals of widely occurring woody vegetation burnt by wild fires within the catchment. By plotting the frequency distribution of all ^{14}C ages based on charcoal irrespective of their sample position versus those based on shell, an interesting palaeo-environmental picture

emerges. According to the timeline, ages based on shell closely succeed those based on charcoal (Fig. 3). From this it can be inferred that intervals marked by droughts and wildfires (charcoal) are followed by floods (shells) that entrain and redistribute the age proxies along with readily erodible sediments.

Flinders Silts luminescence dates

Optically stimulated luminescence dating takes advantage of the characteristic of common mineral grains, such as quartz sand, to accumulate stored energy in their crystal lattices, which is released on exposure to light. Consequently, sediments buried in a depositional event gradually acquire an

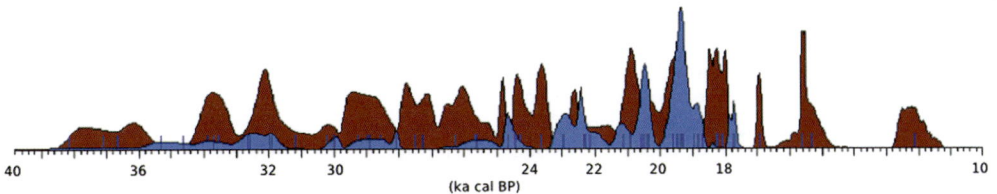

Fig. 3. Frequency distribution of calibrated ^{14}C ages based on charcoal pieces (brown) and shell (blue) on a linear scale from 40–10 ka cal. BP.

increasing energy that can be freed and measured in the laboratory by stimulation. 'Light-sensitive' electrons can be released by optical stimulation (Huntley *et al.* 1985). The energy absorbed by the sample mineral depends on its exposure to radioactivity from the decay processes of naturally occurring radioactive elements within the sediments and on cosmic irradiation, which together make up the environmental dose rate (D_r). Except for the oldest Hookina Floodplain samples, the D_r was remarkably similar across all samples discussed, probably a reflection of their common aeolian source (Table 2). At the altitudes of the study area, cosmic-ray contribution is largely a function of overburden thickness (Prescott & Hutton 1994) and is trivial relative to the radioactive contribution of the fine-grained sediments. The largest uncertainty in the calculation of accurate D_r is water content. Water attenuates the ionizing radiation more than constituent mineral grains of the sedimentary unit (Aitken 1998). Although present-day water contents can readily be measured in the laboratory and adequately taken into account in the final D_r calculation, they do not necessarily represent the average water content within the sediments over its time of burial (Li *et al.* 2008). Measured present-day water contents range from c. 2.5 to 10% across all recently collected samples and, where re-sampled, were found to vary up to c. 5% between seasons. Hence, these values were used to calculate a maximum and a minimum D_r estimate for each sample, and to determine a 'most likely estimate' by averaging the D_r estimate for the measured present-day water content with that obtained with 5% water content (Fig. 4 and Table 2).

Quartz grains for this study were sampled, extracted and prepared using standard techniques (see, e.g. Huntley *et al.* 1993). The intensity of the luminescence signal increases with the radiation dose in a nonlinear way, differing for every subsample of quartz grains. To translate the measured natural luminescence signal into an accurate laboratory equivalent-dose (D_e), the aliquot-specific response to a sequence of laboratory-administered radiation doses is established by the single-aliquot regenerative-dose protocol (SAR), comparing initially measured natural OSL signals with equivalent sensitivity-corrected signals from irradiation-dose responses performed on the same aliquot (Murray & Wintle 2000). The SAR protocol allows for internal checks to exclude deficient aliquots based on their recycling ratio, test dose errors, background difference and curve-fit (Wintle & Murray 2006). For the Flinders Silts samples, we used the rejection criteria recommended by Jacobs *et al.* (2006). Significant variations between D_e values of multiple aliquots of a sample reflect either insufficient exposure to daylight during the transport process or post-depositional introduction of more recently bleached sediments by infiltration, desiccation cracking and bioturbation. Only in exceptional geomorphological circumstances are all grains sufficiently exposed to reset their pre-existing luminescence signals. Particularly in fluvial and colluvial environments, a potentially large fraction retain prior D_e values. Averaging such 'poorly bleached' sample D_e values will result in a potentially significant OSL age overestimation (Olley *et al.* 1998). It follows that it is crucial to establish the variability in distribution of D_e values statistically. A prerequisite for this is small aliquot sizes, to minimize the effect of averaging. With the development of luminescence readers designed for rapid single-grain measurements (Duller *et al.* 1999), this can now be achieved on the basis of single grains. However, only a small number of quartz grains produce a valid signal (Duller 2004), so that similar results are obtained by reducing the aliquot size to the smallest 'effective' number of grains or 'small aliquots'. This is demonstrated by OSL sample WL07-FP 6, for which D_e values were independently established from single grains and small aliquots of roughly 15 grains (Tables 1 and 2).

Consequently, the application of appropriate statistical age models to infer the depositional population is crucial to the interpretation of the data. The variability in D_e values is best displayed in radial plots by plotting aliquots as points with their precision against the *x*-axis (increasing towards the right) and their D_e value against the *y*-axis (Olley & Reed 2003). The latter is scaled in terms of standard deviations (SDs) from a given central value or 'radial line' (Table 2, final column). More precise D_e values plot towards the right and statistically concordant ages fall within a band of two SDs about the radial line (Galbraith 1990). Different radial lines correspond to different central D_e values and OSL ages. Populations where more than 5% of the constituent aliquots fall outside the two SDs band are termed 'over-dispersed'. In the present study, the number of populations and the degree of over-dispersion is statistically determined by the best data fit (i.e. the lowest Bayesian information criterion value (BIC) while observing the maximum log-likelihood (L_{max}) to avoid overfitting), using the Finite Mixture Model (FMM) (Galbraith & Green 1990; Galbraith 2005). The resulting discrete populations are presented both numerically and as radial plots (Table 2). This approach allows the evaluation of any D_e populations within a stratigraphic context as either 'depositional', 'inherited' or 'post-depositional'. In this study, only discrete D_e populations exceeding 25% are discussed in terms of 'ages'; the remainder is accounted for as outliers. Some caution needs to be exercised in interpreting 'inherited' populations as proxies indicative of residual ages of the reworked sediments, for any previous D_r environments are likely to have differed from the present. However, the environmental dose rates for the Flinders Silts samples are very similar and cosmic-ray contributions (and variations) minor. According to results of the FMM analysis, two samples are best described by a single population (K1887, K1888), for which the Central Dose Model (CDM) was employed (Galbraith *et al.* 1999; Galbraith 2005; Table 2). In the most recent chronological studies of the Namib Silts, the Minimum Age Model (MAM; Olley *et al.* 1998; Galbraith 2005) was used (Bourke *et al.* 2003; Srivastava *et al.* 2006), revising and shifting a previously established LGM chronostratigraphy based on ^{14}C ages (Vogel 1982) to the Holocene. To allow for a comparison, results returned by this age model are reported as well (Table 1). However, large unsystematic discrepancies to otherwise orderly and paired internal chronostratigraphies reveal the inadequacy of the MAM in the context of alluvial loess sedimentation.

Chronostratigraphy of the Flinders Silts

The description of stratigraphic sections follows a transect from the western piedmont plain across into the Brachina catchment to the Wilkawillina catchment on the eastern side, which is approximately in line with prevailing contemporary (Schwerdtfeger & Curran 1996) and palaeo-wind directions dominated by the winter westerlies (Jennings 1968; Shulmeister *et al.* 2004; Fig. 1). Whereas the overall thickness of the reworked loess deposits decreases in a downwind direction, most evident in a comparison between the western and eastern piedmont plains, the actual depth of the silts in the stratigraphic sections is a complex function of the size of the upstream catchment, the local topography and the degree of incision.

Hookina Silts

Hookina Creek is deeply entrenched into a broad, gently rolling floodplain. A sequence of stacked calcareous palaeosols is exposed along vertical banks of unconsolidated fine-grained alluvium with near-horizontal bedding planes and localized cross-stratified sandy bands. Earlier attempts to date these regional Bca-horizons, which can be traced laterally for hundreds of metres, faced the constraints of radiocarbon dating of pedogenic carbonate (Williams 1973; for discussion of the limitations and assumptions involved see Williams & Polach 1971; Callen *et al.* 1983; Fontes & Gasse 1989) and sample quantities required for conventional radiocarbon dating (Williams 1982). Three

stratigraphic sections along the thalweg of the trunk channel are presented and OSL-dated (Fig. 5), establishing the onset and termination of Hookina Silts aggradation and the age of the parent material of the major palaeosols. HK07-D is the most upstream stratigraphic section from the western piedmont plain within the Hookina catchment and is less than 2 km west of the range front. Its surface is concordant with the highest and oldest floodplain and exhibits the uppermost well-developed Bca-horizon, which is also developed within both downstream sections. Approximately 6 km downstream, stratigraphic section HK07-L functions as a spatial link between HK07-D and HK07-M, another 1 km downstream and separated by a partially loess-draped quartzite ridge.

The 8.5 m high vertical cliff face of HK07-D hosts a sequence of at least four Bca-horizons (Fig. 6) that extend laterally for *c.* 500 m. Each horizon is topped by *c.* 30 cm thick continuous red (e.g. 5YR 5/6 dry–4/6 moist) sheets consisting of sand-sized mud aggregates with lenses of rounded and platy pebbles of variable lithology, hereafter referred to as 'chromatic bands' to include similar colours. The uppermost band mantles the present surface and exhibits a platy structure with rootlets and insect burrows. The buried chromatic bands are more compacted, of blocky to prismatic structure, and in places have slickensides and organic streaks. All, including the chromatic surface drape, are associated with discontinuous gravel. A central unit between *c.* 100 and 375 cm is sandier, with poorly preserved cross-stratification and intercalated silt drapes. A thin gravel sheet at *c.* 230 cm depth forms a local disconformity, in places scouring up to *c.* 30 cm into the underlying sediments. The imbricated, clast-supported mostly platy gravel indicates flow directions from the NE to NNE. The three OSL ages from the basal Bca-horizon and the two chromatic bands topping the well-developed central Bca-horizons indicate rapid aggradation of the host material between 84.6 ± 7.0 ka (K1887) and 82.5 ± 13.5 ka (K1889), although the large error ranges potentially extend that interval to *c.* 20 ka. The second sample from the base (K1888) is close to saturation and the few valid D_e aliquots are likely to reflect only an inherited age of 119.7 ± 12.8 ka. The base of the parent material of the uppermost Bca is dated to 19.3 ± 1.6 ka (K1890), with a large inherited population suggesting a source of reworked material last deposited 38.9 ± 2.7 ka.

Derived from the foot of the ruins of the old Hookina Hotel, abandoned half a century ago, HK07-L displays a distinct 'twin chromatic band' associated with and resting on a prominent composite Bca-horizon (Fig. 7). This band, at a depth of *c.* 275–325 cm (Fig. 6), peters out a few hundred metres upstream but can be traced laterally for

Table 2. *OSL data: environmental dose rate data (D_r) for variable water contents and corresponding equivalent*

Sample codes (and designation)	Irradiation data (TSAC and XRS)		D_r (H_2O % Gy ka^{-1})	
K1890 (HK07-D5) small aliquots	U (ppm)	1.67 ± 0.40	2.5	2.51 ± 0.14
	Th (ppm)	6.95 ± 1.31	5.1	2.44 ± 0.14
	K (%)	1.50 ± 0.05	10	2.32 ± 0.13
	Cosmic D_r (μGy a^{-1})	192 ± 19.2	AV	2.57 ± 0.10
K1889 (HK07-D4) small aliquots	U (ppm)	0.95 ± 0.49	2.5	2.64 ± 0.18
	Th (ppm)	11.19 ± 1.65	3.7	2.61 ± 0.17
	K (%)	1.56 ± 0.05	10	2.44 ± 0.16
	Cosmic D_r (μGy a^{-1})	138 ± 13.8	AV	2.59 ± 0.17
K1888 (HK07-D2) small aliquots	U (ppm)	2.06 ± 0.59	2.5	3.24 ± 0.21
	Th (ppm)	10.02 ± 1.96	3.4	3.27 ± 0.21
	K (%)	2.04 ± 0.06	10	3.02 ± 0.20
	Cosmic D_r (μGy a^{-1})	128 ± 12.8	AV	3.21 ± 0.21
K1887 (HK07-D1) small aliquots	U (ppm)	2.56 ± 0.39	2.5	3.38 ± 0.14
	Th (ppm)	10.13 ± 1.31	5.9	3.51 ± 0.15
	K (%)	2.18 ± 0.07	10	3.23 ± 0.14
	Cosmic D_r (μGy a^{-1})	100 ± 10.0	AV	3.40 ± 0.15

dose data (D_e)

FMM/CDM D_e pop. (Gy/relative prop. %)		OD	BIC	Radial plots
99.82 ± 5.84	63%	0.12	11.85	
49.60 ± 3.55	37%			
–	–			
–	–			
213.43 ± 31.84	76%	0.17	4.18	
274.03 ± 82.05	24%			
–	–			
–	–			
384.35 ± 32.86	100%	0	–	
–	–			
–	–			
–	–			
287.44 ± 20.29	100%	0	–	
–	–			
–	–			
–	–			

(*Continued*)

Table 2. *Continued*

Sample codes (and designation)	Irradiation data (TSAC and XRS)		D_r (H_2O % Gy ka^{-1})	
K1892 (HK07-L4) small aliquots	U (ppm)	2.39 ± 0.27	2.5	3.49 ± 0.11
	Th (ppm)	9.08 ± 0.90	7.9	3.29 ± 0.11
	K (%)	2.204 ± 0.066	10	3.22 ± 0.11
	Cosmic D_r (μGy a^{-1})	175 ± 17.5	AV	3.34 ± 0.11
K1891 (HK07-L2) small aliquots	U (ppm)	2.28 ± 0.42	2.5	3.06 ± 0.16
	Th (ppm)	8.36 ± 1.39	6.0	2.95 ± 0.15
	K (%)	1.85 ± 0.06	10	2.83 ± 0.14
	Cosmic D_r (μGy a^{-1})	164 ± 16.4	AV	3.01 ± 0.10
AdGL-08006 (HK07-M5) small aliquots	U (ppm)	2.01 ± 0.73	2.5	3.55 ± 0.24
	Th (ppm)	14.76 ± 2.43	7.4	3.37 ± 0.23
	K (%)	2.02 ± 0.06	10	3.28 ± 0.22
	Cosmic D_r (μGy a^{-1})	201 ± 20.1	AV	3.42 ± 0.23
AdGL-08005 (HK07-M3) small aliquots	U (ppm)	2.55 ± 0.30	1.9	2.38 ± 0.11
	Th (ppm)	6.27 ± 0.98	2.5	2.36 ± 0.11
	K (%)	1.32 ± 0.04	10	2.18 ± 0.10
	Cosmic D_r (μGy a^{-1})	112 ± 11.2	AV	2.34 ± 0.10

FMM/CDM D_e pop. (Gy/relative prop. %)		OD	BIC	Radial plots
100.97 ± 4.46	92%	0.09	1.88	
38.86 ± 4.27	*8%*			
–	–			
–	–			
108.86 ± 8.16	90%	0.20	11.12	
258.20 ± 68.22	*10%*			
–	–			
–	–			
61.61 ± 3.59	54%	0.15	58.22	
28.18 ± 2.27	33%			
180.34 ± 27.57	*9%*			
9.45 ± 2.04	*5%*			
292.64 ± 18.35	90%	0.15	3.23	
200.15 ± 40.85	10%			
–	–			
–	–			

(Continued)

Table 2. *Continued*

Sample codes (and designation)	Irradiation data (TSAC and XRS)		D_r (H_2O % Gy ka^{-1})	
AdGL-08004 (HK07-M1) small aliquots	U (ppm)	2.11 ± 0.57	2.5	2.95 ± 0.19
	Th (ppm)	11.22 ± 1.90	5.3	2.86 ± 0.19
	K (%)	1.73 ± 0.05	10	2.72 ± 0.18
	Cosmic D_r (μGy a^{-1})	81 ± 8.1	AV	2.86 ± 0.19
AdGL-08003 (BRA07-G4) small aliquots	U (ppm)	3.23 ± 0.28	2.3	3.33 ± 0.11
	Th (ppm)	8.70 ± 0.93	2.5	3.33 ± 0.11
	K (%)	1.98 ± 0.06	10	3.07 ± 0.10
	Cosmic D_r (μGy a^{-1})	145 ± 14.45	AV	3.28 ± 0.11
AdGL-08002 (BRA07-G2) small aliquots	U (ppm)	3.18 ± 0.29	2.5	3.16 ± 0.11
	Th (ppm)	10.56 ± 0.94	6.6	3.02 ± 0.10
	K (%)	1.77 ± 0.05	10	2.91 ± 0.10
	Cosmic D_r (μGy a^{-1})	64 ± 6.4	AV	3.05 ± 0.10
AdGL-08001 (BRA07-G 1) small aliquots	U (ppm)	2.66 ± 0.37	2.5	3.26 ± 0.13
	Th (ppm)	11.91 ± 1.23	3.8	3.21 ± 0.13
	K (%)	1.91 ± 0.06	10	3.00 ± 0.12
	Cosmic D_r (μGy a^{-1})	52 ± 5.2	AV	3.19 ± 0.13

FMM/CDM D_e pop. (Gy/relative prop. %)		OD	BIC	Radial plots
309.90 ± 24.76	57%	0.20	36.91	
65.78 ± 7.43	25%			
162.45 ± 28.33	18%			
–	–			
52.89 ± 3.85	74%	0.15	12.26	
87.79 ± 14.49	26%			
–	–			
–	–			
93.55 ± 7.69	47%	0.05	8.64	
118.45 ± 12.86	43%			
249.64 ± 37.12	10%			
–	–			
103.07 ± 3.21	68%	0.05	20.26	
160.15 ± 12.78	18%			
68.94 ± 5.56	14%			
–	–			

(Continued)

Table 2. *Continued*

Sample codes (and designation)	Irradiation data (TSAC and XRS)		D_r (H$_2$O % Gy ka^{-1})	
CLW-95/4 (WL07-FP6-SG) single grains	U (ppm) Th (ppm) K (%) Cosmic D_r (μGy a^{-1})	2.33 $\pm$ 0.33 8.68 $\pm$ 1.09 2.06 $\pm$ 0.06 202 $\pm$ 20.2/190 $\pm$ 19.0	2.5 9.4 10 AV	3.36 $\pm$ 0.13 3.13 $\pm$ 0.12 3.11 $\pm$ 0.12 3.26 $\pm$ 0.09
AdGL-07009 (WL07-FP6-SA) small aliquots	U (ppm) Th (ppm) K (%) Cosmic D_r (μGy a^{-1})	2.33 $\pm$ 0.33 8.68 $\pm$ 1.09 2.06 $\pm$ 0.06 202 $\pm$ 20.2/190 $\pm$ 19.0	2.5 9.7 10 AV	3.36 $\pm$ 0.13 3.12 $\pm$ 0.12 3.11 $\pm$ 0.12 3.26 $\pm$ 0.09
AdGL-07008 (WL07-FP5) small aliquots	U (ppm) Th (ppm) K (%) Cosmic D_r (μGy a^{-1})	2.59 $\pm$ 0.25 10.45 $\pm$ 0.45 2.10 $\pm$ 0.05 189 $\pm$ 18.9/160 $\pm$ 16.0	2.5 8.4 10 AV	3.49 $\pm$ 0.09 3.27 $\pm$ 0.09 3.22 $\pm$ 0.09 3.33 $\pm$ 0.09
AdGL-07007 (WL07-FP3) small aliquots	U (ppm) Th (ppm) K (%) Cosmic D_r (μGy a^{-1})	2.21 $\pm$ 0.32 10.42 $\pm$ 1.08 2.19 $\pm$ 0.07 179 $\pm$ 17.9/90 $\pm$ 9.0	2.5 2.5 10 AV	3.44 $\pm$ 0.13 3.44 $\pm$ 0.13 3.17 $\pm$ 0.12 3.24 $\pm$ 0.09

FMM/CDM D_e pop. (Gy/relative prop. %)		OD	BIC	Radial plots
55.63 ± 5.03	83%	0.22	82.98	
24.12 ± 3.62	9%			
76.12 ± 37.35	8%			
–	–			
56.57 ± 5.12	63%	0.1	11.85	
80.25 ± 7.95	37%			
–	–			
–	–			
83.01 ± 3.93	91%	0.10	2.19	
167.61 ± 30.19	9%			
–	–			
–	–			
126.05 ± 8.62	69%	0.05	6.00	
73.71 ± 3.93	31%			
–	–			
–	–			

(Continued)

Table 2. *Continued*

Sample codes (and designation)	Irradiation data (TSAC and XRS)		D_r (H$_2$O % Gy ka^{-1})	
AdGL-07006 (WL07-FP1) small aliquots	U (ppm) Th (ppm) K (%) Cosmic D_r (μGy a^{-1})	2.19 $\pm$ 0.34 10.21 $\pm$ 1.12 1.78 $\pm$ 0.05 114 $\pm$ 11.4/70 $\pm$ 0.7	2.5 3.0 10 AV	3.04 $\pm$ 0.13 3.02 $\pm$ 0.13 2.80 $\pm$ 0.12 3.02 $\pm$ 0.05
AdGL-08007 (WL07-FP0) small aliquots	U (ppm) Th (ppm) K (%) Cosmic D_r (μGy a^{-1})	2.54 $\pm$ 0.35 7.50 $\pm$ 1.14 2.02 $\pm$ 0.06 100 $\pm$ 10.0/60 $\pm$ 0.6	0.1 2.5 10 AV	3.17 $\pm$ 0.13 3.07 $\pm$ 0.13 2.83 $\pm$ 0.12 3.07 $\pm$ 0.13
CLW-95/5 (WL07-G6) single grains	U (ppm) Th (ppm) K (%) Cosmic D_r (μGy a^{-1})	2.05 $\pm$ 0.34 10.45 $\pm$ 1.15 2.35 $\pm$ 0.07 187 $\pm$ 18.7	2.5 3.9 10 AV	3.69 $\pm$ 0.14 3.63 $\pm$ 0.14 3.40 $\pm$ 0.13 3.61 $\pm$ 0.14
CLW-95/3 (WL07-G5) single grains	U (ppm) Th (ppm) K (%) Cosmic D_r (μGy a^{-1})	2.80 $\pm$ 0.32 8.11 $\pm$ 1.05 2.15 $\pm$ 0.06 178 $\pm$ 17.8	2.5 4.8 10 AV	3.47 $\pm$ 0.13 3.38 $\pm$ 0.12 3.20 $\pm$ 0.12 3.38 $\pm$ 0.12

FMM/CDM D_e pop. (Gy/relative prop. %)		OD	BIC	Radial plots
217.69 ± 17.42	77%	0	3.87	
137.53 ± 11.63	23%			
–	–			
–	–			
143.61 ± 12.88	59%	0.10	7.38	
202.11 ± 21.61	41%			
–	–			
–	–			
59.48 ± 4.41	67%	0.15	24.42	
132.83 ± 15.75	33%			
–	–			
–	–			
99.21 ± 6.51	61%	0.15	76.54	
61.93 ± 4.06	39%			
–	–			
–	–			

(*Continued*)

Table 2. *Continued*

Sample codes (and designation)	Irradiation data (TSAC and XRS)		D_r (H_2O % Gy ka^{-1})	
CLW-95/2 (WL07-G3) single grains	U (ppm) Th (ppm) K (%) Cosmic D_r (μGy a^{-1})	1.88 ± 0.33 9.56 ± 1.10 2.18 ± 0.07 153 ± 15.3	2.5 3.0 10 AV	3.35 ± 0.13 3.33 ± 0.13 3.09 ± 0.12 3.30 ± 0.13
CLW-95/1 (WL07-G1) single grains	U (ppm) Th (ppm) K (%) Cosmic D_r (μGy a^{-1})	1.55 ± 0.45 10.35 ± 1.49 1.93 ± 0.06 104 ± 10.4	2.5 9.8 10 AV	3.04 ± 0.16 2.80 ± 0.15 2.80 ± 0.15 2.88 ± 0.16
AdGL-07002 (CAS06-1 3) small aliquots	U (ppm) Th (ppm) K (%) Cosmic D_r (μGy a^{-1})	1.54 ± 0.37 14.22 ± 1.25 1.92 ± 0.05 115 ± 11.5	2.5 3.3 10 AV	3.360 ± 0.137 3.330 ± 0.136 3.105 ± 0.126 3.300 ± 0.135
AdGL-07001 (CAS06-1 1) small aliquots	U (ppm) Th (ppm) K (%) Cosmic D_r (μGy a^{-1})	1.99 ± 0.34 9.33 ± 1.12 1.44 ± 0.04 205 ± 20.5	2.5 3.3 10 AV	2.588 ± 0.121 2.574 ± 0.120 2.388 ± 0.111 2.546 ± 0.119

Discrete D_e populations are calculated using the Finite Mixture Model (FMM) 'fmix.s' (Galbraith 2005), and numerically listed by their Bayesian information criterion value (BIC). Where D_e values fall into one discrete population, the Central Dose Model (CDM) against the *x*-axis (increasing towards the right), and D_e values plotted against the *y*-axis (bands indicating 2 SD from a given central in regular font, whereas D_e populations consisting of <25% are shown in italics and treated as outliers and not converted into of 10% and an average (AV) between the measured present-day and a long-term water content of 5% in order to reduce the seasonal impact of

FMM/CDM D_e pop. (Gy/relative prop. %)		OD	BIC	Radial plots
79.03 ± 2.31	64%	0.03	15.28	
107.59 ± 4.21	27%			
45.60 ± 3.62	8%			
–	–			
124.91 ± 3.36	59%	0	24.98	
69.00 ± 4.38	32%			
46.07 ± 4.66	9%			
–	–			
76.43 ± 1.84	54%	0.02	2.40	
99.98 ± 4.38	39%			
173.63 ± 28.70	7%			
–	–			
97.08 ± 4.97	51%	0.06	−1.25	
74.78 ± 4.10	49%			
–	–			
–	–			

relative proportion in the sample. The number of D_e populations and the over-dispersion factor (OD) are largely determined by the lowest 'cdose.s' (Galbraith 2005) is employed. All aliquots or grains are presented as radial plots; that is, as points with precision values plotted value or 'radial line') (Olley & Reed 2003). The D_e population interpreted to relate to the last deposition of the sediment is shown ages. Different D_r estimates are presented for water contents as measured in the laboratory, a minimum value of 2.5%, a maximum value sample collection.

FLINDERS SILTS   OSL Age-Depth Plot

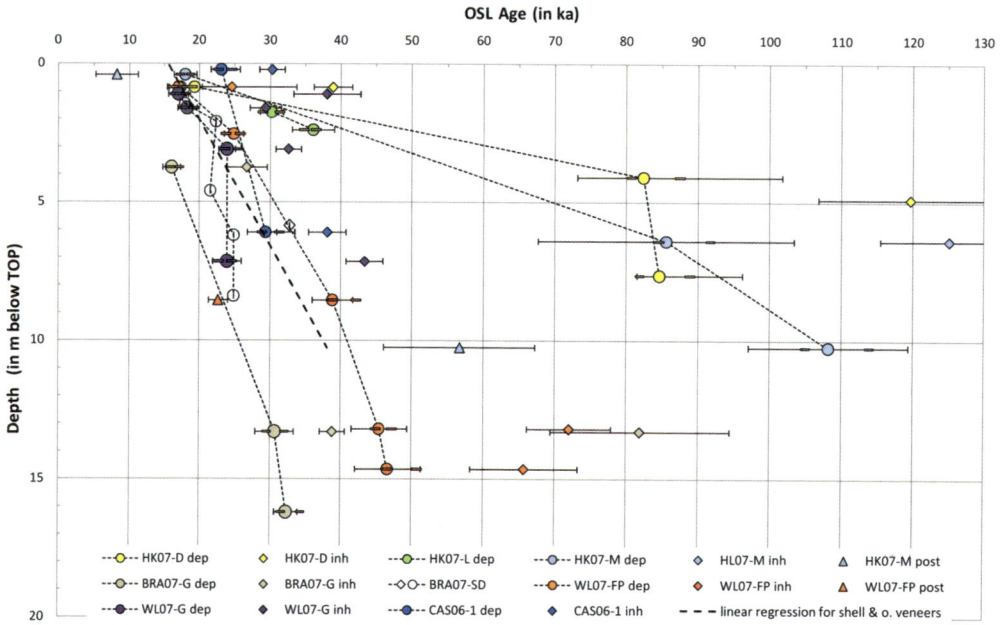

Fig. 4. Flinders Silts OSL data, colour-coded by stratigraphic section. Age estimates referring to the last depositional event are depicted as circles flanked by rectangles indicating the impact of changes in water content for 2.5% and 10%. Significant 'inherited' D_e populations are plotted as diamonds and significant 'post-depositional' D_e populations as triangles. The trend line (linear regression) for AMS-dated veneers of organic detritus and carbonate shells (excluding two Holocene samples) is indicated for comparison with Figure 3.

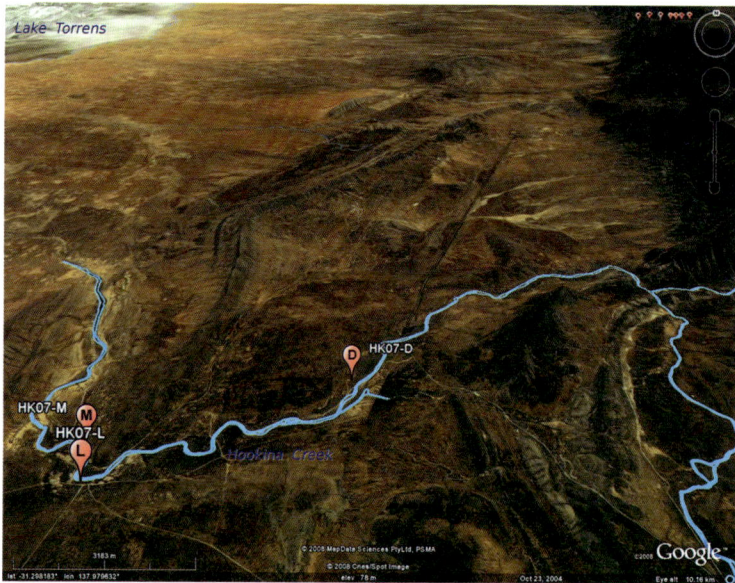

Fig. 5. Hookina Silts stratigraphic sections as seen from the south at an altitude of *c.* 10.16 km on Cnes/Spot imagery (23/10/2004) projected on the SRTM digital elevation model (DEM; elevation exaggeration factor 3) in Google Earth 4.3, with Lake Torrens playa to the west and the location of the Brachina Silts stratigraphic sections in the background. Locations of all sections within the Hookina catchment are indicated by red pointers with letters corresponding to section names (see also Fig. 1).

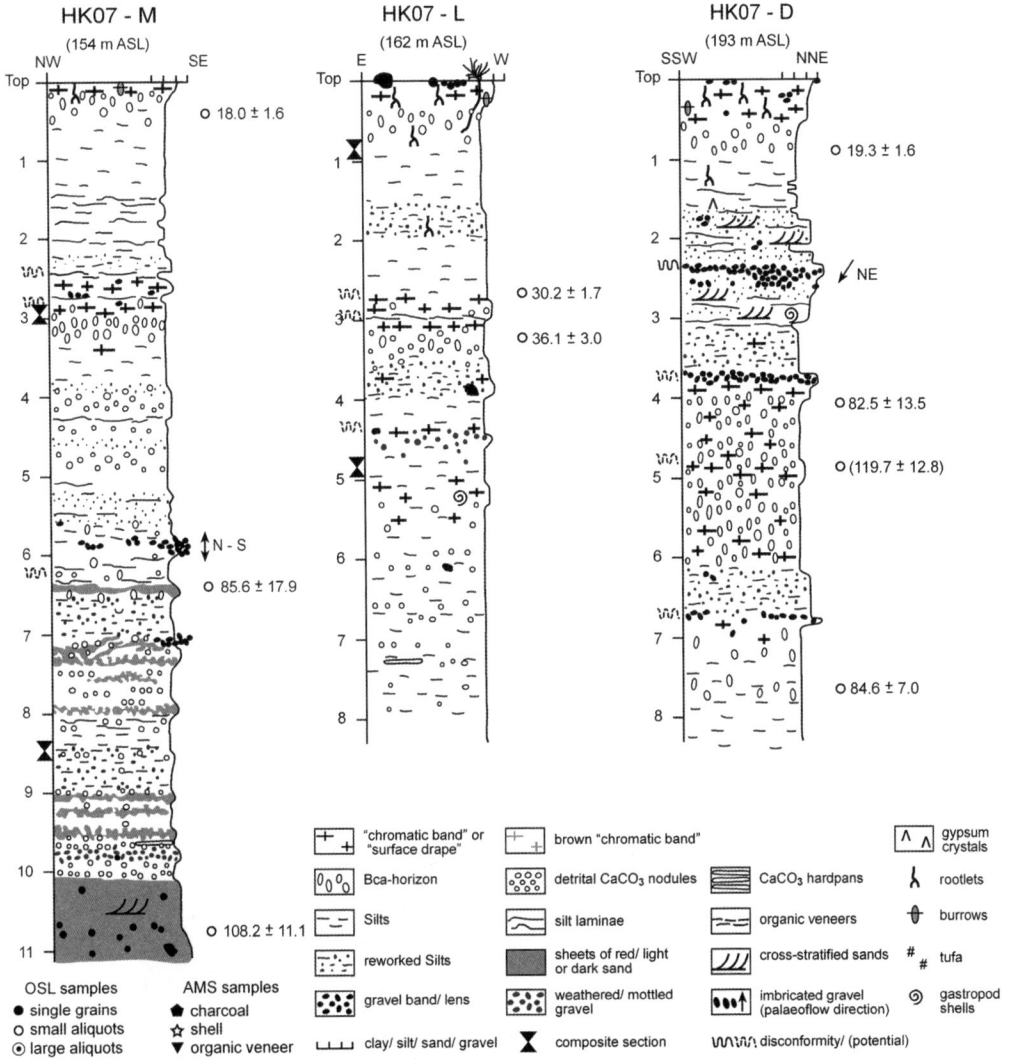

Fig. 6. Hookina Silts stratigraphic sections; main lithostratigraphic units and chronostratigraphy.

Fig. 7. Hookina Floodplain looking towards the Range front and stratigraphic section HK07-L.

more than 1 km downstream, where it terminates in stratigraphic section HK07-M. The lower band is dated to 36.1 ± 3.0 ka (K1891) and the upper, separated by a light carbonate-indurated silt drape, to 30.2 ± 1.7 ka (K1892). Abundant artefacts from the former hotel are incorporated in the uppermost chromatic surface drape.

The freshly collapsed 11 m high vertical stratigraphic section HK07-M presents the most complete record from the Hookina Floodplain. Here, the Hookina Silts rest on fluvial sands and gravel, and continue without obvious erosional breaks towards the top of the section, level with the highest floodplain surface (Fig. 6). The lowermost *c.* 4 m are layered, comprising multiple sheets of rolled detrital calcareous nodules (transported nodular calcrete) from reworked Bca-horizons, which alternate with seven distinct *c*. 20 cm thick undulating red bands consisting of well-sorted fine sand and silt. In rare instances, these bands are associated with narrow gravel-filled chutes and overlie *in situ* calcareous rhizocretions. Sheets of nodular calcrete occur at present on and downstream of erosional surfaces truncating the uppermost calcareous Bca-horizon as the result of rare overbank flood events. No modern analogue for the red bands was observed, but colour and modal sizes match the fine component of dune fields that extend from Lake Torrens towards the range front (Fig. 1). The fluvial sands below the silts are dated to the onset of the last glacial cycle at 108.2 ± 11.1 ka (AdGL-08004) and the uppermost red band was deposited at 85.6 ± 17.9 ka (AdGL-08005), with a dominant inherited component with an age of 125.1 ka ± 9.6 ka. The lowermost sample (AdGL-08004) was taken less than 1 m below a large cut-and-fill structure from which large mottled root casts extend and its minor post-depositional age of 23.0 ± 3.0 ka is likely to reflect rapid early LGM-infilling with reworked fine-grained material. Deposition of the topmost Bca-horizon is dated to 18.0 ± 1.6 ka (AdGL-08006) and the sample contains a significant post-depositional

population of 8.3 ± 0.9 ka, attributed to bioturbation and pedogenesis, possibly linked to the formation of the Bca-horizon.

Brachina Silts

The fine-grained valley-fill formations within the Brachina Creek catchment form an intramontane floodplain remnant upstream of the narrow Brachina Gorge that cuts through a sequence of resistant ridges (Fig. 8). They were initially mapped and described as lacustrine deposits (Callen & Reid 1994; Cock *et al.* 1999). A theodolite survey and dating programme in 2000 established that they form a gently sloping surface parallel to older rock-cut terraces and the present thalweg, extending into the gorge. This led to their interpretation as alluvial wetland deposits (Williams *et al.* 2001). Additional dating, Differential Global Positioning System (DGPS)-surveys, detailed sedimentological descriptions, and geophysical and geochemical analyses of six stratigraphic sections have led to the present reinterpretation of the Brachina Silts as a succession of two intramontane floodplains. Areas upstream of bedrock constrictions, tributary mouths and protected embayments back-flooded during major precipitation events, resulting in laminated slackwater deposits. Three stratigraphic sections are

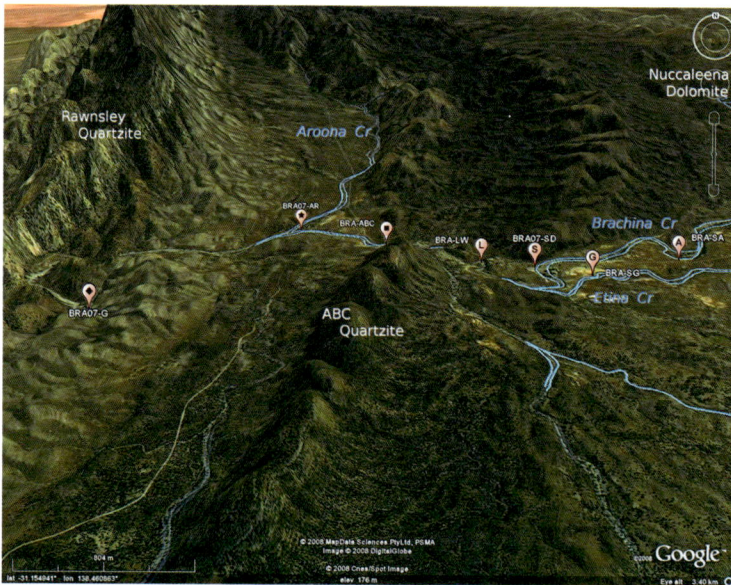

Fig. 8. Brachina Silts stratigraphic sections as seen from the south at an altitude of *c*. 3.4 km on Cnes/Spot imagery (23/10/2004) projected on the SRTM DEM (elevation exaggeration factor 3) in Google Earth 4.3, with Aroona Creek joining in from the north between the steep ABC Quartzite and Rawnsley Quartzite ranges. Locations of all sections within the Brachina catchment are indicated by red pointers with letters corresponding to section names (see also Fig. 1).

presented from the Brachina Floodplain upstream of the ABC Range: (1) BRA-SA situated upstream where the floodplain is confined by a rise of the Nuccaleena Dolomite Formation; (2) BRA-SG halfway along Etina Creek, which according to imbrication patterns of cut-and-fill structures embedded within the floodplain may have persisted as a main line of flow throughout the floodplain aggradation; (3) BRA07-SD, which is within a slight depression c. 300 m to the west of the trunk channel at the foot of the steep eastern valley slope in an area backflooded by the constriction posed by the ABC Range. BRA07-SD is entrenched by the present channel of Brachina Creek (Haberlah *et al.* 2010). Three stratigraphic sections follow downstream from remnants within the Brachina Gorge. BRA-LW is within the deeply entrenched trunk channel near the eastern mouth of the gorge. Downstream of the constriction formed by the quartzite ABC Range, Brachina Creek meets the wide Aroona valley, which is eroded into shales of the Bunyeroo Formation. Stratigraphic section BRA07-AR is located where Aroona Creek joins Brachina Creek. Finally, just upstream of the steeply dipping Rawnsley Quartzite, Brachina Creek makes a sharp turn into an embayment eroded by three minor tributaries. Within this back-flooded area, the c. 18 m high terrace remnant BRA07-G is preserved.

BRA-SA is a c. 6 m sequence of Brachina Silts and gravel (Fig. 9). Aggradation rates correlate well with the two other sections from the Brachina

Floodplain. Two features set it apart, however: the more oxidized (red) colour of the silts (7.5YR 5/8 dry–4/6 moist), and three to four generations of intercalated widespread cut-and-fill structures of clast-supported gravels. The lowermost is below fine-grained sediments dated to 24.8 ± 0.1 ka cal. BP (Beta-84141), followed by two generations of chutes cutting into material aged between 22.2 ± 0.2 ka cal. BP (OZC709) and 23.0 ± 0.2 ka cal. BP (OZC704). The uppermost gravel spreads out as a continuous sheet on the silts. The sequence is topped by another 1 m of fine-grained sediments with a Bca-horizon developed below a chromatic surface drape (5YR 6/6 dry–4/6 moist).

The uppermost 2 m have been dated in detail in the downstream section BRA-SG and show continuous aggradation until 18.1 ± 0.2 ka cal. BP (Wk-6548). BRA-SG is marked by multiple thin layers of detrital calcareous casts of sedges and tufa clasts. Here, the Brachina Silts are part of an overall greyish unit that increases in thickness towards the range front, and indicates waterlogged conditions (Cock *et al.* 1999). However, as the undisturbed sequence of ^{14}C dates based on gastropod shells indicates, bioturbation must be assumed minimal and the growth of the now fossilized vegetation was restricted.

BRA07-SD is further downstream and across the floodplain in a slight depression at the base of a bedrock slope (Fig. 10). This stratigraphic section is distinctly layered to laminated in its

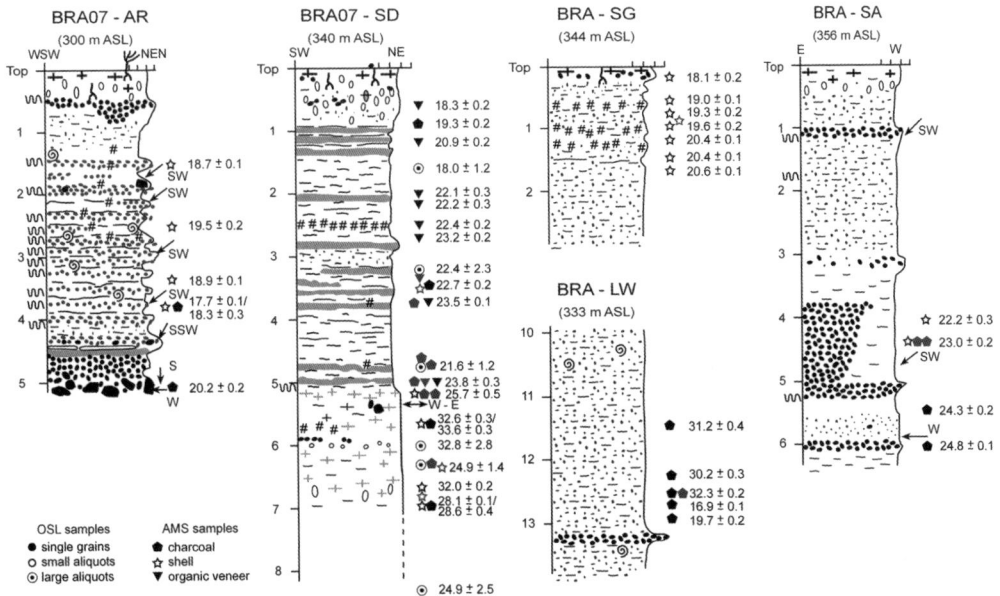

Fig. 9. Brachina Silts stratigraphic sections; main lithostratigraphic units and chronostratigraphy. (For legend see Fig 6.)

Fig. 10. The layered and laminated slackwater stratigraphic section BRA07-SD.

upper 5 m and massive and more chromatic in its lower part (Fig. 9) (Haberlah *et al.* 2010). The sharp boundary between these units is marked by a *c.* 10–20 cm thick, brown (7.5YR 5/2 dry–4/2 moist) chromatic band previously interpreted as a 'palaeosol' (Williams *et al.* 2001). The microfossil record of BRA07-SD indicates a major palaeo-environmental change at and above the 'palaeosol', from a large diversity of unbroken freshwater species of ostracods and gastropods to a lower diversity reflecting high-energy (fragmented shells), more saline and intermittent conditions (Glasby *et al.* 2007). Despite the close spacing of ^{14}C dates, it proved impossible to quantify the hiatus between the two aggradational units with certainty. Age estimates based on the lowermost veneer of organic detritus indicate an onset of laminated aggradation at 23.8 ± 0.3 ka cal. BP (SSAMS ANU-4205). Older ages further up the sequence are based on charcoal pieces. The top of the lower unit (i.e. the material of the 'palaeosol') is dated to 25.7 ± 0.5 ka cal. BP (OZJ909) based on shell, followed by large charcoal pieces returning ages between 32.2 ± 0.3 ka cal. BP (SSAMS ANU 2036) and 33.6 ± 0.4 ka cal. BP (OZJ904). Associated shells are dated to 32.0 ± 0.2 ka cal. BP (OZJ907) and 32.6 ± 0.6 ka cal. BP (OZJ908), suggesting rapid aggradation. This inference is supported by the OSL age of 32.8 ± 2.8 ka (AdGL-96003) based on large aliquots. However, a number of indicators bring into question the depositional significance of these age proxies; two further OSL samples below return unequivocally ages of 24.9 ± 1.4 ka (AdGL-96006) and 24.9 ± 2.5 ka (AdGL-96002), which in more water-saturated conditions could prove a slight underestimate (Williams *et al.* 2001). AdGL-96003 is sampled from a layer marked by transported nodular calcrete most probably derived from former Bca-horizons stripped from the slope or upstream remnants of a previous generation of floodplain. This suggests that the older ages are inherited and the sediment is reworked. Finally, two paired ^{14}C ages from

shell and small charcoal pieces collected at the present base return ages of $28.6\ 0 \pm 0.4$ ka cal. BP (SSAMS ANU-2039) and 28.1 ± 0.1 cal. BP (SSAMS ANU-1811), respectively.

The layered upper sequence of BRA07-SD aggraded episodically throughout the LGM. At intervals, discrete light bands consisting of silt and fine sand subdivide darker laminated units. These light bands are continuous for over 100 m and can be traced towards the trunk channel to the SW, increasing in thickness and finally being associated with gravel and cross-stratification. They are topped by undulating couplets of grey fine-grained sediments, organic veneers and streaks of tufa. The lower 3 m of the layered upper unit aggraded prior to *c.* 22 ka (Fig. 9). Towards 19.3 ± 0.2 ka cal. BP (SSAMS ANU-4109), a final compact succession of light bands was deposited. In contrast to underlying sediments, it is affected by pedogenesis and bioturbation extending from the uppermost Bca-horizon. The layered and laminated sequence terminates at 18.3 ± 0.2 ka cal. BP (SSAMS ANU-4107) with lenses of gravel and is partially mantled by a chromatic surface drape.

Situated near the mouth of Brachina Gorge within the trunk channel is BRA-LW, a section of Brachina Silts entrenched to more than 13.5 m (Fig. 9). Four ^{14}C ages based on charcoal pieces collected from within 1 m return ages ranging from 30.2 ± 0.3 ka cal. BP (Wk-6562) to 32.3 ± 0.2 ka cal. BP (Beta-96171), similar to those in BRA07-SD but 6 m lower within the sequence. However, samples closer to the base dated to 19.7 ± 0.2 ka cal. BP (Wk-6564) and 16.9 ± 0.1 ka cal. BP (Wk-6561) again call into question the extent to which these reflect the depositional event or the age of reworked sediments.

The Brachina Silts spread across the second intramontane floodplain where the Brachina Creek debouches from the narrow gap through the ABC Range Quartzite Formation and is met by Aroona Creek. BRA07-AR is situated immediately upstream of the junction of the Brachina trunk channel with the Aroona Creek tributary (Fig. 8). The stratigraphic section consists of a succession of weathered gravel bands intercalated with laminated silt couplets (Fig. 9). The gravel is derived from local shales of the Bunyeroo Formation and interspersed with clasts of tufa from benches that crop out upstream. The fine-grained sediments reflect multiple fluxes of suspension load with vertical fining-upward trends from sand to silt draped by veneers of organic flotsam (Haberlah & McTainsh 2011). At the base, almost level with the present creek bed, large pebbles and cobbles of variable lithology were excavated and identified as traction load of the Brachina Creek and dated to 20.2 ± 0.2 ka cal. BP (OZK517). According to five ages based on shell and charcoal, the overlying fine sequence aggraded rapidly between 19.5 ± 0.2 ka cal. BP (OZK002) and 17.7 ± 0.1 ka cal. BP (SSAMS ANU-1812). Repeated age inversions indicate that the material is from reworked sediments. The $c.$ 40 m wide section marked by the mottled gravel bands appears to be a cut-and-fill formed within a floodplain that may be older.

Less than 3 km downstream, Brachina Creek is once more confined to a narrow gorge cutting through the Rawnsley Quartzite. Approaching the range front, it makes a sharp turn and floods into an embayment eroded by three minor creeks coming out of the Wonoka Limestone and Siltstone (Fig. 8). A complex terrace remnant of Brachina Silts up to $c.$ 18 m high (Williams *et al.* 2001) is dated close to its downstream termination. Two OSL ages from the base of BRA07-G put the onset of the fine-grained aggradation to 32.4 ± 1.7 ka (AdGL-08001) and 30.7 ± 2.7 ka (AdGL-08002). Both samples are only partially bleached with a significant inherited D_e population dating to 38.9 ± 4.4 ka (AdGL-08002; see Table 1), possibly indicating reworked sediments of similar ages to those discussed for sections further upstream. Shells bracketed by the OSL samples return ages of 34.6 ± 0.4 ka cal. BP (ANU BG27) and 28.9 ± 4.4 ka cal. BP (ANU BG42). The upper part of the sequence consists of silts intercalated with alluvial fan matrix-supported gravels from a

tributary valley. Around 16.1 ± 1.3 ka (AdGL-08003), the gravels spread out across the section.

Wilkawillina Silts

The central Flinders Ranges form a large eroded anticline structurally controlled by the Oraparinna Diapir (Callen & Reid 1994; Fig. 1). As a result, a similar sequence of wide valleys and steep ridges to those described from the Brachina Creek catchment occurs on the eastern side of the Ranges. Here, in mirror image to the Brachina and Etina Creeks, the Second Plain and Moodlatanna Creeks converge in a fine-grained floodplain and are joined by Mt Billy Creek upstream of Wilkawillina Gorge (Fig. 11). Four stratigraphic sections from the Wilkawillina Floodplain are described and dated at similar geomorphological sites with the aim of testing to what extent aggradational and erosional events were synchronous across different catchments. From the distal position within the floodplain, WL08-UFP is from within a bedrock embayment in the immediate vicinity of the present confluence with the Second Plain Creek, and situated less than 200 m upstream of a bedrock constriction cutting through a resistant ridge of the ABC Range Quartzite (Callen & Reid 1994). The depositional setting is in many ways comparable with that of BRA07-SD, and both sections display a number of stratigraphic similarities.

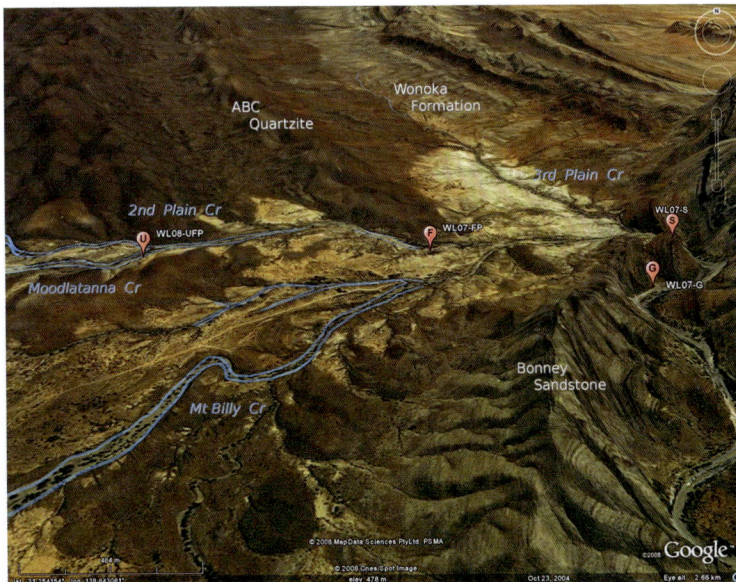

Fig. 11. Wilkawillina Silts stratigraphic sections as seen from the south at an altitude of $c.$ 2.66 km projected on the SRTM DEM (elevation exaggeration factor 3) in Google Earth 4.3, with the Second Plain Creek and Mt Billy Creek coming in from the west upstream the Wilkawillina Gorge. Locations of all sections within the Wilkawillina catchment are indicated by red pointers with letters corresponding to section names (see also Fig. 1).

Stratigraphic section WL07-FP is a further 1500 m downstream, and is situated at the downstream end of an interfluve flowing from Mt Billy Creek and joining the trunk channel ahead of a less pronounced ridge of ABC Range Quartzite. The composite section is marked by multiple generations of gravel sheets and lies in a comparable landscape setting (interfluve) to BRA-SA and BRA-SG. Further downstream, the Wilkawillina Creek cuts through the Bonney Sandstone but fails to breach the massive Rawnsley Quartzite. At the gorge entrance, the Third Plain Creek, occupying a valley incised within the Wonoka Formation, joins the trunk channel. Here, in a mirror image to BRA07-AR, a slackwater sequence with bands of mottled gravel is situated at section WL07-S. In contrast, this section occurs opposite the largely eroded confluence in a back-flooded embayment (Fig. 11). Finally, as an analogue to BRA07-G, the remnant of WL07-G lies in a protected embayment within Wilkawillina Gorge.

With more than 12 m of vertical exposure and well-preserved laminations in the lower sequence, WL08-UFP has the potential to complement and extend the high-resolution record of BRA07-SD (Fig. 12) (Haberlah *et al.* 2010). Perhaps the most apparent commonality between the two is the presence of distinct brown (10YR 5/3 dry–4/4 moist) chromatic bands or horizons, displaying transitional lower but clear-cut upper boundaries associated with marked colour changes between underlying and overlying units. WL08-UFP exhibits three such 'palaeosols' varying in thickness and maturity, the best-defined at *c.* 750–765 cm (Fig. 13). Shells collected above this chromatic band marking the onset of a more chromatic unit returned an age of

30.0 $\pm$ 0.2 ka cal. BP (Wk23524), and shells below from the Bca-horizon of 22.9 0 $\pm$ 0.2 ka cal. BP (Wk23467). A discrete pocket of intact shells from the lower of the closely spaced chromatic bands at *c.* 855–890 cm returned an age of 35.3 $\pm$ 0.4 ka cal. BP (Wk23525). Disregarding the shells from the Bca-horizon, the ages of the closely spaced chromatic bands compare well with OSL ages obtained from the 'twin chromatic band' from the Hookina Creek Floodplain. Both chromatic bands are associated with angular clasts and pebbles derived from local outcrops, and increase in thickness and angularity towards the bedrock slope. The fine-grained sediments underlying the chromatic bands are cemented by carbonate, in places forming large *in situ* calcareous rhizocretions, but less so below the uppermost at *c.* 440–460 cm. This chromatic band marks the boundary between the lower more chromatic unit and an overlying grey unit, and, in terms of colour association and depth below the top, resembles the 'palaeosol' of BRA07-SD (Williams *et al.* 2001). The minimum age for the end of the fine-grained aggradational regime on the distal floodplain comes from the age of 15.6 $\pm$ 0.1 ka cal. BP (Wk23465) from the matrix-supported gravel sheets overlying the Wilkawillina Silts. A chromatic surface drape (5YR 5/6 dry–4/6 moist) with embedded Aboriginal artefacts overlies the sequence. Large intact snail shells sampled from it return an early Holocene age of *c.* 8.2 $\pm$ 0.1 ka cal. BP (Wk23464).

The *c.* 15 m thick composite section of WL07-FP occurs at the downstream end of the largest interfluve of the Wilkawillina Floodplain. Its base rests on gravels and cobbles embedded in a matrix of fluvial sands (Fig. 13). The lowermost

Fig. 12. Stratigraphic section WL08-UFP from the distal Wilkawillina Floodplain.

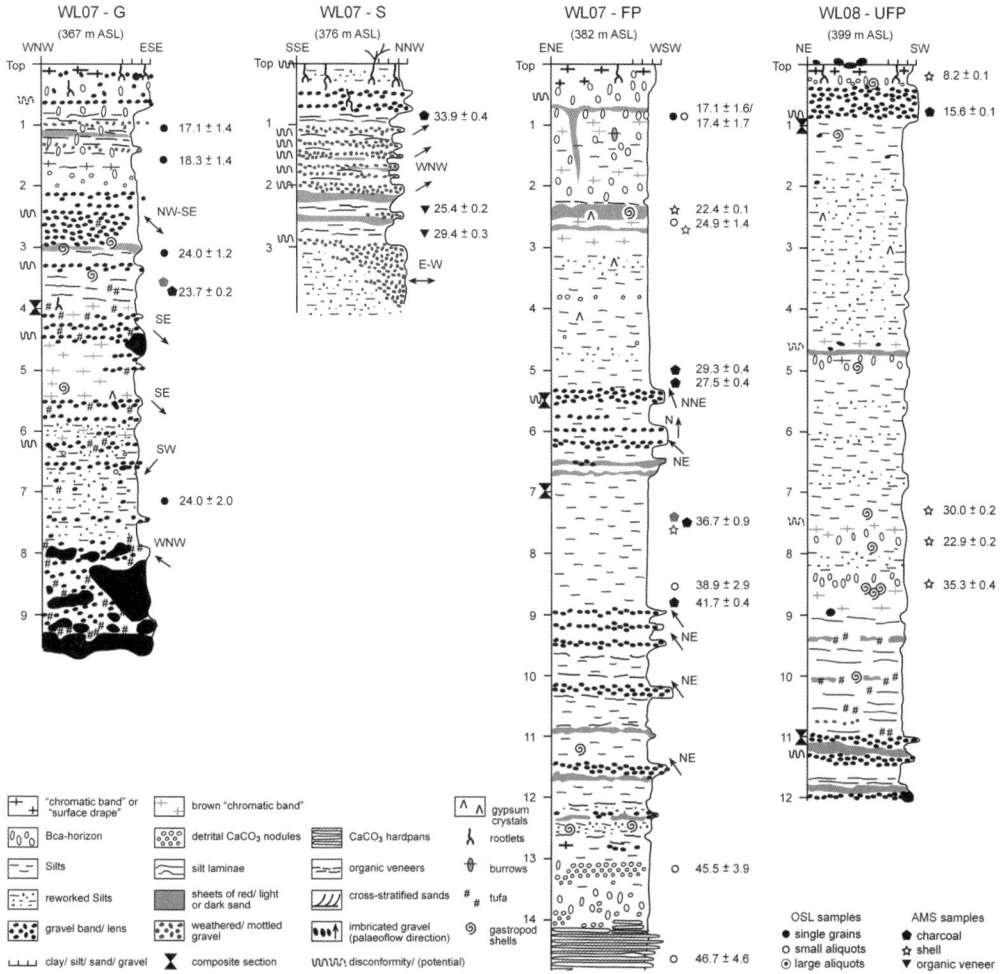

Fig. 13. Wilkawillina Silts stratigraphic sections; main lithostratigraphic units and chronostratigraphy.

fine-grained sediments contain multiple hardpans of carbonate and return an OSL age of 46.7 ± 4.6 ka (AdGL-08007), followed by 45.5 ± 3.9 ka (AdGL-07006) with a dominant residual signal of 72.1 ± 5.9 ka from the overlying lowermost Bca-horizon. Above, a sequence of pristine light-coloured silts is marked by sheets of gravel that terminate a few metres downstream. Further upstream, multiple generations of clast-supported gravel-filled chutes testify to persistent flow directions from the NE across the floodplain, and to continuous aggradation of the silts. Over two intervals, the sheets of gravel spread out. The termination of the lower gravel sheet is associated with charcoal dated to 41.7 ± 0.4 ka cal. BP (OZK010). Subsequently, fine-grained sediments aggraded between 38.9 ± 2.9 ka (AdGL-07007) and 36.7 ± 0.9 ka cal. BP (OZK013) (excluding an unlikely young shell

age). Interestingly, the OSL age here comprises a small, but well-bleached, population of grains indicating an age of 22.7 ± 1.4 ka (Tables 1 and 2). The most likely explanation is that at this time, when no more gravel was transported across the interfluve, the trunk channel incised to this depth. The second generation of gravel sheets is covered by >5 m of exclusively fine-grained sediments preserved in the form of insular terrace remnants. These consist of grey silts, layered in appearance and displaying continuous light bands of silt and fine sand. Two pieces of charcoal collected from the base of this unit returned ages of 29.3 ± 0.4 ka cal. BP (SSAMS ANU-2033) and 27.5 ± 0.4 ka cal. BP (SSAMS ANU-2032). Above, a close sequence of horizontally laminated silt and fine-sand bands, hosting many intact gastropods, was dated to 22.4 ± 0.1 ka cal. BP (SSAMS ANU-1813). A pronounced intercalated red, light band returned an OSL age of 24.9 ± 1.4 ka (AdGL-07008). The termination in a similar red, light silt and fine-sand band from which deep desiccation cracks extend was dated by two independent OSL samples; one based on single grains (CLW-95/4) and the other on small aliquots (AdGL-07009). The determined ages of 17.1 ± 1.6 ka and 17.4 ± 1.7 ka are identical within error ranges. The same applies to an inherited early LGM age (however, of significant proportion only in AdGL-07009), suggesting reworking of upstream sediments.

Section WL07-S is banked against the steep ridge of the Bonney Sandstone, at the downstream termination of the Wilkawillina Floodplain where the channel is constricted to less than *c*. 30 m. It is situated opposite the largely eroded confluence of the trunk channel with its final major tributary within an embayment filled by many metres of grey silts. From the level of the present channel, closely spaced couplets of mottled gravel and laminated fine-grained sediments alternate (Fig. 13), comparable in appearance with those of BRA07-AR (Fig. 9). The age of a basal veneer of organic detritus is 29.4 ± 0.3 ka calBP (SSAMS ANU-2229), followed by 25.4 ± 0.2 ka calBP (SSAMS ANU-2227). A piece of charcoal extracted from the uppermost preserved veneer returned an age of 33.9 ± 0.4 ka calBP (SSAMS ANU-2001), probably inherited.

Within a protected embayment around the first bend within Wilkawillina Gorge lies the layered terrace remnant WL07-G (Fig. 14). The sequence was excavated down to bedrock that in places is capped by thick carbonate (tufa) benches incorporating metre-sized boulders. The lower *c*. 4 m of the section consists of *c*. 25% of thin bands of reworked fine-grained sediments. These alternate with matrix-supported bands of gravel, in places incorporating large slabs of limestone that slid down from the adjacent slope into the lower overall more chromatic unit (Fig. 13). Imbrication

Fig. 14. Stratigraphic section WL07-G within the first embayment of Wilkawillina Gorge.

patterns indicate flow directions to the SW, similar to the present-day system. The single-grain OSL analysis for the base of the silt aggradation (CLW-95/1) returned two dominant D_e populations (Tables 1 and 2): the first aged at 43.4 ± 2.6 ka and the second at 24.0 ± 2.0 ka. The lower unit terminates in a brown chromatic band (10YR 5/2 dry–4/2 moist) upon which rests a lighter greyish and distinctly layered unit. This upper unit contains small lenses of fine clast-supported gravel, in places displaying herringbone cross-stratification indicating flow in and out of the embayment. The base of this unit was dated on two separate pieces of charcoal to 23.7 ± 0.2 ka cal. BP (OZK019, OZK020), an age confirmed by a single-grain OSL sample of 24.0 ± 1.2 ka (CLW-95/2). The OSL sample was taken from a more chromatic (2.5Y 6/4 dry–4/4 moist) unit marked by rolled detrital calcareous nodules probably derived from redistributed slope mantles from the embayment, an inference supported by a significant inherited age of 32.6 ± 1.8 ka. The more massive fine-grained sediments give way to a unit of layered silts topped by thick veneers of organic detritus, the base of which is single-grain OSL dated to 18.3 ± 1.4 ka (CLW-95/3), followed by the topmost discrete light band deposited at 17.1 ± 1.4 ka (CLW-95/5).

Discussion of the alluvial loess record from the Flinders Ranges

The onset of the aggradation of proximal dust-derived alluvium in the Flinders Ranges is recorded from the western piedmont plain from stratigraphic sections less than 50 km downwind of a major potential source of the material: the Lake Torrens playa. At this stage, the low resolution of OSL dates only allows us to state that between the end of the last Interglacial optimum and *c*. 83 ka a *c*. 4 m thick sequence of fine-grained sediments was widely deposited over the Hookina Creek Floodplain. Towards the end of this interval, the upstream section (HK07-D) records multiple pronounced Bca-horizons associated with chromatic bands. At the same time, the downstream section (HK07-M) records fluvial activity and reworking of dunes and calcareous palaeosols. As for the nature of the chromatic bands, also described as 'palaeosols' (Williams 1973, 1982; Williams *et al.* 2001), and in the case of the Namib Silts interpreted as 'cambic horizons' (Brunotte *et al.* 2009), the chromatic surface drapes that mantle most of the present floodplain surfaces may provide a modern analogue. A number of characteristics clearly distinguish both from the underlying silts: their red colour, characteristic of oxidation; association with small pebbles;

and their degree of aggregation (subplasticity). Typically, but not in all cases, they overlie Bca-horizons, which on first impression suggests that they represent the equivalent A-horizon. However, observations on the formation of the modern analogue indicate a different origin. Local land-owners describe how after heavy rainfall the surface 'levels itself out' and mounds and trenches are reduced to a common surface. As for the source of the material, after major flood events such as the most recent one in February 2007, large parts of the trunk channel were draped by silts entrained from the slumping banks. With desiccation, these may have been rapidly entrained by gusts of wind and recycled in the form of aggregated dust over the floodplain. Support for this hypothesis comes from ruins of buildings, such as the abandoned Hookina Hotel. Wherever the roof is missing, the interior is covered by a similar thin chromatic sediment drape. The associated gravel is surface lag left by overbank floods and often concentrated linearly by runoff in the waning stage. Hence, such chromatic bands or 'recycled silts' reflect intervals characteristic of both surface stability and fluvial erosion. When buried, these principally aeolian sheets make excellent marker horizons.

Within the Ranges, the onset of the aggradation of the Flinders Silts is dated to c. 47 ka (WL07-FP). However, inherited ages indicate that the fine-grained sediments are possibly reworked material last deposited c. 70 ka. Between c. 47 ka and c. 37 ka, close to 7–8 m of light-coloured fine-grained sediments intercalated with gravel sheets were deposited within the Wilkawillina catchment.

The next youngest succession of chromatic bands is closely spaced and dated to c. 36 ka and c. 30 ka, respectively (HK07-L and WL08-UFP). Other stratigraphic sections that span this interval record fluvial activity reflected in cross-stratified sands (HK07-D) and multiple extensive gravel sheets and chutes (WL07-FP). Where these eolian accessions mantled bedrock slopes, in many cases they subsequently became incorporated as slope-wash in the valley-fills. The chromatic lower unit of BRA07-SD with sheets of carbonate is likely to reflect this interval. At section WL07-FP, this took place between c. 29 ka and c. 25 ka. The 6 m thick channel fill of stratigraphic section CAS06-1 in the upper Parachilna Creek catchment (Fig. 1) records a similar rapid depositional regime, the onset of which is dated to 29.4 ± 2.1 ka (AdGL07001, Table 1). Intercalated and often matrix-supported sheets of gravel within the gorges, recorded in WL07-S and BRA07-G, point to fluvial redistribution of sediment in the lead-up to the LGM. The brown chromatic band ('palaeosol') in BRA07-SD, the c. 440–460 cm chromatic band in WL08-UFP (undated), and the twin bands of WL07-FP

suggest an interval of surface stability prior to c. 24 ka. This was followed regionally by rapid aggradation of fine-grained layered and laminated unweathered sediments. According to the c. 4 m succession of laminated slackwater deposits in BRA07-SD, this occurred during the course of at least 12 high-magnitude and numerous smaller floods (Haberlah et al. 2010), most pronounced between c. 24 ka and c. 21 ka and continuing until c. 18 ka. To what extent these choked the gorges remains to be established by higher resolution dating of WL07-G. Over the Hookina Floodplain, c. 2 m aggraded during the LGM proper (HK07-L), and large channels were rapidly filled with silts (HK07-M). According to results from BRA07-AR and BRA-LW, subsequent incision within the gorges is likely to have reached almost the present level at c. 20 ka. Headward erosion during the peak of the LGM is also reflected further upstream by gravel chutes and sheets (BRA-SA) incising into and spreading across the floodplain, possibly indicating sediment starvation as its cause. This interval gave way to rapid aggradation of the trunk channel, which raised the base level for tributaries recorded in slackwater deposition from both sides of the Ranges (BRA07-AR, BRA-LW and WL07-G). All floodplain sections document near synchronous final aggradation of silts between c. 19 ka and c. 17 ka. Within the gorges and distal parts of the floodplain, aggradation of a fast decreasing fine component and increasing gravel influx persisted until c. 16 ka (BRA07-G, WL07-G, WL08-UFP). After that, the termination of this fine-grained aggradational regime by incision was final.

Conclusions

The alluvial loess record from the Flinders Ranges presents a largely continuous palaeo-environmental record from semi-arid southern Australia for the period between the last interglacial and the Deglacial. Based on 124 ages (94 AMS and 30 OSL) from various catchments, regional synchronicity of periods of aggradation, surface stability and incision can be inferred, reflecting major changes of the palaeo-environment. The current chronological resolution is highest between the lead-up to the Last Glacial Maximum and the early Deglacial. The LGM presents itself as an interval characterized by a highly variable climate, which is in accordance with other recent studies from Australia (Petherick et al. 2008a, b, 2009) and New Zealand (Newnham et al. 2007). Periods of relative surface stability reflected in two closely spaced intervals of pedogenic activity between c. 36 and 30 ka were followed by widespread erosion and reworking.

Another shorter-lived interval of stability prior to *c.* 24 ka gave way to a climate in which large amounts of proximal dust were deposited across the catchments. These loess mantles were redistributed episodically and transported downstream by at least a dozen high-magnitude and numerous smaller floods (Haberlah *et al.* 2010). The peak of the LGM at *c.* 20 ka was marked locally by rapid incision within the gorges and gravel alluviation on the floodplains. Fine-grained aggradation resumed (or continued) at a slower rate after this. The termination of this aggradational regime *c.* 18–16 ka was final and culminated in episodic incision. The Late Pleistocene floodplain remnants are today deeply entrenched and in process of being laterally eroded by ephemeral traction load streams.

Future work analysing independent palaeo-environmental proxies collected from the stratigraphic sections will test and refine this model of late Quaternary landscape evolution. A major unanswered question concerns the extent to which these intervals of aggradation, surface stability and erosion reflect changes in the precipitation regime, the vegetation cover and the sediment supply. At this stage we can only state with confidence that the influx of large quantities of loess must be considered as an important variable.

Our approach to dating a variety of buried materials allows us to critically assess their utility as proxies to infer the timing of deposition. Radiocarbon dates on charcoal proved to be problematic proxies in an environment prone to sediment redistribution, given the potential of charcoal to survive multiple episodes of reworking. Overall, the dated charcoal within the catchment reflects times of locally abundant woody vegetation burnt by wildfires. The most reliable approximation of depositional ages came from OSL dating. Partial bleaching during fluvial transport is common, but given the homogeneous nature of the Flinders Silts this is not a disadvantage. By applying the Finite Mixture Models (Galbraith 2005), multiple ages can be obtained from many samples relating to 'inherited', 'depositional' and 'post-depositional' bleaching histories of the sampled quartz grains.

The inferred scenario of LGM floods redistributing loess mantles, choking narrow gorges and causing widespread back-flooding that resulted in laminated slackwater deposits and ephemeral wetlands is new for Australia. Abnormal wet features within the generally more arid LGM landscape have so far been answered with the minevaporal model proposed by Galloway (1965). However, there is an increasing realization that droughts and floods go together (Lancaster 2002), and that semi-arid landscapes are to a great extent shaped by such events during climatic deterioration

towards drier and cooler conditions (Zielhofer & Faust 2008). The rapid aggradation of the fine-grained valley-fills in the Flinders Ranges prompts us to ask whether short-lived flood events might not explain the widespread beach ridges of the playa lakes in the flanking basins. Our chronostratigraphic approach could also provide a useful working model for similar fine-grained valley-fills outside Australia, such as the Namib Silts and the Sinai Silts.

Global context

This contribution is one of an increasing number from both hemispheres demonstrating that the LGM, as defined by Mix *et al.* (2001), was far more complex than earlier workers had realized (e.g. Farrera *et al.* 1999), as were the times immediately before and after it. A further set of issues that has attracted considerable recent attention is the question of whether global climatic changes were driven primarily by changes in North Atlantic ocean circulation and whether or not such changes were synchronous from pole to equator across both hemispheres. Results from an ice core from Dronning Maud Land, Antarctica ('EDML' core), at comparable resolution to previously published Greenland ice core records, showed a one-to-one coupling between all Antarctic warm events and Greenland Dansgaard–Oeschger cold events, consistent with the bi-polar seesaw, perhaps reflecting a reduction in meridional overturning in the Atlantic (EPICA Community Members 2006). A recent compilation of [10]Be exposure dates for the onset of major retreat in mid-latitude LGM mountain glaciers gave a mean age of 17.3 ± 0.5 ka for the Southern Hemisphere and 17.4 ± 0.5 ka for the Northern Hemisphere, showing that the retreat was synchronous in both hemispheres (Schaefer *et al.* 2006). The onset of glacier retreat therefore coincided with the onset of postglacial warming revealed in the Antarctic high-resolution EPICA Dome C ice core record but with a cooling trend in the Greenland GISP 2 ice core, where warming did not begin until the onset of the Bölling–Allerød (B–A) interstadial event at 14.7 ka.

What is now needed to advance palaeoclimatic research in Australia is a series of well-dated proxy records along a set of transects from temperate south to tropical north, with the capacity to provide seasonal temperature and precipitation signals.

We thank CRC LEME (Cooperative Research Centre for Landscape Environments and Mineral Exploration), the Australian Research Council (ARC grant DP0559577) and the Australian Institute of Nuclear Science and Engineering (AINSE Grants 96/192R, 99/001 and 07/160) for generous financial support. Thanks also go to the Flinders

Ranges National Park authorities for research permits and friendly accommodation, station owners R. Spears and Rex for letting us work on their land, and to J. von dem Bussche, T. Brookman and L. Parham for assisting in the field. In particular, we are indebted to T. Pietsch for volunteering to run all single-grain samples at the luminescence laboratory of CSIRO Land and Water in Canberra, J. Prescott for kindly providing the facilities, encouragement and supervision at the Adelaide luminescence laboratory, and S. Fallon from the Australian National University Radiocarbon Dating Laboratory for preparing most of the labour-intensive radiocarbon samples based on organic veneers. Finally we would like to thank the two anonymous reviewers for their valuable comments, and the Australian and New Zealand Geomorphology Group together with the Geological Society of London for funding the first colour figure in the hardcopy version of this paper.

References

AITKEN, M. J. 1998. *An Introduction to Optical Dating.* Oxford University Press, Oxford.

BLONG, R. J. & GILLESPIE, R. 1978. Fluvially transported charcoal gives erroneous ^{14}C ages for recent deposits. *Nature*, **271**, 739–741.

BOURKE, M. C., CHILD, A. & STOKES, S. 2003. Optical age estimates for hyper-arid fluvial deposits at Homeb, Namibia. *Quaternary Science Reviews*, **22**, 1099–1103.

BOWLER, J. M. 1978. Quaternary climate and tectonics in the evolution of the Riverine Plain, southeastern Australia. *In*: DAVIES, J. L. & WILLIAMS, M. A. J. (eds) *Landform Evolution in Australasia.* Australian National University Press, Canberra, 70–112.

BRUNOTTE, E., MAURER, B., FISCHER, P., LOMAX, J. & SANDER, H. 2009. A sequence of fluvial and aeolian deposits (desert loess) and palaeosoils covering the last 60 ka in the Opuwo basin, Kaokoland/Kunene region, Namibia) based on luminescence dating. *Quaternary International*, **196**, 71–85.

CALLEN, R. A. & REID, P. W. 1994. *Geology of the Flinders Ranges National Park.* South Australian Geological Survey, Special Map 1:75 000.

CALLEN, R. A., WASSON, R. J. & GILLESPIE, R. 1983. Reliability of radiocarbon dating of pedogenic carbonate in the Australian arid zone. *Sedimentary Geology*, **35**, 1–14.

COCK, B. J., WILLIAMS, M. A. J. & ADAMSON, D. A. 1999. Pleistocene Lake Brachina: a preliminary stratigraphy and chronology of lacustrine sediments from the central Flinders Ranges, South Australia. *Australian Journal of Earth Sciences*, **46**, 61–69.

CROCKER, R. L. 1946. Post-Miocene climatic and geologic history and its significance in relation to the genesis of the major soil types of South Australia. *Council for Scientific and Industrial Research Bulletin*, **193**, 5–65.

DART, R. C., BAROVICH, K. M., CHITTLEBOROUGH, D. J. & HILL, S. M. 2007. Calcium in regolith carbonates of central and southern Australia: its source and implications for the global carbon cycle. *Palaeogeography, Palaeoclimatology, Palaeoecology*, **249**, 322–334.

DULLER, G. A. T. 2004. Luminescence dating of Quaternary sediments: recent advances. *Journal of Quaternary Science*, **19**, 183–192.

DULLER, G. A. T., BOTTER-JENSEN, L., MURRAY, A. S. & TRUSCOTT, A. J. 1999. Single grain laser luminescence (SGLL) measurements using a novel automated reader. *Nuclear Instruments and Methods in Physics Research*, **155**, 506–514.

EITEL, B., BLÜMEL, W. D., HÜSER, K. & MAUZ, B. 2001. Dust and loessic alluvial deposits in northwestern Namibia (Damaraland, Kaokoveld): sedimentology and palaeoclimatic evidence based on luminescence data. *Quaternary International*, **76**, 57–65.

EITEL, B., KADEREIT, A., BLÜMEL, W. D., HÜSER, K. & KROMER, B. 2005. The Amspoort Silts, northern Namib Desert (Namibia): formation, age and palaeoclimatic evidence of river-end deposits. *Geomorphology*, **64**, 299–314.

EPICA COMMUNITY MEMBERS 2006. One-to-one coupling of glacial climate variability in Greenland and Antarctica. *Nature*, **444**, 195–198.

FAIRBANKS, R. G., MORTLOCK, R. A. ET AL. 2005. Radiocarbon calibration curve spanning 0 to 50 000 years BP based on paired ^{230}Th/^{234}U/^{238}U and ^{14}C dates on pristine corals. *Quaternary Science Reviews*, **24**, 1781–1796.

FARRERA, I., HARRISON, S. P. ET AL. 1999. Tropical climates of the last glacial maximum: a new synthesis of terrestrial palaeoclimatic data. I. Vegetation, lake levels and geochemistry. *Climate Dynamics*, **15**, 823–856.

FERGUSON, C. W., HUBER, B. & SUESS, H. E. 1966. Determination of the age of Swiss lake dwellings as an example of dendrochronologically-calibrated radiocarbon dating. *Zeitschrift für Naturforschung, Teil A*, **21**, 1173–1177.

FONTES, J.-C. & GASSE, F. 1989. On the ages of humid Holocene and late Pleistocene phases in north Africa—remarks on 'Late Quaternary climatic reconstruction for the Maghreb (North Africa)' by P. Rognon. *Palaeogeography, Palaeoclimatology, Palaeoecology*, **70**, 393–398.

GALBRAITH, R. F. 1990. The radial plot: graphical assessment of spread in ages. *Nuclear Tracks and Radiation Measurements*, **17**, 207–214.

GALBRAITH, R. F. 2005. *Statistics for Fission Track Analysis.* Chapman & Hall–CRC, Boca Raton, FL.

GALBRAITH, R. F. & GREEN, P. F. 1990. Estimating the component ages in a finite mixture. *Nuclear Tracks and Radiation Measurements*, **17**, 197–206.

GALBRAITH, R. F., ROBERTS, R. G., LASLETT, G. M., YOSHIDA, H. & OLLEY, J. M. 1999. Optical dating and multiple grains of quartz from Jinmium rock shelter, northern Australia: part 1, experimental design and statistical models. *Archaeometry*, **41**, 339–364.

GALLOWAY, R. W. 1965. Late Quaternary climates in Australia. *Journal of Geology*, **73**, 603–618.

GENTILLI, J. 1986. Climate. *In*: JEANS, D. N. (ed.) *Australia—A Geography. The Natural Environment.* Sydney University Press, Sydney, 14–48.

GINGELE, F. X. & DE DECKKER, P. 2005. Late Quaternary fluctuations of palaeoproductivity in the Murray Canyons area, south Australian continental margin.

Palaeogeography, Palaeoclimatology, Palaeoecology, **220**, 361–373.

GLASBY, P., WILLIAMS, M. A. J., McKIRDY, D., SYMONDS, R. & CHIVAS, A. R. 2007. Late Pleistocene environments in the Flinders Ranges, Australia: preliminary evidence from microfossils and stable isotopes. *Quaternary Australasia*, **24**, 19–28.

GOOSSENS, D. 1988. The effect of surface curvature on the deposition of loess: a physical model. *Catena* **15**, 179–194.

GOUDIE, A. S. 1972. Climate, weathering, crust formation, dunes and fluvial features of the central Namib Desert, near Gobabeb, South West Africa. *Madoqua II*, **1**, 15–31.

HABERLAH, D. 2006. Depositional models of late Pleistocene fine-grained valley-fill formations in the Flinders Ranges, SA. *In*: FITZPATRICK, R. W. & SHAND, P. (eds) *Regolith 2006—Consolidation and Dispersion of Ideas*. Cooperative Research Centre for Landscape Environments and Mineral Exploration, Western Australia, Perth, 122–126.

HABERLAH, D. 2007. A call for Australian loess. *Area*, **39**, 224–229.

HABERLAH, D. 2008. Response to Smalley's discussion of 'A call for Australian loess'. *Area*, **40**, 135–136.

HABERLAH, D. & McTAINSH, G. H. 2011. Quantifying particle aggregation in sediments. *Sedimentology*, in press.

HABERLAH, D., WILLIAMS, M. A. J., HILL, S. M., HALVERSON, G. & GLASBY, P. 2007. A terminal last glacial maximum (LGM) loess-derived palaeoflood record from South Australia? *Quaternary International*, **167–168**, 150.

HABERLAH, D., WILLIAMS, M. A. J. *ET AL.* 2010. Loess and floods: High-resolution multi-proxy data of Last Glacial Maximum (LGM) slackwater deposition in the Flinders Ranges, semi-arid South Australia. *Quaternary Science Reviews*, **29**, 2673–2693.

HEINE, K. & HEINE, J. T. 2002. A paleohydrologic reinterpretation of the Homeb Silts, Kuiseb River, central Namib Desert (Namibia) and paleoclimatic implications. *Catena*, **48**, 107–130.

HESSE, P. P. & McTAINSH, G. H. 2003. Australian dust deposits: modern processes and the Quaternary record. *Quaternary Science Reviews*, **22**, 2007–2035.

HOFFMAN, D. L., BECK, J. W., RICHARDS, D. A., SMART, P. L., MATTEY, D. P., PATERSON, B. A. & HAWKESWORTH, C. J. 2008. Atmospheric radiocarbon variation between 44 and 28 ka based on U-series dated speleothem. *Geophysical Research Abstracts*, **10**, SRef-ID: 1607-7962/gra/EGU2008-A-10064.

HUA, Q., JACOBSEN, G. E., ZOPPI, U., LAWSON, E. M., WILLIAMS, A. A., SMITH, A. M. & McGANN, M. J. 2001. Progress in radiocarbon target preparation at the ANTARES AMS Centre. *Radiocarbon*, **43**, 275–282.

HUGHEN, K., SOUTHON, J., LEHMAN, S., BERTRAND, C. & TURNBULL, J. 2006. Marine-derived ^{14}C calibration and activity record for the past 50 000 years updated from the Cariaco Basin. *Quaternary Science Reviews*, **25**, 3216–3227.

HUNTLEY, D. J., GODFREY-SMITH, D. I. & THEWALT, M. L. W. 1985. Optical dating of sediments. *Nature*, **313**, 105–107.

HUNTLEY, D. J., HUTTON, J. T. & PRESCOTT, J. R. 1993. The stranded beach-dune sequence of south-east Australia: a test of thermoluminescence dating, 0–800 ka. *Quaternary Science Reviews*, **12**, 1–20.

ISSAR, A. S. & BRUINS, H. J. 1983. Special climatological conditions in the deserts of Sinai and the Negev during the latest Pleistocene. *Palaeogeography, Palaeoclimatology, Palaeoecology*, **43**, 63–72.

ISSAR, A. S. & ECKSTEIN, Y. 1969. The lacustrine beds of Wadi Feiran, Sinai: their origin and significance. *Israel Journal of Earth Sciences*, **18**, 29–32.

JACOBS, Z., DULLER, G. A. T. & WINTLE, A. G. 2006. Interpretation of single grain De distributions and calculation of De. *Radiation Measurements*, **41**, 264–277.

JENNINGS, J. N. 1968. A revised map of the desert dunes of Australia. *Australian Geographer*, **10**, 408–409.

KEMP, J. & SPOONER, N. A. 2007. Evidence for regionally wet conditions before the LGM in southeast Australia: OSL ages from a large palaeochannel in the Lachlan Valley. *Journal of Quaternary Science*, **22**, 423–427.

LANCASTER, N. 2002. How dry was dry?—Late Pleistocene palaeoclimates in the Namib Desert. *Quaternary Science Reviews*, **21**, 769–782.

LEOPOLD, M., VÖLKEL, J. & HEINE, K. 2006. A ground penetrating radar survey of late Holocene fluvial sediments in northwest Namibian river valleys: characterisation and comparison. *Journal of the Geological Society, London*, **163**, 923–936.

LI, B., LI, S. & WINTLE, A. G. 2008. Overcoming environmental dose rate changes in luminescence dating of waterlain deposits. *Geochronometria*, **30**, 33–40.

MACKELLAR, D. 1971. *The Poems of Dorothea Mackellar*. Rigby, Adelaide.

MIX, A. C., BARD, E. & SCHNEIDER, R. 2001. Environmental processes of the ice age: land, oceans, glaciers (EPILOG). *Quaternary Science Reviews*, **20**, 627–657.

MURRAY, A. S. & WINTLE, A. G. 2000. Luminescence dating of quartz using an improved single-aliquot regenerative-dose protocol. *Radiation Measurements*, **32**, 57–73.

NANSON, G. C., PRICE, D. M. *ET AL.* 2008. Alluvial evidence for major climate and flow regime changes during the middle and late Quaternary in eastern central Australia. *Geomorphology*, **101**, 109–129.

NEWNHAM, R. M., LOWE, D. J., GILES, T. & ALLOWAY, B. V. 2007. Vegetation and climate of Auckland, New Zealand, since *ca*. 32 000 cal. yr ago: support for an extended LGM. *Journal of Quaternary Science*, **22**, 517–534.

OLLEY, J. M. & REED, M. 2003. *Radial Plot 1.3*. CSIRO Land & Water, Canberra.

OLLEY, J. M., CAITCHEON, G. & MURRAY, A. S. 1998. The distribution of apparent dose as determined by optically stimulated luminescence in small aliquots of fluvial quartz: implications for dating young sediments. *Quaternary Geochronology*, **17**, 1033–1040.

PETHERICK, L. M., McGOWAN, H. A. & MOSS, P. 2008a. Climate variability during the last glacial maximum in eastern Australia: evidence of two stadials? *Journal of Quaternary Science*, **23**, 787–802.

PETHERICK, L. M., McGOWAN, H. A., MOSS, P. T. & KAMBER, B. S. 2008b. Late Quaternary aridity and dust transport pathways in eastern Australia. *Quaternary Australasia*, **25**, 2–11.

PETHERICK, L. M., McGOWAN, H. A. & KAMBER, B. S. 2009. Reconstructing transport pathways for late Quaternary dust from eastern Australia using the composition of trace elements of long traveled dusts. *Geomorphology*, **105**, 67–79.

PREISS, W. V. 1987. The Adelaide geosyncline—late Proterozoic stratigraphy, sedimentation, palaeontology and tectonics. *Geological Survey of South Australia, Bulletin*, **53**, 438.

PREISS, W. V. 1999. *Parachilna, South Australia.* 1:250 000 Geological Series—Explanatory Notes. Primary Industries and Resources SA, Adelaide.

PRESCOTT, J. R. & HUTTON, J. T. 1994. Cosmic ray contributions to dose rates for luminescence and ESR dating: large depths and long-term time variations. *Radiation Measurements*, **23**, 497–500.

QUIGLEY, M. C., SANDIFORD, M. & CUPPER, M. L. 2007. Distinguishing tectonic from climatic controls on range-front sedimentation. *Basin Research*, **19**, 491–505.

REIMER, P. J., BAILLIE, M. G. L. ET AL. 2004. Intcal04 atmospheric radiocarbon age calibration, 26–0 cal kyr bP. *Radiocarbon*, **46**, 1029–1058.

REVEL-ROLLAND, M., DE DECKKER, P. ET AL. 2006. Eastern Australia: a possible source of dust in East Antarctica interglacial ice. *Earth and Planetary Science Letters*, **249**, 1–13.

RÖGNER, K., KNABE, K., ROSCHER, B., SMYKATZ-KLOSS, W. & ZÖLLER, L. 2004. Alluvial loess in the Central Sinai: occurrence, origin, and palaeoclimatological consideration. *In*: *Paleoecology of Quaternary Drylands.* Lecture Notes in Earth Sciences, **102**, 79–99.

SCHAEFER, J. M., DENTON, G. H. ET AL. 2006. Nearsynchronous interhemispheric termination of the last glacial maximum in mid-latitudes. *Science*, **312**, 1510–1513.

SCHWERDTFEGER, P. & CURRAN, E. 1996. Climate of the Flinders Ranges. *In*: DAVIES, M., TWIDALE, C. R. & TYLER, M. J. (eds) *Natural History of the Flinders Ranges.* Royal Society of South Australia Occasional Publications, **7**, 63–75.

SHULMEISTER, J., GOODWIN, I. ET AL. 2004. The southern hemisphere westerlies in the Australasian sector over the last glacial cycle: a synthesis. *Quaternary International*, **118–119**, 23–53.

SMITH, R. M. H., MASON, T. R. & WARD, J. D. 1993. Flash-flood sediments and ichnofacies of the late Pleistocene Homeb Silts, Kuiseb River, Namibia. *Sedimentary Geology*, **85**, 579–599.

SPRIGG, R. C. 1979. Stranded and submerged sea-beach systems of southeast South Australia and the aeolian desert cycle. *Sedimentary Geology*, **22**, 53–96.

SRIVASTAVA, P., BROOK, G. A., MARAIS, E., MORTHEKAI, P. & SINGHVI, A. K. 2006. Depositional environment and OSL chronology of the Homeb silt deposits, Kuiseb River, Namibia. *Quaternary Research*, **65**, 478–491.

TALBOT, M. R., HOLM, K. & WILLIAMS, M. A. J. 1994. Sedimentation in low gradient desert margin systems: a comparison of the late Triassic of north-west Somerset (England) and the late Quaternary of east–central Australia. *In*: ROSEN, M. R. (ed.) *Paleoclimate and Basin Evolution of Playa Systems.* Geological Society of America, Special Papers, **289**, 97–117.

TUNIZ, C., BIRD, J. R., HERZOG, G. F. & FINK, D. 1998. *Accelerator Mass Spectrometry: Ultrasensitive Analysis for Global Science*, 1st edn. CRC Press, Boca Raton, FL.

VOGEL, J. C. 1982. The age of the Kuiseb River silt terrace at Homeb. *Palaeoecology of Africa*, **15**, 201–209.

WALKER, M. 2005. *Quaternary Dating Methods: An Introduction.* Wiley, Chichester.

WARD, J. D. 1987. *The Cenozoic Succession in the Kuiseb Valley, Central Namib Desert.* Geological Survey of South West Africa/Namibia Memoir, **9**.

WENINGER, B. & JÖRIS, O. 2008. A ^{14}C age calibration curve for the last 60 ka: the Greenland–Hulu U/Th timescale and its impact on understanding the middle to upper Paleolithic transition in western Eurasia. *Journal of Human Evolution*, **55**, 772–781.

WENINGER, B., JÖRIS, O. & DANZEGLOCKE, U. 2008. *CalPal-2007. Cologne Radiocarbon Calibration & Palaeoclimate Research Package.* World Wide Web Address: http://www.calpal.de/.

WILLIAMS, D. L. G. 1982. *Late Pleistocene vertebrates and palaeo-environments of the Flinders and Mount Lofty Ranges.* PhD thesis, Flinders University of South Australia, Adelaide.

WILLIAMS, G. E. 1973. Late Quaternary piedmont sedimentation, soil formation and paleoclimates in arid South Australia. *Zeitschrift für Geomorphologie Neue Folge*, **17**, 102–125.

WILLIAMS, G. E. & POLACH, H. A. 1971. Radiocarbon dating of arid-zone calcareous paleosols. *Geological Society of America Bulletin*, **82**, 3069–3086.

WILLIAMS, M. A. J. & NITSCHKE, N. 2005. Influence of wind-blown dust on landscape evolution in the Flinders Ranges, South Australia. *South Australian Geographical Journal*, **104**, 25–36.

WILLIAMS, M. A. J., DE DECKKER, P., ADAMSON, D. A. & TALBOT, M. R. 1991. Episodic fluviatile, lacustrine and aeolian sedimentation in a late Quaternary desert margin system, central western New South Wales. *In*: WILLIAMS, M. A. J., DE DECKKER, P. & KERSHAW, A. P. (eds) *The Cainozoic in Australia: A Re-appraisal of the Evidence.* Geological Society of Australia Special Publication, **18**, 258–287.

WILLIAMS, M. A. J., PRESCOTT, J. R., CHAPPELL, J., ADAMSON, D., COCK, B., WALKER, K. & GELL, P. 2001. The enigma of a late Pleistocene wetland in the Flinders Ranges, South Australia. *Quaternary International*, **83–85**, 129–144.

WILLIAMS, M. A. J., NITSCHKE, N. & CHOR, C. 2006. Complex geomorphic response to late Pleistocene climatic changes in the arid Flinders Ranges of South Australia. *Géomorphologie: Relief, Processus, Environnement*, **4**, 249–258.

WINTLE, A. G. & MURRAY, A. S. 2006. A review of quartz optically stimulated luminescence characteristics and their relevance in single-aliquot regeneration dating protocols. *Radiation Measurements*, **41**, 369–391.

ZIELHOFER, C. & FAUST, D. 2008. Mid- and late Holocene fluvial chronology of Tunisia. *Quaternary Science Reviews*, **27**, 580–588.

Eroding Australia: rates and processes from Bega Valley to Arnhem Land

ARJUN M. HEIMSATH[1]*, JOHN CHAPPELL[2] & KEITH FIFIELD[3]

[1]*School of Earth and Space Exploration, Arizona State University, Tempe, AZ 85287, USA*

[2]*Research School of Earth Sciences, Australian National University, Canberra, ACT 0200, Australia*

[3]*Research School of Physical Sciences and Engineering, Australian National University, Canberra, ACT 0200, Australia*

*Corresponding author (e-mail: Arjun.Heimsath@asu.edu)

Abstract: We report erosion rates determined from *in situ* produced cosmogenic [10]Be across a spectrum of Australian climatic zones, from the soil-mantled SE Australian escarpment through semi-arid bedrock ranges of southern and central Australia, to soil-mantled ridges at a monsoonal tropical site near the Arnhem escarpment. Climate has a major effect on the balance between erosion and transport and also on erosion rate: the highest rates, averaging $35 \, \text{m Ma}^{-1}$, were from soil-mantled, transport-limited spurs in the humid temperate region around the base of the SE escarpment; the lowest, averaging about $1.5 \, \text{m Ma}^{-1}$, were from the steep, weathering-limited, rocky slopes of Kings Canyon and Mt Sonder in semi-arid central Australia. Between these extremes, other factors come into play including rock-type, slope, and recruitment of vegetation. We measured intermediate average erosion rates from rocky slopes in the semi-arid Flinders and MacDonnell ranges, and from soil-mantled sites at both semi-arid Tyler Pass in central Australia and the tropical monsoonal site. At soil-mantled sites in both the SE and tropical north, soil production generally declines exponentially with increasing soil thickness, although at the tropical site this relationship does not persist under thin soil thicknesses and the relationship here is 'humped'. Results from Tyler Pass show uniform soil thicknesses and soil production rates of about $6.5 \, \text{m Ma}^{-1}$, supporting a longstanding hypothesis that equilibrium, soil-mantled hillslopes erode in concert with stream incision and form convex-up spurs of constant curvature. Moreover, weathering-limited slopes and spurs also occur in the same region: the average erosion rate for rocky sandstone spurs at Glen Helen is $7 \, \text{m Ma}^{-1}$, similar to the Tyler Pass soil-mantled slopes, whereas the average rate for high, quartzite spurs at Mount Sonder is $1.8 \, \text{m Ma}^{-1}$. The extremely low rates measured across bedrock-dominated landscapes suggest that the ridge–valley topography observed today is likely to have been shaped as long ago as the Late Miocene. These rates and processes quantified across different, undisturbed landscapes provide critical data for landscape evolution models.

It is widely held that different climatic regions are characterized by different geomorphological processes and landforms. Being a tectonically quiet continent, Australia might be expected to express, almost purely, the effect of climate on landscape evolution, modulated only by lithology and ancient structures. From the temperate eastern highlands through the monsoonal north to the arid centre, the continent presents a broad suite of climatic zones and associated landforms, ranging from soil-mantled ranges to stony mesas and inselbergs. However, the climate of Australia changed greatly during its northward drift over the last few tens of millions of years (e.g. see Fujioka & Chappell 2010), and whether the landforms seen today were formed under climates like those of today depends on their rates of geomorphological change, governed by the erosional processes acting upon them. Erosion rates determined from *in situ* cosmogenic nuclides at some sites in Australia are very low (*c.* $1 \, \text{m Ma}^{-1}$), such as from residual hills in the semi-arid zone (Bierman & Caffee 2002). If such rates were widespread, landscapes with local relief of tens to hundreds of metres would be unlikely to have evolved under climates similar to those of today.

We explore relationships between erosion rate and geomorphological form using cosmogenic nuclide measurements of erosion rates in undisturbed hilly landscapes with local relief ranging to several hundred metres, in several climatic provinces across Australia, including the temperate SE, the monsoonal tropics and the semi-arid centre. We determined erosion rates using cosmogenic [10]Be and [26]Al in samples collected across spurs

From: BISHOP, P. & PILLANS, B. (eds) *Australian Landscapes.* Geological Society, London, Special Publications, **346**, 225–241. DOI: 10.1144/SP346.12 0305-8719/10/$15.00 © The Geological Society of London 2010.

and slopes, as well as average erosion rates from samples of stream and river sediments.

Concepts and methods

Landforms evolve by surface lowering, caused by loss or erosion of rock (including weathered rock). In this study, measurements of cosmogenic radionuclides (^{10}Be and ^{26}Al) are used to estimate erosion rates of both soil-mantled and bare rock surfaces [reviews of methods used here have been given by Nishiizumi *et al.* (1993), Bierman (1994), Bierman & Steig (1996), Granger *et al.* (1996), Gosse & Phillips (2001), and Cockburn & Summerfield (2004)]. As shown by Lal (1991), *in situ* cosmogenic nuclide content is inversely related to erosion rate:

$$N(z) = \frac{P_0 e^{-\mu z}}{\lambda + \mu \varepsilon} \qquad (1)$$

where $N(z)$ is nuclide content (atoms per gram) at depth z, P_0 is surface ($z = 0$) production rate of the nuclide of interest at the latitude and altitude of the sample site, $\mu = \rho/\Lambda$ where ρ is rock density and Λ is mean free path of cosmic rays, λ is the radionuclide decay constant, and ε is erosion rate. It is important to note that equation (1) assumes that erosion proceeds smoothly at the grain or small fragment scale and at a constant rate. Taking into account the effects of latitude, altitude and topographic shielding on P_0, equation (1) is widely used to calculate rates of erosion and soil production. The approaches to soil-mantled and rocky surfaces are, however, different, and results for both cases may be distorted by the effects of climatic shifts, and are summarized briefly.

Soil-mantled slopes. Where soil is produced through weathering of the underlying rock and soil is lost by soil transport at the same rate as it is produced, erosion equates to soil production. Where transport can be characterized as diffusion-like creep, soil thickness depends on slope curvature and, provided that factors affecting creep such as soil biota and vegetation cover remain unchanged, the balance between soil production and transport should be in steady state and soil thickness at any point should remain constant (Dietrich *et al.* 1995; Heimsath *et al.* 1997, 1999). (It should be noted that, as land surfaces are lowered, slope curvature may slowly change and steady state cannot persist indefinitely, but is approximated for metre-scale lowering where the radius of slope curvature is large relative to soil thickness.) When steady-state conditions hold, the rate of erosion/soil production can be determined from measurements of cosmogenic

nuclides in weathered rock at soil base, using equation (1) with $z =$ soil thickness. One criterion for steady state is that soil depth is negative-exponentially dependent on curvature (Heimsath *et al.* 1997, 1999); another is that the rate of lowering at a point remains constant, which can be tested by measuring cosmogenic nuclide profiles down the sides of emergent bedrock tors, if present (Heimsath *et al.* 2000).

Rocky slopes. Unless rocky surfaces are eroding smoothly by solution or grain-by-grain loss, surface lowering tends to occur by intermittently shedding of joint-controlled blocks or exfoliation slabs. The apparent erosion rate calculated by equation (1) for a surface sample will vary according to the time elapsed since the surface was exposed after loss of a prior block: for example, if N_u is the nuclide content at the upper surface of a block when it falls off, the nuclide content N_b of the surface exposed immediately after it fell is

$$N_b = N_u e^{-\mu L} \qquad (2)$$

where L is block thickness. For jointed rocks with joint spacing L, the nuclide content of most surfaces should lie in the range N_b to N_u. Moreover, the mean nuclide content, $\overline{N}$, of a block at the moment of falling is

$$\overline{N} = N_u \frac{(1 - e^{-\mu L})}{\mu L}. \qquad (3)$$

Numerical evaluation of stepwise erosion by successive block losses shows that the mean value of nuclide content in a block closely approximates the uniform-rate value in equation (1), for blocks up to several metres thick. Provided that block break-up is statistically uniform, the average content of newly fallen blocks (or sediment derived from them) will represent the average erosion rate and, having measured $\overline{N}$, the expected range of values for a set of random samples from a blocky slope can be estimated from equations (2) and (3) together (also see Reinhardt *et al.* 2007).

Effects of climatic shifts. Major climatic changes, such as occurred widely and repeatedly throughout the Quaternary Era, can nullify the assumption of uniform erosion. For example, slopes that today are soil-mantled may have had negligible soil cover under Late Pleistocene arid or periglacial conditions, in which case the cosmogenic nuclide content will be higher than expected under steady-state soil cover and the erosion rate calculated using present soil depth will be falsely low. Conversely, if slopes that today are bare rock were previously mantled by soil or sediment, the erosion rate calculated for a bare surface will be too high. In short, where these scenarios are likely, apparent erosion rates for soil-mantled slopes should be taken as minimum estimates, and those for bare rocky surfaces as maximum estimates.

Field samples. We collected three types of sample for evaluating erosion rates: rock samples a few centimetres thick from exposed bedrock surfaces; saprolite from immediately beneath soil base in soil-mantled landforms; and sediment samples from both low-order channels and slope-foot colluvium. Only quartz-bearing rocks were sampled, from which purified quartz was prepared for measurements of ^{10}Be and ^{26}Al, following standard methods. Unless otherwise stated, cosmogenic nuclides were measured with the ANU 14UD Pelletron accelerator as described by Fifield (1999). ^{9}Be carrier blanks gave ^{10}Be $< 3 \times 10^{-15}$ atoms g^{-1}, $< 0.1\%$ of the concentrations in our field samples. Separate aliquots of purified quartz from the same specimen showed good reproducibility (Table 1, sample P199), and field reproducibility also was good: samples taken 1.5 m apart at a waterfall crest gave statistically equivalent results (Table 1, Br299U and Br299L), as did two samples taken 7 m apart at a low granite dome, eroding by centimetre-scale exfoliation (MD1 and MD3). Field sampling patterns varied with site, and are outlined in site descriptions, below.

Field sites

We sampled moderate to steep ranges in different Australian climatic zones, including the humid southeastern escarpment and highlands, the semi-arid Flinders Ranges in southern Australia and MacDonnell Ranges in central Australia, together with the Arnhem escarpment in the tropical monsoonal north. Those in the SE and Arnhem Land

have previously been described in detail whereas results from the remainder are reported here for the first time, some of which will be the subject of later, more detailed papers. No major post-Palaeozoic orogenic movements have occurred in any of these regions, although the SE is subject to slow isostatic uplift (Lambeck & Stephenson 1985; Persano *et al.* 2002, 2005) together with late Cretaceous uplift associated with Tasman Sea rifting (Hayes & Ringis 1973). Evidence for late Cenozoic thrusting and uplift has been reported from the Flinders Ranges (Quigley *et al.* 2007*a*, *b*). However, the time and depth scales recorded by cosmogenic nuclide concentrations are small compared with long-term landscape evolution of the Australian interior, which is more fruitfully addressed by thermochronological and palaeomagnetic methods [Stewart *et al.* 1986; Belton *et al.* 2004; Pillans 2007; and see reviews by Kohn *et al.* (2002) and Vasconcelos *et al.* (2008)].

Our field sites fall into two broad groups: soil-mantled, transport-limited slopes, which were examined across all climatic zones, and bedrock-dominated, weathering-limited slopes, which were sampled only in the semi-arid ranges.

Soil-mantled sites

Southeastern Australia: Bega–Bredbo region. We summarize here results reported previously in detailed studies at two soil-mantled sites in SE NSW, one from spurs beside Nunnock River (NR) at the base of the passive-margin escarpment, inland from the town of Bega (Heimsath *et al.* 2000) and the other from Frogs Hollow (FH) on tablelands above the escarpment, *c.* 900 m above sea level (asl) (Heimsath *et al.* 2001*a*). Two further sites near NR have since been studied (see Fig. 1): one lies in lowlands *c.* 25 km SE of NR, 200 m asl ('Snug' site), and the other is a spur on the escarpment above NR that rises to the crest at

Table 1. *Repeatability tests*

Sample	Material	Prod. rate	Exp. factor	^{10}Be (10^6 atoms g^{-1})	E (m Ma^{-1})
(a) Same sample, different aliquots of quartz					
P199	Hard sandstone slope	6.89	0.95	0.22 ± 0.02	16.35 ± 1.46
P199	(repeat)	6.89	0.95	0.21 ± 0.02	17.06 ± 1.68
(b) Paired samples from same field site					
Br229U	Quartzite knickpoint	6.22	0.8	0.24 ± 0.03	11.17 ± 1.44
Br299L	Quartzite knickpoint	6.22	0.8	0.19 ± 0.02	14.07 ± 1.51
MD1	Exfoliating granite	7	1	1.56 ± 0.15	2.21 ± 0.24
MD3	Exfoliating granite	7	1	1.69 ± 0.14	2.01 ± 0.18

All errors propagated to erosion rate, E. Production factor scales nuclide prodution for slope, soil cover, elevation and latitude. Exp. factor is the correction factor applied for local exposure.

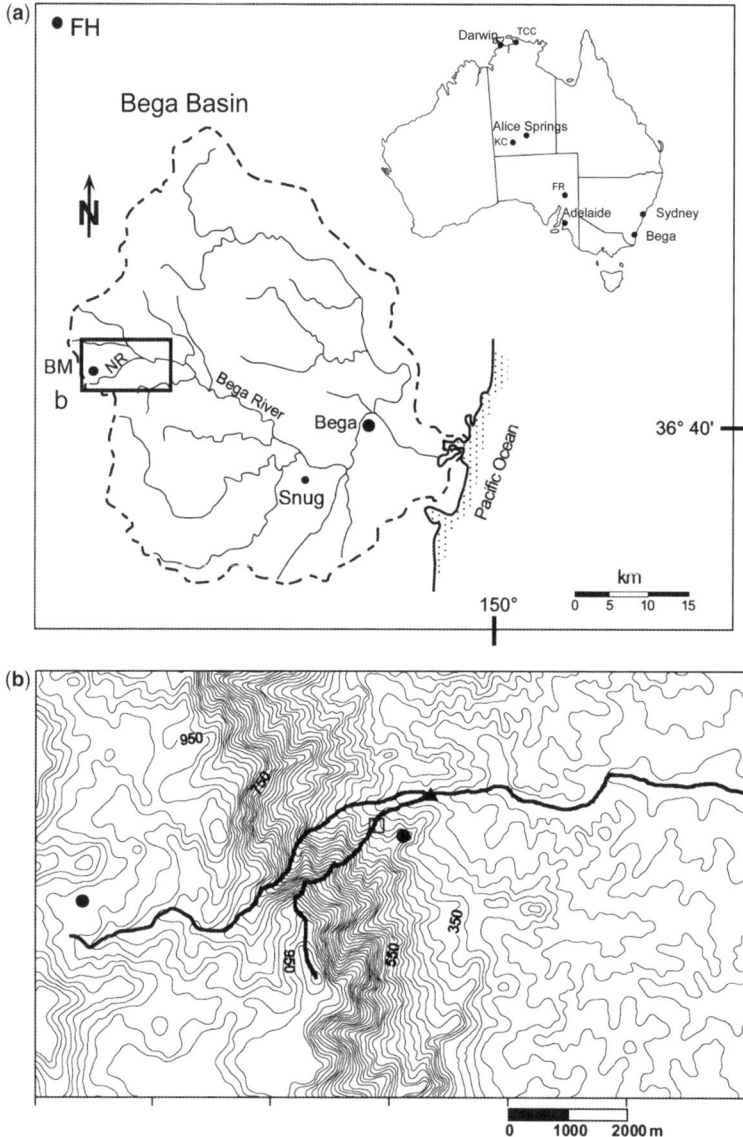

Fig. 1. (**a**) Bega Basin outline with field sites of previous work: FH, Frogs Hollow; BM, Brown Mountain; NR, Nunnock River; Snug, Snugburra. Inset shows outline map of Australia with cities nearest field sites of this paper. (**b**) Topography spanning the escarpment in the area shown by the rectangle in (a) with the transects used for sample collection shown. Contour interval is 20 m from the 1:100 000 digital data (Australian Surveying and Land Information Group) from Bega (sheet 8824) map. Modified from Heimsath *et al.* (2000) and showing locations used for that study (open square is approximate survey area; black dot on ridge crest is where the tor samples were collected; black triangle shows initial Nunnock River stream sediments). Other Australia sites: FR, Flinders Range; KC, Kings Canyon; TCC, Tin Camp Creek.

Brown Mountain (BM site). Cosmogenic nuclide determinations from all four sites together with an account of landscape evolution have been reviewed by Heimsath *et al.* (2006). Chemical weathering across the escarpment has been examined by Burke *et al.* (2009), whereas Braun *et al.* (2001),

Heimsath *et al.* (2002, 2005) and Kaste *et al.* (2007) examined the physical mixing processes at NR.

In summary, the physiography of these sites is as follows. The NR site comprises convex spurs at Nunnock River, a bedrock channel that rises *c.* 1000 m above sea level in the tablelands and

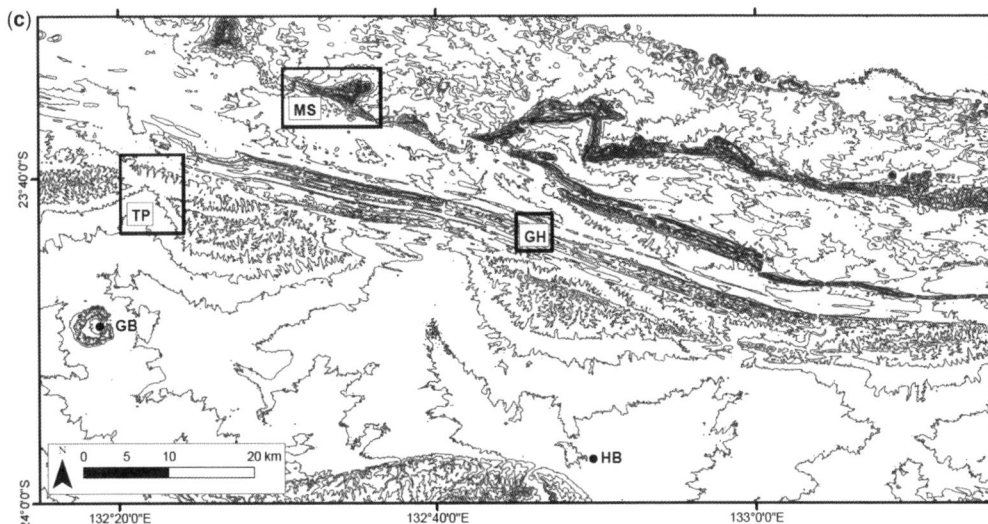

Fig. 1. (*Continued*) Other field site areas (TP, Tyler Pass; GH, Glen Helen; MS, Mount Sonder; shown in Figures 2, 6 and 7, respectively) are located by boxes in (**c**). The eastern border of the map is 70 km from Alice Springs. GB, Gosse Bluff; HB, Hermannsburg.

steeply descends the escarpment (Fig. 1a and b). The convex spurs are soil mantled and steepen downslope to intervening channels, which pass through very coarse, consolidated debris-flow deposits. Bedrock cliffs are prominent near the escarpment top, although they are less dramatic than those to the north in the New England Fold Belt described by Weissel & Seidl (1997, 1998), and are eroding by block failure on the escarpment face and by thin exfoliation and sheet erosion on top. Little sediment has accumulated beneath the cliffs and it appears that the upper region of the escarpment is weathering limited. From the escarpment base to the coast, the landscape is dominated by rolling hills of Late Silurian–Early Devonian granodiorite. Inland, tableland and ranges, dominated by Ordovician–Silurian metasediments and Devonian granites with extensive Tertiary basalt flows (Richardson 1976), lie c. 800–1500 m asl and typically have relatively gentle, soil-mantled slopes (<25°), punctuated by frequent tors. Rainfall at the coastal lowlands site (Snug: 200 m asl, c. 900 mm a^{-1}) and at escarpment base (NR: c. 400 m; c. 1200 mm a^{-1}), is higher than on the highlands (FH: c. 900 m; c. 700 mm a^{-1}), and is distributed throughout the year (Richardson 1976; Australian Bureau of Meteorology: www.bom.gov.au/climate/averages). We refer to these areas as humid, temperate lowlands (NR and Snug) and cool highlands (FH and Brown Mt).

Central Australia: Tyler Pass. The western MacDonnell Ranges, west of Alice Springs in central Australia, are dominated by steep bedrock ridges

in folded Palaeozoic and Precambrian quartzites, metasediments and gneisses. Unusual in this landscape, soil-mantled convex-up spurs occur on late Palaeozoic lithified conglomerate at Tyler Pass, about 150 km west of Alice Springs (Figs 1c and 2). Local relief of the spurs seen in Figure 2 is c. 20 m; hillslope lengths are c. 100 m, and slopes reach c. 25° as they approach dry bedrock channels that lie between them. The soil is strongly bioturbated and its thickness is almost uniform (c. 0.5 m). Tape and clinometer measurements showed that slope increases steadily downslope from ridge crests, indicating constant curvature. This geometry and constant-thickness soil suggest that ridge profiles have reached equilibrium.

Cosmogenic nuclide samples were collected from weathered rock below the soil base, and sand and pebbles were collected from local deposits in first- and second-order channels dissecting the hills, and from the dry bed of the main channel draining south from Tyler Pass.

Arnhem escarpment: Tin Camp Creek. The site is a soil-mantled basin in hard Precambrian sandstone, in the region of the Arnhem escarpment. A ridge, partly soil-mantled, bisects the basin: the soils are mostly red loamy earths and shallow gravelly loam with some micaceous silty yellow earths and minor solodic soils on the alluvial flats (Riley & Williams 1991). Bedrock exposures across the hills are significantly less weathered than saprolite beneath the soil mantle (Wells *et al.* 2008). Vegetation is open dry-sclerophyll forest interspersed

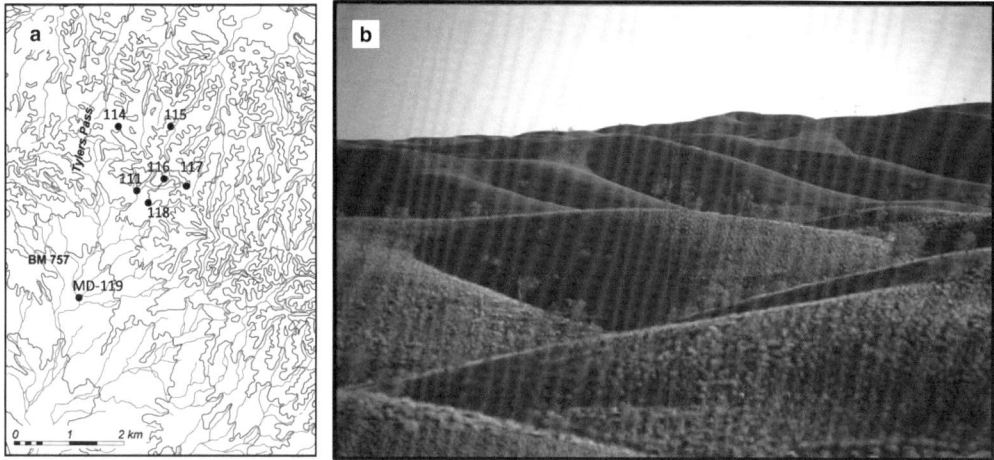

Fig. 2. (a) Tyler Pass field area. Contours and ephemeral streams extracted from Australian Surveying and Land Information Group (AUSLIG) draft map, 5350: Gosse Bluff. Elevation at pass is 815 m; 20 m contour intervals. MD sample numbers label approximate sample locations. MD-119 is at 23°41.795'S, 132°21.246'E. (b) Photograph of the soil-mantled hills of Tyler Pass.

with seasonal tall grassland; tree-throw and burrowing animals contribute to soil transport, and feral water-buffalo and pigs have caused local erosion (Riley & Williams 1991; Hancock *et al.* 2000; Townsend & Douglas 2000; Saynor *et al.* 2004, 2009; Staben & Evans 2008). Rainfall is *c.* 1400 mm a^{-1}, falling almost entirely between October and April, typically in short, high-intensity storms. The climate is classed as monsoon–tropical.

Tin Camp Creek is undisturbed by colonial land use and, as it closely resembles the terrane at Energy Resources Australia Ranger uranium Mine (ERARM), where geomorphological processes and rates are of high interest, is a site of intensive study. Fieldwork was carried out by A.H. in collaboration with G. Hancock of the Office of the Supervising Scientist, Jabiru, and D. Fink of the Australian Nuclear Science and Technology Organisation (Heimsath *et al.* 2009). Our sampling traversed the central ridge, with samples of both exposed and unweathered bedrock as well as weathered rock from below a range of soil depths, together with catchment sediment samples, to assess the spatial variability of erosion rates across a landscape that is used as a reference site by the uranium mining industry (Hancock & Evans 2006; Hancock *et al.* 2008*a*, *b*; Fig. 3).

Bedrock-dominated landscapes

Flinders Ranges. The Flinders Ranges in South Australia, flanked by the great salt playas of lakes Frome, Eyre and Torrens, are dominated by Late Proterozoic rocks, moderately to steeply dipping,

from very hard quartzites through sandstones and limestones to comparatively friable quartzose mudstones. The Ranges rise to over 600 m and are structurally controlled, varying from high quartzite ridges to rounded hills and cuestas in cyclic sandstone–shale–limestone sequences. Seismic activity and evidence of late Cenozoic thrusting suggest continuing tectonic uplift (Quigley *et al.* 2007*a*, *b*), although the rate has yet to be determined and probably is slow. Regional climate is semi-arid, with mean annual precipitation ranging from 250 mm a^{-1} at Port Augusta to over 400 mm a^{-1} in the Ranges.

Our study area centres on Brachina and Parachilna Creeks, bedrock channels that flow westward from the diapiric core of the Ranges through cuestas and rounded ridges of cyclic sandstone and mudstone, and then pass through deep, steep-sided gorges cut in hard sandstone and quartzite. Slope erosion is dominated by shedding of joint-controlled blocks. There is no continuous detrital cover, and the slopes are weathering-limited. Rock surfaces are reddened by iron oxides, suggesting relatively slow slope retreat, and falling blocks tend to disintegrate during their downslope passage. West of the gorge, the wide, boulder-strewn channels form fans and dwindle as they head towards playa-lake Torrens.

We collected samples around the gorges of Brachina and Parachilna Creeks (Fig. 4), from steep to precipitous slopes in jointed quartzite (Brachina gorge) and sandstone (both gorges), to determine erosion rates of weathering-limited slopes and also to assess the variation of cosmogenic nuclides in

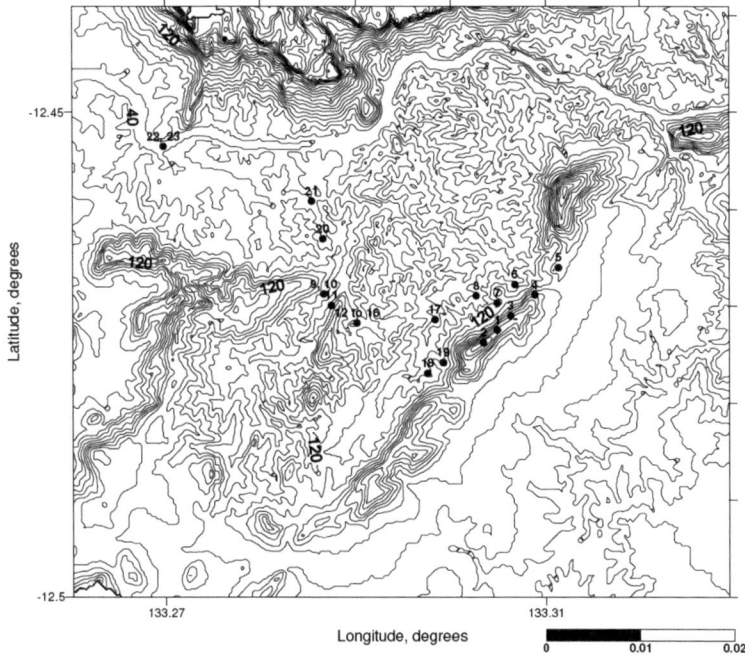

Fig. 3. Contour map and sample locations for Tin Camp Creek (TCC) field area of Heimsath *et al.* (2009). Topography extracted from regional digital elevation model (e.g. Hancock & Evans 2006); 10 m contour intervals. Sample locations shown with black dots and (TC) number of Heimsath *et al.* (2009). Samples 22 and 23 mark the main drainage of TCC.

point samples from joint-controlled rocky slopes. We also sampled quartzite gravel and bedrock knickpoints in low-order channels.

Central Australia: Kings Canyon. Carved in gently dipping Late Palaeozoic sandstone (Bagas 1988), the bedrock-dominated inner drainage basins in Kings Canyon National Park include regularly spaced, convex-up bedrock ridges with relatively steep slopes (Fig. 5) that are remarkably similar in form to steep, soil-mantled landscapes of coastal Oregon (Heimsath *et al.* 2001b; Roering *et al.* 2001), despite the fact that Kings Canyon lies in arid, tropical central Australia and receives only *c.* 260 mm a^{-1} of rainfall.

The regularly spaced convex ridges pose a geomorphological puzzle. Evolution of convex ridges under a soil mantle is expected with diffusive soil creep and depth-dependent soil production (Heimsath *et al.* 1997, 2000), but it is unclear how these forms could develop in weathering-limited bedrock landscapes, eroding by processes acting on steep rectilinear and concave slopes of the Flinders Ranges. Conjecturing that the convex ridges may have formed under a soil mantle during wetter climate in the past, we hypothesize

that present-day erosion rates are inconsistent with evolution of the observed topography, and collected samples downslope as shown in Figure 5.

Central Australia: Mt Sonder and western MacDonnell Ranges. The MacDonnell Ranges are dominated by steep, structurally controlled bedrock ranges: quartzites form high, precipitous rectilinear slopes, whereas sandstone and other rocks tend to be dissected in parallel, convex-up rounded spurs similar to those of Kings Canyon. The climate is semi-arid, similar to that of Kings Canyon, with *c.* 2–300 mm a^{-1} highly variable rainfall. Spinifex (*Triodea*) and other grasses together with intermittent acacia and eucalyptus shrubs and trees mantle the stony landscape. In addition to Tyler Pass, described above, we selected two other field sites to assess the effect of lithology on erosion rates: one at Glen Helen in Hermannsburg Sandstone (Figs 1c and 6) and the other on Mt Sonder in Heavitree Quartzite (Figs 1c and 7). Both are bedrock-dominated; erosion processes are similar to those in the Flinders Ranges, outlined above, and both are characterized by rounded bedrock ridges with no soil mantle. Despite the clear differences in rock hardness, the two sites are morphologically

Fig. 4. Photograph of the Flinders Range blocky cliff exposure used to calibrate a bedrock erosion model. Samples collected from above and below the bedrock steps shown in the photograph were compared with average erosion rates from low-order stream sands draining catchments adjacent to this ridge.

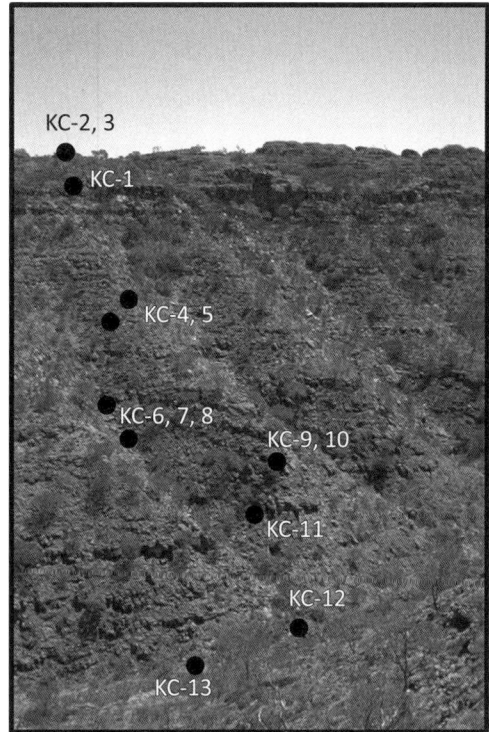

Fig. 5. Photograph of the bedrock ridge–valley topography of Kings Canyon with approximate sample locations shown with black filled circles. Samples were collected from the 'beehive'-shaped features of the Mereenie Sandstone at the top of the ridge, down the axis of the central ridge of the photograph, with some samples from the active channels.

similar, with maximum spur-crest slopes of *c.* 35°. Local relief is *c.* 200 m at the Glen Helen site (Fig. 6) and ranges from 250 m to >500 m at Mt Sonder (Fig. 7).

Results

Soil-mantled landscapes

Rates of soil production. Soil production and transport are considered to be in steady state at the NR and Snug sites in humid southeastern Australia. In their intensive study at the NR site, Heimsath *et al.* (2000) showed that soil thickness is negative-exponentially dependent on slope curvature, and we have confirmed a similar relationship at Snug. Moreover, cosmogenic nuclide concentrations in granite tors at NR reveal uniform-rate lowering of the surrounding soil (Heimsath *et al.* 2000). Cosmogenic measurements from beneath soil base robustly define an exponential decline of

the soil production with increasing soil thickness; as reported previously

$$\varepsilon(H) = 53 \pm 2e^{-(0.022 \pm 0.001)H} \qquad (4)$$

where $\varepsilon(H)$ is soil production rate (m Ma^{-1}) and H is soil thickness (cm). Results from Brown Mountain and FH also accord with this relationship (Fig. 8a), although it must be noted that the FH site at 900 m asl probably is not in steady state, having almost certainly been affected by periglacial solifluction (Heimsath *et al.* 2001*a*).

Turning to tropical Australia, measurements of soil production and soil thickness at Tin Can Creek (TCC) in the monsoonal far north (Heimsath *et al.* 2009) overlap completely with those from our southeastern sites, as Figure 8 shows, except that the TCC rates are lower where soil thickness approaches zero and suggest a 'humped' production function; Wilkinson *et al.* (2005) proposed a similar

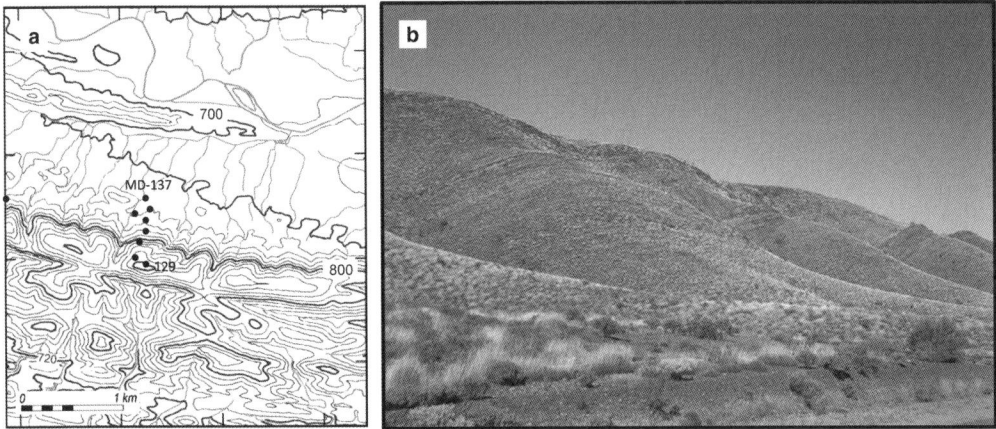

Fig. 6. (**a**) Glen Helen field area. Contours and ephemeral streams extracted from Defense Imagery and Geospatial, 1:100 000 map Series R621, Sheet 5550: MacDonnell Ranges; 20 m contours. MD sample numbers label approximate sample locations. MD-130 sampled near 129 and MD-137 is at 23°42.501′S, 132°46.443′E. (**b**) Photograph of the ridge from the Alice Springs–Tyler Pass Road, looking SW. Despite the soil-mantled appearance, these hills are almost entirely bedrock, albeit a fractured and weathered bedrock able to support ample vegetation.

relationship in the Blue Mountains, SE Australia. That landscapes with such different climates show such similar soil production functions seems surprising but suggests that similar processes, presumably dominantly biotic, are instrumental in producing and transporting soil. Measurements of rates of soil-particle migration would be informative, similar to those reported from NR by Heimsath *et al.* (2002). In contrast, soil production on spurs at Tyler Pass in semi-arid central Australia is significantly lower: rates of 6.6–6.7 m Ma^{-1} were obtained at five pits, under soil thicknesses of 28–35 cm (Table 1; Fig. 8). For comparable thicknesses at both the monsoonal and southern sites, the rate is 30–35 m Ma^{-1}. Moreover, the mean

erosion rate at Tyler Pass, determined from sand and pebble samples from nearby channels, is also 6–7 m Ma^{-1} (Table 2). We presume the difference between semi-arid Tyler Pass and both the monsoonal and the southeastern sites reflects the effect of soil moisture on both chemical weathering and biotic activity.

Exposed rock at soil-mantled sites. Local exposures of bedrock occur at our granitic soil-mantled sites in SE Australia, and at the monsoonal tropical sandstone site (TCC), either as upstanding tors (emergent corestones) or as flat exposures that pass beneath soil at their edges: the data are represented in Figure 8 as points at zero soil thickness. Erosion

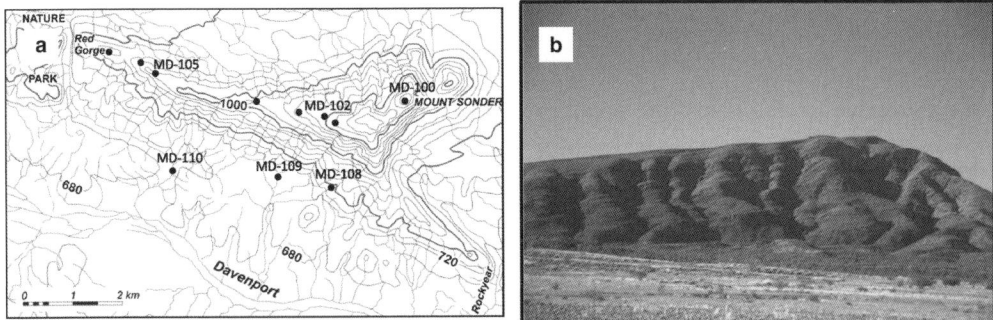

Fig. 7. (**a**) Sonder field area. Contours and ephemeral streams from Australian Surveying and Land Information Group (AUSLIG) draft map, 5450, Hermannsburg; 40 m contours. MD sample numbers label selected approximate sample locations. MD-100 is at 23°34.860′S, 132°34.238′E. (**b**) Photograph of Mt Sonder from the Alice Springs–Tyler Pass Road, looking NE. The highest point (*c.* 1320 m elevation) is not visible.

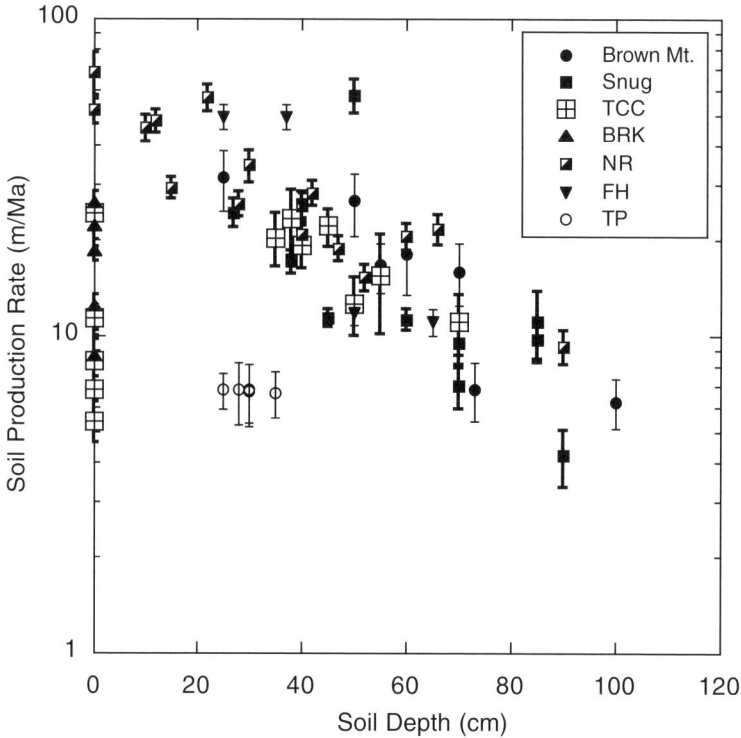

Fig. 8. Soil production rates plotted against overlying soil thickness for all four field sites across the southeastern escarpment (NR, Nunnock River; FH, Frogs Hollow), as well as the seven soil production rate samples collected from beneath soil at Tin Camp Creek (TCC). Additionally, soil production rates from the five Tyler Pass samples, with little variation in soil depth, are plotted with open circles.

rates determined from these exposures show a considerable range ($5-60$ m Ma^{-1}), some being higher and others much lower than results from nearby soil-covered sites. Erosion is by grain-by-grain disintegration at all these sites and the wide spread of results is not attributable to episodic shedding of coarse blocks, but reflects the position of each sample relative to the nearby soil. As Heimsath *et al.* (2000, 2001*a*, *b*) described, rates similar to values from beneath thin soil generally came from flat exposures at soil level, whereas the low erosion rates were obtained from the sides and tops of tors.

Topographic variation of erosion rates. Our erosion-rate measurements from the escarpment that rises from NR to Brown Mountain are strongly altitude dependent. The dataset includes eight samples from small, zero- and first-order streams draining the escarpment, 10 samples from beneath shallow soil on a downhill line on a prominent spur, three samples from granite tor-tops from the escarpment crest, and three samples from the tops of granite cliffs along the steep escarpment face. Rates at a given elevation are consistent between

soil-mantled samples, sediments and bedrock samples. These 24 samples together with the average erosion rate from NR show a strong trend of increasing erosion rate with decreasing altitude (Fig. 9a), from *c.* 3 m Ma^{-1} at the escarpment top to *c.* 35 m Ma^{-1} near the base. There is no relationship between erosion rate and either local slope or basin-averaged slope, which may be due to the lack of relationship between soil thickness and slope as discussed further by Burke *et al.* (2009). Extrapolated to million year time scales, these results imply that the escarpment scarp becomes steeper with time, unless corrected by some process not measured by these data. Not only is it unlikely that the escarpment will become indefinitely steeper but long-term constancy of form is supported by apatite fission-track and (U–Th)/He data (Persano *et al.* 2002, 2005). It seems likely that the very coarse diamictites observed in the ravines near the escarpment base (Heimsath *et al.* 2000, 2006) represent large mass-movements that episodically reverse the trend suggested by Figure 9a.

The TCC data from the tropical monsoonal north show a rather similar pattern. The relief is less than

Table 2. *Erosion rates from cosmogenic nuclide concentrations*

Sample	Weight (g)	Carrier (µg)	^{10}Be (10^6 atoms g^{-1})	Prod. rate	Elev. (m)	E (m Ma^{-1})	Min. exp. (ka)
Kings Canyon							
KC-1	27.49	447	17.052 ± 0.886	6.54	789	0.24 ± 0.03	2607
KC-2	30.34	383	22.379 ± 1.045	6.56	793	0.18 ± 0.02	3411
KC-3	29.18	383	14.124 ± 0.612	6.55	791	0.24 ± 0.03	2156
KC-4	30.07	428	9.719 ± 0.412	6.36	750	0.39 ± 0.04	1528
KC-5	28.82	442	3.792 ± 0.188	6.29	735	0.99 ± 0.12	603
KC-6	30.57	465	1.509 ± 0.178	6.20	715	2.43 ± 0.51	243
KC-7 S	28.45	424	2.109 ± 0.163	6.16	705	1.73 ± 0.27	342
KC-8 P	30.19	434	13.907 ± 0.555	6.13	700	0.26 ± 0.03	2269
KC-9	29.53	398	1.237 ± 0.115	6.08	687	2.97 ± 0.54	204
KC-10	29.62	416	1.880 ± 0.136	6.07	685	1.95 ± 0.30	310
KC-11	31.09	411	1.501 ± 0.148	6.02	675	2.32 ± 0.44	249
KC-12 S	29.92	412	1.026 ± 0.067	5.86	638	3.38 ± 0.51	175
KC-13 P	29.75	469	1.797 ± 0.289	5.85	635	1.93 ± 0.53	307
Mt. Sonder							
MD-100	30.08	445	2.189 ± 0.162	9.42	1320	2.00 ± 0.23	232
MD-101	29.76	469	3.675 ± 0.266	8.90	1240	1.51 ± 0.13	413
MD-102	30.62	458	2.505 ± 0.419	8.53	1180	1.79 ± 0.35	294
MD-103	29.32	456	3.235 ± 0.384	8.21	1125	1.58 ± 0.21	394
MD-104	29.62	440	1.428 ± 0.168	7.67	1030	2.77 ± 0.48	186
MD-105	30.60	478	7.026 ± 0.399	7.21	945	0.76 ± 0.06	974
MD-106	30.64	457	2.992 ± 0.193	7.19	940	1.60 ± 0.15	416
MD-107	29.95	437	1.555 ± 0.115	7.03	910	2.51 ± 0.32	221
MD-108 S	29.88	433	5.605 ± 0.254	6.75	770	0.80 ± 0.07	830
MD-109 S	30.83	425	1.299 ± 0.111	6.75	700	2.84 ± 0.42	192
MD-110 S	28.54	300	4.001 ± 0.286	6.60	675	1.47 ± 0.12	606
Tyler Pass							
MD-111 S	30.15	409	0.636 ± 0.102	6.47	792	6.14 ± 1.30	98
MD-112 P	31.37	387	0.671 ± 0.120	6.48	794	5.83 ± 1.34	104
MD-114 (35 cm)	29.66	417	0.618 ± 0.071	6.56	810	6.61 ± 1.08	94
MD-115 (30 cm)	29.44	417	0.283 ± 0.049	6.51	801	6.67 ± 1.49	44
MD-116 (25 cm)	20.73	447	0.211 ± 0.016	6.53	804	6.76 ± 0.84	32
MD-117 (28 cm)	30.39	352	0.360 ± 0.061	6.51	799	6.75 ± 1.48	55
MD-118 (30 cm)	30.80	393	0.488 ± 0.079	6.61	822	6.75 ± 1.41	74
MD-119 S	28.52	359	0.498 ± 0.041	6.33	760	7.12 ± 1.35	64
MD-119 P	31.39	437	0.624 ± 0.148	6.33	761	5.68 ± 1.68	99
Glen Helen							
MD-129	29.35	397	1.362 ± 0.183	7.00	900	3.20 ± 0.55	217
MD-130	29.89	329	2.245 ± 0.307	7.00	900	1.94 ± 0.34	357
MD-131	29.70	390	0.287 ± 0.075	6.95	890	14.34 ± 4.16	46
MD-132	30.65	394	0.820 ± 0.135	6.70	838	4.99 ± 1.04	132
MD-133	29.48	379	0.343 ± 0.047	6.35	765	11.00 ± 2.12	56
MD-135	30.17	366	0.350 ± 0.062	6.15	720	10.51 ± 2.52	58
MD-136	29.40	367	0.867 ± 0.166	6.12	715	4.29 ± 1.10	143
MD-137	29.78	338	0.543 ± 0.075	6.06	700	6.68 ± 1.35	90
Flinders Range							
Blocky quartzite slopes							
Br8U2	33.74	454	1.140 ± 0.130	6.22	400	2.2 ± 0.28	183
Br9i	31.83	365	0.600 ± 0.050	6.22	400	4.36 ± 0.41	96
F108P	32.17	456	0.570 ± 0.040	6.40	380	5.92 ± 0.44	89
FR9	31.22	460	0.550 ± 0.040	7.20	460	6.31 ± 0.46	76
FR11	16.76	309	0.590 ± 0.070	7.04	445	6.37 ± 0.8	84
Br8U1	30.37	374	0.320 ± 0.040	6.22	400	8.32 ± 0.97	51
FR10	31.29	383	0.290 ± 0.020	6.94	350	10.82 ± 0.87	42
FR12	31.45	380	0.350 ± 0.030	7.04	470	10.81 ± 0.96	50
FR7	23.09	306	0.290 ± 0.060	6.89	400	11.37 ± 2.7	42

(Continued)

Table 2. *Continued*

Sample	Weight (g)	Carrier (µg)	^{10}Be (10^6 atoms g^{-1})	Prod. rate	Elev. (m)	E (m Ma^{-1})	Min. exp. (ka)
Quartzite sediments							
F109s	21.25	316	0.900 ± 0.050	6.40	350	3.66 ± 0.22	141
P499	31.10	356	0.790 ± 0.050	7.04	380	5.12 ± 0.31	112
Br399	30.11	457	0.660 ± 0.040	7.14	375	5.15 ± 0.33	92
F109P	23.32	300	0.540 ± 0.040	6.40	350	6.27 ± 0.49	84
F110s	30.51	366	0.530 ± 0.050	7.96	325	7.6 ± 0.75	67
Rocky sandstone slopes							
P399	30.62	312	0.520 ± 0.024	7.14	390	7.31 ± 0.35	73
FR4	29.98	350	0.410 ± 0.060	6.89	400	9.1 ± 1.39	60
P299	31.58	353	0.270 ± 0.040	7.14	375	14.07 ± 2.37	38
P199	31.28	354	0.220 ± 0.020	6.89	350	16.35 ± 1.46	32
P199	31.10	356	0.210 ± 0.020	6.89	350	17.06 ± 1.68	30
FR5	30.55	344	0.150 ± 0.020	6.89	400	21.91 ± 3.54	22

All errors propagated to erosion rate, E; Min. exp. is the minimum exposure age to cosmic-ray flux. Production factor scales nuclide prodution for slope, soil cover, elevation and latitude. Calculated using Cronus online calculator (http://hess.ess.washington.edu/math/). S and P, sand and pebbles, respectively, referring to catchment-averaged detrital sediment samples. [For Tin Camp Creek data see Heimsath *et al.* (2009)]. For Bega Valley data see Heimsath *et al.* (2000, 2001*a*, 2006).

that of the southeastern escarpment (200 m v. 800 m) but the physiography is similar, with soil-mantled convex spurs below rocky cliffs, topped by ridge-and-plateau terrain. Figure 9b plots erosion rates v. altitude, determined from soil-mantled spurs, exposed bedrock, and sediments from small channels. Excepting two sediment samples from the lowlands, the results suggest decreasing erosion rate with increasing elevation, from *c.* 20 m Ma^{-1} at 80 m asl to *c.* 5 m Ma^{-1} at 230 m asl. As with the southeastern escarpment, the pattern implies that slopes around TCC become progressively steeper through time, unless the trend is reversed by processes not apparent today.

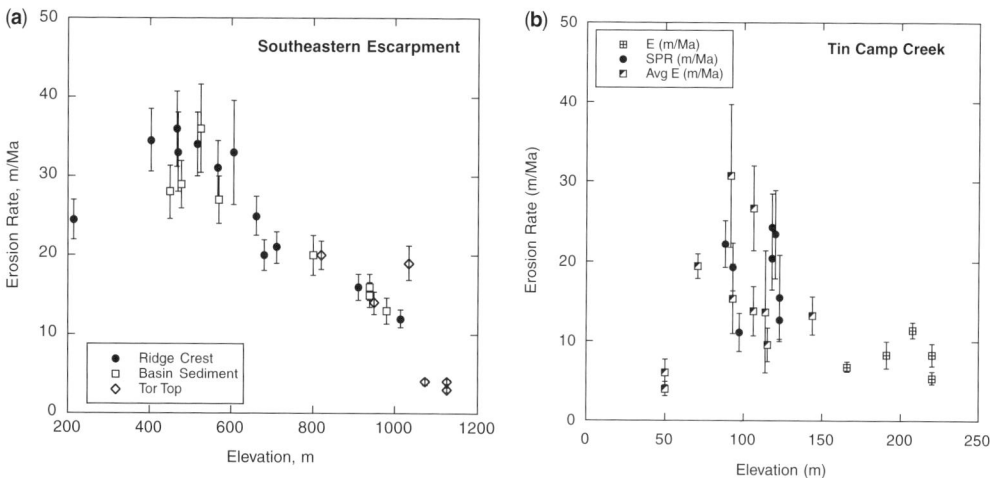

Fig. 9. (**a**) Erosion rates from ridge crest (filled black circles), small catchment (open squares) and tor top (open diamonds) samples plotted as a function of elevation across the southeastern escarpment face. For most of the samples, the ridge-crest and small catchment samples were adjacent for similar elevations. Average values for the soil production rates at the scarp base (NR) and the coastal lowland (Snug) sites are included. (**b**) Soil production (filled black circles), average erosion (half-filled squares) and point-specific bedrock erosion (open squares with crosses) plotted against elevation for the Tin Camp Creek field site. Lowest elevation samples are from the Tin Camp Creek, draining the entire field area, as well as the extensive stony highlands. Adapted from Heimsath *et al.* (2009).

Bedrock-dominated landscapes

Erosion in the steep, weathering-limited bedrock slopes and ridges that we examined in the Flinders and MacDonnell Ranges is dominated by joint-controlled break-up on stratified and schistose rock surfaces, and by exfoliation and granular disintegration on massive, coarsely crystalline plutonic rocks. Clasts and grains drift downslope, disintegrating further as they go and lodging temporarily in joint-angle pockets but, on the slopes we sampled, not in sufficient quantities to form talus slopes or colluvial mantles. Vegetation patches of herbs, grasses and shrubs pattern the slopes, rooted in loose gravelly sand. Drainage is focused by steep bedrock channels, with discontinuous alluvial deposits of mixed sand, gravel and angular boulders, interrupted by knickpoints. Although physiographically similar, the Flinders and Mac-Donnell Ranges yielded somewhat different results, and are summarized separately.

Erosion rates on different rock-types. (i) Flinders Ranges. We sampled *in situ* rock surfaces on steep, coarsely jointed quartzite slopes (joint spacing 60 cm) (nine samples), sediments derived from these slopes (five samples), and rocky sandstone slopes (five samples). Results are listed in Table 2. Erosion rates calculated from ^{10}Be measurements for quartzite samples varied: surfaces that appeared to have been exposed by slip-off of overlying blocks gave the highest apparent rates (FR7, FR10, and FR12) whereas surfaces judged to have been long exposed, from their state of red-weathering, gave the lowest apparent rates (Br8U2, BR9i). The mean of the measurements from rock surfaces is 7.4 m Ma^{-1}; the range (2.2–11.4 m Ma^{-1}) is similar to that predicted using equations (2) and (3) with block size $L = 60$ cm (1.8–9.5 m Ma^{-1}). Standard deviation and standard errors are ± 3.2 and ± 1.1, respectively. The mean rate from quartzite sand and gravel channel sediments is 5.6 m Ma^{-1} (standard error ± 0.7 m), statistically similar to the average for the *in situ* samples, suggesting that (1) the average erosion rate obtained from the *in situ* samples is a reasonable estimate, and (2) exposure to cosmic rays as weathered clasts move from the rocky slopes into stream sediments is minor, relative to *in situ* exposure before break-out from the slope; that is, the travel time is relatively short for these landscapes.

Erosion rates from *in situ* sandstone samples are significantly higher than from the quartzite, ranging from 7.3 to 21.9 m Ma^{-1} with a mean of 14 $\pm$ 2 m Ma^{-1}. Extrapolation back over tens of millions of years suggests that sandstone ridge-crests should be substantially lower than quartzite crests. Although this is locally the case in the neighbourhood of

Brachina and Parachilna gorges, it is not so throughout the Flinders Ranges as a whole, suggesting continuing, slow differential uplift. Quigley *et al.* (2007a, b) reached a similar conclusion at faulted terrain in the central Flinders Ranges.

(ii) MacDonnell Ranges and Kings Canyon. Long, convex-up spurs in jointed bedrock were sampled in the western MacDonnell Ranges, in quartzite at Mt Sonder and sandstone at Glen Helen, and in sandstone at Kings Canyon. Surface erosion rates measured at Mount Sonder are consistently low, ranging from 0.7 to 2.8 m Ma^{-1} (eight samples: Table 2) with a mean of 1.8 $\pm$ 0.2 m Ma^{-1}. Three sediment samples from small bedrock channels give essentially the same result: 1.7 $\pm$ 0.6 m Ma^{-1}. Surface erosion rates from the Glen Helen sandstone site are both higher and more scattered, ranging from 1.9 to 14.3 m Ma^{-1} (six samples) with a mean of 6.6 $\pm$ 2.0 m Ma^{-1}. In contrast, erosion rate measurements from Kings Canyon sandstone site are low, ranging from 0.18 to 2.97 m Ma^{-1} (nine samples), with a mean of 1.3 $\pm$ 0.4 m Ma^{-1}. Four sediment samples from small bedrock channels give a similar result: 1.8 $\pm$ 0.6 m Ma^{-1}.

Topographic variation of erosion rate

In contrast with the high, soil-mantled escarpment in humid SE Australia, where erosion rate declines markedly with increasing altitude, we found no regular variation of erosion rate with altitude at our sites in semi-arid central Australia. Although the variability inherent in point measurements from coarsely jointed surfaces will tend to mask trends, we consider that overall there are sufficient measurements for any trends to be apparent, if present. Allowing for inherent scatter, erosion rates at Mt Sonder, which has both the greatest relief and the most measurements, effectively are the same at all altitudes, from the base at *c.* 700 m asl to the top at 1350 m asl (Fig. 10a). Results from Glen Helen are too scattered for any trend to be revealed (Fig. 10b), whereas at Kings Canyon, erosion rates appear to drop in a step-like manner, from *c.* 2 m Ma^{-1} below *c.* 720 m asl to <0.5 mMa^{-1} above 720 m (Fig. 10c).

Discussion

Ranking the sites in ascending order in terms of erosion rates (means and standard errors in m Ma^{-1}), we have: Kings Canyon sandstone (1.5 $\pm$ 0.3); Mt Sonder quartzite (1.8 $\pm$ 0.2); Tyler Pass conglomerate (6.5 $\pm$ 0.2); Flinders Ranges quartzite (6.7 $\pm$ 0.7); Glen Helen sandstone (7.1 $\pm$ 1.5); Flinders Ranges sandstone (14 $\pm$ 2);

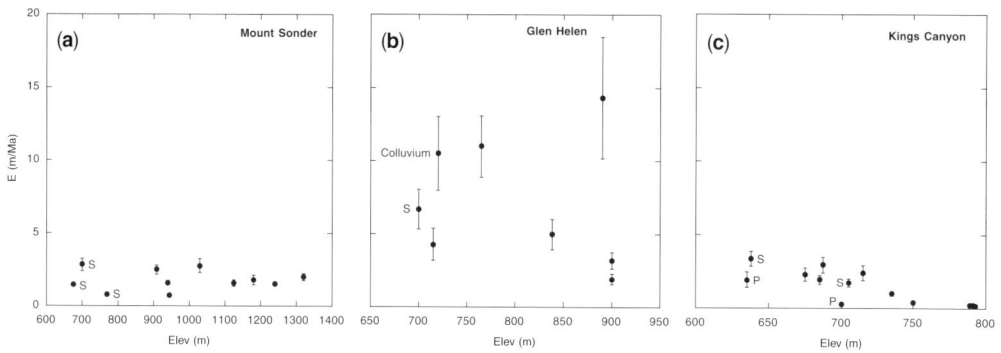

Fig. 10. Bedrock erosion rates from point-specific samples, as well as average erosion rates from sediment samples, plotted as functions of sample elevation. (**a**) Rates from Mount Sonder are roughly equal, with the lowest erosion rates from sampled tor surfaces on the ridge crest. The three lowest elevation samples are catchment-averaged rates from sands collected in different drainages (Fig. 7a). (**b**) Rates from the Glen Helen ridge transect show greater scatter. S, sand draining the ridges in first-order drainage. (**c**) Rates from Kings Canyon range from almost zero at the top to average rates from the sediments roughly equivalent to the Mount Sonder field area, roughly 100 km north. Sand (S) samples show higher erosion rates than pebble (P) samples collected at same location.

monsoonal north, TCC sandstone (14 ± 4); SE highlands, Brown Mt and FH granodiorite (14 ± 3); SE lowlands, NR and Snug granodiorite (35 ± 6). A number of factors affecting erosion rate emerge from this ranking: climate, presence or absence of soil, and rock type. Climatic history may also be a factor.

A climatic influence seems obvious (see Fig. 11). The average rates are lowest at Kings Canyon and Mt Sonder in semi-arid central Australia and are comparable with those reported for other semi-arid Australian sites, including inselbergs on the Eyre Peninsula in South Australia (Bierman & Caffee 2002) and the Davenport Range in the Northern Territory (Belton *et al.* 2004). In contrast, the rates are about 20 times higher at the NR and Snug sites in the temperate SE. Moreover, the low and high erosion rates parallel the absence or presence of a soil mantle: at the low end (the weathering-limited case), the supply of weathered detritus is less than the capacity of sliding, rolling and wash processes to move it downslope; at the high end (the transport-limited case), the supply exceeds the capacity of these simple processes and, as a result, a detrital blanket ('soil') builds up until downslope motion of the entire mantle equilibrates with supply, which in turn is mediated by the dependence of soil production on soil thickness [as shown by equation (4)]. Thus, climate, or, more exactly, moisture and temperature, governs both weathering rate and the balance of erosional processes, at least at the semi-arid and humid temperate ends of the spectrum.

Between these extremes, the effects of climate on weathering rates and presence or absence of soil are more blurred. Although rock hardness and jointing density appear to have an effect, as shown by the

difference between neighbouring rocky quartzite and sandstone slopes in the Flinders Ranges (6.7 ± 0.7 v. $14 \pm 2 \text{ m Ma}^{-1}$), other factors as well as climate influence both erosion rate and presence or absence of soil. For example, Tyler Pass and Glen Helen sites are convex-up spurs with similar mean erosion rates (6.5 and 7.1 m Ma^{-1}), but TP is soil-mantled whereas GH is not: the difference cannot be attributed to climate, which is the same

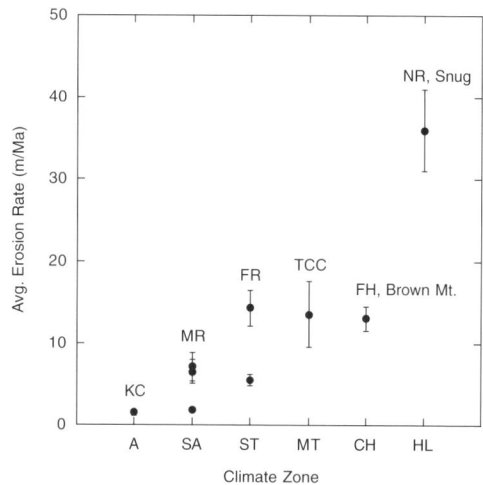

Fig. 11. Average erosion rate plotted against the climate zone characterizing each field site and labelled with the field site represented. A, arid (KC, Kings Canyon); SA, semi-arid (MR, MacDonnell Range); ST, semi-arid–temperate (FR, Flinders Range); MT, monsoonal tropic (TCC, Tin Camp Creek); CH, cool highland (FH, Frogs Hollow); HL, humid lowland (NR, Nunnock River).

at both sites. Furthermore, three other sites have similar mean erosion rates (c. 14 m Ma^{-1}) but different climates and cover; one is the bare, rocky Flinders Ranges sandstone, whereas two others are soil-mantled: TCC sandstone in the monsoonal north and the SE highland sites at Brown Mt and FH. We conclude that the threshold between weathering- and transport-limited slopes must depend not only on the rate of production of weathered detritus but also on local factors such as slope angle, runoff intensity and vegetation recruitment. However, without further studies at the field sites, we cannot resolve these effects.

Climatic history may be a confounding factor, if climate shifts led to changes of detrital or soil cover, which may be the case at some of our climatically mid-range sites on Late Quaternary time scales. In the SE highlands, for example, it is likely that soil locally was stripped during Late Quaternary glacial periods by periglacial processes (Heimsath et al. 2001a) that were active in the SE highlands, in which case erosion rates calculated using present-day soil thicknesses would be minimum estimates. Conversely, if sites that today are bare rock were blanketed by aeolian dust or sand in Late Quaternary times, as Williams et al. (2001) suggested for parts of the Flinders Ranges, then the calculated erosion rates will be maximum estimates.

We turn to the question of whether the observed erosional processes, acting at the measured rates, have generated the landforms observed today. This is strongly supported by our detailed studies at the soil-mantled lowland sites in humid southeastern Australia, where erosion rates are relatively rapid (Heimsath et al. 2000, 2002) and arguably is the case at the monsoonal tropical TCC site, but seems unlikely at the central Australian sites that have very low erosion rates. With a rate of 1–2 m Ma^{-1}, surface lowering since Late Miocene times would amount to only 10–20 m, which is small compared with the scale of the major spurs at Mt Sonder, for example. Surface conditions have changed over the last several million years at all our sites in the semi-arid region (as reviewed by Fujioka & Chappell 2010). In the Miocene, continuous forest cover throughout Australia gave way in the central regions to sclerophyll woodland and savannah; alkaline lakes developed accompanied by dissection and alluvial fan activity, until, from Late Pliocene to mid-Pleistocene times, soil-covered plains were transformed to stony deserts and, as aridity deepened, playas and extensive dunefields came into being. Thus, we suggest that the convex-up spurs, widespread throughout the MacDonnell Ranges and still soil-mantled today at Tyler Pass, evolved under soil cover that was lost as regional aridity deepened, perhaps in early or mid-Pleistocene times.

Conclusions

We examined a wide range of moderate to steep upland landscapes across Australia to quantify erosion rates and processes and to evaluate their dependence on climate. At first order, the slopes are either bedrock-dominated, weathering-limited surfaces or soil-mantled, transport-limited surfaces. We found robust support for the exponential decline of soil production rates with increasing soil thickness across the passive margin exposed in the Bega Valley in the humid SE. A soil-mantled site within the Arnhem escarpment in tropical monsoonal northern Australia showed a similar relationship with similar rates of soil production, except that the rate apparently declines at minimal soil thickness, suggesting a 'humped' soil production function (Heimsath et al. 2009).

Results from bedrock-dominated landscapes in semi-arid central Australia, mostly in hard sandstone, and a soil-mantled area at Tyler Pass, include very low erosion rates of less than 2 m Ma^{-1} on long, steep slopes at Mt Sonder and Kings Canyon, but range to c. 7 m Ma^{-1} on both bedrock spurs at Glen Helen and soil-mantled spurs at Tyler Pass. Comparable results were obtained from steep, rocky slopes in the semi-arid Flinders Ranges in southern Australia. Such slow rates of erosion suggest that the ridge–valley topography observed today was probably shaped under dramatically different climatic conditions, potentially during the late Miocene or early Pliocene.

The suite of results from different field sites indicates that erosion rates generally increase with increasing precipitation and decreasing temperature. However, to more completely quantify the potential relationship between erosion and climate we must include chemical weathering processes and a more rigorous quantification of palaeoclimates across the field sites. This compilation of results across Australia also emphasizes the robustness of cosmogenic nuclides to quantify millennial-scale erosion rates. Although these rates, and the different processes quantified across such different landscapes, offer a striking comparison of undisturbed landscapes, they do not capture the impact of recent anthropogenic land use. The rates do, however, provide critical parameters to help drive landscape evolution models seeking to explain the long-term evolution of the Earth's surface.

Cosmogenic nuclide measurements and field support were funded by The Research School of Earth Sciences (ANU), the US National Science Foundation (NSF) with postdoctoral support and grants DEB-0128995 and EAR-0239655 to A.M.H. Results from Arnhem Land were generated with D. Fink at the Australian Nuclear Science and Technology Organisation (ANSTO) and G. Hancock at the Department of Civil, Surveying and Environmental

Engineering, University of Newcastle. S. Selkirk used her magic wand to turn illegible 'blue line' 1:100 000 maps into the field maps for the MacDonnell Ranges field sites, and M. Zoldak made the regional topographic map. P. Bishop and an anonymous reviewer helped improve the manuscript with their comments.

References

BAGAS, L. 1988. *Geology of Kings Canyon National Park.* Northern Territory Department of Mines and Energy, Report, **4**.

BELTON, D. X., BROWN, R. W., KOHN, B. P., FINK, D. & FARLEY, K. A. 2004. Quantitative resolution of the debate over antiquity of the central Australian landscape: implications for the tectonic and geomorphic stability of cratonic interiors. *Earth and Planetary Science Letters*, **219**, 21–34.

BIERMAN, P. R. 1994. Using *in situ* produced cosmogenic isotopes to estimate rates of landscape evolution; a review from the geomorphic perspective. *Journal of Geophysical Research, B, Solid Earth and Planets*, **99**, 13 885–13 896.

BIERMAN, P. R. & CAFFEE, M. 2002. Cosmogenic exposure and erosion history of Australian bedrock landforms. *Geological Society of America Bulletin*, **114**, 787–803.

BIERMAN, P. & STEIG, E. J. 1996. Estimating rates of denudation using cosmogenic isotope abundances in sediment. *Earth Surface Processes and Landforms*, **21**, 125–139.

BRAUN, J., HEIMSATH, A. M. & CHAPPELL, J. 2001. Sediment transport mechanisms on soil-mantled hillslopes. *Geology*, **29**, 683–686.

BURKE, B. C., HEIMSATH, A. M., CHAPPELL, J. & YOO, K. 2009. Weathering the escarpment: chemical and physical rates and processes, southeastern Australia. *Earth Surface Processes and Landforms*, doi: 10.1002/esp.1764.

COCKBURN, H. A. P. & SUMMERFIELD, M. A. 2004. Geomorphological applications of cosmogenic isotope analysis. *Progress in Physical Geography*, **28**, 1–42.

DIETRICH, W. E., REISS, R., HSU, M.-L. & MONTGOMERY, D. R. 1995. A process-based model for colluvial soil depth and shallow landsliding using digital elevation data. *Hydrological Processes*, **9**, 383–400.

FUJIOKA, T. & CHAPPELL, J. 2010. History of Australian aridity: chronology in the evolution of arid landscapes. *In*: BISHOP, P. & PILLANS, B. (eds) *Australian Landscapes*. Geological Society, London, Special Publications, **346**, 121–139.

GOSSE, J. C. & PHILLIPS, F. M. 2001. Terrestrial *in situ* cosmogenic nuclides: theory and application. *Quaternary Science Reviews*, **20**, 1475–1560.

GRANGER, D. E., KIRCHNER, J. W. & FINKEL, R. 1996. Spatially averaged long-term erosion rates measured from *in situ*-produced cosmogenic nuclides in alluvial sediment. *Journal of Geology*, **104**, 249–257.

HANCOCK, G. R. & EVANS, K. G. 2006. Gully position, characteristics and geomorphic thresholds in an undisturbed catchment in northern Australia. *Hydrological Processes*, **20**, 2935–2951.

HANCOCK, G. R., WILLGOOSE, G. R., EVANS, K. G., MOLIERE, D. R. & SAYNOR, M. J. 2000. Medium term erosion simulation of an abandoned mine site using the SIBERIA landscape evolution model. *Australian Journal of Soil Research*, **38**, 249–263.

HANCOCK, G. R., LOUGHRAN, R. J., EVANS, K. G. & BALOG, R. M. 2008*a*. Estimation of soil erosion using field and modelling approaches in an undisturbed Arnhem Land catchment, Northern Territory, Australia. *Geographical Research*, **46**, 333–349.

HANCOCK, G. R., LOWRY, J. B. C., MOLIERE, D. R. & EVANS, K. G. 2008*b*. An evaluation of an enhanced soil erosion and landscape evolution model: a case study assessment of the former Nabarlek uranium mine, Northern Territory, Australia. *Earth Surface Processes and Landforms*, **33**, 2045–2063.

HAYES, D. E. & RINGIS, J. 1973. Seafloor spreading in the Tasman Sea. *Nature*, **243**, 454–458.

HEIMSATH, A. M., DIETRICH, W. E., NISHIIZUMI, K. & FINKEL, R. C. 1997. The soil production function and landscape equilibrium. *Nature*, **388**, 358–361.

HEIMSATH, A. M., DIETRICH, W. E., NISHIIZUMI, K. & FINKEL, R. C. 1999. Cosmogenic nuclides, topography, and the spatial variation of soil depth. *Geomorphology*, **27**, 151–172.

HEIMSATH, A. M., CHAPPELL, J., DIETRICH, W. E., NISHIIZUMI, K. & FINKEL, R. C. 2000. Soil production on a retreating escarpment in southeastern Australia. *Geology*, **28**, 787–790.

HEIMSATH, A. M., CHAPPELL, J., DIETRICH, W. E., NISHIIZUMI, K. & FINKEL, R. C. 2001*a*. Late Quaternary erosion in southeastern Australia: a field example using cosmogenic nuclides. *Quaternary International*, **83–85**, 169–185.

HEIMSATH, A. M., DIETRICH, W. E., NISHIIZUMI, K. & FINKEL, R. C. 2001*b*. Stochastic processes of soil production and transport: erosion rates, topographic variation, and cosmogenic nuclides in the Oregon Coast Range. *Earth Surface Processes and Landforms*, **26**, 531–552.

HEIMSATH, A. M., CHAPPELL, J., SPOONER, N. A. & QUESTIAUX, D. G. 2002. Creeping soil. *Geology*, **30**, 111–114.

HEIMSATH, A. M., FURBISH, D. J. & DIETRICH, W. E. 2005. The illusion of diffusion: field evidence for depth-dependent sediment transport. *Geology*, **33**, 949–952.

HEIMSATH, A. M., CHAPPELL, J., FINKEL, R. C., FIFIELD, L. K. & ALIMANOVIC, A. 2006. Escarpment erosion and landscape evolution in southeastern Australia. *In*: WILLETT, S. D., HOVIUS, N., BRANDON, M. T. & FISHER, D. M. (eds) *Tectonics, Climate and Landscape Evolution*. Geological Society of America, Special Papers, **398**, 173–190, doi: 10.1130/2006.2398(10).

HEIMSATH, A. M., FINK, D. & HANCOCK, G. R. 2009. The 'humped' soil production function: eroding Arnhem Land, Australia. *Earth Surface Processes and Landforms*, doi: 10.1002/esp.1859.

KASTE, J., HEIMSATH, A. M. & BOSTICK, B. C. 2007. Short-term soil mixing quantified with fallout radionuclides. *Geology*, **35**, 243–246.

KOHN, B. P., GLEADOW, A. J. W., BROWN, R. W., GALLAGHER, K., O'SULLIVAN, P. B. & FOSTER, D. A.

2002. Shaping the Australian crust over the last 300 million years: insights from fission track thermotectonic imaging and denudation studies of key terranes. *Australian Journal of Earth Sciences*, **49**, 697–717.

LAL, D. 1991. Cosmic ray labeling of erosion surfaces: *in situ* nuclide production rates and erosion models. *Earth and Planetary Science Letters*, **104**, 424–439.

LAMBECK, K. & STEPHENSON, R. 1985. Post-orogenic evolution of a mountain range; south-eastern Australian highlands. *Geophysical Research Letters*, **12**, 801–804.

NISHIIZUMI, K., KOHL, C. P. *ET AL.* 1993. Role of *in situ* cosmogenic nuclides ^{10}Be and ^{26}Al in the study of diverse geomorphic processes. *Earth Surface Processes and Landforms*, **18**, 407–425.

PERSANO, C., STUART, F. M., BISHOP, P. & BARFOD, D. N. 2002. Apatite (U–Th)/He age constraints on the development of the great escarpment on the southeastern Australian passive margin. *Earth and Planetary Science Letters*, **200**, 79–90.

PERSANO, C., STUART, F. M., BISHOP, P. & DEMPSTER, T. J. 2005. Deciphering continental breakup in eastern Australia using low-temperature thermochronometers. *Journal of Geophysical Research*, **110**, B12405, doi: 10.1029/2004JB003325.

PILLANS, B. 2007. Pre-Quaternary landscape inheritance in Australia. *Journal of Quaternary Science*, **22**, 439–447.

QUIGLEY, M., SANDIFORD, M., FIFIELD, K. & ALIMANOVIC, A. 2007*a*. Bedrock erosion and relief production in the northern Flinders Ranges, Australia. *Earth Surface Processes and Landforms*, **32**, 929–944.

QUIGLEY, M., SANDIFORD, M., FIFIELD, K. & ALIMANOVIC, A. 2007*b*. Landscape responses to intraplate tectonism: quantitative constraints from ^{10}Be nuclide abundances. *Earth and Planetary Science Letters*, **261**, 120–133.

REINHARDT, L. J., HOEY, T. B., BARROWS, T. T., DEMPSTER, T. J., BISHOP, P. & FIFIELD, L. K. 2007. Interpreting erosion rates from cosmogenic radionuclide concentrations measured in rapidly eroding terrain. *Earth Surface Processes and Landforms*, **32**, 390–406.

RICHARDSON, S. J. 1976. *Geology of the Michelago 1:100 000 sheet 8726*. Geological Survey of New South Wales, Department of Mineral Resources and Development, Canberra.

RILEY, S. J. & WILLIAMS, D. K. 1991. Thresholds of gullying, Arnhem Land, Northern Territory. *Malaysian Journal of Tropical Agriculture*, **22**, 133–143.

ROERING, J. J., KIRCHNER, J. W. & DIETRICH, W. E. 2001. Hillslope evolution by nonlinear, slope-dependent transport: steady state morphology and equilibrium

adjustment timescales. *Journal of Geophysical Research—Solid Earth*, **106**, 16 499–16 513.

SAYNOR, M. J., ERSKINE, W. D., EVANS, K. G. & ELIOT, I. 2004. Gully initiation and implications for management of scour holes in the vicinity of Jabiluka mine, Northern Territory, Australia. *Geografiska Annaler Series A—Physical Geography*, **86**, 191–203.

SAYNOR, M. J., STABEN, G. *ET AL.* 2009. *Impact of Cyclone Monica on catchments within the Alligator Rivers Region. Field Survey Data.* Supervising Scientist, Darwin, Internal Report, **557**.

STABEN, G. & EVANS, K. G. 2008. Estimates of tree canopy loss as a result of Cyclone Monica, in the Magela Creek catchment, northern Australia. *Austral Ecology*, **33**, 562–569.

STEWART, A. J., BLAKE, D. H. & OLLIER, C. D. 1986. Cambrian river terraces and ridgetops in central Australia: oldest peristing landforms? *Science*, **233**, 758–761.

TOWNSEND, S. A. & DOUGLAS, D. D. 2000. The effect of three fire regimes on stream water quality, water yield and export coefficients in a tropical savanna (northern Australia). *Journal of Hydrology*, **229**, 118–137.

VASCONCELOS, P. M., KNESEL, K. M., COHEN, B. E. & HEIM, J. A. 2008. Geochronology of the Australian Cenozoic: a history of tectonic and igneous activity, weathering, erosion, and sedimentation. *Australian Journal of Earth Sciences*, **55**, 865–914.

WEISSEL, J. K. & SEIDL, M. A. 1997. Influence of rock strength properties on escarpment retreat across passive continental margins. *Geology*, **25**, 631–634.

WEISSEL, J. K. & SEIDL, M. A. 1998. Inland propagation of erosional escarpments and river profile evolution across the southeast Australian passive continental margin. *In*: TINKLER, K. J. & WOHL, E. E. (eds) *Rivers Over Rock: Fluvial Processes in Bedrock Channels*. Geophysical Monograph, American Geophysical Union, **107**, 189–206.

WELLS, T., WILLGOOSE, G. R. & HANCOCK, G. R. 2008. Modeling weathering pathways and processes of the fragmentation of salt weathered quartz–chlorite schist. *Journal of Geophysical Research—Earth Surface*, **113**, F01014, doi: 10.1029/2006JF000714.

WILKINSON, M. T., CHAPPELL, J., HUMPHREYS, G. S., FIFIELD, K., SMITH, B. & HESSE, P. 2005. Soil production in heath and forest, Blue Mountains, Australia: influence of lithology and palaeoclimate. *Earth Surface Processes and Landforms*, **30**, 923–934.

WILLIAMS, M., PRESCOTT, J., CHAPPELL, J., ADAMSON, D., COCK, B., WALKER, K. & GELL, P. 2001. The enigma of a late Pleistocene wetland in the Flinders Ranges, South Australia. *Quaternary International*, **83–85**, 129–144.

Tectonic geomorphology of Australia

MARK C. QUIGLEY[1]*, DAN CLARK[2] & MIKE SANDIFORD[3]

[1]*Department of Geological Sciences, University of Canterbury, Christchurch 8014, New Zealand*

[2]*Geoscience Australia, GPO Box 378, Canberra, ACT, 2601, Australia*

[3]*School of Earth Sciences, University of Melbourne, Melbourne, Victoria 3010, Australia*

*Corresponding author (e-mail: mark.quigley@canterbury.ac.nz)

Abstract: The Australian continent is actively deforming in response to far-field stresses generated by plate boundary interactions and buoyancy forces associated with mantle dynamics. On the largest scale (several 10^3 km), the submergence of the northern continental shelf is driven by dynamic topography caused by mantle downwelling along the Indo-Pacific subduction system and accentuated by a regionally elevated geoid. The emergence of the southern shelf is attributed to the progressive movement of Australia away from a dynamic topography low. On the intermediate scale (several 10^2 km), low-amplitude (*c.* 100–200 m) long-wavelength (*c.* 100–300 km) topographic undulations are driven by (1) anomalous, smaller-scale upper mantle convection, and/or (2) lithospheric-scale buckling associated with plate boundary tectonic forcing. On the smallest scale (10^1 km), fault-related deformation driven by partitioning of far-field stresses has modified surface topography at rates of up to *c.* 170 m Ma^{-1}, generated more than 30–50% of the contemporary topographic relief between some of Australia's highlands and adjacent piedmonts, and exerted a first-order control on long-term (10^4–10^6 a) bedrock erosion. Although Australia is often regarded as tectonically and geomorphologically quiescent, Neogene to Recent tectonically induced landscape evolution has occurred across the continent, with geomorphological expressions ranging from mild to dramatic.

Australia is one of the lowest, flattest, most arid, and most slowly eroding continents on Earth. The average elevation of the continent is only *c.* 330 m above sea level (asl), maximum local topographic relief is everywhere <1500 m (defined by elevation ranges with 100 km radii) and two-thirds of the continent is semi-arid to arid. With the exception of localized upland areas in the Flinders and Mt Lofty Ranges and the Eastern Highlands, bedrock erosion rates are typically ≤ 1–10 m Ma^{-1} (Wellman & McDougall 1974; Bishop 1984, 1985; Young & MacDougall 1993; Bierman & Caffee 2002; Belton *et al.* 2004; Chappell 2006). Despite this apparent geomorphological longevity, Australia has had a dynamic Neogene to Recent tectonic history. Australia has migrated more than 3000 km to the NNE at a rate of 6–7 cm a^{-1} over the past 45 Ma as part of the Indo-Australian Plate (Fig. 1) (DeMets *et al.* 1990, 1994; Bock *et al.* 2003), making it the fastest moving continent since the Eocene (Sandiford 2007). Australia experiences a relatively high level of seismicity for a supposedly 'stable' intraplate continental region (Johnston *et al.* 1994; Johnston 1996) and has a rich record of Neogene and Quaternary faulting (e.g. Andrews 1910; Fenner 1930, 1931; Miles 1952; Beavis 1960; Hills 1961; Sandiford 2003*b*; Quigley *et al.* 2006; Hillis *et al.* 2008). Thus,

although large parts of the Australian landscape are likely to be ancient (e.g. Twidale 1994, 2000) features reflecting recent landscape rejuvenation in response to Neogene to Recent plate motion and associated intraplate deformation are widespread (Sandiford 2003*a, b*; Sandiford *et al.* 2009). Quantitative landscape analysis indicates that even some apparently ancient landscapes have undergone kilometre-scale burial and denudation through Phanerozoic in the arid continental interior (Belton *et al.* 2004).

In this paper, we reveal how intraplate tectonism and mantle processes have contributed to the Neogene to Recent geomorphological evolution of the Australian continent. We summarize geological evidence for distinct modes of deformation and speculate on how each deformation mode has influenced the spatial and temporal evolution of Australia's coastlines, uplands, interior basins and fluvial systems. Although pre-Pliocene fault-related tectonic activity in onshore Australia was widespread in the Tertiary (e.g. Raiber & Webb 2008) we restrict most of our discussion here to faulting during the Pliocene to Recent interval, as these faults are typically considered 'active' in the Australian context (Sandiford 2003*b*). We contend that Neogene to Recent tectonism, particularly in the last *c.* 5 Ma, and mantle processes have exerted an

From: BISHOP, P. & PILLANS, B. (eds) *Australian Landscapes.* Geological Society, London, Special Publications, **346**, 243–265. DOI: 10.1144/SP346.13 0305-8719/10/$15.00 © The Geological Society of London 2010.

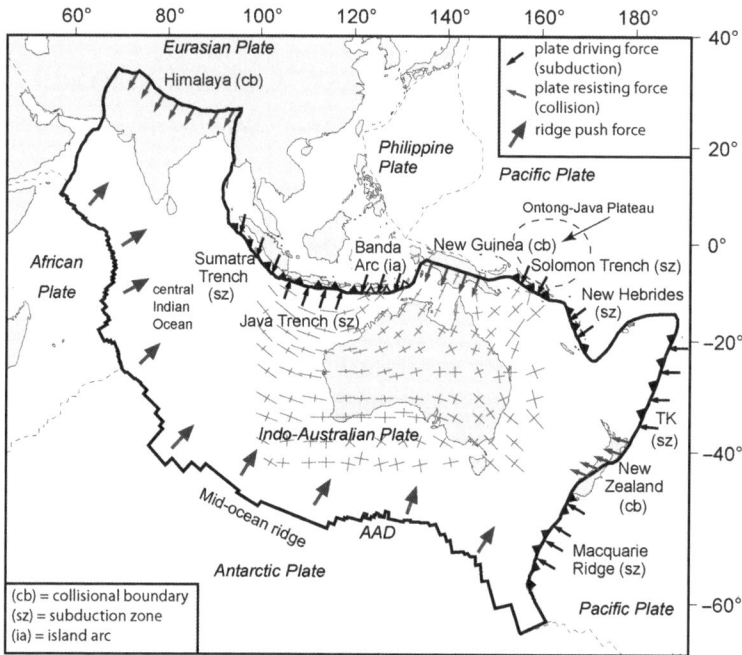

Fig. 1. Indo-Australian Plate with plate boundary forces and orientation of modelled maximum and minimum horizontal stresses used in the finite-element stress modelling of Reynolds *et al.* (2003). Much of the southern part of the continent has an east–west-oriented maximum horizontal compressive stress oriented at a high angle to the NNE-oriented plate velocity vector. Filled triangles along plate boundary indicate the direction of subduction; open triangles delineate the Banda Arc. TK, Tonga–Kermadec Trench; AAD, Australian–Antarctic discordance (from Reynolds *et al.* 2003; Hillis *et al.* 2008).

influence on the geomorphological evolution of the Australian landscape, and must be considered when interpreting contemporary Australian landforms.

The Australian intraplate stress field: characteristics, age and origin

Indo-Australian Plate (IAP) motion is driven principally by the 'pull force' associated with subduction zones in the Indonesia region and is resisted by continent–continent collision in the Himalayan, New Zealand and New Guinea orogens (Fig. 1; Coblentz *et al.* 1995, 1998; Sandiford *et al.* 2004). The Australian *in situ* stress field, as inferred from earthquake focal mechanism solutions, borehole breakouts and fracture arrays (Hillis *et al.* 2008), is unusual for a plate interior in that it is characterized by maximum horizontal compressive stress azimuths (σ_{Hmax}) oriented at a high angle to the NNE-oriented plate velocity vector. σ_{Hmax} varies from roughly east–west in western and south–central Australia, to NE–SW in northern, central and central–east Australia, to NNE–SSW in NE Australia, and to NW–SE in SE Australia (Fig. 1).

The modern *in situ* stress field has been extrapolated back to the terminal Miocene–early Pliocene (10–6 Ma) by comparing σ_{Hmax} trends with palaeo-stress orientations inferred from kinematic investigations of Plio-Quaternary faults (Fig. 2a; Sandiford 2003*b*; Celerier *et al.* 2005; Quigley *et al.* 2006; Green 2007) and structural–stratigraphic relationships in Neogene strata (Fig. 2b, c; Dickinson *et al.* 2001, 2002; Sandiford 2003*b*). For instance, sedimentary basins in SE Australia (e.g. the Gippsland and Otway Basins) underwent significant inversion at *c.* 8–6 Ma (Dickinson *et al.* 2002; Sandiford 2003*b*; Sandiford *et al.* 2004). Reverse faults preserved in the offshore record parallel Plio-Quaternary structures in onshore basins (Sandiford 2003*b*; Sandiford *et al.* 2004), and up to 1 km of stratigraphic section has been locally removed on reverse fault-bounded topographic highs, suggesting that a significant episode of tectonic uplift and accompanying erosion occurred in this time interval (Dickinson *et al.* 2001; Sandiford 2003*b*). Sandiford *et al.* (2004) attributed the NW–SE-oriented σ_{Hmax} in SE Australia to increased intraplate stress levels associated with the increased IAP–Pacific Plate coupling and uplift of the

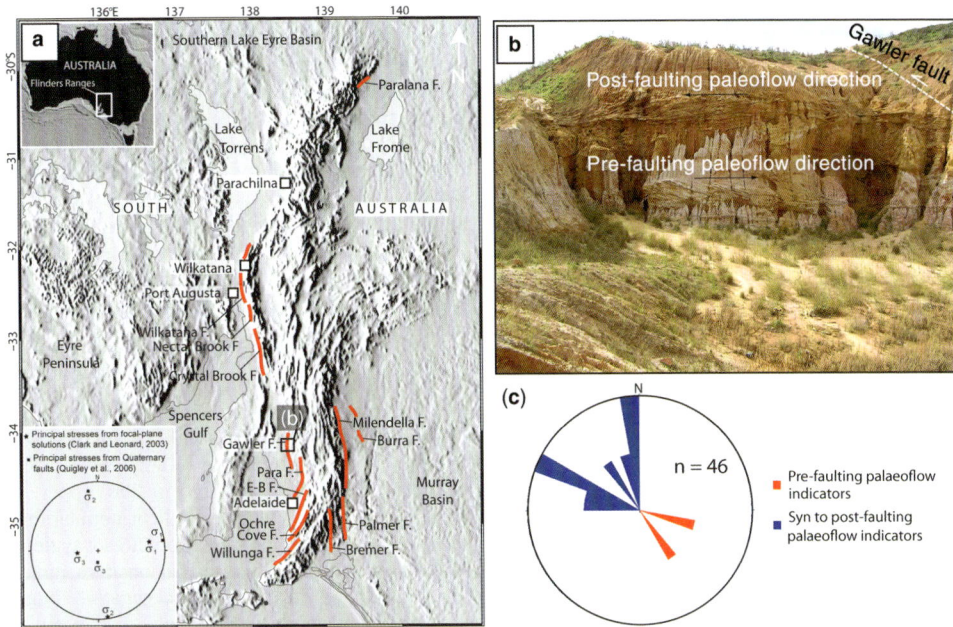

Fig. 2. (**a**) Flinders Ranges DEM with locations of Quaternary faults. E-B F., Eden–Burnside Fault. Further details have been given by Sandiford (2003b). Lower inset shows consistency between principal stresses estimated from earthquake focal mechanisms and those determined from geological studies of Quaternary faults (specifically, the Wilkatana, Burra and Mundi Mundi Faults) (from Quigley *et al.* 2006). (**b**) Cross-sectional view of alluvial sediments in the footwall of the Gawler Fault looking NE, showing a Mid- to Late Miocene reversal in drainage flow direction from towards the fault (SE) to parallel to the fault (north to NE), postulated to be the result of fault activity (from Green 2007). (**c**) Rose diagram showing palaeoflow directions obtained from cross-bed orientations and clast imbrication in the section shown in (b). The change in palaeoflow should be noted, from SE-directed prior to Gawler Fault initiation to NE-directed during or after fault growth, implying drainage reversal as a result of faulting (from Green 2007).

Southern Alps in New Zealand at *c.* 5–10 Ma (Wellman 1979; Batt & Braun 1999).

In central southern Australia, stress conditions consistent with the modern stress field may have been established somewhat earlier. Mid- to late Miocene alluvial sequences west of the Mt Lofty Ranges reveal drainage direction shifts from eastward-directed palaeo-flow towards impending uplifts to northward-directed palaeo-flow adjacent to north–south-oriented reverse fault scarps, implying that tectonic uplift associated with roughly east–west compression caused drainage reorganization by the terminal Miocene (Fig. 2b, c; Green 2007). At Sellicks Beach, subvertical Cambrian sedimentary rocks are unconformably overlain by steeply west-dipping Oligocene sedimentary rocks that are, in turn, overlain by moderately west-dipping Oligo-Miocene sedimentary rocks and, finally, gently west-dipping Pleistocene gravels (Lemon & McGowran 1989). This series of progressive unconformities, ranging in age from possibly Late Eocene or Early Oligocene through the Miocene, Pliocene and Pleistocene suggests that the processes governing tectonic tilting of

these sequences have been active since at least the Oligocene (Lemon & McGowran 1989). The trace of a locally exposed ESE-dipping reverse fault lies west of these outcrops, implying that movement on this fault may have been responsible for the observed structural geometry. Computer modelling of the structural relationships is consistent with formation as a result of fault-propagated folding (Lemon & McGowran 1989) in response to roughly east–west shortening, although the fault geometry is not well constrained. The kinematic consistency between Oligocene(?) deformation and more recent tectonism implies that the same tectonic stresses may have driven these deformation regimes. Deformation is likely to have occurred in discrete pulses as opposed to being continuous (Dyksterhuis & Müller 2008). On the basis of numerical stress modelling, Dyksterhuis & Müller (2008) proposed that periods of roughly east–west compressional tectonism may have affected the Flinders Ranges in the Eocene (beginning at *c.* 55 Ma) and from *c.* 12 Ma to the present, separated by an early to mid-Miocene period of tectonic quiescence.

In NW Australia, transpressional deformation and uplift of parts of the NW Shelf initiated in the interval 11–5.5 Ma, with distinct deformation pulses recognized at *c.* 8 Ma and *c.* 3 Ma (Packham 1996; Keep *et al.* 2000, 2002). Reverse faulting and growth of fault propagation anticlines in the Carnarvon Basin continue to the present, as indicated by the uplift of Pleistocene marine terraces (van de Graaff *et al.* 1976).

In summary, the bulk of structural–stratigraphic evidence suggests that the modern, intraplate stress field was firmly entrenched by the mid- to late Miocene (Dickinson *et al.* 2001; Sandiford *et al.* 2004; Hillis *et al.* 2008), with the possibility that deformation within a regime similar to the modern tectonic regime began as early as the Eocene or Oligocene in south–central Australia (Lemon & McGowran 1989). Plate-scale finite-element modelling of the IAP intraplate stress field replicates σ_{Hmax} trends by balancing plate driving forces (slab-pull and ridge-push), plate resisting forces (compression associated with the Indo-Asian collisional front in the Himalayas, the New Guinea fold-and-thrust belt, and the New Zealand Southern Alps) and tractions induced by mantle flow (Coblentz *et al.* 1995, 1998). Using inferences drawn from stress modelling studies, Sandiford *et al.* (2004) attributed the 10–6 Ma onset of active tectonism in SE Australia to the synchronous development of transpression along the IAP–Pacific Plate boundary segments of southern New Zealand, the Puyseguer Trench and the Macquarie Ridge (Walcott 1998; Massell *et al.* 2000). However, it is also worth noting that the onset of deformation in the central Indian Ocean at around this time has been attributed to increases in stress levels propagated from the Himalayan–Tibet system (Molnar *et al.* 1993; Martinod & Molnar 1995). These results imply that a significant component of the intraplate stress field relates to forces exerted from orogenic belts up to several thousands of kilometres from the plate interior (Coblentz *et al.* 1995, 1998; Reynolds *et al.* 2003; Sandiford *et al.* 2004). Hence, the IAP appears to have responded with increasing intraplate compression to a complex evolving plate boundary scenario over the last 10 Ma, and the onset of faulting at specific intraplate locations probably reflects rising stress levels related to the combination of all plate boundary forcings (e.g. Dyksterhuis & Müller 2008).

The mantle, the geoid, and dynamic topography

The mean shape of the Earth contains long-wavelength (up to 0.5 of Earth's circumference), low-amplitude (*c.* 300 m) deviations from a perfect mathematical ellipsoid that relate to variations in density distribution within the deep mantle. This varying gravitational equipotential surface is called the geoid and is best approximated by global mean sea level. Where a relative density deficiency exists the geoid (sea level) will dip below the mean ellipsoid, and where a relative density surplus exists the geoid (sea level) will rise above the mean ellipsoid.

The term dynamic topography refers to the deflection of the solid surface of the Earth resulting from buoyancy forces associated with thermal convection in the viscous mantle. The wavelength of dynamic topography relates to the depth and scale of convection. Whole-mantle convection is likely to produce undulations on a similar scale to geoid undulations (several thousand kilometres wavelength, ± 500 m amplitude; Kaban *et al.* 2005), whereas shallow convection in the upper sublithospheric mantle would result in undulations of the order of several hundred kilometres wavelength. Mantle upwelling will result in positive dynamic topography and downwelling in negative dynamic topography. Dynamic topography has been used elsewhere to explain large-scale subsidence patterns on continental platforms and in sedimentary basins (Mitrovica *et al.* 1989; Gurnis 1992; Stern *et al.* 1992; Russell & Gurnis 1994; Pysklywec & Mitrovica 1999; Wegmann *et al.* 2007). Modelling studies predict that dynamic topographic undulations may reach several kilometres in amplitude (Lithgow-Bertelloni & Gurnis 1997).

For several reasons, the Australian continent is arguably the best natural laboratory on Earth for investigating the topographic and geomorphological effects of geoid variations and dynamic topography through the Neogene. On its voyage towards an active subduction realm and away from a mid-ocean ridge, Australia has traversed a region of the geoid with a present-day change in height from -20 m in the Southern Ocean to $+80$ m in Melanesia (Sandiford 2007). This voyage has resulted in the extensive preservation of palaeo-shorelines that provide a datum of long-wavelength changes in the position of the continent relative to palaeo-sea level. The Australian coastline is beyond the flexural response wavelength (*c.* 200–300 km) of active plate boundaries, and thus crustal deformation is not driven by plate flexure associated with subduction. Long- to intermediate-wavelength changes in surface topography and geomorphology can thus been interpreted within the context of geoid anomalies and dynamic topography (Sandiford 2007; Sandiford & Egholm 2008; Sandiford *et al.* 2009), although offshore sedimentation may have contributed some component to the subsidence by enhancing crustal flexure (e.g. Pazzaglia & Gardner 1994).

Long-wavelength (10^3 km) deformation

Characteristics

The width of the continental shelf and the distribution of Cenozoic marine and nearshore sediments around the Australian margin are highly asymmetric (Fig. 3). The southern continental shelf has a characteristic width of c. 100 km and ranges from 20 to 200 km wide. Eocene to Quaternary marine deposits

are preserved up to several hundred kilometres inland, and up to 250 m above current sea level (Sandiford 2007). The northern shelf is almost everywhere >200 km wide and is locally as broad as 500 km. Eocene to Quaternary marine deposits are almost entirely situated offshore up to 50 m below sea level (Veevers 2000), indicating that northern shelf subsidence has exceeded the progressive Neogene eustatic sea-level fall of c. 100–150 m (Sandiford 2007). The variations in

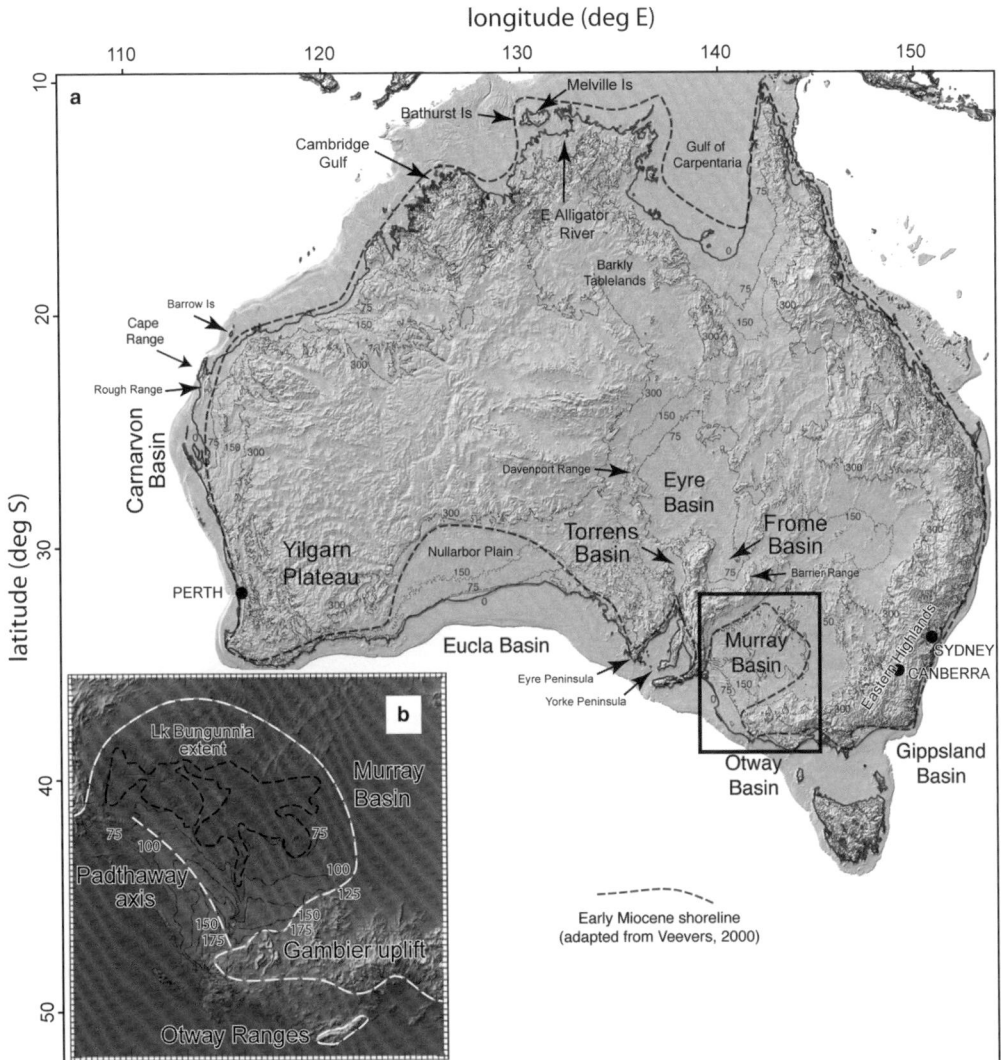

Fig. 3. (**a**) Shaded relief image of the Australian continent and its continental shelf (at elevations greater than −250 m) derived from Geoscience Australia's 9 arc second 'bathytopo' dataset. Contours are shown for 75, 150 and 300 m. The thick dashed line shows the approximate position of Early Miocene shorelines (from Veevers 2000; Sandiford 2007). (**b**) Shaded relief image of the Murray Basin showing the Padthaway and Gambier uplifts, and the former extent of Lake Bungunnia. The 25 m topographic contours across the Padthaway axis are also marked (from Sandiford *et al.* 2009).

shelf thickness and distribution of marine deposits between the northern and southern Australian continental shelves imply a differential vertical displacement between SW and north Australia of *c.* 300 m since the late Eocene. The asymmetric pattern of Eocene to Pliocene shorelines is similarly found in last interglacial shoreline elevations, which tend to be elevated along the southern margin by several metres relative to the northern margin (Murray-Wallace & Belperio 1991; Belperio *et al.* 2002), implying that the forces driving this asymmetry continue to be active.

Long-wavelength variations in marine shoreline elevation are also evident across the southern part of the continent. Early Neogene shorelines decrease eastward from *c.* 250 m asl in the western Eucla Basin to *c.* 100 m asl in the eastern Eucla Basin over a distance of *c.* 1000 km, implying as much as *c.* 150 m of post mid-Neogene (<15 Ma), long-wavelength differential vertical displacement (Figs 3 & 4). This implies an uplift of *c.* 100–150 m for the western Eucla Basin when corrected for Neogene eustatic sea levels. Pliocene marine sequences in the Eucla Basin are restricted to

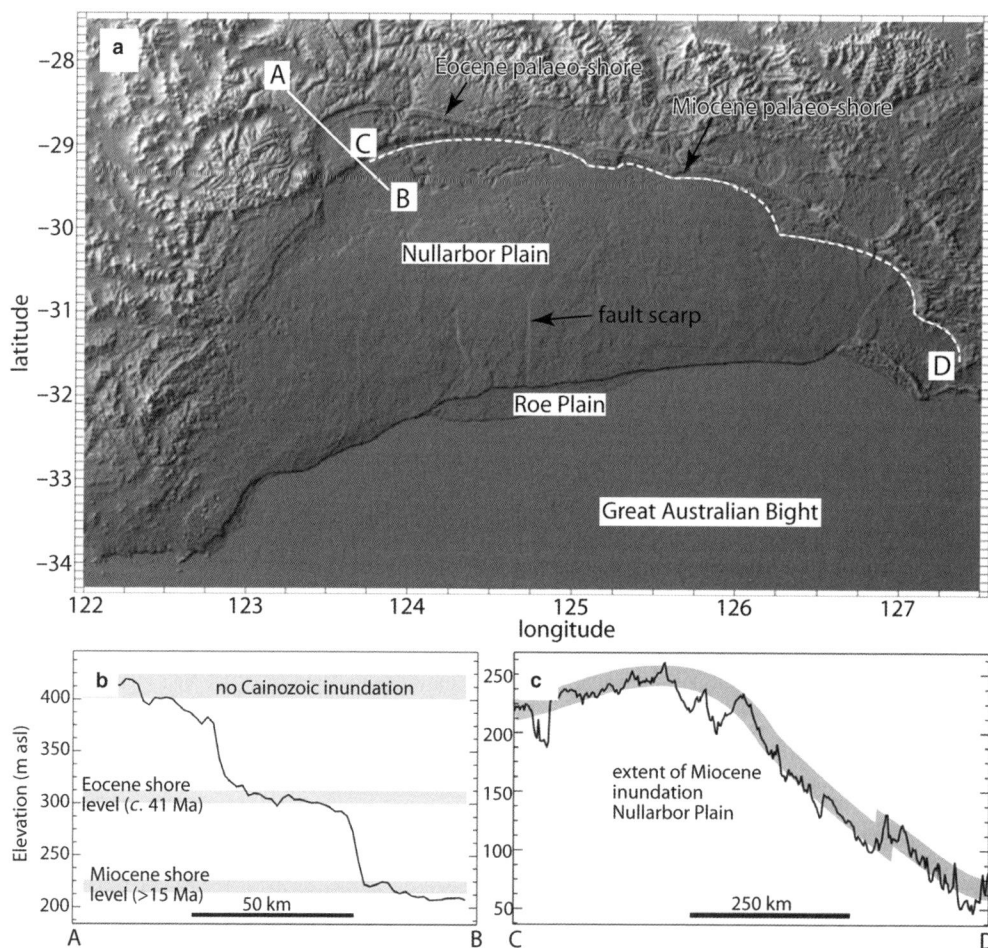

Fig. 4. (**a**) Shaded relief of the Nullarbor Plain showing various palaeo-shoreline features of Eocene to mid-Miocene age (>15 Ma) along its northern margin. The Roe Plain is a Pliocene surface. North–south-trending faults crossing the Nullarbor Plain have cumulative throws of up to 30 m (from Sandiford *et al.* 2009). (**b**) Section A–B normal to the palaeo-shorelines in the northwestern part of the Eucla Basin showing inundation levels. (**c**) Section C–D parallel to the northern margin of the Nullarbor Plain showing the interpreted limit of Miocene inundation. The lowest frequency component (indicated by the grey band), at wavelengths of order 103 km, implies a deep mantle origin related to dynamic topography.

elevations of ≤40 m asl and are nowhere present on the surface of the Nullarbor Plain, implying that post-Miocene sea levels never encroached onto land surfaces above c. 70 m asl. In the central Murray Basin, Miocene and Pliocene strand-lines are at equivalent elevations (c. 70 m asl), implying minimal Mio-Pliocene basin uplift. The consistency between Pliocene strandline elevations between these regions suggests that the western Eucla is uplifted by c. 180 m relative to the Murray Basin, and most of this uplift is constrained to the 15–5 Ma interval.

Origin and rates

The apparent sinking of the northern part of the Australian continent inferred from the absence of onshore Cenozoic shorelines and record of stratal onlap implies that the northern margin has subsided at higher rates than the long-term, Neogene eustatic sea-level fall of c. 100–150 m (Sandiford 2007). This subsidence reflects the progressive transport of the Australian continent towards a geoid high and dynamic topography low associated with the active subduction realm to the north. Separating the geoid from dynamic topography fields is challenging; however, if one assumes that the present-day geoid provides an adequate representation of the geoid in the past, then the northern margin of the continent has traversed a geoid gradient of c. 30–40 m over the last 15 Ma (Sandiford 2007), consistent with an instantaneous rate of change in geoid height along the northern continental margin of c. $+2$ m Ma^{-1} (Sandiford & Quigley 2009). The geoid effect accounts for c. 20–40% of the c. 100–150 m of total subsidence, leaving a remaining signal of c. 60–110 m to be accounted for by the dynamic topographic effect associated with downward tilting of the continental shelf in the direction of plate motion plus any effects of subsidence caused by sediment loading.

The uplift of the SW part of the continent is thought to relate to the progressive northward movement of the continent away from a dynamic topography low (Sandiford 2007; Sandiford & Quigley 2009). The dynamic topography low south of Australia is indicated by the anomalously low elevation of the mid-ocean ridge along the Australian–Antarctic discordance (AAD) (Fig. 1), attributed to the presence of a relict of a former slab residing above the mantle transition zone (Gurnis et al. 1998). Movement of the continent away from this anomalous mantle is responsible for a dynamic topographic uplift of up to c. 200 m in amplitude between 15 and 5 Ma, equivalent to a rate of c. 20 m Ma^{-1} (Sandiford 2007; Sandiford et al. 2009). The instantaneous rate of change in geoid height along the southern continental margin

varies between -0.2 and $+1$ m Ma^{-1}, with the lowest rates in the SW part of the continent (Sandiford & Quigley 2009).

The combined effect of northern shelf submergence and southwestern shelf emergence has resulted in a c. 300 m differential continental 'tilting' over the last 15 Ma at a rate of c. 10–20 m Ma^{-1} (Sandiford 2007). Although the rate of dynamic topographic forcing is several orders of magnitude lower than maximum eustatic sea-level changes, the finite amplitude of dynamic topography is larger than the eustatic variation, at least since the Neogene. In this respect, Australia's tectonic voyage across geoid undulations and changes in the dynamic topography may be as important as climate-driven sea-level fluctuations when considering the long-term geomorphological evolution of the continent.

Geomorphological implications

The influence of persistent downward displacement of the northern Australian land surface has resulted in low stream gradients (<1 m per 10 km) in major river systems draining to the north (e.g. Alligator River, NT, Fig. 3). Nearshore surface elevations are generally restricted to <50 m within 30 km from the northern coastline. Offshore bathymetry between Australia and Papua New Guinea is almost entirely <100 m deep and is punctuated by numerous large offshore islands (Melville, Bathurst; Fig. 3).

Upward displacement of the SSW part of Australia has also affected stream development. The formation of cave systems in the Nullarbor may relate in part to dynamic topographic uplift as ground water tables lowered to keep pace with regional base-level lowering (Webb & James 2006). The formation of early Middle Eocene stranded inset valleys (palaeochannels) on the Yilgarn Plateau (Fig. 3) has been attributed to lowered geomorphological base level and increased stream gradients associated with slow surface uplift (de Broeckert & Sandiford 2005). Longitudinal profiles of west-flowing streams, including the Avon River NE of Perth (Fig. 5), show convex-up 'disequilibrium' longitudinal profiles, implying that incision has been outpaced by relative base-level fall (Fig. 5). It is highly improbable that these profiles relate to escarpment formation, which took place in the Late Eocene (Beard 2003). Late Quaternary incision rates derived from cosmogenic ^{10}Be and ^{26}Al dating of bedrock surfaces at the base of the Avon River (c. 3.5 m Ma^{-1}) are slower than estimated dynamic topography uplift rates of 10–20 m Ma^{-1} (see above), implying that the formation and maintenance of steepened stream longitudinal profiles may be attributed to upstream

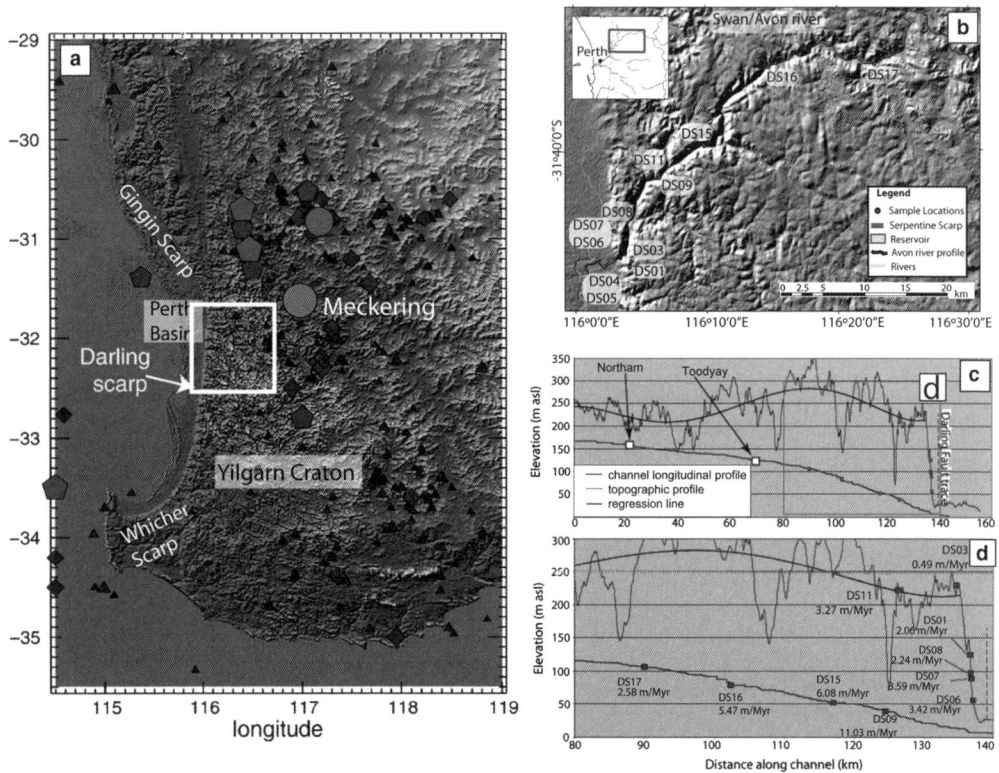

Fig. 5. (a) Seismicity of western Australia, showing absence of earthquakes along the Darling escarpment.
(b) Location of cosmogenic nuclide samples (DS01–17) along the Avon river section. (c) Longitudinal profile
of the river showing convex-up 'disequilibrium' profile that deviates from a concave-up graded equilibrium profile.
(d) Enlargement of box in (c) showing sample locations and erosion rates. The decay of erosion rate estimates in the
Avon River away from the range front should be noted (from Jakica et al. 2010).

propagation of knickpoints formed during slow, regional base-level lowering associated with continental-scale tilting (Jakica et al. 2010). In contrast to the northern Australian coastline, sea cliffs are abundant and often result in nearshore elevations of >50 m in height (Nullarbor Plain). Offshore islands are small and rare, and bathymetry increases steeply away from the coast. Topographic effects associated with long-wavelength deformation may thus be responsible for shaping several features of the contemporary landscape, including coastline geomorphology and stream profiles.

Intermediate-wavelength (10^2–10^3 km) deformation

Characteristics

Regions of intermediate-wavelength (100–1000 km), low-amplitude (100–200 m) topographic undulations have developed in several parts of the

continent and are distinct from the long-wavelength signal described above. The southern Australian volcanic field (Joyce 1975) is associated with a zone of subdued regional uplift, with a characteristic uplift of c. 60 m, maximum uplift of c. 240 m, and wavelength of c. 100 km evidenced by warping of Pliocene and Quaternary beach ridge systems (Wallace et al. 2005; Sandiford et al. 2009). This zone of uplift can be traced for c. 400 km west from Melbourne, where it bifurcates into the Padthaway and Gambier uplifts (Fig. 3b).

The region encompassing the Flinders Ranges and adjacent basins (Frome, Torrens; Fig. 3) also displays evidence for intermediate-wavelength deformation (Celerier et al. 2005; Quigley et al. 2007c). Plio-Quaternary alluvial fans on the eastern and northwestern side of the ranges are commonly uplifted and incised proximal to the range front (Coats 1973; Sandiford 2003b), with the fan-breaching streams emptying 30–50 km outboard of the range front in the Frome and Torrens basins (Figs 3 & 6). On the eastern side of the ranges,

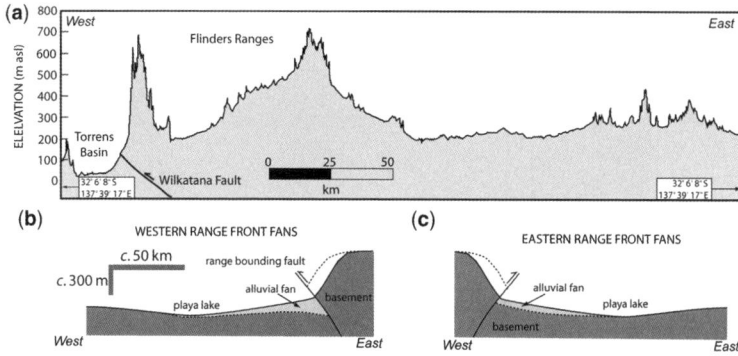

Fig. 6. (a) East–west topographic cross-section across the central Flinders Ranges. Length of section is 275 km. (b) Schematic cross-section of the western range front, showing geometry of basement–alluvium interface. Base of fans dips gently away from ranges distal to range front and towards ranges close to range front, indicative of flexural subsidence in response to loading superimposed on broader scale domal uplift. (c) Schematic cross-section of the eastern range front. Basement–alluvium interface dips gently away from range front, a geometry that Cèlérier *et al.* (2005) attributed to long-wavelength flexural buckling of the lithosphere in the eastern Flinders Ranges (from Quigley *et al.* 2007c).

and parts of the western side of the ranges (e.g. Parachilna), the underlying basement–alluvium interface dips gently away from the ranges towards the basins, inconsistent with the architecture of flexurally controlled footwall basins, where the unconformity between the alluvial fans and the underlying basement typically dips towards the ranges (Celerier *et al.* 2005). The regional distribution of basinward-dipping top-of-basement surfaces implicates a component of low-amplitude (100–200 m), intermediate-wavelength (c. 200 km) regional deformation (Celerier *et al.* 2005; Quigley *et al.* 2007c). In other parts of the western side of the ranges (e.g. Wilkatana), the unconformity between the alluvial fans and the underlying basement dips towards the ranges (Quigley *et al.* 2007c). This implies that the footwall architecture varies along the western range front as a result of differing amounts of flexural footwall subsidence that probably relates to differences in sediment yield and range front faulting histories (Quigley *et al.* 2007c). The Torrens Basin, which is currently separated from the sea in Spencer Gulf by a sill of only c. 30–40 m asl, contains no evidence for marine incursions in the form of palaeo-shorelines or marine sediment (Johns 1968). Pliocene sea-level highstands should have encroached into these low-lying regions if the present surface topography was static since the Pliocene, implying that absolute subsidence of the order of several tens of metres must have occurred in this region in the late Neogene following Miocene to Pliocene eustatic sea-level highs (Quigley *et al.* 2007c).

Intermediate-wavelength topographic development is also evident in the Lake Eyre region. The Eyre basin, with a present-day minimum elevation

of 10 m below sea level (bsl) is confined by a sill at c. 80 m asl from the Torrens Basin to the south (Fig. 7). This drainage divide is marked by arcuate strandline deposits of the large, late Miocene palaeo-lake Billa Kalina (Callen & Cowley 1998). At its maximum size of c. 15 000 km^2 this lake would seem to require a catchment that included much of the present Eyre and Torrens basins. The difference in elevation between the base of the Billa Kalina deposits (c. 120 m asl) and Lake Eyre (c. 15 m asl) implies a post-late Miocene topographic inversion of c. 135 m with a wavelength of c. 250 km (Sandiford *et al.* 2009). The Eyre Basin is fringed to the west by the Davenport Ranges. Intermediate-wavelength (c. 50 km) gentle folding is indicated by subtle undulation (c. 200 m) of the top-of-basement surface, currently expressed as an exhumed, warped and incised fossil planation surface forming the Mt Margaret Plateau on the top of the Davenport Ranges (Wopfner 1968; Johnson 2004). The plateau surface contains Cretaceous gravel lags at elevations of 400 m asl within the Davenport ranges that correlate with rocks exposed at elevations of <150 m asl along the eastern range front, indicating >250 m of post-Cretaceous uplift. Wopfner (1968) attributed this to Quaternary fault-related uplift (see next section); however, the broad wavelength of the doming suggests that it may relate to gentle folding.

Origin and rates

Although a few faults appear to displace basalts of the Newer Volcanics (e.g. the Rowsley Fault), significant episodes of faulting between c. 6–4 Ma (Paine *et al.* 2004) and 2–1 Ma (Sandiford 2003a)

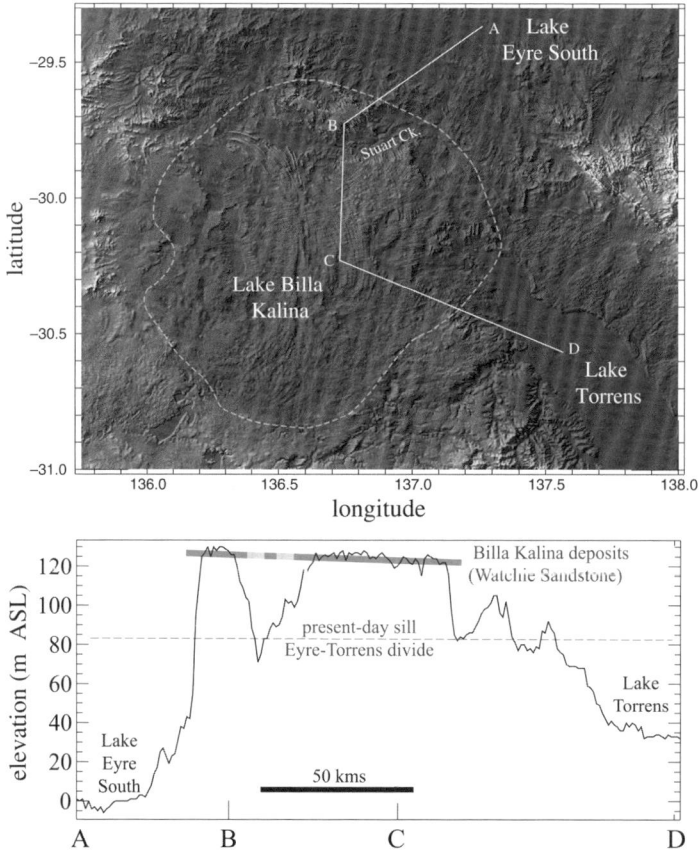

Fig. 7. (**a**) Shaded relief image of Billa Kalina Basin, South Australia, showing distinctive arcuate palaeo-shoreline features of Lake Billa Kalina of probable Miocene age. The inferred approximate maximum extent of the lake is indicated by the dashed line, based on limits of relict palaeo-shorelines. (**b**) Topographic profile along line A–B–C–D. It should be noted that Lake Billa Kalina is now perched on the present drainage divide between the Torrens and Eyre Basins, implying significant topographic inversion since the lake formation (from Sandiford *et al.* 2009).

do not appear to have significantly affected the southern Australian volcanic field. The Padthaway and Gambier uplifts are not associated with any significant faulting and the Padthaway uplift axis is almost parallel to the regional NW–SE compression direction defined by *in situ* stresses (Fig. 1). These observations imply that uplift is unlikely to be associated with contractional tectonic faulting or buckling at these locations. Known neotectonic faults from the Murray Basin (e.g. the Morgan, Hamley and Danyo Faults, and faults underlying the Neckarboo and Iona Ridges) universally strike towards the north and NE (Sandiford 2003*b*). The coincidence of mild yet regionally extensive basaltic volcanism with uplift of the volcanic field led Demidjuk *et al.* (2007) to conclude that surface uplift was driven by upper mantle upwelling associated with small-scale, edge-driven convection beneath the Australian plate. Uplift of Quaternary

beach ridges on the Mount Gambier Coastal Plain over the last 780 ka indicates a surface uplift rate of *c.* 75 m Ma^{-1} (Belperio 1995). Associated gentle subsidence in the more internal parts of the Murray Basin several hundred kilometres to the north is interpreted to have occurred above the downwelling part of the convection circuit (Demidjuk *et al.* 2007).

Conversely, the intermediate-wavelength pattern of deformation associated with Flinders Ranges and surrounding anomalously low basins (such as the Torrens and Frome basins) has been attributed to lithospheric-scale (10^2 km) tectonically induced buckling (Celerier *et al.* 2005). This interpretation is based both on the topography of the top-of-basement surface and the observed positive coherence between topography and Bouguer gravity fields, implying that the Flinders Ranges are not supported by a crustal root but rather have risen in response to lithospheric folding. The Indian Ocean provides an analogous system, whereby buckling of the oceanic lithosphere beginning at around 8 Ma ago has been attributed to increases in intraplate stress levels caused by an increase in plate forcing induced by the rise in the Himalayan–Tibetan orogen at this time (e.g. Martinod & Molnar 1995). Similar scale buckling of the continental lithosphere perpendicular to the regional σ_{Hmax} is associated with localized seismicity in intraplate India (Vita-Finzi 2004). For the Flinders Ranges, the notion of lithospheric-scale buckling along a roughly north–south axis is consistent with the prevailing east–west σ_{Hmax} trend in this part of the continent.

The pattern of topography in the Lake Eyre region is more puzzling. The bullseye-shaped Eyre topographic depression is similar to other intracontinental basins associated with mantle downwelling (e.g. Hudson Bay, Canada; Wanganui Basin, New Zealand) (Mitrovica et al. 1989; Stern et al. 1992). However, the presence of folded and faulted Miocene alluvial strata along the western flank of Lake Eyre (Waclawik et al. 2008) and the long-wavelength folding of the basement surface in the Davenport Ranges suggest that regional compression has affected this region and thus provides an equally feasible mechanism for intermediate-wavelength deformation (e.g. Celerier et al. 2005; Vita-Finzi 2004).

Geomorphological implications

Intermediate-wavelength deformation has influenced the geomorphological evolution of uplands, basins, and alluvial systems at the sites described above. Uplift associated with the Padthaway high is suggested to have dammed the Murray River, forming a large palaeo-lake (Lake Bungunnia) several hundred kilometres to the north (Fig. 3b; Stephenson 1986). Lake formation and subsequent abandonment, and the consequent changes in stream base level, are likely to have had a strong influence on stream gradient evolution and aggradation–dissection patterns. Intermediate-wavelength deformation in south–central Australia has influenced the geomorphological evolution of broad upland systems (Flinders and Davenport Ranges) and adjacent basins (Eyre, Frome, Torrens) through effects on the spatial and temporal evolution of regional topography. The migration of a major basin depocentre from Billa Kalina to Eyre has resulted in significant drainage reversals in the region; for instance, former Billa Kalina southwestward-younging shorelines (formed by shoreline retreat) are now incised by east- and NE-flowing streams (Fig. 7). Broad uplift of the Flinders Ranges and Davenport Ranges relative to Lakes Frome, Torrens and Eyre has shifted the location of depocentres further from the range fronts, resulting in the incision of alluvial fans proximal to the ranges and final deposition of sediment more distal from the ranges (Williams 1973; Quigley et al. 2007c; J. Bowler, pers. comm.). Marine shoreline evolution has also been strongly influenced by intermediate-wavelength deformation. The lack of marine sedimentation at elevations of 30–40 m between Torrens Basin and Spencer Gulf (Fig. 3) contrasts with the presence of Pliocene strandlines at >200 m asl in parts of the Murray Basin, implying a regime characterized by both long-wavelength post-Pliocene subsidence and uplift. Continued slow subsidence between Spencer Gulf and Torrens coupled with sea-level rise may allow northward penetration of the sea into this region (Quigley et al. 2007c).

Short-wavelength (10^1 km) deformation

Characteristics

Insights into short-wavelength, fault-related deformation have been obtained from historical seismicity and faulting and the prehistoric 'neotectonic' geological record of faulting. Much of Australia's seismicity is concentrated into four distinct zones; the SW, NW, Flinders and SE seismic zones (Fig. 8). Calculated earthquake focal mechanism solutions indicate reverse faulting associated with roughly NW–SE horizontal compression throughout the SE seismic zone (Allen et al. 2005; Nelson et al. 2006). Flinders seismic zone mechanisms indicate strike-slip and reverse mechanisms with a broadly east–west horizontal compression (Hillis & Reynolds 2000; Clark & Leonard 2003; Hillis & Reynolds 2003). SW seismic zone mechanisms indicate reverse faulting mechanisms with east–west-oriented maximum horizontal stress (Denham et al. 1979; Clark & Leonard 2003). The NW seismic zone is typified by a NW–SE-oriented maximum horizontal stress orientation and predominantly strike-slip mechanisms (Hillis et al. 2008).

Seven instrumentally recorded earthquakes have produced surface ruptures, all occurring within the last 40 years (Fig. 9a). Earthquakes ranged in magnitude from 5.6 to 6.8, and produced scarps of 3.5–37 km length and 0.4–2.5 m height (Clark & McCue 2003). All historical scarps contain a dominant reverse-slip component with subsidiary oblique-slip. However, all historical fault scarps occurred in farming and/or grazing areas with low relief and thus their geomorphological impacts on the natural landscape have been short-lived, with rapid scarp degradation (Machette et al. 1993; Clark & McCue 2003; Crone et al. 2003).

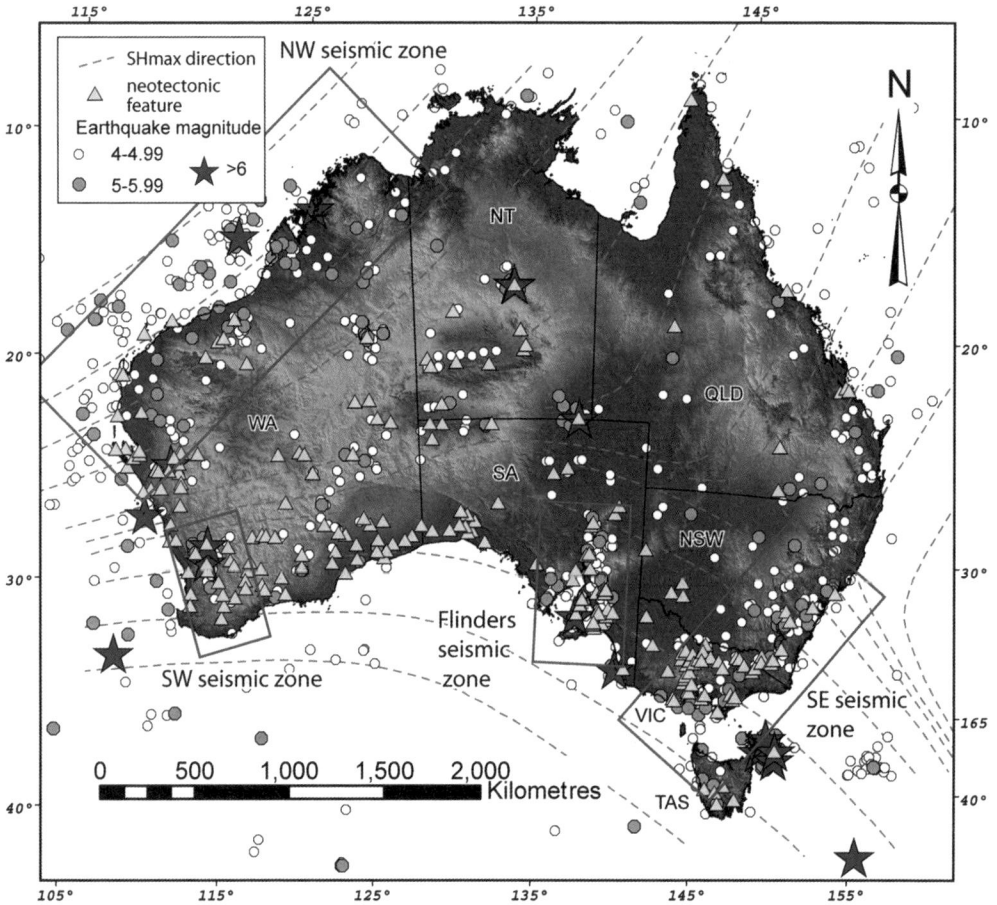

Fig. 8. Distribution of M>4 historical earthquake epicentres and identified neotectonic features in Australia. Earthquake epicentre and neotectonics data are courtesy of Geoscience Australia. Clustering of earthquake epicentre distributions indicates four seismic zones; NW seismic zone, SW seismic zone, Flinders seismic zone, and SE seismic zone. Maximum horizontal stress directions (σ_{Hmax}) after Hillis & Reynolds (2003) are shown.

Geological studies have identified several hundred faults across the continent with demonstrable Pliocene and/or Quaternary displacements (Fig. 9; Clark & McCue 2003; Crone *et al.* 2003; Sandiford 2003*b*; Twidale & Bourne 2004; Clark 2006, 2009; Quigley *et al.* 2006; Clark *et al.* 2010). The clearest evidence for active tectonic deformation in the Australian landscape is found in the fault-related landscapes around upland systems in SE and south–central parts of the continent (Sandiford 2003*b*). The Flinders and Mount Lofty Ranges of South Australia are bounded to the east and west by reverse faults that thrust Proterozoic and/or Cambrian basement rocks above Quaternary sediment (Fig. 9c–f). Faults in this region with documented Pliocene to Quaternary displacements include the Wilkatana, Burra, Milendella, Para, Paralana, Willunga, Cambrai, Morgan,

Gawler, Clarendon, Eden, Bremer, and Ediacara (Fig. 2a; Binks 1972; Williams 1973; Bourman & Lindsay 1989; Sandiford 2003*b*; Quigley *et al.* 2006). Fault-propagation folding is well developed in Miocene strata within the footwall of the Willunga and Gawler Faults (Fig. 9c; Lemon & McGowran 1989; Green 2007). The Barrier Ranges are flanked to the west by the Mundi Mundi and Kantappa Faults, both which displace late Quaternary strata (Quigley *et al.* 2006). The eastern edge of the Davenport Ranges is defined by the Mt Margaret and Levi Faults, which have been interpreted to have tens to hundreds of metres of Quaternary offset (Wopfner 1968). Miocene sediments forming the adjacent Lake Eyre alluvial plain have also been folded and faulted (Waclawik *et al.* 2008). The Eyre Peninsula consists of a series of en echelon reverse faults that displace Miocene to Quaternary stratigraphy (Fig. 9b; Miles 1952; Dunham 1992; Hutton et al. 1994; Crone et al. 2003). A series of scarps on the Yorke Peninsula have been attributed to Quaternary faulting as they offset a Pleistocene calcrete horizon (Crawford 1965). In SE Australia, upland systems such as the Otway and Strzelecki Ranges have similarly been affected by Pliocene to Quaternary reverse faulting. On the northern flanks of the Otway Range, in southern Victoria, the remnants of a Pliocene strandplain rise c. 120 m over a series of ENE–WSW-trending faults and monoclines to elevations of c. 250 m (Sandiford 2003a; Sandiford et al. 2004). Along with its correlatives in the Murray and Gippsland Basins (Holdgate et al. 2003; Wallace et al. 2005), this strandplain developed during falling sea levels following a 6 Ma highstand at c. 65 m above present-day sea level (Brown & Stephenson 1991), implying almost 200 m of fault-related tectonic uplift. In the Eastern Highlands of northern Victoria, Palaeozoic gneiss has been thrust some 160 m over Quaternary alluvium along the Tawonga Fault (Beavis 1960; Hills 1975). Elsewhere in the highlands, Palaeozoic rocks are similarly thrust over Miocene and younger strata along the Khancoban–Yellow Bog, Kiewa, Adaminaby, Lake George, and Shoalhaven Faults (e.g. Beavis & Beavis 1976; Abel 1985; Sharp 2004; Twidale & Bourne 2004). Uplift along the Cadell Fault has been dated as latest Pleistocene (Bowler & Harford 1966; Bowler 1978; Rutherfurd & Kenyon 2005; Clark et al. 2007). The Waratah Fault at Cape Liptrap in SE Victoria has displaced a last interglacial (c. 125 ka) marine terrace by up to 5.1 m (Gardner et al. 2009). The Lake Edgar Fault in Tasmania has incurred three surface-rupturing events with average displacements of c. 2.5–3 m in the last c. 50 ka, with the most recent event occurring at c. 18–17 ka (Clark et al. 2010). Kinematic studies of many of these faults indicate that faulting occurred in response to roughly east–west- to SE–NW-oriented palaeo σ_{Hmax} consistent with focal mechanism and contemporary stress data, implying that Neogene to Quaternary faulting is linked with the modern tectonic regime (Sandiford 2003b; Quigley et al. 2006; Gardner et al. 2009).

The relationship between Neogene to Quaternary faulting and uplifted topography is less clear in western Australia. Although some of the largest recorded earthquakes and the largest moment release occurs in the SW seismic zone (Leonard 2008; Braun et al. 2009), the only feature with appreciable relief in this part of the continent is the Darling Scarp, which is historically aseismic and contains no evidence for Quaternary displacement (Sandiford & Egholm 2008). Faults with demonstrated Quaternary displacement, including

the Meckering (Fig. 9a), Calingiri, Cadoux, Hyden, Lort River, Dumbleyung and Mt Narryer Faults, are associated with discrete scarps <2.5–5 m high (Williams 1979; Gordon & Lewis 1980; Lewis et al. 1981; Crone et al. 2003; Twidale & Bourne 2004; Estrada 2009). The absence of evidence for continuing fault-related relief generation on geological time scales in this part of the continent implies that seismic activity has only recently commenced or that, unlike SE and south–central Australia, is not localized on discrete structures at geological time scales. The sparse palaeoseismological dataset available is most consistent with the latter interpretation.

In the NW seismic zone, a series of asymmetric anticlines (e.g. Cape Range, Barrow, Rough Range anticlines) have developed as fault propagation folds above blind reverse faults (Hocking 1988; Hillis et al. 2008). The growth of fault propagation anticlines is generally dated as Miocene and younger in the Carnarvon Basin (e.g. Barber 1988; Hearty et al. 2002). Concentrations of earthquake epicentres (Geoscience Australia online earthquake catalogue: www.ga.gov.au), and emerged Pleistocene marine terraces on the Cape Range anticline and anticlinal folds in offshore Plio-Quaternary seafloor sediments (van de Graaff et al. 1976), indicate that deformation has continued to the present.

All of the faults described above for which kinematics can be resolved involve either purely dip-slip reverse movement or oblique-reverse movement, indicating that crustal thickening and uplift occurs across the Australian continent. Estimates of Quaternary earthquake magnitudes (M), based on fault rupture lengths, single-event displacements and inferred ranges of hypocentral rupture depth, range from M = 5.8 to 7.3 (Clark & McCue 2003; Quigley et al. 2006). The data are consistent with magnitude estimates for the largest recorded Australian earthquakes (Meeberiee, WA 1941, M = 7.3; Meckering, WA 1968, M = 6.8; Tennant Creek, NT 1988; M = 6.7). Estimates of surface-rupturing earthquake recurrence intervals from faults with multiple displacements range from 1 in 22 ka to ≥1 in 83 ka (Crone et al. 2003; Quigley et al. 2006; Clark et al. 2008).

Origin and rates

Seismic strain rates have been calculated for the Australian continent using historical seismicity. Assuming a maximum earthquake magnitude of M = 7, the maximum seismic strain rate calculated for Australia's seismic zones is c. 10^{-16} s^{-1}, and the continent-averaged strain rate is no more than c. 10^{-17} s^{-1} (Celerier et al. 2005; Leonard 2008; Braun et al. 2009). A bulk strain rate of 10^{-16} s^{-1} in uniaxial compression implies a total shortening

Fig. 9. Field photographs of selected historical and prehistoric faults across the Australian continent, from west to east. (**a**) Meckering Fault scarp in Western Australia, formed on 14 October 1968 during the M_S 6.8 Meckering earthquake. View is to the south. The fault has a westward heave of 2.44 m, a dextral lateral component of movement of 1.54 m, and a throw of 1.98 m (Dentith *et al.* 2009). Photograph courtesy of K. McCue and the Australian Earthquake Engineering Society. (**b**) Murninnie Fault on the eastern Eyre Peninsula, South Australia. View to the south. The fault plane is oriented 170°, dips 60° to the west, and thrusts Proterozoic basement gneiss over Pleistocene(?) sandstones (main image) and Pliocene clay (inset), indicating that the most recent event occurred no earlier than the Pleistocene. (**c**) Gawler Fault east of the Mt Lofty Ranges, South Australia. View to the east. The fault plane strikes NNE, dips 40–60° to the east, and thrusts Cambrian siltstone over late Miocene(?) gravels. Fault-propagated folding in the Miocene strata is evident from the folded bedding. (**d**) Wilkatana Fault in the central Flinders Ranges, South Australia. View to the north. The fault strikes NNW, dips 40–60°E, and thrusts Neoproterozoic quartzite over Pleistocene gravels with a reverse-oblique total slip of up to 15 m (Quigley *et al.* 2006). The last major earthquake on this fault occurred *c.* 30 ka ago. (**e**) Milendella Fault in the eastern Mt Lofty ranges, South Australia. View to the south. The fault strikes north, dips *c.* 50°W, and juxtaposes metamorphosed Cambrian rocks in the hanging wall with a footwall composed of a

of c. 250 m Ma^{-1} across a c. 100 km wide zone, such as the Flinders seismic zone (Sandiford & Quigley 2009). Such shortening could be accounted for with slip on 10 faults each accommodating about 25 m of horizontal slip per million years, equating to 25 m Ma^{-1} of vertical uplift for 45° dipping faults (Flinders Fault dips range from c. 30 to 80°; Celerier et al. 2005; Quigley et al. 2006). Accounting for some aseismic slip, actual fault slips may be even larger, and suggest slip rates to the closest order of magnitude in the range between 10^1 and 10^2 m Ma^{-1} in the most active parts of the continent.

Fault slip rates derived from cumulative displacements of Pliocene and Quaternary sediment in SE and south–central Australia range from 20 to 150 m Ma^{-1} (Sandiford 2003b; Quigley et al. 2006). Displacement of Neogene strata in the Otway Ranges has accumulated in the last 5–6 Ma (Sandiford et al. 2004) giving time-averaged displacement on bounding faults c. 40–50 m Ma^{-1}. Penultimate earthquakes on the historically active Marryat Creek and Meckering Faults occurred >100 ka ago, equating to rates of <10–35 m Ma^{-1} (Machette et al. 1993). The Rosedale Fault in Gippsland is purported to have experienced a slip rate of 50–80 m Ma^{-1} over the Early to Middle Pleistocene (Holdgate et al. 2003). However, fault slip rates determined from single fault exposures are difficult to assess because of the tendency of intracontinental faulting to cluster in time and space, resulting in under- or overestimation depending on the time of the last event (Crone & Machette 1997; Crone et al. 1997, 2003). For example, faults in the Wilkatana area of the central Flinders Ranges incurred upwards of 15 m of cumulative slip as a result of three or more earthquakes since c. 67 ka ago, equating to a c. 67 ka to Recent rate (c. 225 m Ma^{-1}) that is significantly more rapid than the Pliocene to Recent rate (20–50 m Ma^{-1}).

There is some indication that the temporal clustering behaviour emerging from single fault studies may be symptomatic of a larger picture of the more or less continuous tectonic activity from late Miocene to recent being punctuated by 'pulses'

of activity in specific deforming regions. For example, major deformation episodes are constrained to the interval 6–4 Ma in SW Victoria (Paine et al. 2004) and 2–1 Ma in the Otway Ranges (Sandiford 2003a). An episode of deformation ceased at 1.0 Ma in the offshore Gippsland Basin although it continued onshore until c. 250 ka (Holdgate et al. 2003). Holdgate et al. (2008) presented evidence from the SE Highlands that resurrected the idea of a punctuated post-Eocene Kosciusko Uplift event (see Browne 1967) that continued into the Late Pliocene and possibly the Pleistocene. It is possible that this event might relate to the pulse of deformation seen in SE Australian offshore basins in the interval 10–5 Ma associated with the reorganization of the crustal stress field to its present configuration (Dickinson et al. 2001, 2002; Sandiford et al. 2004; Hillis et al. 2008). The Lake George Fault near Canberra (Singh et al. 1981; Abel 1985), and faults of the Lapstone Structural Complex near Sydney (Fig. 3) may also have accumulated much of their Neogene displacement in this event (Clark 2010). Indeed, palaeomagnetic data indicate that folding and uplift relating to the Lapstone Structural Complex had largely ceased by the late Pliocene (Bishop et al. 1982 with age recalculated by Pillans 2003). In contrast, Gardner et al. (2009) obtained slip rates on the Waratah Fault of 10–40 m Ma^{-1} for displacements across both 125 ka and Pliocene marine terraces, so indicating that there may be some fault systems that do not exhibit highly variable long-term slip rates.

The factors that localize seismic activity and associated fault-related deformation within the Australian continent have been addressed in a number of studies. Fault reactivation of pre-existing structural weaknesses such as ancient fault zones almost certainly plays a fundamental role (Crone et al. 2003; Dentith & Featherstone 2003; Quigley et al. 2006). Faults commonly occur along geological boundaries such as inherited lithotectonic boundaries (e.g. Wilkatana and Roopena–Ash Reef Faults within the Lake Torrens Rift Zone) and topographic boundaries such as range fronts (Flinders and Mt. Lofty Ranges). It is unclear how surface topography influences near-surface fault

---

Fig. 9. ($Continued$) sliver of Lower Miocene limestone and a Quaternary gravel and clay sequence that elsewhere contains the Brunhes–Matuyama palaeomagnetic reversal at c. 780 ka BP (Sandiford 2003b). A metamorphic foliation can be traced across a wedge of colluvial material derived from the hanging wall immediately above the footwall, implying that this colluvium may have resulted from post-seismic collapse of the hanging wall with a fault displacement of up to 7 m. The total post-Early Miocene throw on the Milendella Fault is at least c. 60–90 m (Sandiford 2003b). (**f**) Burra Fault east of the central Flinders Ranges, South Australia. View to north. The fault strikes north, dips c. 30°W, and thrusts Neoproterozoic tillite over Quaternary gravels. The total fault displacement is c. 4 m and predates c. 83 ka (Quigley et al. 2006). (**g**) Lake Edgar Fault scarp and related geomorphological features in southern Tasmania. View to the south. The fault cross-cuts a periglacial alluvial fan and has incurred three surface-rupturing events with average displacements of c. 2.5–3 m in the last c. 50 ka, with the most recent event occurring at c. 18–17 ka (from Clark et al. 2010).

geometry, if at all. However, significant along-strike variations in fault dip are common (e.g. Milendella Fault, Wilkatana Fault).

Celerier *et al.* (2005) showed how variations in both absolute abundance and depth of heat-producing elements provide a plausible thermal control on lithospheric strength that helps localize deformation in the Flinders Ranges. The Flinders Ranges form part of a zone of anomalous surface heat flow (Neumann *et al.* 2000), with an average surface heat flow of 85 mW m^{-2} reflecting unusually elevated heat production in the Proterozoic basement. Celerier *et al.* (2005) concluded that the uppermost mantle beneath the Flinders seismic zone may be 50–100 °C hotter than surrounding zones because of the distribution of heat production within the crust, thus providing a thermally weak zone that is prime for focusing deformation. Sandiford & Egholm (2008) argued that the general correspondence between elevated earthquake activity and proximity to the edge of the continent indicates that thermal effects associated with the ocean–continent transition may also help control the pattern of active deformation in the SW seismic zone. They showed that the effects of steady-state lateral heat flow across transitional lithosphere can produce a weakening effect 100–200 km inboard of the ocean–continent transition. The anisotropic distribution of intraplate deformation is thus best explained by focusing of far-field stress into pre-existing fault zones, regions of enhanced crustal heat flow, craton boundaries, and regions associated with Moho temperature variations (Sandiford & Egholm 2008).

Geomorphological implications

Recurring earthquake activity has resulted in several hundreds of metres of fault displacement in parts of SE Australia, such as the Flinders, Mt Lofty, and Otway Ranges (Bourman & Lindsay 1989; Tokarev *et al.* 1999; Sandiford 2003*a*, *b*) and in parts of the Eastern Highlands (Hills 1975; Abel 1985; Sharp 2004) over the last 5–10 Ma. Quigley *et al.* (2007*b*) suggested that, although fault slip rates are modest compared with those of active faults in plate boundary settings, rates of relative surface uplift (i.e. uplift of mountain summit surfaces relative to adjacent piedmonts) may be comparable with those of more tectonically active environments because of the extremely low rates of bedrock erosion in Australian uplifted terrains. For instance, summit surfaces in the central Flinders Ranges have been uplifted relative to flanking piedmonts by more than 12 m in the last 70 ka, equivalent to a rate of *c.* 170 m Ma^{-1}, as a result of activity on range bounding faults. If such rates were sustained over longer time-scales, the

present-day relief characterizing much of the Flinders Ranges could have been created in as little as 3 Ma. Taking into account the intermittent nature of faulting in Australia, Sandiford (2003*b*) and Quigley *et al.* (2006) suggested that 30–50% of the present-day elevation of the Flinders Ranges relative to adjacent piedmonts has developed in the last 5 Ma. Inter-seismic subsidence may have played a role in the depression of topography along the western Flinders Range front, where the Torrens Basin has subsided several tens of metres since the terminal Miocene (Quigley *et al.* 2007*c*). At certain time intervals, some Australian uplands may thus have been uplifted more than mountain belts in tectonically active plate boundary settings such as New Zealand, the Cascadia accretionary margin, and Taiwan because the latter are balanced by rapid rates of erosion and may thus be in topographic steady state (Willet & Brandon 2002), whereas rock uplift has occasionally exceeded erosion in isolated, tectonically active, semi-arid to arid parts of Australia (e.g. Quigley *et al.* 2007*b*).

Coseismic surface displacements can be demonstrated to have exerted an influence on the evolution of many Australian stream systems and, consequently, intra-catchment relief. Surface rupturing along the Wilkatana Fault in the Flinders Ranges differentially uplifted the upstream reach of stream-beds, forming waterfalls (knickpoints) in longitudinal stream profiles (Quigley *et al.* 2006). Subsequent headward retreat of knickpoints at rates of >100 m Ma^{-1} into catchment systems was invoked to explain the valley-in-valley geomorphology upstream of the Wilkatana Fault (Quigley *et al.* 2007*a*). Stream incision rates are similar to bedrock uplift rates, implying that fault-generated uplift has played a major role in modulating intra-catchment relief production rates, presumably as a result of tectonic increases in stream profile steepness and stream erosive power.

Coseismic stream damming is evident at a number of sites across Australia. Perhaps the most spectacular example is found in the Echuca district of Victoria, where Late Pleistocene surface rupturing along the Cadell Fault diverted and temporarily dammed the flow of the Murray River, forming Lake Kanyapella (Bowler & Harford 1966; Bowler 1978; Rutherfurd & Kenyon 2005; Stone 2006). The river eventually formed northern and southern drainages that deflected around the uplifted fault block (Bowler & Harford 1966; Rutherfurd & Kenyon 2005; Clark *et al.* 2007). The lower Murray River is inferred to have once flowed NW continuously across the Mt Lofty Ranges along what is now the Broughton River, depositing sediments into the Broughton delta in the Spencer Gulf [Williams & Goode 1978; but see also Harris *et al.* (1980) for a contrary viewpoint]. Post-Eocene tectonic uplift of the Mt Lofty Ranges is inferred to have diverted the Murray to the south, forming the conspicuous modern-day diversion adjacent to the Morgan Fault (Williams & Goode 1978). Coseismic ponding of streams occurred along the Ash Reef Fault in the Eyre Peninsula (Hutton *et al.* 1994).

Coseismic ground shaking has also played a role in shaping Australia's geomorphology. Coseismic landslides have been identified in the Flinders Ranges (Quigley *et al.* 2007a) and are interpreted as major contributors to catchment sediment fluxes throughout the region. Quigley *et al.* (2007a)

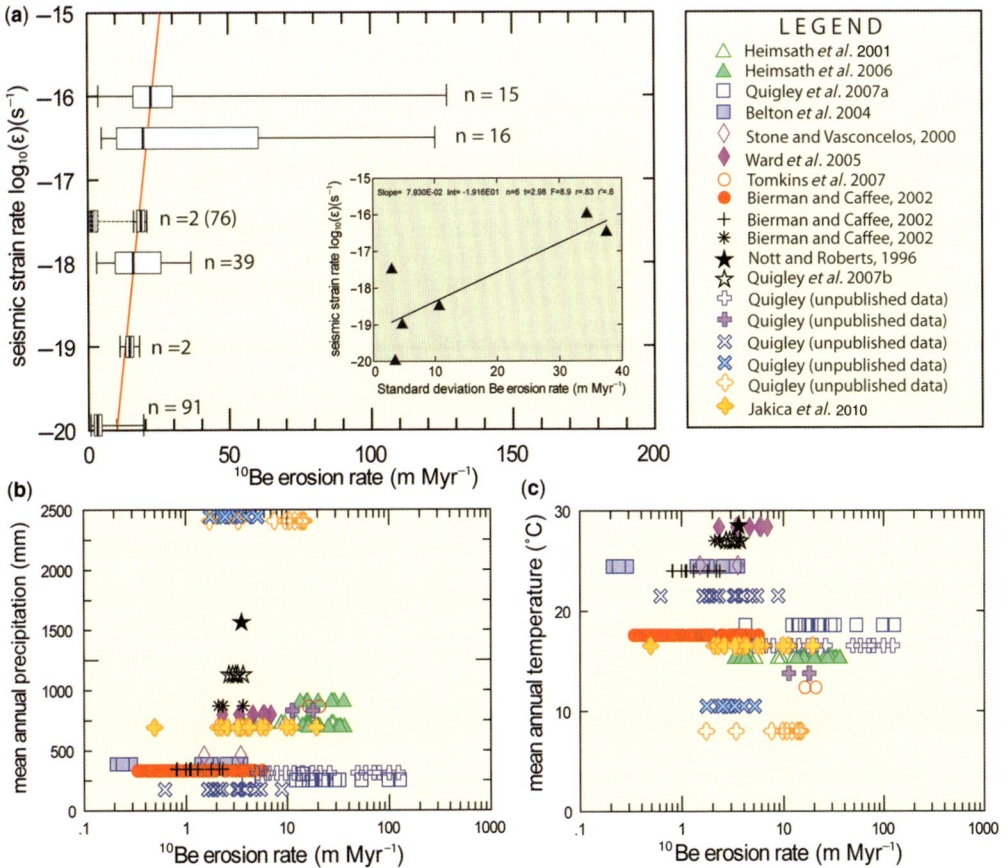

Fig. 10. Cosmogenic ^{10}Be ($t_{1/2} = 1.5$ Ma) is produced *in situ* in near-surface materials as a result of high-energy cosmic-ray reactions. Concentrations can be measured using accelerator mass spectrometry and interpreted in terms of surface exposure histories and/or bedrock erosion rates. (**a**) Whisker-box plot of seismic strain rate (from Braun *et al.* 2009) v. erosion rate (determined from cosmogenic ^{10}Be analyses) for Australian bedrock outcrops across regional tectonic gradients. Outer ticks represent sample minimum and maximum, box boundaries define lower and upper quartiles, and middle line indicates the median erosion rate. *n*, number of samples per suite. For seismic strain rate of $\log_{10} - 17.5$, two datasets are shown; inselberg samples (which are relatively unjointed) are discounted from the second, higher erosion rate grouping to compare similar rock masses with consistent joint spacing throughout this analysis. The best-fit line through the data shows an increase in bedrock erosion rate with increasing seismic strain rate, suggesting a first-order tectonic control on bedrock erosion. Qualitatively, increased erosion rates also appear to correlate with increased neotectonic activity (Fig. 8). The inset shows an increase in standard deviation of bedrock erosion for data populations of increasing seismic strain rate. We interpret this to reflect increased erosion rates and erosion rate variability in bedrock landscapes subjected to intermittent large earthquakes and coseismic shaking. (**b**) Mean annual precipitation v. bedrock erosion rate, showing the absence of any clear correlation between these parameters across large precipitation gradients. (**c**) Mean annual temperature v. bedrock erosion rate, showing the absence of any clear correlation between these parameters across large mean annual temperature gradients. The results of (b) and (c) suggest that over 10 ka–1 Ma time scales, climate variability plays a minimal role in determining erosion rate variability for Australian bedrock outcrops relative to tectonic forcing.

speculated that the large volume of Late Pleistocene sediment accumulated in the Wilkatana fans resulted from reworking of intra-catchment sediment derived from coseismic ground shaking. On the basis of anomalously low cosmogenic ^{10}Be concentrations in bedrock and alluvium derived from fault-affected catchments, Quigley *et al.* (2007*a*) suggested that the effects of tectonic perturbations (e.g. anomalously high hillslope sediment volume release as a result of coseismic landslides and anomalously steep longitudinal stream profiles) may reside within Australian landscapes for *c.* 30 ka or longer. Cosmogenic nuclide concentrations thus appear to provide an informative tool on the degree of neotectonism that has affected Australian bedrock landscapes (Fig. 10). The aseismic Darling escarpment, which has not been affected by late Quaternary tectonism, yields high cosmogenic ^{10}Be concentrations and appears to be in topographic steady state (Jakica *et al.* 2010) whereas the Flinders Ranges landscapes that have been uplifted in the late Quaternary yield low ^{10}Be concentrations (Quigley *et al.* 2007*a*, *c*). Seismic shaking has produced fractured bedrock slabs or 'A-tents' at several sites in South Australia (Twidale & Bourne 2000; Twidale & Campbell 2005), providing an observable connection between earthquake activity and bedrock fracturing. The earthquake recurrence interval and magnitude of coseismic ground shaking associated with seismic events is likely to exert considerable control over the amount of bedrock available for transport during subsequent erosional events (e.g. floods; Quigley *et al.* 2007*c*). Kink-bands in *c.* 120 ka BP interglacial cemented dune limestones near Cape Liptrap provide additional evidence for coseismic rock mass deformation in the Quaternary (Sandiford 2003*b*; Gardner *et al.* 2009).

Preliminary investigations of the relationships between bedrock erosion, determined from cosmogenic ^{10}Be concentrations, and extrinsic variables including seismic strain rate, mean annual precipitation, mean annual temperature and temperature range suggest that tectonic activity, as opposed to climate, exerts the dominant control on bedrock erosion rate across the Australian continent (Fig. 10). Intriguingly, a continent-wide inventory of cosmogenic nuclide erosion rate determinations indicates that bedrock erosion rates are low (0.5–5 m Ma^{-1}) with low variability across large climatic gradients, implying that on time scales of 10–100 ka, climate appears to play a minimal role in modulating bedrock erosion rates (Fig. 10; Quigley *et al.* 2007*a*; Quigley 2008). Conversely, erosion rates are highly variable (*c.* 3–130 m Ma^{-1}) in regions of elevated seismic activity (high seismic strain rate) relative to regions of relatively low seismic activity (low seismic strain rate)

(Fig. 10; Quigley *et al.* 2007*a*, *c*; Jakica *et al.* 2010). We interpret this dataset to indicate that coseismic rock fracturing and transport during infrequent earthquakes is the primary driver of long-term erosion in bedrock landscapes. Although this is perhaps not a surprising conclusion to be drawn from tectonically active regions (Riebe *et al.* 2001*a*, *b*), it is perhaps more remarkable for a mild intraplate setting such as Australia.

Conclusions

Many of Australia's landscapes have developed in response to active geodynamic processes, including long-wavelength continental tilting, intermediate-wavelength folding, and short-wavelength fault displacement of the surface. These processes have extensively modified the coastline of Australia, strongly influenced patterns of marine inundation, and influenced the geometry of many of Australia's streams, hillslopes, basins, and uplands over the last 5–10 Ma. New techniques such as cosmogenic nuclide analysis have allowed bedrock erosion rates, surface uplift rates, and relief production rates to be quantified in selected localities. When combined with structural and stratigraphic studies, these results confirm that the geomorphology of parts of the Australian continent has been changed dramatically in the Late Cenozoic in response to active tectonic processes. Although the modern intraplate tectonic regime is characterized by relatively infrequent large earthquakes compared with plate boundary regions, the correlation between bedrock erosion rates and seismic activity suggests that an active tectonic regime continues to exert a profound influence on landscape evolution in parts of the great southern continent.

We thank P. Bishop, T. Gardner and J. Webb for reviews that improved the quality of this paper. Our views on the Australian landscape have profited greatly through discussions with M. Williams, J. Chappell, S. Hill, J. Bowler, K. Fifield, D. Fink, P. Bierman and many others. Some of the unpublished cosmogenic nuclide data were obtained using ANSTO–University of Melbourne collaborative research grants in 2008 and 2009. D.C. publishes with the permission of the CEO of Geoscience Australia.

References

ABEL, R. S. 1985. *Geology of the Lake George Basin, N.S.W.* Bureau of Mineral Resources, Geology and Geophysics Record, **1985/4**.

ALLEN, T. I., GIBSON, G. & CULL, J. P. 2005. Stress-field constraints from recent intraplate seismicity in southeastern Australia. *Australian Journal of Earth Sciences*, **52**, 217–229.

ANDREWS, A. C. 1910. Geographical unity of Eastern Australia in Late and Post Tertiary time. *Proceedings* of the Royal Society of New South Wales, **54**, 251–350.

BARBER, P. 1988. The Exmouth Plateau deep water frontier: a case study. In: PURCELL, P. G. & PURCELL, R. R. (eds) The North West Shelf, Australia. Proceedings of the Petroleum Exploration Society of Australia Symposium. PESA, Perth, 173–187.

BATT, G. E. & BRAUN, J. 1999. The tectonic evolution of the Southern Alps, New Zealand: insights from fully thermally coupled dynamical modeling. Geophysical Journal International, **136**, 403–420.

BEARD, J. S. 2003. Palaeodrainage and the geomorphologic evolution of passive margins in Southwestern Australia. Zeitschrift für Geomorphologie, **47**, 273–288.

BEAVIS, F. C. 1960. The Tawonga Fault, northeast Victoria. Proceedings of the Royal Society of Victoria, **72**, 95–100.

BEAVIS, F. C. & BEAVIS, J. C. 1976. Structural geology in the Kiewa region of the Metamorphic Complex, North-East Victoria. Proceedings of the Royal Society of Victoria, **88**, 66–75.

BELPERIO, A. P. 1995. Quaternary. In: DREXEL, J. F. & PREISS, W. V. (eds) The Geology of South Australia, Volume 2, The Phanerozoic. Geological Survey of South Australia Bulletin, **54**, 219–281.

BELPERIO, A. P., HARVEY, N. & BOURMAN, R. P. 2002. Spatial and temporal variability in the Holocene sea-level record of the South Australian coastline. Sedimentary Geology, **150**, 153–169.

BELTON, D. X., BROWN, R. W., KOHN, B. P., FINK, D. & FARLEY, K. A. 2004. Quantitative resolution of the debate over antiquity of the central Australian landscape: implications for the tectonic and geomorphic stability of cratonic interiors. Earth and Planetary Science Letters, **219**, 21–34.

BIERMAN, P. R. & CAFFEE, M. 2002. Cosmogenic exposure and erosion history of Australian bedrock landforms. Geological Society of America Bulletin, **114**, 787–803.

BINKS, P. J. 1972. Late Cainozoic uplift of the Ediacara Range, South Australia. Proceedings of the Australian Institute of Minerals and Metallurgy, **243**, 47–55.

BISHOP, P. 1984. Modern and ancient rates of erosion of central eastern N.S.W. and their implications. In: LOUGHRAN, R. J. (compiler) Drainage Basin Erosion and Sedimentation. University of Newcastle and Soil Conservation Service of NSW, Newcastle, 35–42.

BISHOP, P. 1985. Southeast Australian Late Mesozoic and Cenozoic denudation rates. A test for late Tertiary increases in continental denudation. Geology, **13**, 469–482.

BISHOP, P., HUNT, P. & SCHMIDT, P. W. 1982. Limits to the age of the Lapstone Monocline, N.S.W.; a palaeomagnetic study. Journal of the Geological Society of Australia, **29**, 319–326.

BOCK, Y., PRAWIRODIRDJO, L. ET AL. 2003. Crustal motion in Indonesia from Global Positioning System measurements. Journal of Geophysical Research, **108**, 2367, doi: 10.1029/2001JB000324.

BOURMAN, R. P. & LINDSAY, J. M. 1989. Timing, extent and character of late Cainozoic faulting on the eastern margin of the Mt Lofty Ranges, South Australia. Transactions of the Royal Society of South Australia, **113**, 63–67.

BOWLER, J. M. 1978. Quaternary climate and tectonics in the evolution of the Riverine Plain, south eastern Australia. In: WILLIAMS, M. A. J. (ed.) Landform Evolution in Australasia. ANU Press, Canberra, 70–112.

BOWLER, J. M. & HARFORD, L. B. 1966. Quaternary tectonics and the evolution of the Riverine Plain near Echuca, Victoria. Geological Society of Australia Journal, **13**, 339–354.

BRAUN, J., GESTO, F., BURBIDGE, D., CUMMINS, P., SANDIFORD, M., GLEADOW, A. & KOHN, B. 2009. Constraints on the current rate of deformation and surface uplift of the Australian continent from a new seismic database. Australian Journal of Earth Sciences, **56**, 99–110.

BROWN, C. M. & STEPHENSON, A. E. 1991. Geology of the Murray Basin, Southeastern Australia. Bureau of Mineral Resources, Geology & Geophysics Australia Bulletin, **235**.

BROWNE, W. R. 1967. Geomorphology of the Kosciusko block and its north and south extensions. Proceedings of the Linnean Society of New South Wales, **92**, 117–144.

CALLEN, R. A. & COWLEY, W. M. 1998. Billla Kalina Basin. In: DREXEL, J. F. & PREISS, W. V. (eds) The Geology of South Australia, Volume 2, the Phanerozoic. South Australia Geological Survey Bulletin, **54**, 195–198.

CÈLÈRIER, J., SANDIFORD, M., HANSEN, D. L. & QUIGLEY, M. 2005. Modes of active intraplate deformation, Flinders Ranges, Australia. Tectonics, **24**, doi: 10.029/2004&C001679.

CHAPPELL, J. 2006. Australian landscape processes measured with cosmogenic nuclides. In: PILLANS, B. (ed.) Regolith Geochronology and Landscape Evolution. CRC LEME, Perth, 19–26.

CLARK, D. 2006. Neotectonics-based intraplate seismicity models and seismic hazard. Paper presented at Australian Earthquake Engineering Society Meeting, Albury, NSW. World Wide Web Address: http://www.aees.org.au/Proceedings/2006_Papers/069_Clark.pdf.

CLARK, D. 2009. Potential geologic sources of seismic hazard in the Sydney Basin: proceedings volume of a one day workshop, April 2005. Geoscience Australia Record, **2009/011**.

CLARK, D. 2010. Identification of Quaternary scarps in southwest and central west Western Australia using DEM-based hill shading: application to seismic hazard assessment and neotectonics. In: Special Publication of the International Journal of Remote Sensing (in press).

CLARK, D. & MCCUE, K. 2003. Australian palaeoseismology: towards a better basis for seismic hazard estimation. Annals of Geophysics, **46**, 1087–1105.

CLARK, D., VAN DISSEN, R., CUPPER, M., COLLINS, C. & PRENDERGAST, A. 2007. Temporal clustering of surface ruptures on stable continental region faults: a case study from the Cadell Fault scarp, south eastern Australia. Paper presented at Australian Earthquake Engineering Society Conference, Wollongong. World Wide Web Address: http://www.aees.org.au/Proceedings/2007_Papers/17_Clark%2c_Dan.pdf.

CLARK, D., DENTITH, M., WYRWOLL, K. H., YANCHOU, L., DENT, V. & FEATHERSTONE, C. 2008. The Hyden fault scarp, Western Australia: paleoseismic evidence for repeated Quaternary displacement in an intracratonic setting. *Australian Journal of Earth Sciences*, **55**, 379–395.

CLARK, D. J. & LEONARD, M. 2003. Principal stress orientations from multiple focal plane solutions: new insight into the Australian intraplate stress field. *In*: HILLIS, R. R. & MULLER, D. (eds) *Evolution and Dynamics of the Australian Plate*. Geological Society of Australia and Geological Society of America Joint Special Publication, **22**, 91–105.

CLARK, D. J., CUPPER, M., SANDIFORD, M. & KIERNAN, K. 2010. Style and timing of late Quaternary faulting on the Lake Edgar Fault, southwest Tasmania, Australia: implications for hazard assessment in intracratonic areas. *In*: *Paleoseismology*. Geological Society of America, Special Papers (in press).

CLOETINGH, S. & WORTEL, R. 1986. Stress in the Indo-Australian plate. *Tectonophysics*, **132**, 49–67.

COATS, R. P. 1973. *Copley map sheet. South Australia*. Geological Survey of South Australia. Geological Atlas 1:250 000 Series, Sheet **SH54-09**.

COBLENTZ, D. D., SANDIFORD, M., RICHARDSON, R. M., ZHOU, S. & HILLIS, R. 1995. The origins of the intraplate stress field in continental Australia. *Earth and Planetary Science Letters*, **133**, 299–309.

COBLENTZ, D. D., ZHOU, S., HILLIS, R. R., RICHARDSON, R. M. & SANDIFORD, M. 1998. Topography boundary forces and the Indo-Australian intraplate stress field. *Journal of Geophysical Research*, **103**, 919–938.

CRAWFORD, A. R. 1965. *The geology of Yorke Peninsula. South Australia*. Geological Survey of South Australia Bulletin, **39**.

CRONE, A. J. & MACHETTE, M. N. 1997. The temporal variability of surface-faulting earthquakes in stable continental regions; a challenge to seismic-hazard assessments. *Geological Society of America, Abstracts with Programs*, **29**, 71.

CRONE, A. J., MACHETTE, M. & BOWMAN, J. R. 1997. Episodic nature of earthquake activity in stable continental regions revealed by palaeoseismicity studies of Australian and North American quaternary faults. *Australian Journal of Earth Sciences*, **44**, 203–214.

CRONE, A. J., DE MARTINI, P. M., MACHETTE, M. N., OKUMURA, K. & PRESCOTT, J. R. 2003. Paleoseismicity of two historically quiescent faults in Australia: implications for fault behavior in stable continental regions. *Bulletin of the Seismological Society of America*, **93**, 1913–1934.

DE BROECKERT, P. & SANDIFORD, M. 2005. Buried inset-valleys in the eastern Yilgarn Craton, Western Australia: geomorphology, age, and allogenic control. *Journal of Geology*, **113**, 471–493.

DEMETS, C., GORDON, R. G., ARGUS, D. F. & STEIN, S. 1990. Current plate motions. *Geophysical Journal International*, **101**, 425–478.

DEMETS, C., GORDON, R., ARGUS, D. & STEIN, S. 1994. Effect of recent revisions to the geomagnetic reversal time scale on estimates of current plate motions. *Geophysical Research Letters*, **21**, 2191–2194.

DEMIDJUK, Z., TURNER, S., SANDIFORD, M., GEORGE, R., FODEN, J. & ETHERIDGE, M. 2007. U-series isotope and geodynamic constraints on mantle melting processes beneath the Newer Volcanic Province in South Australia. *Earth and Planetary Science Letters*, **261**, 517–533.

DENHAM, D., ALEXANDER, L. G. & WOROTNICKI, G. 1979. Stresses in the Australian crust; evidence from earthquake and *in situ* stress measurements. *BMR Journal of Australian Geology & Geophysics*, **4**, 289–295.

DENTITH, M. C. & FEATHERSTONE, W. E. 2003. Controls on intra-plate seismicity in southwestern Australia. *Tectonophysics*, **376**, 167–184.

DENTITH, M. C., CLARK, D. & FEATHERSTONE, W. E. 2009. Aeromagnetic mapping of Precambrian geological structures that controlled the 1968 Meckering earthquake: implications for intraplate seismicity in Western Australia. *Tectonophysics*, **475**, 544–553.

DICKINSON, J. A., WALLACE, M. W., HOLDGATE, G. R., DANIELS, J., GALLAGHER, S. J. & THOMAS, L. 2001. Neogene tectonics in SE Australia: implications for petroleum systems. *APPEA Journal*, **41**, 37–52.

DICKINSON, J. A., WALLACE, M. W., HOLDGATE, G. R., GALLAGHER, S. J. & THOMAS, L. 2002. Origin and timing of the Miocene–Pliocene unconformity in southeast Australia. *Journal of Sedimentary Research*, **72**, 288–303.

DUNHAM, M. H. R. 1992. *The geomorphological nature and age of the linear escarpments of northeast Eyre Peninsula*. BSc Hons thesis, University of Adelaide.

DYKSTERHUIS, S. & MÜLLER, R. D. 2008. Cause and evolution of intraplate orogeny in Australia. *Geology*, **36**, 495–498.

ESTRADA, B. 2009. *Neotectonic and Palaeoseismological Studies in the Southwest of Western Australia*. PhD thesis, University of Western Australia, Perth.

FENNER, C. 1930. The major structural and physiographic features of South Australia. *Transactions of the Royal Society of South Australia*, **54**, 1–36.

FENNER, C. 1931. *South Australia. A Geographical Study*. Whitcombe and Tombs, Melbourne.

GARDNER, T., WEBB, J. *ET AL.* 2009. Episodic intraplate deformation of stable continental margins: evidence from Late Neogene and Quaternary marine terraces, Cape Liptrap, Southeastern Australia. *Quaternary Science Reviews*, **28**, 39–53.

GORDON, F. R. & LEWIS, J. D. 1980. The Meckering and Calingiri earthquakes October 1968 and March 1970. *Western Australia Geological Survey Bulletin*, **126**, 229.

GREEN, M. 2007. *The Structure and Kinematics of the Concordia Fault: Constraining the Age of the Indo-Australian Plate Stress Field*. BSc Hons thesis, University of Melbourne.

GURNIS, M. 1992. Dynamic topography. *In*: NIERENBERG, W. A. (ed.) *Encyclopedia of Earth System Science, Vol. 2*. Academic Press, San Diego, CA, 105–109.

GURNIS, M., MUELLER, R. D. & MORESI, L. 1998. Dynamics of Cretaceous vertical motion of Australia and the Australian–Antarctic discordance. *Science*, **279**, 1499–1504.

HARRIS, W. K., LINDSAY, J. M. & TWIDALE, C. R. 1980. Possible western outlet for an ancient Murray River in South Australia. 2. A discussion. *Search*, **11**, 226–227.

HEARTY, D. J., ELLIS, G. K. & WEBSTER, K. A. 2002. Geological history of the western Barrow Sub-basin: implications for hydrocarbon entrapment at Woollybutt and surrounding oil and gas fields. *In*: KEEP, M. & MOSS,

S. J. (eds) *The Sedimentary Basins of Western Australia 3. Proceedings of the Petroleum Exploration Society of Australia Symposium.* PESA, Perth, 577–598.

HEIMSATH, A. M., CHAPPELL, J., DIETRICH, W. E., NISHIIZUMI, K. & FINKEL, R. C. 2001. Late Quaternary erosion in southeastern Australia: a field example using cosmogenic nuclides. *Quaternary International*, **83–85**, 169–185.

HEIMSATH, A. M., CHAPPELL, J., FINKEL, R. C., FIFIELD, K. & ALIMANOVIC, A. 2006. Escarpment erosion and landscape evolution in southeastern Australia. *In*: WILLET, S. D., HOVIUS, N., BRANDON, M. T. & FISHER, D. M. (eds) *Tectonics, Climate, and Landscape Evolution.* Geological Society of America, Special Publication, **398**, 173–190.

HILLIS, R. & REYNOLDS, S. 2003. *In situ* stress field of Australia. *In*: HILLIS, R. R. & MULLER, D. (eds) *Evolution and Dynamics of the Australian Plate.* Geological Society of Australia, Special Publication, **22**, 101–113.

HILLIS, R. R. & REYNOLDS, S. D. 2000. The Australian stress map. *Journal of the Geological Society, London*, **157**, 915–921.

HILLIS, R. R., SANDIFORD, M., REYNOLDS, S. D. & QUIGLEY, M. C. 2008. Present-day stresses, seismicity and Neogene-to-Recent tectonics of Australia's 'passive' margins: intraplate deformation controlled by plate boundary forces. *In*: JOHNSON, H., DORÉ, A. G., GATLIFF, R. W., HOLDSWORTH, R., LUNDIN, E. R. & RITCHIE, J. D. (eds) *The Nature and Origin of Compression in Passive Margins.* Geological Society, London, Special Publications, **306**, 71–90.

HILLS, E. S. 1961. Morphotectonics and geomorphological sciences with special reference to Australia. *Quarterly Journal of the Geological Society of London*, **117**, 77–89.

HILLS, E. S. 1975. *Physiography of Victoria; An Introduction to Geomorphology.* Whitcombe & Tombs, Melbourne.

HOCKING, R. M. 1988. Regional geology of the northern Carnarvon Basin. *In*: PURCELL, P. G. & PURCELL, R. R. (eds) *North West Shelf Symposium, Aug. 10–12, 1988.* PESA, Perth, WA, 97–114.

HOLDGATE, G. R., WALLACE, M. W., GALLAGHER, S. J., SMITH, A. J., KEENE, J. B., MOORE, D. & SHAFIK, S. 2003. Plio-Pleistocene tectonics and eustacy in the Gippsland Basin, southeast Australia: evidence from magnetic imagery and marine geological data. *Australian Journal of Earth Sciences*, **50**, 403–426.

HOLDGATE, G. R., WALLACE, M. W., GALLAGHER, S. J., WAGSTAFF, B. E. & MOORE, D. 2008. No mountains to snow on: major post-Eocene uplift of the East Victoria Highlands; evidence from Cenozoic deposits. *Australian Journal of Earth Sciences*, **55**, 211–234.

HUTTON, J. T., PRESCOTT, J. R., BOWMAN, J. R., DUNHAM, M. N. E., CRONE, A. J., MACHETTE, M. N. & TWIDALE, C. R. 1994. Thermoluminescence dating of Australian palaeoearthquakes. *Quaternary Geochronology*, **13**, 143–147.

JAKICA, S., QUIGLEY, M., SANDIFORD, M., CLARK, D., FIFIELD, K. & ALIMANOVIC, A. 2010. Geomorphic evolution of the Darling Scarp, Western Australia: constraints from cosmogenic ^{10}Be. *Earth Surface Processes and Landforms.* doi: 10.1002/esp.2058.

JOHNS, R. K. 1968. Investigation of Lake Torrens, report of investigations—South Australia. *Geological Survey of South Australia Report of Investigation*, **31**, 1–62.

JOHNSON, D. 2004. *The Geology of Australia.* Cambridge University Press, Cambridge.

JOHNSTON, A. C. 1996. Moment magnitude assessment of stable continental earthquakes, Part 2: historical seismicity. *Geophysical Journal International*, **125**, 639–678.

JOHNSTON, A. C., COPPERSMITH, K. J., KANTER, L. R. & CORNELL, C. A. 1994. *The earthquakes of stable continental regions.* Electric Power Research Institute Report, **TR102261V1**.

JOYCE, E. B. 1975. Quaternary volcanism and tectonics in southeastern Australia. *Bulletin of the Royal Society of New Zealand*, **13**, 169–176.

KABAN, M., SCHWINTZER, P. & REIGBER, C. 2005. Dynamic topography as reflected in the global gravity field. *In*: REIGBER, C., LÜHR, H., SCHWINTZER, P. & WICKERT, J. (eds) *Earth Observation with CHAMP: Results from Three Years in Orbit.* Springer, Berlin.

KEEP, M., BISHOP, A. & LONGLEY, I. 2000. Neogene wrench reactivation of the Barcoo sub-basin, northwest Australia: implications for Neogene tectonics of the northern Australia margin. *Petroleum Geoscience*, **6**, 211–220.

KEEP, M., CLOUGH, M. & LANGHI, L. 2002. Neogene tectonic and structural evolution of the Timor Sea region, NW Australia. *In*: KEEP, M. & MOSS, S. (eds) *The Sedimentary Basins of Western Australia III, Proceedings West Australian Basins Symposium.* PESA, Perth, 341–353.

LEONARD, M. 2008. One hundred years of earthquake recording in Australia. *Bulletin of the Seismological Society of America*, **98**, 1458–1470.

LEMON, N. M. & MCGOWRAN, B. 1989. *Structural Development of the Willunga Embayment, St. Vincent Basin, South Australia.* National Centre for Petroleum Geology and Geophysics, Adelaide (unpublished).

LEWIS, J. D., DAETWYLER, N. A., BUNTING, J. A. & MONCRIEFF, J. S. 1981. *The Cadoux Earthquake.* Western Australia Geological Survey Report, **1981/11**.

LITHGOW-BERTELLONI, C. & GURNIS, M. 1997. Cenozoic subsidence and uplift of continents from time-varying dynamic topography. *Geology*, **25**, 735–738.

MACHETTE, M. N., CRONE, A. J. & BOWMAN, J. R. 1993. Geologic investigations of the 1986 Marryat Creek, Australia, earthquake—implications for paleoseismicity in stable continental regions. *US Geological Survey Bulletin*, **2032-B**, 29.

MARTINOD, J. & MOLNAR, P. 1995. Lithospheric folding in the Indian Ocean and the rheology of the oceanic plate. *Bulletin de la Société Géologique de France*, **166**, 813–821.

MASSELL, C. G., COFFIN, M. F. *ET AL.* 2000. Neotectonics of the Macquarie Ridge Complex, Australia–Pacific plate boundary. *Journal of Geophysical Research*, **105**, 13 457–13 480.

MILES, K. R. 1952. Tertiary faulting in northeastern Eyre Peninsula. *Transactions of the Royal Society of South Australia*, **75**, 89–96.

MITROVICA, J. X., BEAUMONT, C. & JARVIS, G. T. 1989. Tilting of continental interiors by the dynamical effects of subduction. *Tectonics*, **8**, 1078–1094.

MOLNAR, P., ENGLAND, P. & MARTINOD, J. 1993. Mantle dynamics, the uplift of the Tibetan Plateau, and the Indian monsoon. *Reviews of Geophysics*, **31**, 357–396.

MURRAY-WALLACE, C. V. & BELPERIO, A. P. 1991. The last interglacial shoreline in Australia; a review. *Quaternary Science Reviews*, **10**, 441–461.

NELSON, E., HILLIS, R. *ET AL.* 2006. Present-day state-of-stress of southeast Australia. *APPEA Journal*, **46**, 283–305.

NEUMANN, N., SANDIFORD, M. & FODEN, J. 2000. Regional geochemistry and continental heat flow: Implications for the origin of the South Australian heat flow anomaly. *Earth and Planetary Science Letters*, **183**, 107–120.

NOTT, J. & ROBERTS, R. G. 1996. Time and process rates over the past 100 m.y.: a case for dramatically increased landscape denudation rates during the late Quaternary in northern Australia. *Geology*, **24**, 883–887.

PACKHAM, G. 1996. Cenozoic SE Asia; reconstructing its aggregation and reorganisation. *In*: HALL, R. & BLUNDELL, D. J. (eds) *Tectonic Evolution of Southeast Asia*. Geological Society, London, Special Publications, **106**, 123–152.

PAINE, M. D., BENNETTS, D. A., WEBB, J. A. & MORLAND, V. J. 2004. Nature and extent of Pliocene strandlines in southwestern Victoria and their application to Late Neogene tectonics. *Australian Journal of Earth Sciences*, **51**, 407–422.

PAZZAGLIA, F. & GARDNER, T. 1994. Late Cenozoic flexural deformation of the middle U.S. Atlantic passive margin. *Journal of Geophysical Research*, **99**, 12 143–12 157.

PILLANS, B. 2003. Dating ferruginous regolith to determine seismic hazard at Lucas Heights, Sydney. *In*: ROACH, I. C. (ed.) *Advances in Regolith: Proceedings of the CRC LEME Regional Regolith Symposia, 2003*. CRC LEME, Bentley, Western Australia, 324–327.

PYSKLYWEC, R. N. & MITROVICA, J. X. 1999. The role of subduction-induced subsidence in the evolution of the Karoo Basin. *Journal of Geology*, **107**, 155–164.

QUIGLEY, M. 2008. Active tectonics, climate change, and surface processes: insights from continent-wide paleoseismic and cosmogenic nuclide datasets. *Geological Society of New Zealand Annual Conference Geosciences '08*, Wellington, NZ, 25 November 2008.

QUIGLEY, M. C., CUPPER, M. L. & SANDIFORD, M. 2006. Quaternary faults of southern Australia: palaeoseismicity, slip rates and origin. *Australian Journal of Earth Sciences*, **53**, 285–301.

QUIGLEY, M., SANDIFORD, M., ALIMANOVIC, A. & FIFIELD, L. K. 2007a. Landscape responses to intraplate tectonism: quantitative constraints from [10]Be abundances. *Earth and Planetary Science Letters*, **261**, 120–133.

QUIGLEY, M., SANDIFORD, M., FIFIELD, K. & ALIMANOVIC, A. 2007b. Bedrock erosion and relief production in the northern Flinders Ranges, Australia. *Earth Surface Processes and Landforms*, **32**, 929–944.

QUIGLEY, M. C., SANDIFORD, M. & CUPPER, M. L. 2007c. Distinguishing tectonic from climatic controls on range-front sedimentation. *Basin Research*, doi: 10.1111/j.1365-2117.2007.00336.x.

RAIBER, M. & WEBB, J. 2008. Tectonic control of Tertiary deposition in the Streatham Deep-Lead System in western Victoria. *Australian Journal of Earth Sciences*, **55**, 493–508.

REYNOLDS, S. D., COBLENTZ, D. & HILLIS, R. R. 2003. Influences of plate-boundary forces on the regional intraplate stress field of continental Australia. *In*: HILLIS, R. R. & MULLER, R. D. (eds) *Evolution and Dynamics of the Australian Plate*. Geological Society of Australia Special Publication, **22**, 59–70.

RIEBE, C. S., KIRCHNER, J. W., GRANGER, D. E. & FINKEL, R. C. 2001a. Strong tectonic and weak climatic control of long-term chemical weathering rates. *Geology*, **29**, 511–514.

RIEBE, C. S., KIRCHNER, J., GRANGER, D. & FINKEL, R. 2001b. Minimal climatic control of erosion rates in the Sierra Nevada, California. *Geology*, **29**, 447–450.

RUSSELL, M. & GURNIS, M. 1994. The planform of epeirogeny: Vertical motions of Australia during the Cretaceous. *Basin Research*, **6**, 63–76.

RUTHERFURD, I. D. & KENYON, C. 2005. Geomorphology of the Barmah–Millewa forest. *Proceedings of the Royal Society of Victoria*, **117**, 23–39.

SANDIFORD, M. 2003a. Geomorphic constraints on the late Neogene tectonics of the Otway Ranges. *Australian Journal of Earth Sciences*, **50**, 69–80.

SANDIFORD, M. 2003b. Neotectonics of southeastern Australia: linking the Quaternary faulting record with seismicity and *in situ* stress. *In*: HILLIS, R. R. & MULLER, D. (eds) *Evolution and Dynamics of the Australian Plate*. Geological Society of Australia Special Publication, **22**, 101–113.

SANDIFORD, M. 2007. The tilting continent: a new constraint on the dynamic topographic field from Australia. *Earth and Planetary Science Letters*, **261**, 152–163.

SANDIFORD, M. & EGHOLM, D. L. 2008. Enhanced intraplate seismicity along continental margins: Some causes and consequences. *Tectonophysics*, **457**, 197–208.

SANDIFORD, M. & QUIGLEY, M. C. 2009. Topo-Oz: insights into the various modes of intraplate deformation in the Australian continent. *Tectonophysics*, doi: 10.1016/j.tecto.2009.01.028.

SANDIFORD, M., WALLACE, M. & COBLENTZ, D. 2004. Origin of the *in situ* stress field in southeastern Australia. *Basin Research*, **16**, 325–338.

SANDIFORD, M., QUIGLEY, M., DE BROEKERT, P. & JAKICA, S. 2009. Tectonic framework for the Cainozoic cratonic basins of Australia. *Australian Journal of Earth Sciences*, **56**, 5–18.

SHARP, K. R. 2004. Cenozoic volcanism, tectonism and stream derangement in the Snowy Mountains and northern Monaro of New South Wales. *Australian Journal of Earth Sciences*, **51**, 67–83.

SINGH, G., OPDYKE, N. D. & BOWLER, J. M. 1981. Late Cainozoic stratigraphy, magnetic chronology and vegetational history from Lake George, N.S.W. *Journal of the Geological Society of Australia*, **28**, 435–452.

STEPHENSON, A. E. 1986. Lake Bungunnia; a Plio-Pleistocene megalake in southern Australia.

Palaeogeography, Palaeoclimatology, Palaeoecology, **57**, 137–156.

STERN, T. A., QUINLAN, G. M. & HOLT, W. E. 1992. Basin formation behind an active subduction zone: three-dimensional flexural modelling of Wanganui Basin, New Zealand. *Basin Research*, **4**, 197–214.

STONE, J. & VASCONCELOS, P. 2000. Studies of geomorphic rates and processes with cosmogenic isotopes—examples from Australia. *Goldschmidt 2000 Journal of Conference Abstracts*, **5**, 961.

STONE, T. 2006. *Late Quaternary Rivers and Lakes of the Cadell Tilt Block Region, Murray Basin, Southeastern Australia*. PhD thesis, University of Melbourne.

TOKAREV, V., SANDIFORD, M. & GOSTIN, V. 1999. Landscape evolution in the Mount Lofty Ranges: Implications for regolith development. *Paper presented at Regolith '98; Australian Regolith and Mineral Exploration; New Approaches to an Old Continent; 3rd Australian Regolith Conference*, Kalgoorlie, Western Australia, 2–9 May 1998.

TOMKINS, K. M., HUMPHREYS, G. S. ET AL. 2007. Contemporary v. long term denudation along a passive plate margin; the role of extreme events. *Earth Surface Processes and Landforms*, **32**, 1013–1031.

TWIDALE, C. R. 1994. Gondwanan (Late Jurassic and Cretaceous) palaeosurfaces of the Australian Craton. *Palaeogeography, Palaeoclimatology, Palaeoecology*, **112**, 157–186.

TWIDALE, C. R. 2000. Early Mesozoic (?Triassic) landscapes in Australia: evidence, argument, and implications. *Journal of Geology*, **108**, 537–552.

TWIDALE, C. R. & BOURNE, J. A. 2000. Rock bursts and associated neotectonic forms at Minnipa Hill, northwestern Eyre Peninsula, South Australia. *Environmental and Engineering Geoscience*, **2**, 129–140.

TWIDALE, C. R. & BOURNE, J. A. 2004. Neotectonism in Australia: its expressions and implications. *Geomorphologie*, **3**, 179–194.

TWIDALE, C. R. & CAMPBELL, E. M. 2005. *Australian Landforms: Understanding a Low, Flat, Arid and Old Landscape*. Rosenberg, Kenthurst, NSW.

VAN DE GRAAFF, W. J. E., DENMAN, P. D. & HOCKING, R. M. 1976. Emerged Pleistocene marine terraces on Cape Range, Western Australia. *Western Australia Geological Survey Annual Report, 1975*, 62–70.

VEEVERS, J. J. 2000. *Billion-Year Earth History of Australia and Neighbours in Gondwanaland*. GEMOC Press, Sydney.

VITA-FINZI, C. 2004. Buckle-controlled seismogenic faulting in peninsular India. *Quaternary Science Reviews*, **23**, 2405–2412.

WACLAWIK, V. G., LANG, S. C. & KRAPF, C. B. E. 2008. Fluvial response to tectonic activity in an intra-continental dryland setting: The Neales River, Lake Eyre, Central Australia. *Geomorphology*, **102**, doi: 10.1016/j.geomorph.2007.06.021.

WALCOTT, R. I. 1998. Modes of oblique compression: late Cainozoic tectonics of the South Island of New Zealand. *Reviews of Geophysics*, **36**, 1–26.

WALLACE, M. W., DICKINSON, J. A., MOORE, D. H. & SANDIFORD, M. 2005. Late Neogene strandlines of southern Victoria: a unique record of eustasy and tectonics in southeast Australia. *Australian Journal of Earth Sciences*, **52**, 279–297.

WARD, I., NANSON, G., HEAD, L., FULLAGAR, R., PRICE, D. & FINK, D. 2005. Late Quaternary landscape evolution in the Keep River region, northwestern Australia. *Quaternary Science Reviews*, **24**, 1906–1922.

WEBB, J. A. & JAMES, J. M. 2006. Karst evolution of the Nullarbor Plain, Australia. *In*: HARMON, R. S. & WICKS, C. M. (eds) *Karst Geomorphology, Hydrology and Geochemistry—a Tribute Volume to Derek C. Ford and William B. White*. Geological Society of America, Special Papers, **404**, 65–78.

WEGMANN, K. W., ZURBEK, B. D. ET AL. 2007. Position of the Snake River watershed divide as an indicator of geodynamic processes in the greater Yellowstone region, western North America. *Geosphere*, **3**, 272–281.

WELLMAN, P. 1979. On the Cainozoic uplift of the southeastern Australian highland. *Journal of the Geological Society of Australia*, **26**, 1–9.

WELLMAN, P. & MCDOUGALL, I. 1974. Potassium–argon ages on the Cainozoic volcanic rocks of New South Wales. *Journal of the Geological Society of Australia*, **21**, 247–272.

WILLET, S. & BRANDON, M. 2002. On steady states in mountain belts. *Geology*, **30**, 175–178.

WILLIAMS, G. E. 1973. Late Quaternary piedmont sedimentation, soil formation and palaeoclimates in arid South Australia. *Zeitschrift für Geomorphologie*, **17**, 102–123.

WILLIAMS, G. E. & GOODE, A. D. 1978. Possible western outlet for an ancient Murray River in South Australia. *Search*, **9**, 443–447.

WILLIAMS, I. R. 1979. Recent fault scarps in the Mount Narryer area, Byro 1:250 000 sheet. *Western Australia. Geological Survey. Annual Report, 1978*, 51–55.

WOPFNER, H. 1968. Cretaceous sediments on the Mt Margaret Plateau and evidence for neo-tectonism. *Geological Survey of South Australia Quarterly Geological Notes*, **28**, 7–11.

YOUNG, R. W. & MACDOUGALL, I. 1993. Long-term landscape evolution: miocene and modern rivers in southern New South Wales. *Journal of Geology*, **101**, 35–49.

Lithology and the evolution of bedrock rivers in post-orogenic settings: constraints from the high-elevation passive continental margin of SE Australia

PAUL BISHOP[1]* & GEOFF GOLDRICK[2,3]

[1]*School of Geographical and Earth Sciences, East Quadrangle, University of Glasgow, Glasgow G12 8QQ, UK*

[2]*School of Geography and Environmental Science, Monash University, Melbourne, VIC 3168, Australia*

[3]*Present address: Coffs Harbour Senior College, Coffs Harbour, NSW 2450, Australia*

*Corresponding author (e-mail: paul.bishop@ges.gla.ac.uk)

Abstract: Understanding the role of lithological variation in the evolution of topography remains a fundamental issue, especially in the neglected post-orogenic terrains. Such settings represent the major part of the Earth's surface and recent modelling suggests that a range of interactions can account for the presence of residual topography for hundreds of millions of years, thereby explaining the great antiquity of landscapes in such settings. Field data from the inland flank of the SE Australian high-elevation continental margin suggest that resistant lithologies act to retard or even preclude the headward transmission of base-level fall driven by the isostatic response to regional denudation. Rejuvenation, be it episodic or continuous, is 'caught up' on these resistant lithologies, meaning in effect that the bedrock channels and hillslopes upstream of these 'stalled' knickpoints have become detached from the base-level changes downstream of the knickpoints. Until these knickpoints are breached, therefore, catchment relief must increase over time, a landscape evolution scenario that has been most notably suggested by Crickmay and Twidale. The role of resistant lithologies indicates that detachment-limited conditions are a key to the longevity of some post-orogenic landscapes, whereas the general importance of transport-limited conditions in the evolution of post-orogenic landscapes remains to be evaluated in field settings. Non-steady-state landscapes may lie at the heart of widespread, slowly evolving post-orogenic settings, such as high-elevation passive continental margins, meaning that non-steady-state landscapes, with increasing relief through time, are the 'rule' rather than the exception.

Bedrock channels are the 'backbone' of the landscape because they set much of the relief structure of tectonically active landscapes and dictate relationships between relief, elevation and denudation (Howard *et al.* 1994; Whipple & Tucker 1999; Hovius 2000). Bedrock channel evolution is thus the key to understanding landscape history and sediment flux: bedrock trunk channels provide the base level to which the whole drainage net and hillslope processes operate (Wohl & Merritts 2001). The last two decades have seen a blossoming of research on bedrock channels, reviving interest in, and building on, the fundamental research of Hack (1957) and Flint (1974) (Tinkler & Wohl 1998). This recent research has addressed a wide range of aspects of bedrock river morphology and processes, including: the distinction between debris-flow bedrock channels and fluvial bedrock channels (e.g. Stock & Dietrich 2003); the stream-power rule for bedrock river incision and its application in the numerical modelling of landscape evolution (e.g. Howard & Kerby 1983; Howard *et al.* 1994; Whipple & Tucker 1999; Tucker & Whipple 2002; van der Beek & Bishop 2003); refinement of the stream-power rule to take explicit account of sediment (e.g. Sklar & Dietrich 1998, 2001; Turowski *et al.* 2007); bedrock river long profile morphology and development (e.g. Bishop *et al.* 1985; Merritts *et al.* 1994; Whipple *et al.* 2000; Roe *et al.* 2002; Kirby *et al.* 2003; Kobor & Roering 2004; Brocard & van der Beek 2006; Goldrick & Bishop 2007); bedrock river knickpoint processes (Holland & Pickup 1976; Gardner 1983; Hayakawa & Matsukura 2003; Bishop *et al.* 2005; Crosby & Whipple 2006; Frankel *et al.* 2007); and bedrock river incision rates and processes (e.g. Bishop 1985; Burbank *et al.* 1996; Hancock *et al.* 1998; Sklar & Dietrich 1998; Hartshorn *et al.* 2002). The development of cosmogenic nuclide analysis (e.g. Bierman 1994; Bierman & Nichols 2004), with its capability of determining the exposure ages and/or erosion rates of bedrock

From: BISHOP, P. & PILLANS, B. (eds) *Australian Landscapes*. Geological Society, London, Special Publications, **346**, 267–287. DOI: 10.1144/SP346.14 0305-8719/10/$15.00 © The Geological Society of London 2010.

surfaces, has been a fundamentally important break-through in studying bedrock channels (e.g. Burbank *et al.* 1996; Hancock *et al.* 1998; Brocard *et al.* 2003; Reusser *et al.* 2006).

Much of this research has had a focus (explicit or implicit) on tectonically active areas, where high rates of rock uplift, high rates of seismicity, and high (to extreme) rates of precipitation 'drive' bedrock incision (e.g. Burbank *et al.* 1996; Hovius 2000; Hartshorn *et al.* 2002; Dadson *et al.* 2003, 2004). Likewise, steady-state rivers, in which rock uplift is matched by river incision and/or catchment lowering, underpin numerical modelling of bedrock rivers (e.g. Whipple & Tucker 1999) and physical modelling of eroding landscapes (e.g. Bonnet & Crave 2003). Less attention has been paid in recent work to post-orogenic terrains, such as passive continental margins, that are not experiencing continuing tectonically driven rock uplift (Bishop 2007). The focus on steady-state tectonically active terrains and the relative neglect of post-orogenic terrains probably reflect the fewer complexities associated with the conceptualization and numerical modelling of denudation of steady-state terrains (e.g. Whipple & Tucker 1999), and the impressive and eye-catching excitement of tectonically active terrains, where bedrock channel lowering may be physically measured in just 1 year (e.g. Hartshorn *et al.* 2002). On the other hand, the focus on orogenic belts has drawn attention away from understanding bedrock river processes and long-term landscape evolution in non-orogenic and post-orogenic areas, which constitute by far the majority of the Earth's subaerial surface area and which were the focus of research when geomorphology was first emerging as a discipline (Davis 1899, 1902).

Passive margin highland belts are perhaps the single post-orogenic terrain to have continued to receive widespread attention in the last two decades, especially in terms of low-temperature thermochronology (Moore *et al.* 1986; Dumitru *et al.* 1991; Brown *et al.* 2000a, b; Persano *et al.* 2002, 2005), landscape evolution (e.g. Bishop 1985, 1986; Gilchrist & Summerfield 1990; Pazzaglia & Gardner 1994, 2000; Gunnell & Fleitout 1998, 2000; Cockburn *et al.* 2002; Matmon *et al.* 2002; Campanile *et al.* 2008), bedrock river evolution (e.g. Bishop *et al.* 1985; Young & McDougall 1993; Goldrick & Bishop 1995), and numerical modelling (e.g. Kooi & Beaumont 1994; Tucker & Slingerland 1994; van der Beek & Braun 1998; van der Beek *et al.* 1999, 2001). Baldwin *et al.* (2003) used numerical modelling to explore the factors responsible for landscape persistence and the time scales of post-orogenic decay of topography. Their work addressed the formerly widespread viewpoint that landscapes

cannot be much older than the Tertiary and are probably no older than the Pleistocene (e.g. Thornbury 1969), a viewpoint that is demonstrably incorrect in many post-orogenic terrains (Young 1983; Twidale 1998). Baldwin *et al.* (2003) showed that a 'standard' detachment-limited river incision numerical model predicts that orogenic topography will decay to 1% of the original topography at the channel head within only about 1–10 Ma of the cessation of orogenic activity. The addition of isostatic compensation (i.e. rock uplift in response to denudation) increases the decay time to 10–30 Ma, and a switch to transport-limited conditions extends this to 36–90 Ma. Finally, if a critical shear stress for erosion is introduced, along with stochastic variability of flood discharge, the time scale of post-orogenic decay of topography extends to hundreds of millions of years.

Baldwin *et al.* (2003) also identified, but did not assess, other factors that could control the rate of post-orogenic topographic decay, including the slowing of landscape evolution by resistant lithologies. In this mechanism, resistant lithologies control the relaxation time of the long profile in (re-)attaining some form of equilibrium (steady-state) form after perturbation. We focus on this issue here and ask the question: What role does lithology play in bedrock river morphology and landscape evolution in post-orogenic terrains? We examine this issue in the context of the post-orogenic setting of the Lachlan River catchment on the high-elevation passive continental margin of southeastern Australia, by analysing the long profile morphology of the right- and left-bank tributaries of the upper Lachlan River, one of the major streams that drains the inland flank of the SE Australian highlands. Our long profile analysis is based on the DS form of the equilibrium bedrock river long profile (Goldrick & Bishop 2007).

Background: DS plots and long profile analysis and projection

The DS form of the equilibrium long profile plots the logarithm of a reach's slope against the logarithm of that reach's downstream distance; it is a slope–distance equivalent (hence 'DS') of the slope–area plot (e.g. Whipple & Tucker 1999) and takes the following form (Goldrick & Bishop 2007):

$$S = kL^{-\lambda} \text{ or } \ln S = \gamma - \lambda \ln L \qquad (1)$$

where S is channel slope, L is downstream distance, λ is a constant, k is a constant equal to RI_{grade}/il, and γ is equal to $\ln k$. R is some measure of lithological resistance to erosion, I_{grade} is the equilibrium rate of channel incision, i is a constant that describes the proportion of stream power that is expended in incision, and l is a constant.

A key feature of the DS form of the long profile is that, in principle, transient channel steepening, such as a knickpoint propagating in response to base-level fall, can be distinguished from equilibrium channel steepening in response to a more resistant lithology. The latter (equilibrium steepening on more resistant lithology) is indicated on the DS plot by a parallel shift in the plot, whereas a disequilibrium knickpoint plots as disordered outliers on the DS plot (Goldrick & Bishop 2007).

Long profile disequilibrium is commonly generated by a drop in base level and is resolved over time by the upstream passage of a wave of incision (a knickpoint), either as a retreating step that maintains the knickpoint's height and form, or as a step that rotate backwards by 'inclination' or 'replacement' and diffuses away (Gardner 1983; Bishop et al. 2005; Crosby & Whipple 2006; Frankel et al. 2007). At any time between the initiation of a persistent, retreating knickpoint and its arrival at the stream's headwaters the stream can be thought of as consisting of three reaches: the upstream reach, which is yet to be affected by the rejuvenation and may therefore remain graded to the previous base level; the over-steepened reach comprising the knickpoint; and the downstream reach that is graded to the new base level. To compare present and past base levels (and thereby to quantify, for example, the amount of trunk stream incision as a result of base level driven by surface uplift), it is necessary to reconstruct the pre-rejuvenation profile using the upstream (unrejuvenated) reach and to project that reconstructed profile downstream.

With the exception of theory-free curve fitting (e.g. Jones 1924; Hovius 2000), long profile reconstructions and projections used to compare present and ancient profiles and to identify changes in base level have been based on a priori models of the equilibrium long profile. We here use the DS form for such projections, rearranging Goldrick & Bishop's (2007) equation (2) to yield

$$\ln(H_0 - H) = \ln\left(\frac{k}{1-\lambda}\right) + (1-\lambda)\ln L. \quad (2)$$

It follows from equation (2) that the value of λ can be determined from the slope of a least-squares linear regression of the log of downstream distance v. the log of the fall $(H_0 - H)$. Once the value of λ has been determined, the value of k can be determined from the intercept (b) of the same regression:

$$b = \ln\frac{k}{1-\lambda} \text{ or } k = (1-\lambda)e^b. \quad (3)$$

The difficulty in solving for λ and k comes from the need to know the value of H_0, which is not the same as the elevation of the divide (Goldrick & Bishop 2007). This difficulty can be overcome by solving equations (2) and (3) iteratively. Estimates of H_0 can be substituted into equation (2) to yield values of λ that are used in turn to derive values of k. These can then be substituted into equation (2) to give a mathematical description of the long profile. The goodness of fit of each description so generated can be evaluated by comparison with the observed profile using the criterion of the standard error of the estimate of elevation, se_H (Goldrick & Bishop 2007). This procedure is repeated and the best estimate of H_0 determined by converging on the value that yields the best fit between the modelled and observed long profile.

The downstream projection of the equilibrium reach above a knickpoint to estimate the amount of base-level fall is illustrated by a hypothetical example in Figure 1, which shows a stream with a

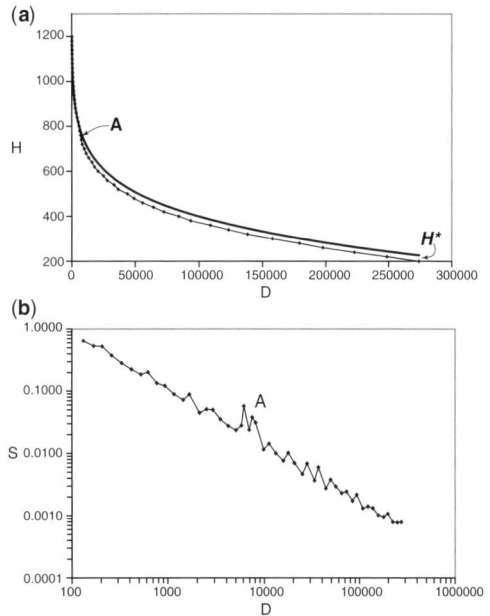

Fig. 1. (a) Long profile of a hypothetical stream (lower line) with $\lambda = 0.9$ and $k = 50$ for all reaches and where the standard error of elevation is 2 m and normally distributed; the reach from $D = 800$ to $D = 720$ is a diffuse knickpoint of 40 m fall generated by the headward propagation of a base-level fall of 40 m. The plot shows the downstream projection of the equilibrium profile upstream of A in accordance with the DS model (upper line); the method of projection has been presented by Goldrick & Bishop (2007). The DS-based estimate of the amount of incision in response to the 40 m base-level fall is 35 m (i.e. $H^* = 35$ m). (b) DS plot of this stream, showing the knickpoint at A.

knickpoint at A and a profile reconstruction and projection based on the DS model of the graded reach upstream of A. The magnitude of the 40 m base-level fall that generated the knickpoint at A can be estimated by projecting the reconstructed equilibrium profile above A to the stream's base level (Goldrick & Bishop 1995, 2007). The difference between the elevation predicted by the reconstructed profile and the actual elevation at any point along the stream is here called H^*, and the value of this difference at base level at the downstream limit of the stream is designated H_D^*. H_D^* is an estimate of the magnitude of stream incision in response to the base-level fall. The DS model prediction of the amount of incision/base-level fall at the downstream limit of the channel in Figure 1 is 35 m whereas the 'true' value is 40 m. The discrepancy of 5 m between the actual base-level drop and the value of H_D^* given by the DS model is mainly a result of the variability resulting from the standard error of 2 m associated with the elevation data. In consequence the calculated values of λ and k are 0.89 and 48, respectively, rather than the 'true' values of 0.9 and 50.

The confidence interval for H_D^* is given by

$$
t \sqrt{\frac{\sum\limits_{j=1}^{n}(H_j - \hat{H}_j)^2}{n-2}\left(\frac{1}{n} + \frac{(L_j - \bar{L})^2}{\sum\limits_{j=1}^{n}L_j^2 - n\bar{L}^2}\right)} \quad (4)
$$

where j is the point for which the confidence interval is to be calculated; n is the total number of points; H_j and $\hat{H}_j$ are the observed and predicted elevations at j, respectively; L_j is the downstream distance of j; and, $\bar{L}$ is the mean value of L (after Ebdon 1985, p. 117). A 95% confidence interval for the example in the preceding paragraph gives $H_D^* = 34 \pm 7$ m, a range that encompasses the actual base-level drop of 40 m. This range is large, relative to the size of the base-level drop, an outcome that emphasizes the sensitivity of long profile projections to errors in the input data, especially when a projection is made over long distances.

The projection of reconstructed profiles to the downstream limit of the stream allows the estimation of total base-level fall to which the stream has responded by incision and knickpoint propagation. Profile projection can also be used to quantify the height of a knickpoint, for which a projection is made from the (graded) section above the knickpoint to the upstream limit of the (graded) section downstream of the knickpoint. In the case of the stream in Figure 1, projection to the upstream limit of the graded reach below A estimates the height of the knickpoint to be 41 ± 2 m.

Denudational isostatic rebound of the Lachlan catchment and long profile adjustment

The Lachlan River catchment drains the inland (western) side of Australia's SE Highlands on the Tasman passive continental margin (Fig. 2). The Lachlan is bounded to the east by the continental divide, to the west by the Cenozoic intracratonic Murray Basin, and to the north and the south by catchments of the Macquarie and Murrumbidgee Rivers, respectively. Downstream of Mandagery Creek, the Lachlan flows across low-gradient interior depositional lowlands before terminating in a very low-gradient inland swamp (O'Brien & Burne 1994). The Murray–Darling system has its base level in the Southern Ocean some 1000 km downstream of Mandagery Creek, the downstream limit of the study area here (Fig. 2). The distance across the low-gradient, swampy, depositional lowlands to the base level in the Southern Ocean means that Cenozoic eustatic sea-level fluctuations are unlikely to have rejuvenated the Lachlan long profile.

Basaltic lavas were erupted into the Lachlan catchment at various times in the Tertiary, providing data on the drainage net at the times of eruption (Bishop et al. 1985; Bishop 1986). K–Ar ages on the basalts, coupled with reconstructions of the valleys into which the basalts flowed, demonstrate that much of the relief and many of the drainage systems of the SE Highlands, including the drainage network of the Lachlan, were established by the Early to Middle Tertiary (Wellman & McDougall 1974; Bishop et al. 1985; Bishop 1986; Young & McDougall 1993). These data notwithstanding, the history of the SE Highlands in New South Wales remains somewhat controversial in detail (Bishop & Goldrick 2000). It has been widely accepted that highlands surface uplift was related to Tasman margin extension and rifting around 100–90 Ma (Veevers 1984; Wellman 1987; Ollier & Pain 1994; O'Sullivan et al. 1995, 1996), which may have 'rejuvenated' a pre-existing topography (e.g. Persano et al. 2005). Few find evidence for active Neogene uplift of this portion of the highlands, notable exceptions being Wellman (1979a, 1987), van der Beek et al. (1995) (but see van der Beek et al. 1999), and Tomkins & Hesse (2004) for the catchment to the north. Recent interpretations of widespread neotectonics in Australia (see Quigley et al. 2010) do not seem relevant to the Lachlan, however, where the known rates of catchment-wide Neogene denudation of the Lachlan (Bishop 1985) and the isostatic equilibrium of the highlands (Wellman 1979b) mean that denudational isostasy can account for the observed

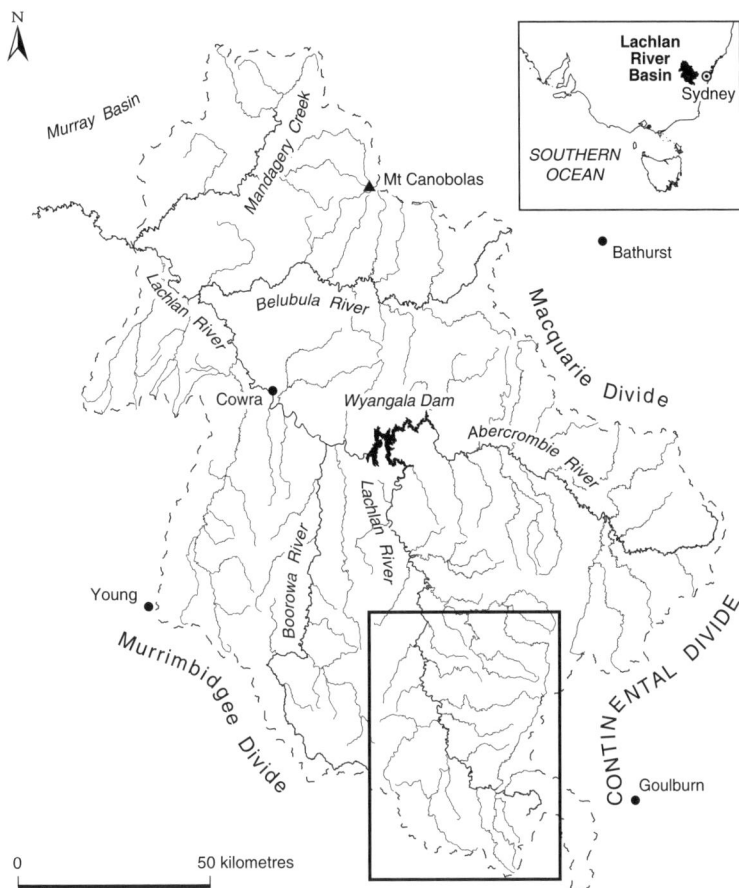

Fig. 2. The Lachlan River's uplands (bedrock) drainage net. Box indicates the area shown in Figure 4.

Neogene rock uplift in the Lachlan (Bishop & Brown 1992, 1993).

The strongest evidence for, and constraint upon, rock uplift along the Lachlan's highlands margin, whatever the mechanism for that uplift is, comes from the 12 Ma old Boorowa basalt (informal name), at the inland edge of the highlands (van der Beek & Bishop 2003; Figs 2 & 3). This basalt overlies fluvial gravels at an elevation 65 m above the modern Lachlan River, indicating a long-term incision rate of about 5 m Ma^{-1}. This incision must approximate the amount of relative base-level fall at the highlands margin, which forms base level for the bedrock Lachlan catchment studied here. In the absence of relative base-level fall at the highlands margin, incision rates at the highlands margin should be close to zero. Indeed, where the base level is the margin between eroding highlands and a depositional basin, as it is in the case of the Lachlan, one might expect an absence of surface

uplift to be accompanied by aggradation of the base level and a gradual up-catchment shift in the margin by back-filling of the highlands valley. In fact, the Lachlan upstream of the Boorowa basalt is characterized by an incised bedrock gorge with terraces up to 35 m above the modern river (Bishop & Brown 1992; van der Beek & Bishop 2003). As noted above, the known rates of denudation of the Lachlan catchment (Bishop 1985) mean that such base-level fall is most reasonably interpreted as the result of isostatic response to catchment-wide denudation (Bishop & Brown 1992, 1993).

The upper Lachlan

Bedrock in the Lachlan catchment consists of meridional belts of Palaeozoic granites and quartz-rich metasediments and silicic volcanic rocks (Fig. 3). The only post-Palaeozoic rocks are the thin Tertiary

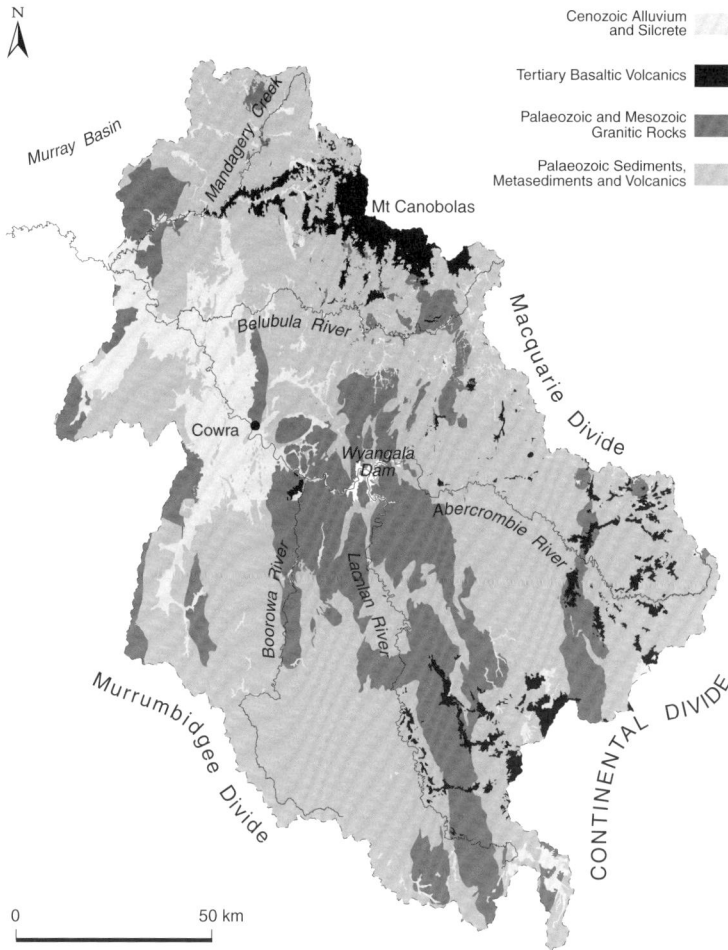

Fig. 3. Geology of the Lachlan's uplands (bedrock) catchment. The Tertiary basaltic volcanic rocks upstream of Cowra at the confluence of the Boorowa and Lachlan Rivers are the 12 Ma Boorowa basalt.

basaltic lavas noted above, and sediments scattered throughout the catchment and sporadic patches of Quaternary alluvium along drainage lines. Even in the areas of apparently more continuous alluvium, the channel bed appears to be everywhere formed in bedrock. Sediment that locally blankets the channel bed reflects catchment disturbance and sediment mobilization over the last century or so, following the introduction of sheep farming by European settlers. Thus, the entire drainage net is essentially formed in bedrock.

The narrow flows of Early Miocene Bevendale Basalt and Wheeo Basalt lie adjacent to the modern streams in the upper Lachlan (Fig. 4; Bishop 1984). These flows thus provide a maximum time limit for stream evolution and have been used several times for the study of evolution of bedrock river long

profiles (Bishop *et al.* 1985; Stock & Montgomery 1999; van der Beek & Bishop 2003). The flows are thin valley-fill basalts, preserving the ancient valleys into which they flowed (Fig. 4). Field relationships, such as the uniform elevation of the upper surfaces of the highest flow remnants, well below the interfluves, and the lack of any basalt remnants at higher elevations, demonstrate that the lavas did not fill their valleys and overtop the interfluves (Bishop *et al.* 1985; Bishop 1986). Post-basaltic stream incision has resulted in relief inversion so that flows now persist as hilltop remnants and ridges that discontinuously preserve the Early Miocene valleys.

The basement geology of the upper Lachlan consists of Ordovician quartz-rich slate and greywacke of the Adaminaby Group surrounding gneissic

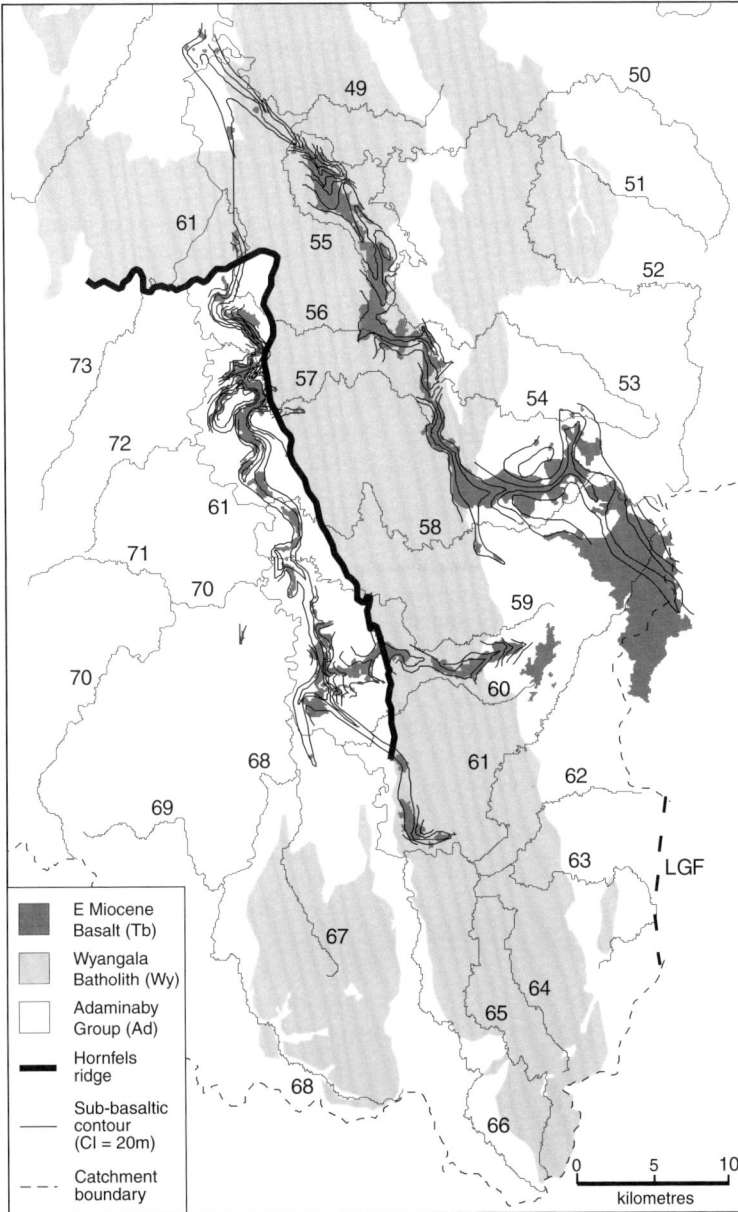

Fig. 4. The drainage network, geology and sub-basaltic contours of the upper Lachlan catchment. Sub-basaltic contours (20 m contour interval) have been taken from figure 8 of Bishop *et al.* (1985) and their elevations decrease westwards and northwards. These contour lines have not been labelled to avoid clutter and because it is their relative spacing and orientation rather than their absolute elevations that are important; likewise, we number (rather than name) the streams to avoid clutter and multiple stream names. The Bevendale Basalt is the southern- and westernmost linear flow remnant [flowing westwards and then northwards adjacent to the Lachlan trunk stream (Stream 61)], and the Wheeo Basalt is the parallel linear flow, to the north and NE of the Bevendale Basalt (Bishop 1984). The broad tabular basalt in the east, at the continental drainage divide, is the Divide Basalt. LGF, Lake George Fault.

granites of the Wyangala Batholith (Fig. 4); all of
these lithologies show meridional foliation. A well-
defined band of locally high relief roughly coincides
with the western parts of the granite, which is
drained by the upper Lachlan's eastern (right-bank)
tributaries, whereas the Ordovician metasediments
drained by the lower-relief western (left-bank)
tributaries are associated with lower relief because
of their lower resistance to erosion. A prominent
ridge of resistant contact metamorphic hornfels
extends for a considerable length along the western
contact of the Ordovician sediments with the granite
(Fig. 4). The modern Lachlan Rivers lies a few kilometres to the west of this ridge, with right-bank
tributaries flowing westward from the granite
through the ridge; basalt remnants demonstrate that
the same situation pertained in the Early Miocene.

Goldrick & Bishop's (2007) preliminary DS
analysis of the right- and left-bank long profiles
showed that the right-bank steepened reaches on
and about the resistant hornfels and granite lithologies are in a state of disequilibrium: the DS profiles
of the right-bank tributaries exhibit two outliers
associated with the resistant hornfels and the granite,
with these DS outliers separating essentially linear
DS reaches that we interpreted to be in equilibrium
(graded). The left-bank tributaries are much less
perturbed by disequilibria and generally grade
smoothly to the trunk stream. We now examine
this asymmetry further, using the DS plots.

The upper Lachlan's right-bank (eastern) tributaries

The right-bank tributaries rise at the continental
divide, descending the eastern flank of the upper
Lachlan's valley to the Lachlan; in doing so, they
cut through the hornfels ridge and across the Bevendale Basalt. The formerly continuous valley-filling
basalt thus now crops out as a series of hill-top cap-
pings isolated from each other by the tributaries.

We digitized the 'blue lines' on the New South
Wales Central Mapping Authority's 1:50 000 topographic sheets (20 m contour interval) to generate
long profiles and DS plots of the perennial streams
of the Lachlan's highlands drainage net (Goldrick
& Bishop 2007). As we noted above, the right-bank
tributaries generally exhibit linear (equilibrium) DS
profiles separated by knickpoints. Streams 59 and 60
are useful exemplars of these long profile characteristics; they are twin lateral streams to the major
east–west remnant of the lava that flowed to the
trunk Lachlan from the continental divide in the
Early Miocene (Bishop 1986) (Figs 5 & 6).

Stream 59 rises on the Adaminaby Group metasediments and flows westward across the Wyangala
Batholith below that east–west basalt cap. It cuts
through the hornfels ridge and the line of the
basalt before joining the trunk stream. The DS plot
shows that, with the exception of a small discontinuity, Stream 59 is graded where it flows across

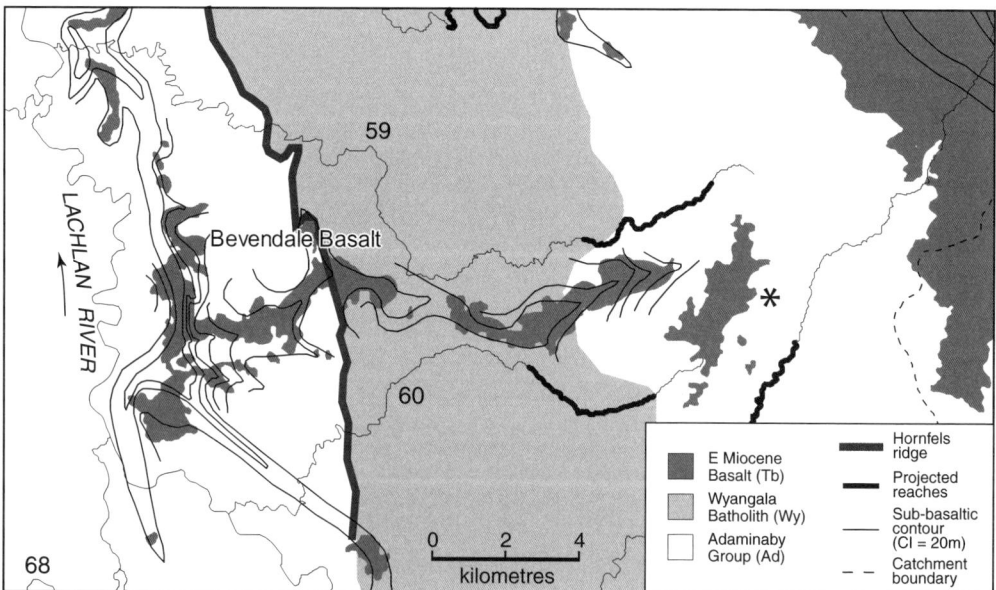

Fig. 5. Detail of drainage network, geology and sub-basaltic contours in the vicinity of Streams 59 and 60. The asterisk marks the western limit of the possible location of the drainage divide for the basalt-filled east–west tributary; the eastern limit for the divide location is the catchment boundary in the east (the continental drainage divide).

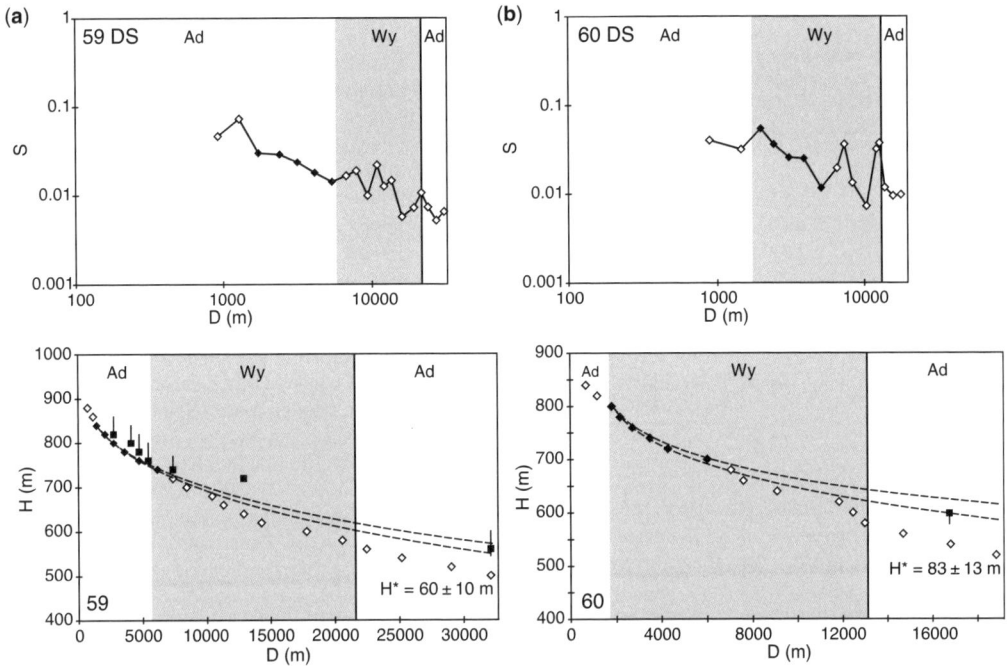

Fig. 6. The DS plot (top) and long profile (bottom) of Stream 59 (left) and Stream 60 (right). Diamonds, contour crossing; filled diamonds, data points used for profile reconstruction; dashed lines show upper and lower confidence limits of projections of reconstructed profiles; squares, elevations of base of basalt (from Fig. 5); fine vertical lines above squares indicate estimated thickness of the basalt flow; fine vertical lines below squares indicate uncertainty of elevation for base of the basalt. Shading and two-letter labels indicate underlying geology [Ad, Adaminaby Group (Palaeozoic quartz-rich metasediments); Wy, Wyangala batholith (Palaeozoic granitic rocks); bold vertical line, hornfels ridge].

the Adaminaby Group sediments (the standard error of the estimate of elevation, $se_H = 1.0$ m). Downstream, on the Wyangala Batholith, the DS plot is marked by high variability and several peaks or knickpoints, including one where the stream crosses the hornfels ridge. The long profile (Fig. 6a) shows that a downstream projection based on the upstream graded reach projects approximately to the base of the basalt where the stream cuts through it. At that point, the projection yields a value of H^* equal to 60 ± 10 m.

Stream 60 is almost the mirror image of Stream 59, rising on the Adaminaby Group metasediments and flowing westward across the granite, through the hornfels ridge, between basalt remnants and then joining the trunk Lachlan. The DS plot (Fig. 6b) shows a low-variability upstream reach that appears graded but the value of se_H is high at 2.8 m. Downstream of that graded reach there is high variability in the DS plot including two well-defined knickpoints, with the second of these located immediately upstream of the hornfels ridge. These knickpoints are clearly evident on the long profile (Fig. 6b). Downstream projection of the graded reach is graded approximately to the

basalt, with a value of H^* equal to 83 ± 13 m where it crosses the line of the basalt.

The profiles of all the eastern right-bank tributaries are broadly similar to those of Streams 59 and 60, being characterized by equilibrium reaches (low variability on the DS plot) interspersed with high-variability disequilibrium reaches. All the profiles show knickpoints near their junction with the trunk stream. The values of H^* associated with these knickpoints are in broad agreement, ranging from 60 ± 10 to 83 ± 13 m, save for the anomalously low, and currently inexplicable, $H = 16 \pm 4$ m for Stream 57 (Table 1). Streams 57 and 58 also have knickpoints further upstream for which the values of H^* at their confluences with the trunk are 122 and 143 ± 10 m, respectively (no uncertainty is given for the projection of the upstream graded reach in Stream 57 because it is based on only three points) (Table 1).

The spacing of the elevation contours of the base of the basalt shows that the north–south portion paralleling the Lachlan preserves the Early Miocene trunk and the east–west portion preserves an Early Miocene tributary, with a notably pronounced steepening upstream of the confluence with the trunk

Table 1. *Values of* H* *for long profile projections to the trunk Lachlan River of the graded reaches upstream of the downstream-most knickpoint in the upper Lachlan's eastern (right-bank) tributaries (see Fig. 4 for location of streams)*

Projected stream	$H*$ (m) at confluence with Lachlan
56	63 ± 4
57	16 ± 4
58	80 ± 5
59	60 ± 10
60	83 ± 13
57 (upstream)	122
58 (upstream)	143 ± 10
Miocene tributary	103 ± 30

(Fig. 4). The sub-basaltic elevation data are sufficiently complete to allow the construction of a DS plot and long profile for the Early Miocene channel down which it flowed. This projection assumes that the headwaters of the basalt-filled channel must have been at least as far east as the easternmost outcrop of the valley-filling basalt (asterisk in Fig. 5) and that the headwaters can have been no further east than the modern continental divide, because the position of the continental divide has been stable since pre-Miocene times (Bishop 1986).

The Miocene basalt-filled channel's DS plot (Fig. 7) shows a marked terminal knickpoint and a steep, but nonetheless linear reach upstream of that. The best fit for a reconstruction of the long profile based on that linear upstream reach is achieved when the Miocene divide is presumed to be at the eastern limit of the valley-filling basalt (asterisk in Fig. 5), but this is a rather poor fit with se_H equal to 7.5 m. Projecting this profile yields a value of $H*$ equal to 103 ± 30 m where

the tributary basalt flow meets the Miocene trunk Lachlan at the elbow in the flow adjacent to the modern Lachlan. In short, and notwithstanding the uncertainties, the Miocene tributary, like most of the modern tributaries, was characterized by considerable disequilibrium and a steep fall to the Miocene trunk.

The knickpoint steepenings, both modern and Miocene times, are broadly spatially associated with granite and/or hornfels, but there is no clear and precise relationship between steepening and geology in general, nor between the terminal knickpoints and the hornfels ridge in particular. Several streams have graded reaches that cross the geological boundaries, whereas others have knickpoints on both the granite and the metasediments. Only in the case of Stream 59 is the terminal knickpoint coincident with the hornfels ridge. The terminal knickpoints of Streams 60 and 58 are upstream of the hornfels ridge on the granite, and the terminal knickpoints of Stream 57 and the Miocene tributary are downstream of the hornfels ridge on the metasediments. The clearest relationship that does emerge from the long profiles analysis is that those streams that flow across the path of the Bevendale Basalt have an upstream reach that is graded approximately to the base of the basalt at the point where they cross the basalt.

The upper Lachlan's left-bank (western) tributaries

Streams 67 and 69 are tributaries of Stream 68 (Fig. 4). The headwaters of Stream 67 flow across granitic rocks of the Wyangala Batholith whereas downstream reaches flow across metasedimentary rocks of the Adaminaby Group. The headwaters of Stream 68 are also on the granite and in its lower reaches it flows across Adaminaby Group metasediments. Stream 69 flows entirely across

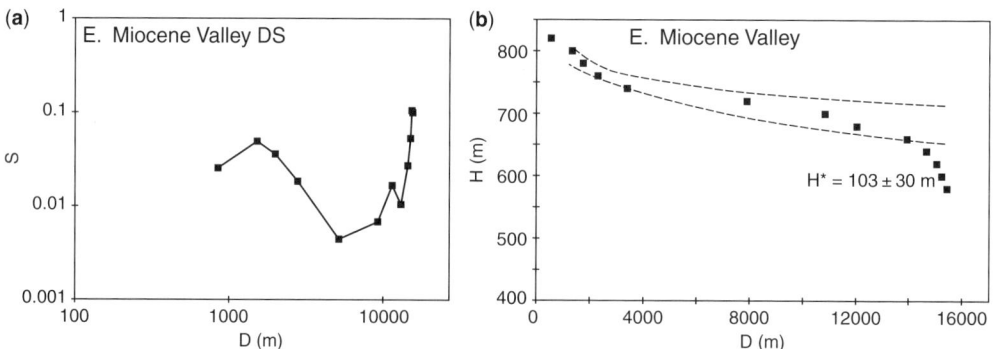

Fig. 7. (**a**) DS plot and (**b**) long profile of the sub-basaltic elevations of the east–west (tributary) Early Miocene lava flow, assuming that the Miocene divide was at the asterisk in Figure 5.

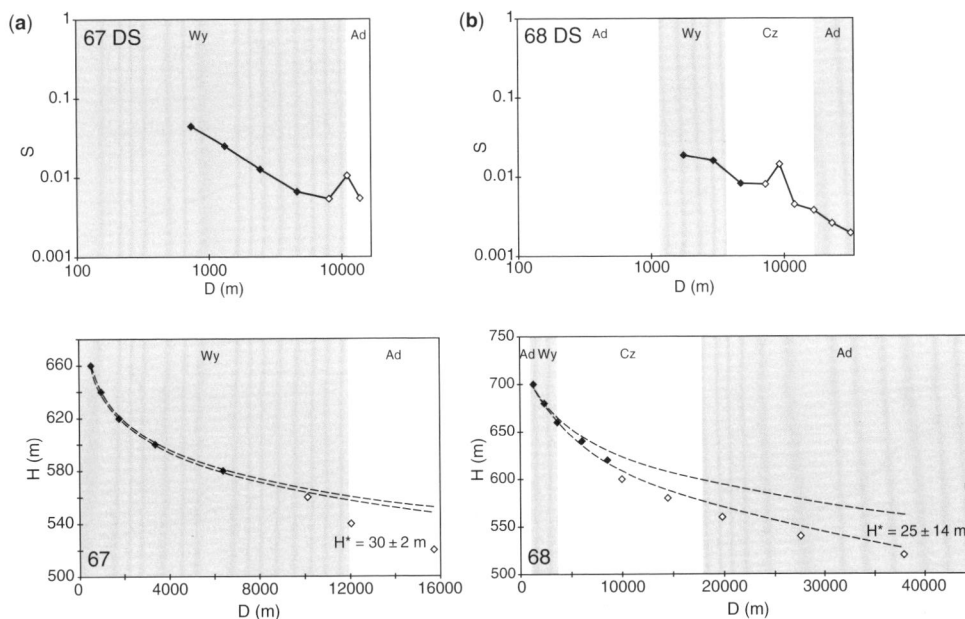

Fig. 8. The DS plot (top) and long profile (bottom) of (**a**) Stream 67 and (**b**) Stream 68. Symbols as in Figure 6; Cz, Cenozoic.

metasedimentary rocks of the Adaminaby Group. The long profiles and DS plots of Streams 67 and 68 (Fig. 8) show that both streams are graded throughout most of their lengths except for the presence of a relatively small, but well-defined, knickpoint on each stream. The values of H^* at the confluence of Streams 67 and 68 are 30 ± 2 m ($se_H = 0.52$ m) and 25 ± 14 m ($se_H = 1.16$ m), respectively. Despite the very large confidence interval associated with the projection of Stream 68, because of the long distance over which it has been projected, the values of H^* are very similar, suggesting that the knickpoints are the result of base-level change rather than lithological variation (see Goldrick & Bishop 1995, 2007). The value of H^* for Stream 68 upstream of the junction with the Lachlan is 26 ± 17 m, about 50 m less than typical values for the eastern tributaries, and the projected profile lies c. 40 m below the basalt.

The DS plot of Stream 69 shows that the stream is not far from graded ($se_H = 2.6$ m) but there is a suggestion of a discontinuity that does not appear to be associated with a lithological change (Fig. 9a). A projection of the upstream reach of Stream 69 ($se_H = 1.27$ m) to the confluence with Stream 68 returns a value of $H^* = 16 \pm 5$ m (Fig. 9a). This is similar to the value of H^* for Stream 68 itself at this point (25 ± 14 m) so that there is a suggestion that the discontinuity along Stream 69 has a similar base-level fall origin to

the knickpoints on Streams 67 and 68, but the steepening on the Stream 69 is more a diffuse 'knick-zone' than a discrete knickpoint.

In summary, the western tributaries of the upper Lachlan River show some evidence of disequilibrium steepening unrelated to lithology, but the degree of steepening along these streams and the corresponding values of H^* (Table 2) are much less than those in the eastern (right-bank) tributaries. Several of the tributaries have comparable values of H^* at their junctions, lending support to the proposal that this steepening is due to a relative fall in base level.

Synthesis

The eastern (right-bank) tributaries are characterized by marked steepening, especially in their distal reaches, whereas the western (left-bank) tributaries have much smoother profiles. The southern tributaries are not treated in detail here, but they are transitional between these two groups, with Streams 61 to 63 similar in form to the eastern tributaries and Streams 64 to 66 similar in form to the western tributaries. The obvious and simplest explanation for the pronounced east–west asymmetry in the tributary long profiles is the lithological influences exerted on the eastern tributary long profiles by the granite–hornfels combination and the basalt. The western tributaries flow mainly across

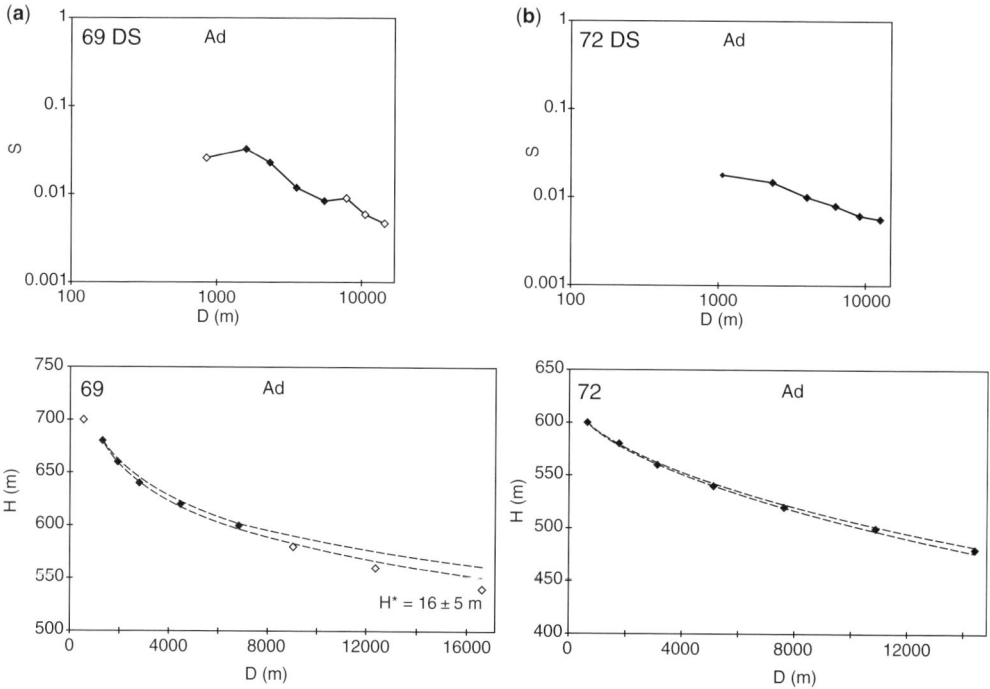

Fig. 9. The DS plot (top) and long profile (bottom) of (**a**) Stream 69 and (**b**) Stream 72. Symbols as in Figure 6.

Table 2. *Values of* H* *for long profile projections of upstream graded reaches of the western (left-bank) tributaries to the upper Lachlan River; and the confluences to which they have been projected*

Projected stream	H^* (m) at confluence	Confluence
67	30 ± 2	67 and 68
68	26 ± 17	67 and 68
68	25 ± 14	68 and 69
69	16 ± 5	68 and 69
73	21 ± 0	73 and 61

Adaminaby Group rocks of moderately low resistance whereas the eastern tributaries flow across a variety of rock types including the more resistant granites of the Wyangala Batholith and the hornfels ridge. The eastern tributaries also all cut through the Bevendale Basalt upstream of their junctions with the trunk stream. Finally, and most importantly, all the eastern tributaries have a reach that is graded approximately to the base of the Bevendale Basalt where the tributaries cross the line of the basalt.

A major perturbation was introduced into the drainage net by the Bevendale Basalt flowing down the east–west tributary and then northward along the Miocene trunk Lachlan. The fact that all the eastern tributaries have a reach that is graded approximately to the base of the Bevendale Basalt suggests the evolutionary history presented schematically in Figure 10. After the eruption, a new trunk stream formed on the western edge of the valley basalt so that the western tributaries maintained an unimpeded path to that new trunk whereas the eastern tributaries flowed across the newly emplaced (and resistant) valley basalt to join the trunk.

The basalt in the trunk would no doubt have introduced some disequilibrium into the profiles of the western streams (flattening them) but in the absence of resistant rocks it is likely that the new trunk and the western tributaries re-attained equilibrium early after the eruption. Disruption of the eastern tributaries must have been considerably greater. The re-establishment of a graded landscape east of the trunk stream would have been inhibited by the greater resistance of the granite, hornfels and the basalt. The influence of the last was particularly important, because the eastern tributaries' low gradients across the top of the basalt, relative to the basalt's lithological resistance, would have led to a differential between the rate at which the trunk and its western tributaries were incising and that at which the eastern tributaries were incising into the basalt. If the attainment of grade in detachment-limited, post-orogenic settings, such as this, is a

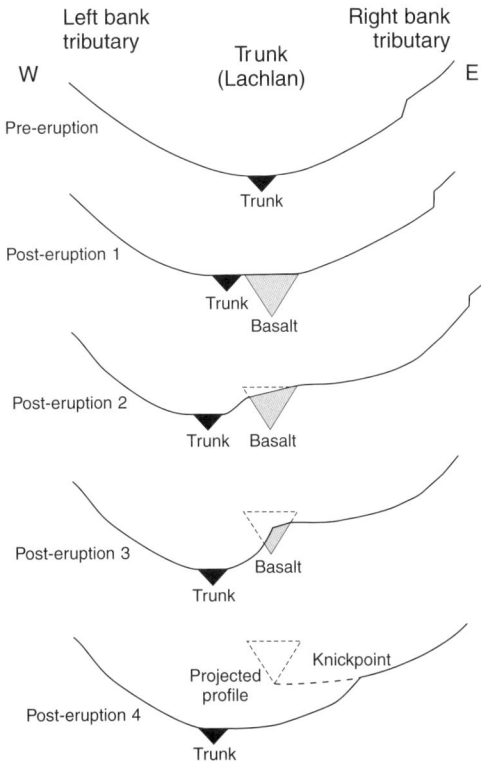

Fig. 10. Schematic illustration of hypothesized post-eruption evolution of tributaries of the upper Lachlan. After the eruption, the trunk stream is re-established on the left side of its basalt-filled valley. Trunk-stream incision generates knickpoints where right-bank tributaries cross the basalt; left-bank tributaries do not have to incise basalt. Long profile projections show that the right-bank tributaries above the knickpoint are graded to the base of the basalt (Fig. 6), whereas the western tributaries did not have to negotiate the basalt and so do not exhibit such grading (Figs 8 & 9, especially Fig. 8b). In the top three panels an earlier knickpoint is also shown propagating up the right-bank tributary.

'bottom-up' process (Bishop 2007) then the basalt would have acted as a local base level with the upstream reaches of the eastern tributaries becoming graded to it. Differential incision rates would have resulted in a steepening of the gradients between the new trunk stream and the basalt (Fig. 10, Post-eruption 2).

Over time, the trunk would have continued to incise at a faster rate than the tributaries on, and upstream of, the basalt so that the height differential between the two would have increased, as would have the gradients across the basalt. Throughout this stage (Fig. 10, Post-eruption 3), the basalt would have acted as a temporary local base level for the

eastern tributaries. Eventually these tributaries would have cut through the basalt, removing the temporary base level provided by the basalt and triggering the upstream migration of a knickpoint (Fig. 10, Post-eruption 4). This stage, which represents the present state of the upper Lachlan drainage net, is characterized by a downstream disequilibrium reach and an upstream reach that is graded approximately to the level of the basalt remnant at the time when it was breached (i.e. the base of the flow).

This model offers the most likely explanation for the gross morphology of the tributaries of the upper Lachlan catchment, with the western tributaries essentially graded to the modern trunk with only minor disequilibrium, and the eastern tributaries characterized by marked downstream disequilibrium steepening and an upstream reach that is graded approximately to the base of the basalt. The data confirm that resistant lithologies can act as temporary local base levels and retard the re-establishment of equilibrium, influencing relaxation times of streams so that resistant reaches can temporarily isolate upstream reaches from the effects of downstream perturbations.

Numerical simulation of the influence of lithology on stream evolution

This interpretation of the role of the basalt in delaying the headward transmission in the right-bank tributaries of base-level fall in the trunk was tested using a 1D finite-difference stream evolution simulation based on the DS form, in which the amount of incision at each iteration is determined by the variables k, λ, L, i and S [see equation (1) above, and Goldrick & Bishop (2007)]. The modelled long profile consists of 100 reaches: reaches 1–72 and 89–100 have a lithological resistance corresponding to $R = 50$, and reaches 73–88 have a doubled lithological resistance of $R = 100$. Each point, x, in the long profile is lowered according to the formula

$$\text{incision} = \frac{il}{R} L_x^{\lambda} \left(\frac{H_x - H_{x+1}}{L_{x+1} - L_x} \right) \qquad (2)$$

where H_x, H_{x+1}, L_x and L_{x+1} are the elevations and downstream distances, respectively of x and $x + 1$. At each iteration the amount of incision is calculated for each reach before the elevations are adjusted.

The development of the profile in the 'posteruption' period is recorded in Figure 11. Figure 11a, b depicts the immediate post-eruption profiles of the tributary, which can be divided into

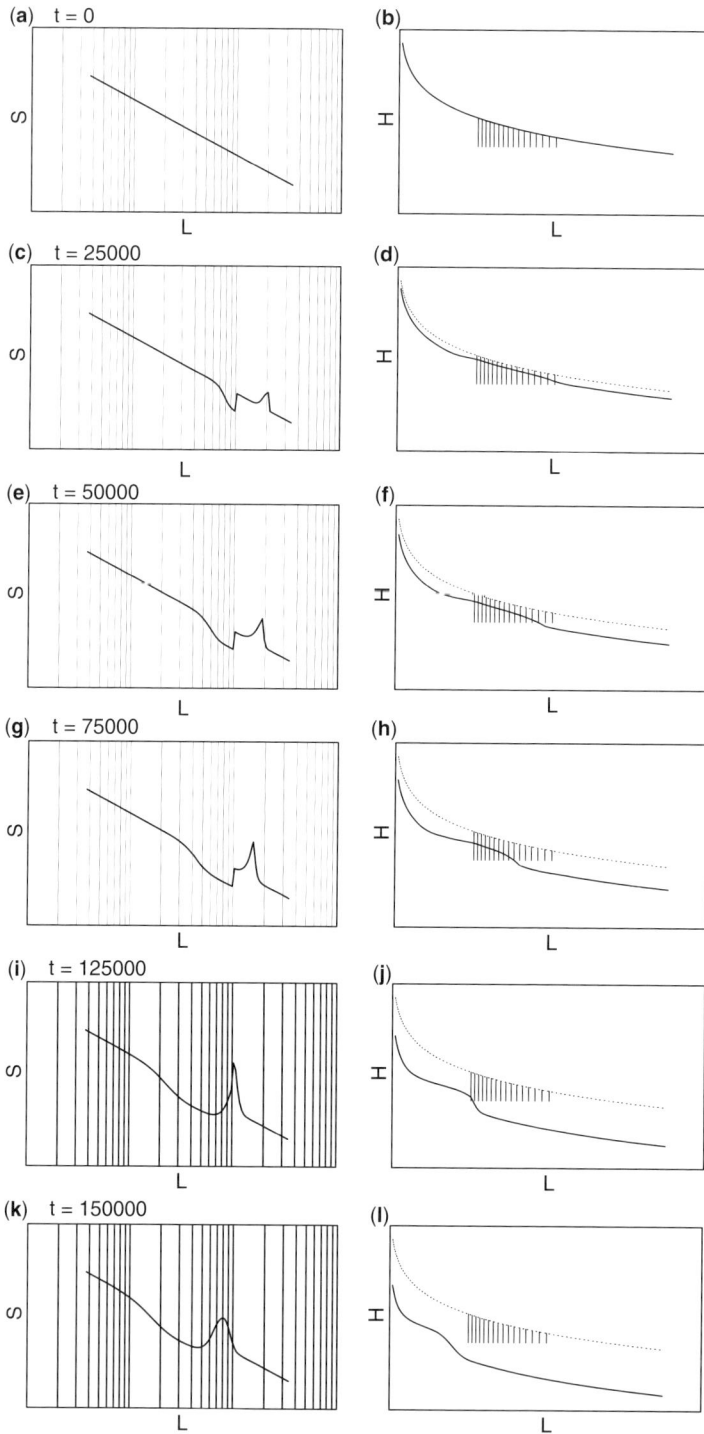

Fig. 11. Numerical simulation of the evolution of the DS plot and long profile of an initially graded stream profile after the emplacement of a segment of resistant lithology. The dashed line on the long profile is the original profile, and the closely spaced vertical lines indicate the extent and thickness of the resistant lithology. t = 0, t = 25 000, etc (upper left of (a), (c), etc) is step number (notional time) in numerical model.

three reaches. The reach downstream of the basalt is graded to the trunk stream, which acts as a base level, and this reach continues to incise at the same rate as the trunk, at the long-term denudation rate driven by denudational isostatic rebound. The reach flowing across the basalt is in disequilibrium because its gradient is low relative to the erosional resistance of the basalt, and it therefore incises at a slower rate than the trunk and the downstream reach. The reach upstream of the basalt is also no longer in equilibrium because the basalt is acting as a temporary local base level that is lowering at a slower rate than the base level to which this reach was previously graded.

After 25 000 iterations (Fig. 11c, d), a clear knickpoint, indicated by a peak on the DS plot, has formed at the downstream edge of the basalt as a result of the rate differential between incision on the country rock and on the basalt. This incision rate differential results in a decrease in the gradients immediately upstream of the basalt, as indicated by low values on the DS plot and the flattening of the long profile immediately upstream of the basalt. After 50 000 iterations the knickpoint has eroded through the downstream edge of the basalt and has become more pronounced (Fig. 11e, f). At the same time gradients upstream have decreased as an increasingly extensive reach has become graded to the upstream edge of the basalt.

After 75 000 iterations an even more pronounced knickpoint has formed and the low-gradient reach has become more extensive. This pattern continues until, at 125 000 iterations, the stream is at the point of breaching the basalt (Fig. 11i, j). After the basalt has been breached, the knickpoint begins to migrate more rapidly upstream and declines ('lies back') so that the peak on the DS plot is less marked and knickpoint becomes laterally more extensive. That is, the knickpoint evolves from a discrete knickpoint into a more diffuse knick-zone. The form of the stream after 150 000 iterations (Fig. 11k, l) is similar to the modern forms of the eastern tributaries of the upper Lachlan, with an upstream reach graded approximately to the base of the basalt, and a knickpoint between this reach and the trunk stream.

The simulation is consistent with the hypothesis that basalt (or any other resistant rock) acting as a temporary local base level can retard the development of the upstream reaches, induce the formation of a knickpoint, and produce the type of profile that is typical of the eastern tributaries.

Discussion

The 'engine' for landscape evolution in the post-orogenic setting of the Lachlan River is denudational isostatic rebound: denudational unloading of the uplands, at rates of $c.$ 5 m Ma^{-1} (Bishop 1985), triggers the isostatic response of rock uplift, which in turn triggers drainage net rejuvenation at the drainage net's base level (in this case the inland edge of the highlands, where the Lachlan flows onto the sedimentary fill of the continental interior) (Bishop & Brown 1992). For reasons that are not yet fully clear, but that probably reflect discontinuous movement on the highland-edge fault(s) on which the denudational isostatic rebound is accommodated, this rejuvenation is evidently discontinuous. Each denudational isostatic uplift 'event' presumably triggers a steepened reach where the Lachlan leaves the highlands around Cowra, and this steepened reach propagates as a knickpoint, headwards along the trunk stream and progressively through the drainage net. The contrast in the long profiles of the upper Lachlan's eastern (right-bank) and western (left-bank) tributaries shows that resistant lithologies act to slow knickpoint propagation, whereas non-resistant lithologies allow rejuvenation to be readily propagated headwards and for graded (equilibrium) long profiles to be re-established (e.g. Fig. 9). There are no indications of the magnitude of each of these rock uplift events, but if the altitudinal separation of each of the terraces in the bedrock reach upstream of the highlands edge (Bishop & Brown 1992) is indicative of the magnitude of the uplift, then each is very substantial. That said, there may be a climatic component to the generation of these terraces and, in any event, the key issues are the triggering of knickpoints at the highlands edge by rock uplift of unknown magnitude as a result of denudational rebound, the propagation of those knickpoints through the drainage net, and the retardation of these knickpoints by resistant lithologies. The values of H^* in the upper Lachlan right-bank tributaries that flow across resistant lithologies integrate all the rebound events for a given time interval.

This lithological retardation of knickpoint propagation is independent of the eruption of basalt into the upper Lachlan's drainage net, as confirmed by the knickpoints in the long profile into which the basalt was extruded at 21 Ma (Fig. 7). In other words, in the absence of the basalt, the lithological resistance of the granite and hornfels still acts to retard knickpoint propagation. That conclusion is confirmed by the fact that Streams 57 and 58 also have upstream knickpoints for which the graded reaches above the knickpoints have values of H^* of 122 and 143 $\pm$ 10 m at their confluences with the trunk (see above). Those graded reaches project to elevations above the basalt and so must predate that basalt. As shown above (Fig. 7), projection to the trunk stream of the long profile of the basalt-filled east–west tributary yields

$H^* = 103 \pm 30$ m and so the graded reaches with $H^* = 122$ and 143 m in Streams 57 and 58 are likely to reflect long profile equilibria prior to 21 Ma, given that they project to a base level in the trunk that is 20–40 m above that to which the basalt-filled Early Miocene tributary long profile projects.

The perturbation caused by the basalt in the trunk stream is, in one sense, a special case, but it has general implications in allowing us (1) to use the instantaneous 'injection' of a resistant lithology into the long profile to assess the role of resistant lithologies in retarding knickpoint propagation, and (2) to place that retardation in a dated time-frame. The basalts have clearly retarded the propagation of post-basaltic rejuvenation into the right-bank tributaries, with the implication that the reaches below the lowermost knickpoint in those right-bank tributaries are still adjusting to post-basaltic rejuvenation (unlike the left-bank tributaries, which have long ago accommodated the earlier post basaltic rejuvenation). That is, H^* in the left-bank tributaries is much less than the 50–80 m of post-basaltic incision along the trunk (which approximates the right-bank H^* values of 60–80 m; Table 1). An important implication of that interpretation is that the left-bank tributaries must be incising at approximately the same rate as the trunk Lachlan whereas the right-bank tributaries must have a range of incision rates along their lengths: between the trunk and the right-bank tributaries' first knickpoint, incision rates must range from the Lachlan's rate in the tributaries' downstream reaches, to (much?) higher rates in the knickpoint reaches, where the delayed rejuvenation is still being accommodated. In the graded reaches of the right-bank tributaries above those downstream-most knickpoints, incision rates are set by the channels' discharges, sediment fluxes and gradients, and not by the trunk stream's rate of incision. In each case, the knickpoint in effect disconnects the graded reach and the trunk Lachlan.

More generally, the data reported here demonstrate the central role that resistant lithologies play in slowing landscape evolution in post-orogenic settings in which landscape response to rock uplift is via bottom-up processes of headward propagating rejuvenation (see Baldwin et al. 2003). The knickpoints on resistant lithologies in the Lachlan's eastern (right-bank) tributaries mean that the upper catchments are not lowering at the same rate as the trunk stream and the left-bank tributaries, and that the relief of the catchment must be increasing. This conclusion is confirmed by the fact that the headwaters of the modern streams rise on the Early Miocene basalts whereas these modern streams have incised below the basalts (Fig. 4; Bishop et al. 1985). We note that the presence of the basalts demonstrates the reality of increasing relief and places a time scale on that relief increase, but it is not a prerequisite for that relief increase.

The numerical modelling of Baldwin et al. (2003) highlights the role of transport-limited conditions in extending the 'life' of post-orogenic terrains to tens and hundreds of millions of years. Those transport-limited conditions mean that river beds consist of a layer of sediment covering the bedrock of the channel bed, shielding it from erosion (see Sklar & Dietrich 1998, 2001). In such transport-limited situations, rivers rarely 'interact' with the substrate and, in effect, do not 'feel' lithological variation (e.g. Brocard & van der Beek 2006). We do not envisage that situation to be the case in the upper Lachlan, which is characterized by low to very low rates of denudation and sediment flux (see Bishop 1985). The knickpoint reaches are not mantled by sediment and so their low rates of erosion are due to a combination of factors including their lithological resistance and the low rates of sediment flux for abrading the bed. In other words, detachment-limited conditions must also be incorporated, and may be a central element, in explanations of the longevity of post-orogenic terrains.

We envisage that the knickpoints are slowed and caught on the steeply dipping hornfels and jointed and foliated granites of the upper Lachlan as has been documented in field and laboratory observations by Frankel et al. (2007). In their experiments, a knickpoint on steeply dipping resistant lithologies evolved by a combination of parallel retreat on the knickpoint face and vertical channel incision on the knickpoint 'top' [although it is unclear whether this channel incision on the knickpoint top is the commonly observed draw-down effect on knickpoints; Haviv et al. (2006)]. Whatever its precise nature, that vertical channel incision must limit the height that knickpoints can attain, and, in any event, knickpoints cannot continue to grow indefinitely because knickpoints of extreme heights will not be stable. The Lachlan data show that knickpoints eventually cut through resistant lithologies and continue their headward propagation. Breaching of the basalts reflects the finite vertical thickness of the basalt (Fig. 10) but the presence of more than one knickpoint in some of the right-bank tributaries (e.g. Streams 57 and 58), with the graded reaches above the upper knickpoints projecting to elevations above the basalt-filled Early Miocene channel, confirms that knickpoints do propagate through the hornfels and into the granites.

Bishop et al. (1985) and Young & McDougall (1993) observed that the passage of the whole of the Neogene (and almost certainly longer) has not resulted in anything that approaches planation of the post-orogenic landscape of SE Australia. Quite the opposite, the data here show increasing relief, much as hypothesized by Crickmay (1974, 1975) and Twidale (1976, 1991), who has been a major champion of Crickmay's 'hypothesis of unequal activity'. This hypothesis proposes that fluvial erosional energy is progressively concentrated in large river valley bottoms and that lower erosion rates in the smaller rivers of the upland areas mean that denudation of upland areas slows (see Bishop 2010). Incision of major channels mean that they become decoupled from slopes. That is, relief amplitude must increase as rivers continue to incise and upstream areas erode more slowly (Crickmay 1975; Twidale 1976). Formalizing this approach in a model of landscape evolution involving increased and increasing relief amplitude (and going beyond the relative sizes of trunk and tributary streams proposed by Crickmay as the key issue), Twidale (1991) highlighted the role of resistant lithologies, structure and different groundwater conditions throughout a drainage basin as the central factors responsible for increasing relief. As Twidale argued: 'water ... is concentrated in and near major channels, for, once a master stream develops, not only surface water but subsurface drainage too gravitates towards it ... [U]plift induces stream incision and water-table lowering, leaving high plains and plateaux perched and dry' (Twidale 1998, p. 663). The retardation of knickpoints on resistant lithologies adds a further dimension to that argument, as does the lack of discharge and sediment supply to incise the bedrock channels and these knickpoints.

Conclusion

The post-orogenic landscape of SE Australia allows the clarification of several major issues in long-term landscape evolution, not least because of its excellent evidence of landscape evolution as recorded by Cenozoic valley-filling basalts. These basalts provide chronologically rigorous evidence of Cenozoic landscape character, long profile morphology and rates of landscape evolution, as well as the opportunity to assess responses to the perturbation of a major resistant lithology being introduced into the drainage network. The long profile morphology of the Lachlan River drainage net demonstrates that continuing rock uplift driven by denudational isostatic rebound is propagated headwards through the drainage net from the inland edge of the bedrock highlands. That rock uplift 'signal' is transmitted more rapidly through those parts of the drainage net formed on less resistant lithologies, such as regionally metamorphosed metasandstones and phyllites, and more slowly on more resistant granites and contact metamorphic hornfels. The basalts themselves also retard landscape response to rock

uplift, but the slowing of such responses occurs on the granites and contact metamorphic rocks, whether the basalts are present or not.

Four general conclusions may be drawn from the study reported here, as follows.

(1) Denudational isostatic rebound is an important and fundamental mechanism for prolonging the time scale for the post-orogenic decay of topography (Bishop & Brown 1992; Baldwin et al. 2003).

(2) Resistant lithologies, and the delay that they exert on the transmission of the signal of rock uplift triggered by denudational isostatic rebound, are further important factors in prolonging the time scale of post-orogenic decay of topography; this second group of factors has hitherto not been evaluated rigorously.

(3) The role of resistant lithologies indicates that detachment-limited conditions are a key to the longevity of (at least some) post-orogenic landscapes. The general importance of transport-limited conditions, as proposed by Baldwin et al. (2003), remains to be evaluated in field settings.

(4) The data demonstrate that the appropriate model for the evolution of landscapes such as that described here is one of spatially unequal activity and increasing relief, as Crickmay (1974, 1975) and Twidale (1991) have emphasized.

The delays in the propagation of knickpoints may persist for considerable periods, reinforced by very low stream power that reflects low discharges in these often semi-arid, intra-continental interiors and their low gradients, as well as low fluxes of sediment to act as erosional 'tools'. In other words, non-steady-state landscapes may lie at the heart of widespread, slowly evolving post-orogenic settings, such as high-elevation passive continental margins, meaning that non-steady landscapes, with increasing relief through time, are the 'rule' rather than the exception on the Earth's surface.

This research was supported by a Monash University Graduate Scholarship and a Research and Travel Grant from the University of Edinburgh Department of Geography (both to G.G.), Australian Research Council grants and a Leverhulme Visiting Fellowship in the University of Edinburgh Department of Geography (both to P.B.), and Monash University. We thank P. van der Beek and C. Pain for their comments, which improved the first draft of this paper. The work reported here is drawn from the PhD theses of the two authors, Bishop's supervised by Martin A. J. Williams and Goldrick's by Bishop. P.B. records here his sincere thanks to M.A.J.W. for inspiration and guidance over an extended period, starting in P.B.'s undergraduate years. Martin's contribution lives on in his students and in their students.

References

BALDWIN, J. A., WHIPPLE, K. X. & TUCKER, G. E. 2003. Implications of the shear stress river incision model for the timescale of postorogenic decay of topography. *Journal of Geophysical Research B*, **108**, 2158, doi: 10.1029/2001JB000550.

BIERMAN, P. R. 1994. Using *in situ* produced cosmogenic isotopes to estimate rates of landscape evolution; a review from the geomorphic perspective. *Journal of Geophysical Research*, **99**, 13 885–13 896.

BIERMAN, P. R. & NICHOLS, K. K. 2004. Rock to sediment—slope to sea with [10]Be—rates of landscape change. *Annual Review of Earth and Planetary Sciences*, **32**, 215–255.

BISHOP, P. 1984. Oligocene and Miocene volcanic rocks and quartzose sediments of the Southern Tablelands, New South Wales: definition of stratigraphic units. *Journal and Proceedings of the Royal Society of N.S.W.*, **117**, 113–117.

BISHOP, P. 1985. Southeast Australian late Mesozoic and Cenozoic denudation rates: a test for late Tertiary increases in continental denudation. *Geology*, **13**, 479–482.

BISHOP, P. 1986. Horizontal stability of the Australian continental drainage divide in south central New South Wales during the Cainozoic. *Australian Journal of Earth Sciences*, **33**, 295–307.

BISHOP, P. 2007. Long-term landscape evolution: linking tectonics and surface processes. *Earth Surface Processes and Landforms*, **32**, 329–365.

BISHOP, P. 2010. Landscape evolution. *In*: GOMEZ, B., BAKER, V. R., GOUDIE, A. S. & ROY, A. G. (eds) *Handbook of Geomorphology*. SAGE Publications, London (in press).

BISHOP, P. & BROWN, R. 1992. Denudational isostatic rebound of intraplate highlands: the Lachlan River valley, Australia. *Earth Surface Processes and Landforms*, **17**, 345–360.

BISHOP, P. & BROWN, R. 1993. Denudational isostatic rebound of intraplate highlands: the Lachlan River valley, Australia. Reply. *Earth Surface Processes and Landforms*, **18**, 753–755.

BISHOP, P. & GOLDRICK, G. 2000. Geomorphological evolution of the East Australian continental margin. *In*: SUMMERFIELD, M. A. (ed.) *Geomorphology and Global Tectonics*. Wiley, Chichester, 227–255.

BISHOP, P., HOEY, T. B., JANSEN, J. D. & ARTZA, I. L. 2005. Knickpoint recession rates and catchment area: the case of uplifted rivers in E Scotland. *Earth Surface Processes and Landforms*, **30**, 767–778.

BISHOP, P., YOUNG, R. W. & McDOUGALL, I. 1985. Stream profile change and long-term landscape evolution: early Miocene and modern rivers of the east Australian highland crest, central New South Wales. *Journal of Geology*, **93**, 455–474.

BONNET, S. & CRAVE, A. 2003. Landscape response to climate change: insights from experimental modeling and implications for tectonic v. climatic uplift of topography. *Geology*, **31**, 123–126.

BROCARD, G. Y., VAN DER BEEK, P. A., BOURLÈS, D. L., SIAME, L. L. & MUGNIER, J.-L. 2003. Long-term fluvial incision rates and postglacial river relaxation time in the French Western Alps from [10]Be dating of alluvial terraces with assessment of inheritance, soil development and wind ablation effects. *Earth and Planetary Science Letters*, **209**, 197–214, doi: 10.1016/S0012-821X(03)00031-1.

BROCARD, G. Y. & VAN DER BEEK, P. A. 2006. Influence of incision rate, rock strength, and bedload supply on bedrock river gradients and valley-flat widths: Field-based evidence and calibrations from western Alpine rivers (southeast France). *In*: WILLETT, S. D., HOVIUS, N., BRANDON, M. T. & FISHER, D. (eds) *Tectonics, Climate, and Landscape Evolution*. Geological Society of America, Special Papers, **398**, 101–126, doi: 10.1130/2006.2398(07).

BROWN, R. W., GALLAGHER, K., GLEADOW, A. J. W. & SUMMERFIELD, M. A. 2000a. Morphotectonic evolution of the South Atlantic margins of Africa and South America. *In*: SUMMERFIELD, M. A. (ed.) *Geomorphology and Global Tectonics*. Wiley, Chichester, 255–281.

BROWN, R. W., SUMMERFIELD, M. A. & GLEADOW, A. J. W. 2000b. Denudational history along a transect across the Drakensberg Escarpment of southern Africa derived from apatite fission track thermochronology. *Journal of Geophysical Research*, **107**, 2350, doi: 10.1029/2001JB000745.

BURBANK, D. W., LELAND, J., FIELDING, E., ANDERSON, R. S., BROZOVIC, N., REID, M. R. & DUNCAN, C. 1996. Bedrock incision, rock uplift and threshold hillslopes in the northwestern Himalayas. *Nature*, **379**, 505–510.

CAMPANILE, D., NAMBIAR, C. G., BISHOP, P., WIDDOWSON, M. & BROWN, R. 2008. Sedimentation record in the Konkan–Kerala basin: implications for the evolution of the Western Ghats and the Western Indian passive margin. *Basin Research*, **20**, 3–22, doi: 10.1111/j.1365-2117.2007.00341.x.

COCKBURN, H. A. P., BROWN, R. W., SUMMERFIELD, M. A. & SEIDL, M. A. 2002. Quantifying passive margin denudation and landscape development using a combined fission-track thermochronology and cosmogenic isotope analysis approach. *Earth and Planetary Science Letters*, **179**, 429–435.

CRICKMAY, C. H. 1974. *The Work of the River*. Macmillan, London.

CRICKMAY, C. H. 1975. The hypothesis of unequal activity. *In*: MELHORN, W. M. & FLEMAL, R. C. (eds) *Theories of Landform Development*. Proceedings of the Sixth Annual Symposia Series, State University of New York, Publications in Geomorphology, 103–109.

CROSBY, B. T. & WHIPPLE, K. X. 2006. Knickpoint initiation and distribution within fluvial networks: 236 waterfalls in the Waipaoa River, North Island, New Zealand. *Geomorphology*, **82**, 16–38.

DADSON, S. J., HOVIUS, N. *ET AL*. 2003. Links between erosion, runoff variability and seismicity in the Taiwan orogen. *Nature*, **426**, 648–651.

DADSON, S. J., HOVIUS, N. *ET AL*. 2004. Earthquake-triggered increase in sediment delivery from an active mountain belt. *Geology*, **32**, 733–736.

DAVIS, W. M. 1899. The geographical cycle. *Geographical Journal*, **14**, 481–504.

DAVIS, W. M. 1902. Base-level, grade and the peneplain. *Journal of Geology*, **10**, 77–111.

DUMITRU, T. A., HILL, K. C. ET AL. 1991. Fission track thermochronology: application to continental rifting of south-eastern Australia. *APPEA Journal*, **31**, 131–142.

EBDON, D. 1985. *Statistics in Geography*. Basil Blackwell, Oxford.

FLINT, J. J. 1974. Stream gradient as a function of order, magnitude and discharge. *Water Resources Research*, **10**, 969–973.

FRANKEL, K. L., PAZZAGLIA, F. J. & VAUGHN, J. D. 2007. Knickpoint evolution in a vertically bedded substrate, upstream-dipping terraces, and Atlantic slope bedrock channels. *Geological Society of America Bulletin*, **119**, 476–486, doi: 10.1130/B25965.1.

GARDNER, T. W. 1983. Experimental study of knickpoint and longitudinal profile evolution in cohesive, homogeneous material. *Geological Society of America Bulletin*, **94**, 664–667.

GILCHRIST, A. R. & SUMMERFIELD, M. A. 1990. Differential denudation and flexural isostasy in formation of rifted-margin upwarps. *Nature*, **346**, 739–742.

GOLDRICK, G. & BISHOP, P. 1995. Differentiating the roles of lithology and uplift in the steepening of bedrock river long profiles: an example from southeastern Australia. *Journal of Geology*, **103**, 227–231.

GOLDRICK, G. & BISHOP, P. 2007. Regional analysis of bedrock stream long profiles: evaluation of Hack's SL form, and formulation and assessment of an alternative (the DS form). *Earth Surface Processes and Landforms*, **32**, 649–671.

GUNNELL, Y. & FLEITOUT, L. 1998. Shoulder uplift of the Western Ghats passive margin, India—a flexural model. *Earth Surface Processes and Landforms*, **23**, 135–153.

GUNNELL, Y. & FLEITOUT, L. 2000. Morphotectonic evolution of the Western Ghats, India. *In*: SUMMERFIELD, M. A. (ed.) *Geomorphology and Global Tectonics*. Wiley, Chichester, 320–368.

HACK, J. T. 1957. Studies of longitudinal stream profiles in Virginia and Maryland. *In*: *Shorter Contributions to General Geology*. US Geological Survey Professional Papers, **294-B**, 45–97.

HANCOCK, G. S., ANDERSON, R. S. & WHIPPLE, K. X. 1998. Beyond power: bedrock river incision process and form. *In*: TINKLER, K. J. & WOHL, E. E. (eds) *Rivers over Rock: Fluvial Processes in Bedrock Channels*. American Geophysical Union, Geophysical Monograph, **107**, 35–60.

HARTSHORN, K., HOVIUS, N., DADE, W. B. & SLINGERLAND, R. L. 2002. Climate-driven bedrock incision in an active mountain belt. *Science*, **297**, 2036–2038.

HAVIV, I., ENZEL, Y., WHIPPLE, K. X., ZILBERMAN, E., STONE, J., MATMON, A. & FIFIELD, L. K. 2006. Amplified erosion above waterfalls and oversteepened bedrock reaches. *Journal of Geophysical Research*, **111**, F04004, doi: 10.1029/2006JF000461.

HAYAKAWA, Y. & MATSUKURA, Y. 2003. Recession rates of waterfalls in Boso Peninsula, Japan, and a predictive equation. *Earth Surface Processes and Landforms*, **28**, 675–684.

HOLLAND, W. N. & PICKUP, G. 1976. Flume study of knickpoint development in stratified sediment. *Geological Society of America Bulletin*, **87**, 76–82.

HOVIUS, N. 2000. Macroscale process systems of mountain belt erosion. *In*: SUMMERFIELD, M. A. (ed.) *Geomorphology and Global Tectonics*. Wiley, Chichester, 77–105.

HOWARD, A. D. & KERBY, G. 1983. Channel changes in badlands. *Geological Society of America Bulletin*, **94**, 739–752.

HOWARD, A. D., DIETRICH, W. E. & SEIDL, M. A. 1994. Modeling fluvial erosion on regional to continental scales. *Journal of Geophysical Research*, **99**, 13 971–13 986.

JONES, O. T. 1924. Longitudinal profiles of the Upper Towy drainage system. *Quarterly Journal of the Geological Society of London*, **80**, 568–609.

KIRBY, E., WHIPPLE, K. X., TANG, W. & CHEN, Z. 2003. Distribution of active rock uplift along the eastern margin of the Tibetan Plateau: inferences from bedrock channel longitudinal long profiles. *Journal of Geophysical Research*, **108**, 2217, doi: 10.1029/2001JB000861.

KOBOR, J. S. & ROERING, J. J. 2004. Systematic variation of bedrock channel gradients in the central Oregon Coast Range: implications for rock uplift and shallow landsliding. *Geomorphology*, **62**, 239–256.

KOOI, H. & BEAUMONT, C. 1994. Escarpment evolution on high-elevation rifted margins: insights derived from a surface process model that combines diffusion, advection, and reaction. *Journal of Geophysical Research*, **99**, 12 191–12 209.

MATMON, A., BIERMAN, P. & ENZEL, Y. 2002. Pattern and tempo of great escarpment erosion. *Geology*, **30**, 1135–1138.

MERRITTS, D. J., VINCENT, K. R. & WOHL, E. E. 1994. Long river profiles, tectonism, and eustacy: a guide to interpreting fluvial terraces. *Journal of Geophysical Research*, **99**, 14 031–14 050.

MOORE, M. W., GLEADOW, A. J. W. & LOVERING, J. F. 1986. Thermal evolution of rifted continental margins: new evidence from fission tracks in basement apatites from southeastern Australia. *Earth and Planetary Science Letters*, **78**, 255–270.

O'BRIEN, P. E. & BURNE, R. V. 1994. The Great Cumbung Swamp—terminus of the low gradient Lachlan River, Eastern Australia. *AGSO Journal of Australian Geology and Geophysics*, **15**, 223–233.

OLLIER, C. D. & PAIN, C. F. 1994. Landscape evolution and tectonics in southeastern Australia. *AGSO Journal of Australian Geology and Geophysics*, **15**, 335–345.

O'SULLIVAN, P. B., KOHN, B. P., FOSTER, D. A. & GLEADOW, A. J. W. 1995. Fission track data from the Bathurst Batholith: evidence for rapid mid-Cretaceous uplift and erosion within the eastern highlands of Australia. *Australian Journal of Earth Science*, **42**, 597–607.

O'SULLIVAN, P. B., FOSTER, D. A., KOHN, B. P. & GLEADOW, A. J. W. 1996. Tectonic implications of early Triassic and middle Cretaceous denudation in the eastern Lachlan Fold Belt, NSW, Australia. *Geology*, **24**, 563–566.

PAZZAGLIA, F. J. & GARDNER, T. W. 1994. Late Cenozoic flexural deformation of the middle U.S. passive margin. *Journal of Geophysical Research*, **99**, 12 143–12 157.

PAZZAGLIA, F. J. & GARDNER, T. W. 2000. Late Cenozoic landscape evolution of the US Atlantic passive margin: insights into a North American Great Escarpment. *In*: SUMMERFIELD, M. A. (ed.) *Geomorphology and Global Tectonics*. Wiley, Chichester, 283–302.

PERSANO, C., STUART, F. M., BISHOP, P. & BARFOD, D. 2002. Apatite (U–Th)/He age constraints on the development of the Great Escarpment on the southeastern Australian passive margin. *Earth and Planetary Science Letters*, **200**, 79–90.

PERSANO, C., STUART, F. M., BISHOP, P. & DEMPSTER, T. 2005. Deciphering continental breakup in eastern Australia by combining apatite (U–Th)/He and fission track thermochronometers. *Journal of Geophysical Research*, **110**, B12405, doi: 10.1029/2004JB003325.

QUIGLEY, M. C., CLARK, D. & SANDIFORD, M. 2010. Tectonic geomorphology of Australia. *In*: BISHOP, P. & PILLANS, B. (eds) *Australian Landscapes*. Geological Society, London, Special Publications, **346**, 243–265.

REUSSER, L., BIERMAN, P., PAVICH, M., LARSEN, J. & FINKEL, R. 2006. An episode of rapid bedrock channel incision during the last glacial cycle, measured with ^{10}Be. *American Journal of Science*, **306**, 69–102.

ROE, G. H., MONTGOMERY, D. R. & HALLET, B. 2002. Effects of orographic precipitation on the concavity of steady-state river profiles. *Geology*, **30**, 143–146.

SKLAR, L. & DIETRICH, W. E. 1998. River longitudinal profiles and bedrock incision models: stream power and the influence of sediment supply. *In*: TINKLER, K. J. & WOHL, E. E. (eds) *Rivers over Rock: Fluvial Processes in Bedrock Channels*. American Geophysical Union, Geophysical Monograph, **107**, 237–260.

SKLAR, L. & DIETRICH, W. E. 2001. Sediment supply, grain size and rock strength controls on rates of river incision into bedrock. *Geology*, **29**, 1087–1090.

STOCK, J. & DIETRICH, W. E. 2003. Valley incision by debris flows: evidence of a topographic signature. *Water Resources Research*, **39**, 1089, doi: 10.1029/2001WR001057.

STOCK, J. & MONTGOMERY, D. R. 1999. Geologic constraints on bedrock river incision using the stream power law. *Journal of Geophysical Research*, **104**, 4983–4993.

THORNBURY, W. D. 1969. *Principles of Geomorphology*. 2nd edn. Wiley, New York.

TINKLER, K. J. & WOHL, E. E. (eds) 1998. *Rivers over Rock: Fluvial Processes in Bedrock Channels*. American Geophysical Union, Geophysical Monograph, **107**.

TOMKINS, K. M. & HESSE, P. P. 2004. Evidence of Late Cenozoic uplift and climate change in the stratigraphy of the Macquarie River valley, New South Wales. *Australian Journal of Earth Sciences*, **51**, 273–290.

TUCKER, G. E. & SLINGERLAND, R. L. 1994. Erosional dynamics, flexural isostasy, and long-lived escarpments: a numerical modeling study. *Journal of Geophysical Research*, **99**, 12 229–12 243.

TUCKER, G. E. & WHIPPLE, K. X. 2002. Topographic outcomes predicted by stream erosion models: sensitivity analysis and intermodel comparison. *Journal of Geophysical Research B*, **107**, doi: 10.1029/2000JB000044.

TUROWSKI, J. M., LAGUE, D. & HOVIUS, N. 2007. Cover effect in bedrock abrasion: a new derivation and its implications for the modeling of bedrock channel morphology. *Journal of Geophysical Research*, **112**, F04006, doi: 10.1029/2006JF000697.

TWIDALE, C. R. 1976. On the survival of palaeoforms. *American Journal of Science*, **276**, 77–94.

TWIDALE, C. R. 1991. A model of landscape evolution involving increased and increasing relief amplitude. *Zeitschrift für Geomorphologie*, **35**, 85–109.

TWIDALE, C. R. 1998. Antiquity of landforms: an 'extremely unlikely' concept vindicated. *Australian Journal of Earth Sciences*, **45**, 657–668.

VAN DER BEEK, P., ANDRIESSEN, P. & CLOETINGH, S. 1995. Morphotectonic evolution of continental rifted margins: inferences from a coupled tectonic–surface processes model and fission track thermochronology. *Tectonics*, **14**, 406–421.

VAN DER BEEK, P. & BISHOP, P. 2003. Cenozoic river profile development in the Upper Lachlan catchment (SE Australia) as a test of quantitative fluvial incision models. *Journal of Geophysical Research*, **108**, 2309, doi: 10.1029/2002JB002125.

VAN DER BEEK, P. & BRAUN, J. 1998. Numerical modelling of landscape evolution on geological time-scales: a parameter analysis and comparison with the southeastern highlands of Australia. *Basin Research*, **10**, 49–68.

VAN DER BEEK, P., BRAUN, J. & LAMBECK, K. 1999. Post-Palaeozoic uplift history of southeastern Australia revisited: results from a process-based model of landscape evolution. *Australian Journal of Earth Sciences*, **46**, 157–172.

VAN DER BEEK, P., PULFORD, A. & BRAUN, J. 2001. Cenozoic landscape evolution in the Blue Mountains (SE Australia): lithological and tectonic controls on rifted margin morphology. *Journal of Geology*, **109**, 35–56.

VEEVERS, J. J. (ed.) 1984. *Phanerozoic Earth History of Australia*. Clarendon Press, Oxford.

WELLMAN, P. 1979*a*. On the Cainozoic uplift of the southeastern Australian highland. *Journal of the Geological Society of Australia*, **26**, 1–9.

WELLMAN, P. 1979*b*. On the isostatic compensation of Australian topography. *BMR Journal of Australian Geology and Geophysics*, **4**, 373–382.

WELLMAN, P. 1987. Eastern Highlands of Australia: their uplift and erosion. *BMR Journal of Australian Geology and Geophysics*, **10**, 277–286.

WELLMAN, P. & MCDOUGALL, I. 1974. Potassium–argon ages on the Cainozoic volcanic rocks of New South Wales, Australia. *Journal of the Geological Society of Australia*, **21**, 247–272.

WHIPPLE, K. X., SNYDER, N. P. & DOLLENMAYER, K. 2000. Rates and processes of bedrock incision by the Upper Ukak River since the 1912 Novarupta ash flow in the Valley of Ten Thousand Smokes, Alaska. *Geology*, **28**, 835–838.

WHIPPLE, K. X. & TUCKER, G. E. 1999. Dynamics of the stream-power incision model: implications for height limits of mountain ranges, landscape response time-scales, and research needs. *Journal of Geophysical Research*, **104**, 17 661–17 674.

WOHL, E. E. & MERRITTS, D. M. 2001. Bedrock channel morphology. *Geological Society of America Bulletin*, **113**, 1205–1212.

YOUNG, R. W. 1983. The tempo of geomorphological change: evidence from southeastern Australia. *Journal of Geology*, **91**, 221–230.

YOUNG, R. W. & McDOUGALL, I. 1993. Long-term landscape evolution: early Miocene and modern rivers in southern New South Wales, Australia. *Journal of Geology*, **101**, 35–49.

Rethinking eastern Australian caves

R. A. L. OSBORNE

Faculty of Education and Social Work, University of Sydney, Sydney, NSW 2006, Australia
(e-mail: armstrong.osborne@sydney.edu.au)

Abstract: There are some 300 bodies of cavernous limestone in eastern Australia, extending from Precipitous Bluff in southeastern Tasmania to the Mitchell Palmer region in north Queensland. These impounded karsts, developed in Palaeozoic limestones of the Tasman Fold Belt System, contain many caves. The caves have a suite of features in common that allows them to be thought of as a major group: the Tasmanic Caves. The Tasmanic Caves include multiphase hypogene caves such as Cathedral Cave at Wellington and multiphase, multiprocess caves such as Jenolan with Carboniferous hypogene and younger paragenetic and fluvial elements. Active hypogene caves occur at Wee Jasper and possibly at five other localities. The Tasmanic Caves are one of the most complex suites of caves in folded Palaeozoic limestones in the world. Field techniques developed to study these caves are now being applied to complex caves in central Europe: in the Czech Republic, Hungary, Poland, Slovakia and Slovenia.

Andrews (1911) recognized that there was a 'Geographical unity of eastern Australia'. Although this may appear to be a quaint early 20th century idea, later geologists recognized the Tasman Fold Belt System as a continental-scale tectonic feature in eastern Australia. Similarly, late 20th century geomorphologists, such as Ollier (1978, 1982), recognized the existence of continental-scale landforms such as the Great Escarpment, running the length of eastern Australia.

The Tasman Fold Belt System in eastern Australia has two Palaeozoic sections, the western Lachlan–Thompson Fold Belt and the eastern New England Fold Belt, both containing cavernous Palaeozoic limestones. If we take a continental-scale view of the karst in these Palaeozoic limestones we find that despite some obvious climatic influences on surface landforms, the caves have many characteristics in common. Caves in Palaeozoic limestones from the southeasternmost tip of Tasmania to northern Queensland are sufficiently similar to be viewed as a cohesive group, the Tasmanic Caves (Fig. 1).

The focus here is on caves, because caves are the prominent features in many eastern Australian karsts. This paper begins by reviewing ideas about eastern Australian caves from 1830 to 1995. The latter part of the paper, dealing with developments in understanding eastern Australian caves from 1996 onwards, concentrates on the author's own work.

The speleological unity of eastern Australia

The Australian Speleological Federation Karst Index Database (http://www.caves.org.au/kid/) lists over 300 areas of cavernous limestone and more than 5000 caves in the Lachlan–Thompson and New England Fold Belts in eastern Australia. All this cavernous limestone is Palaeozoic, and most is early Palaeozoic in age. Much of the limestone is deformed, but extensive dolomitization is restricted to Tasmania and primary gypsum is rare. Most of the limestone outcrops are small, forming impounded fluviokarsts. The most extensive areas of outcrop occur at Chillagoe and Mitchell Palmer in Queensland and at Mole Creek and Junee–Florentine in Tasmania.

Both small- and large-scale karst landforms are generally poorly developed, except at some exceptional localities. Polje are absent, except for some atypical examples at Mole Creek in Tasmania, dolines are relatively uncommon except at Buchan and Mole Creek, and tower karst is well developed only at Chillagoe and Mitchell Palmer in north Queensland. Karren and other small-scale solution forms are often well developed but some of these features are artefacts of forest-clearing and soil erosion (Fig. 2a).

Limestone gorges and canyons, deep by eastern Australian standards, occur in incised areas of the highlands, but these features are more fluvial than karst. Some dry limestone gorges such as those at Colong and Jenolan in New South Wales owe the depth of their incision to the inefficiency of parallel underground drainage.

Many limestone outcrops on the mainland support distinctive floral communities and tiny rainforest refugia have survived in the more incised areas. These floral differences appear to be less distinct in Tasmania, suggesting that at least some of the vegetation changes associated with karst areas

From: BISHOP, P. & PILLANS, B. (eds) *Australian Landscapes*. Geological Society, London, Special Publications,
346, 289–308.   DOI: 10.1144/SP346.15   0305-8719/10/$15.00 © The Geological Society of London 2010.

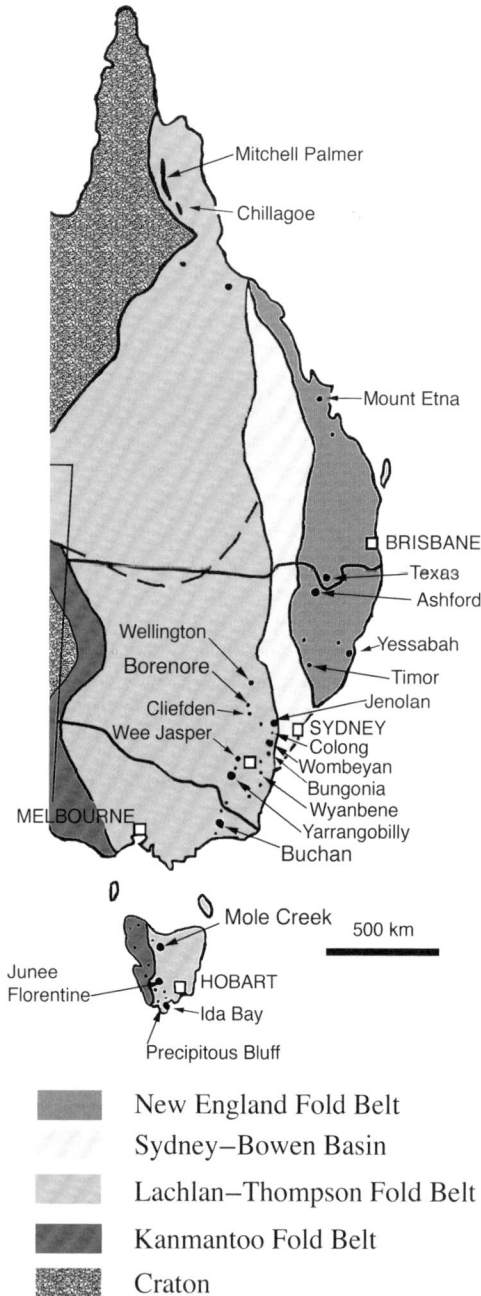

Fig. 1. Karst areas of the Lachlan–Thompson and New England Fold Belts containing the Tasmanic Caves of eastern Australia.

New England Fold Belt

Sydney–Bowen Basin

Lachlan–Thompson Fold Belt

Kanmantoo Fold Belt

Craton

outside Tasmania are due to shelter and water availability for deep-rooted species such as figs and kurrajongs in the karst terrain rather than the plants being specifically calciphile.

Thinking about eastern Australian caves, 1830–1995

Almost all research on the origin and evolution of eastern Australian caves, from its very beginnings in the 1830s until the end of the 1990s, related cave development to contemporary explanations of the evolution of the adjacent surface landscape.

1830–1952: protoscience of eastern Australian caves

The early history of karst science in eastern Australia is essentially the study of caves and bone-bearing sediment in caves. Karst as a term did not appear in the local literature until the 1960s and karst landforms were rarely mentioned before then. The term karst was probably first applied in an Australian context by the Czech, Jiří V. Daneš, in the title of his 1917 paper relating to the Chillagoe karst, 'Karststudien in Australien' (see Jennings 1966). It is interesting to note that, in the title of his paper discussing Daneš' scientific contribution, Jennings referred to 'the Chillagoe Caves District' not the Chillagoe karst.

Scientific study of caves in eastern Australia arguably dates back to Thomas Mitchell's visit to Wellington Caves, NSW in June 1830 to view the recently discovered fossil bone deposits (Mitchell 1838). Mitchell surveyed the main caves, described the stratigraphy of the bone deposits and speculated on whether the cave sediments were deposited by general flooding, local flooding or by marine incursion.

A. M. Thomson, first professor of geology at Sydney University, visited Wellington Caves in September 1869, just over 2 years before his untimely death at the age of 30 in November 1871. Thomson recognized that the caves had been formed 'by the dissolving action of carbonic acid water' and may have been the first worker anywhere to recognize unroofed caves (see Osborne 1991).

For most of the late 19th century, the focus of science in caves was on understanding the palaeontology and taphonomy of fossil bones and the sediments in which they occurred. Little was written about the caves themselves or the surface karst landforms. The tenor of these times can be summed up in the comments of Robert Etheridge Jr. on Kybean Caves that 'From a strictly geological point of view, the deposit of red earth and the stalagmitic floors are the most important points' (Etheridge 1892, p. 24).

Lamont Young produced a three-page report on Wombeyan Caves (Young 1880). Despite writing he had 'little to add' to Young's report, Edgeworth David was probably the first to propose a fluvial

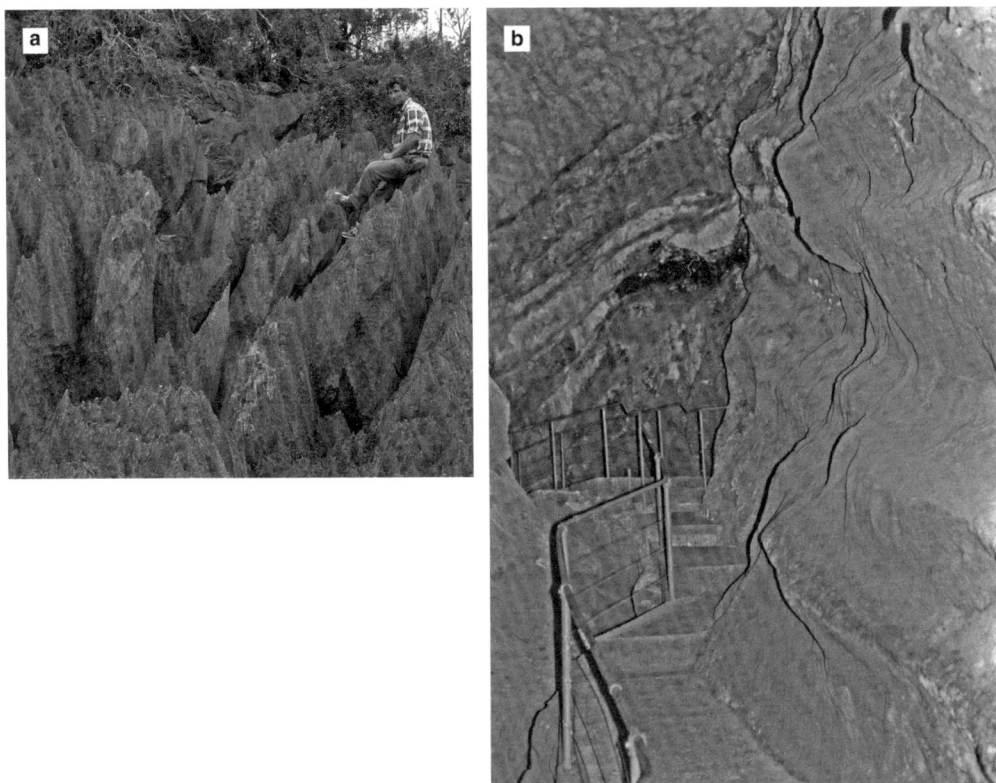

Fig. 2. (**a**) Karren field at Mt Sebastopol west of Yessabah, NSW; (**b**) 'like the "scotia" of the architect', rising and falling notches in the wall of River Cave at Jenolan Caves, NSW.

origin for eastern Australian caves when he stated: 'From their shapes and positions it is evident that these caves, like nearly all similar hollows in limestone rock, are deserted subterranean watercourses' (David 1897, p. 151). This is an important point, as a fluvial mechanism for cave development in eastern Australia was to persist largely unchallenged for almost all of the 20th century.

There was great activity in cave exploration and surveying at the turn of the 20th century as State governments became involved in developing cave tourism. Deep caves were explored, with the 46 m entrance pitch in Drum Cave at Bungonia being descended in 1891 (Nurse 1972). Cave keepers and guides were exploring caves, and professional surveyors, such as Oliver Trickett in NSW, were mapping them (see Middleton 1991). Despite this exploratory and survey activity, there was little or no cave or karst science *per se* in eastern Australia in first half of the 20th century. One outstanding event in this period was a paper on Jenolan Caves' bedrock geology by Sussmilch & Stone (1915), which gave what is probably the first numerical estimate for the age of an eastern Australian cave,

namely, 500 ka for the upper levels of Jenolan Caves based on valley incision after the Kosciusko Epoch (uplift), then thought to have occurred 2 Ma ago. Sussmilch (1923), writing in the guidebook for the Pan-Pacific Science Congress, provided a simple fluvial model for the origin of Jenolan Caves, noting that the youngest caves were at river level. He commented: 'Quite obviously, also, the caves themselves cannot be older than the valley in which they occur' (Sussmilch 1923, p. 15). Subsequent work would bring into question this seemingly innocuous and self-evident assertion.

In his contribution to the Pan-Pacific Science Congress guidebook, Griffith Taylor, Australia's first professor of geography, indicated that he made observations in Jenolan Caves and thought seriously about their origin (Taylor 1923). Taylor recognized five levels in the caves and suggested that they resulted from a combination of erosion of the impounding insoluble beds downstream of the limestone, variation in rainfall and variation in uplift. Cave erosional stages occurred as a result of either or both increased rainfall and/or a halt to uplift. Cave floors and/or tiny slots between the levels represented periods of uplift or low rainfall. Taylor recognized that the steep bedding in the limestone meant that horizontal cave ceilings were unrelated to bedding.

Taylor (1923) also recognized rising and falling notches in the cave walls: 'In many cases the floor of such a channel, or one side of it, has disappeared, leaving hollows like the "scotia" of the architect. At times these "scotias" rise and fall along the side of a cave, showing that the water was under pressure and the erosion was partly due to an upward flow' (Taylor 1923 p. 24). A 'scotia' is an architectural moulding with a concave profile. These features were ignored until the 1980s and were finally recognized as being paragenetic in origin by Osborne (1999a) (Fig. 2b). [Paragenesis is the upward solution of a cave ceiling by water flowing over actively depositing sediment (Renault 1968)].

Colditz (1943) related the development of Wellington Caves to a period of 'rest' from river entrenchment during the late Pliocene. Later workers including Frank (1971) and Francis (1973) also linked cave development at Wellington to varying histories of the development of the Bell River valley, reinforcing the link between underground and surface landscape development.

Laseron (1954) devoted a whole chapter in his popular book, *The Face of Australia*, to Australian caves, which in addition to describing the caves of the Nullarbor, coastal Western Australia and Naracoorte, devoted almost seven pages to eastern Australian caves. Laseron mentioned many eastern Australian caves and proposed that the caves originated as erosion lowered soft strata adjacent to steeply dipping limestone, creating a limestone barrier, which the stream then penetrated forming caves. He related cave development to uplift in the Pliocene. Laseron's diagram, apparently modelled on Jenolan, shows a river producing multiple levels of caves as it penetrates a limestone body.

Taylor (1958) linked his 1923 five-stage fluvial model for Jenolan Caves to the erosion of the Jenolan Valley, the uplift of the Blue Mountains Plateau and to five flats on the monocline at the eastern boundary of the Plateau. Taylor also described Bungonia Gorge, mentioning caves and sinkholes, and borrowed a direct comparison with the Classical Karst from Jennings and King in an untraceable 1954 field guide: 'I am reminded of a wide area of limestone south of Trieste, which is known as the Karst region. Here the scenery of limestone weathering has been carefully studied, and the Karst features of Yugoslavia are found in most limestone districts' (Taylor 1958, p. 188).

1952–1984: the Jennings era, young caves in young landscapes

From his arrival in Australia in 1952 until his death in 1984, Joseph Newell (Joe) Jennings made an outstanding and prodigious contribution to the science of Australian caves and karst, and particularly to the science of eastern Australian caves. Jennings was the first to systematically describe eastern Australian caves, and he introduced karst hydrology and karst process geomorphology to Australia. Although his work in eastern Australia concentrated on caves and karst areas close to Canberra; Burra, Bungonia, Cooleman Plain, Wee Jasper, Wombeyan and Yarrangobilly, Jennings also studied the Mole Creek area in Tasmania and Chillagoe in North Queensland.

The view of the geological establishment about caves at the time was presented to the public in a NSW Geological Survey pamphlet (Anonymous 1968). The following short extracts indicate the view about the age and survival of caves: 'every large cave is probably at least half a million years old'; 'All important limestone caves in the world are less than 10 million years old'; 'Geologically, caves are very short-lived. Only a few million years can intervene between the initiation of a cave and its destruction by roof collapse.'

The journal *Helictite* 'devoted entirely to papers on cave research' commenced publication in 1962 and Jennings became its most prolific author, contributing 30 papers, several with D. I. Smith and A. P. Spate. Jennings' colleagues and students made significant contributions to cave and karst science both generally and in eastern Australia. These included Ollier's review on speleochronology (Ollier 1966) and his extensive work on Pacific island karst (e.g. Ollier & Holdsworth 1969; Ollier 1975), Frank's work on cave sediments (Frank 1969, 1971, 1972a, b, 1973, 1974, 1975) and Gillieson's work on cave sediments and tropical karst (e.g. Gillieson 1981, 1985).

Jennings (1967) discussed three eastern Australian karsts, Wee Jasper, Cooleman Plain and Mole Creek. Jennings noted the importance of structure, the role of rejuvenation and the influence of climate, and speculated on the possible role of palaeoclimate. It is noteworthy that the 10 page description by Jennings (1967) of the Mole Creek karst is still the major published account of this important karst area.

Although Frank's work was mostly within the dominant paradigm, a small section of his second paper on Borenore Caves is prescient of things to come. Commenting on the development of leisegang-banded ferruginous cement in palaeokarst sandstone at Borenore in NSW, Frank remarked: 'these processes take a considerable length of time and probably longer, in fact, than the normal life-span of a cave system' (Frank 1973, p. 36).

Both Jennings and Frank (1971) were aware that many eastern Australian caves did not fit a simple fluvial model, so Jennings (1977) used the term 'nothephreatic' to describe solution by relatively still water below the watertable. Unfortunately, nothephreatic is now generally used to indicate laminar flow in phreatic conduits (Field 1999).

In his detailed study of the now flooded Texas Caves in SE Queensland Grimes (1978) encountered cave morphology similar to that described by Jennings (1977) as nothephreatic. When describing these features Grimes observed that 'there are no scallops to suggest strong currents, nor are there long linear passages of the type found in most stream caves . . . Further there are often no apparent entrances or exits for through flowing streams at the levels of these features' (Grimes 1978, pp. 40–41). These are very similar to the observations that led Osborne (2001a) to develop a new interpretation of eastern Australian cave origins.

For Jennings, caves were young. His views were summed up in his own words 'the probability is that most caves developed during the course of the Quaternary' (Jennings 1983). Whether caves were old or young became the key element of cave and karst science in eastern Australia between 1979 and 1995. Jennings' death in 1984 left a large void in the small cave and karst science community in Australia. Jennings' posthumously published paper on the magnetostratigraphy of cave sediments at Wee Jasper (Schmidt et al. 1984) suggested that the ages of the oldest sections of Punchbowl Cave were between 0.8 and 2.3 Ma. These findings had little impact on the debate about cave ages as the relationship between the dated sediments and the origin of the caves was uncertain.

1979–1995: young caves in old landscapes v. old caves in old landscapes

In the late 1970s and early 1980s, there was an increase in interest and much controversy as attempts were made to relate cave development to the then-new idea that eastern Australian landscapes had significant antiquity (Young 1977, 1982; Bishop 1985, 1986). Eventually, old caves in old landscapes became the accepted idea. The nature of the discourse during this period is illustrated by controversy over the likely ages of Bungonia and Timor Caves in NSW.

Pratt (1964) related cave development at Bungonia to then-current ideas about the history of landform development in eastern Australia. Pratt proposed that the upper level of the caves developed and was filled with sediment during the Miocene

and that deep cave development resulted from rejuvenation following the 'Kosciusko Uplift'. Jennings (1965) discussed Pratt's hypothesis, pointing out some problems, particularly that of relating cave history to general models of landscape development, but did not reject Pratt's ideas out of hand. Jennings et al. (1972) recognized increasing evidence for landscape antiquity, noted the conflict with 'the freshness of the cave forms and lack of demonstrably ancient fill', but decided to 'argue simply that they (the caves) may belong to the Tertiary' (Jennings et al. 1972, p. 143). James et al. (1978), with Jennings as the third author, were more convinced that Bungonia Caves must be young. Although recognizing increasing evidence for the landscape having a pre-Eocene age, they did not conclude that the caves were old. Armed with the caves' 'juvenile appearance' and the possibility that the caves were perched by geological barriers, James et al. concluded: 'most of the caves could be considerably younger than the rejuvenation which formed the gorge. On an interpretation of this type it is no longer necessary to attribute active dynamic phreatic development in caves such as Odyssey to the early Tertiary' (James et al. 1978, p. 61). Thus, the phenomenon of 'young' caves was saved from the 'threat' of ancient landscapes.

Connolly & Francis (1979) in their paper on Timor Caves made the first attempt to link cave development with the then-emerging view that eastern Australian landscapes were ancient. They suggested that the high-level Main Cave formed in the Cretaceous and that three caves at a lower level formed during the Palaeogene. This conclusion was based on the relationships between the landscape and basalt flows dated as Palaeocene–Eocene in age. This paper caused considerable controversy, and Connolly (1983) published a retraction, contending that field relationships between the basalt and the valley floor had been misinterpreted and that dykes intersected by the caves were the same age as the basalt. Francis & Osborne (1983) replied that the dykes had not been examined and that Connolly's new mapping data were inconclusive. Osborne (1986) presented new data supporting Connolly & Francis' (1979) original conclusions about the basalt–landscape relationships. The 'dykes' were petrographically dissimilar to Tertiary basalt flows on the surface, were not dykes but lava flows filling ancient caves, and K/Ar dating showed that they were at least 73.5 Ma old, older than the 53 Ma old lava flows on the surface.

In the 1980s, strongly lithified palaeokarst deposits filling cave-shaped cavities in the bedrock were discovered in eastern Australian caves (Osborne 1984). This and later work showed that karstification and cave development had occurred many times in the geological history of eastern Australia. The discovery of palaeokarst initially had little impact on ideas about the origin and evolution of 'modern' eastern Australian caves. Although palaeokarst was discussed in the review of New South Wales karst by Osborne & Branagan (1988), the intersection of palaeokarst by modern caves was not seen as significant and cave development was directly linked with landscape development.

By the early 1990s, old caves in old landscapes had become the dominant paradigm, culminating in the works of Webb *et al.* (1991), who proposed an Eocene or older origin for Buchan Caves in Victoria, and Osborne (1993*a*), who proposed a late Cretaceous origin for the controversial high-level passages at Bungonia Caves in NSW. Fluvial cave studies continued in the Buchan area into the 1990s with the work of Webb *et al.* (1992) and Fabel *et al.* (1996).

Rethinking eastern Australian caves, 1996–2008

By the mid-1990s, many local workers had abandoned cave and karst research completely. At this time, the author found that whereas the palaeokarst itself began to make more sense, the caves themselves became increasingly inexplicable. It was easy to describe the palaeokarst, but not the caves that intersected it. Some cave passages at Jenolan were cavities left behind after sulphide-bearing palaeokarst deposits were exhumed (Osborne 1993*b*). There was evidence that part of Wyanbene Cave was an exhumed ore body. Sulphide-bearing palaeokarst and low-grade ore bodies were frequently associated with major cave systems, but these materials were thought to be associated with the development of large breakdown chambers, rather than with initial cave formation (Osborne 1996). (The term breakdown is used to describe the processes by which cave ceilings, walls and floors fail, including slab failure, block failure, spalling and heaving, and the rubble produced by these processes.)

The literature on palaeokarst contained a paradox. Although many caves in eastern Australia intersected palaeokarst, there were few reports in the international literature of caves intersecting palaeokarst. Was palaeokarst in caves elsewhere not being recognized, or were eastern Australian caves unusual?

A visit to Europe in 1997 was crucial. In Slovenia, palaeokarst was recognized in surface outcrop, in quarries and in road excavations, but not in caves. The caves seemed unfamiliar; they were like those in the textbooks, but not like those in eastern Australia. This raised another point: cave passage cross-sections in most textbooks assume that bedding was horizontal, but in eastern Australia, the bedding is often steep to vertical and this has not been taken into account. The large keyhole-shaped passages in steeply dipping limestone at Jenolan could not have formed the same way as keyhole passages in horizontal strata. Eastern Australian cave profiles needed to be reinterpreted.

The similarities found in Europe were also surprising. Szemlö-Hegyi Cave in Budapest, the archetypal thermal show cave excavated by rising hot spring water, looks like a giant version of Gaden Cave at Wellington and has many features that eastern Australian workers, following Jennings, had been calling nothephreatic. Visits to other 'unusual' caves, Ochtinská Aragonite Cave in Slovakia, and Zbrazov Aragonite and Mladecske Caves in the Czech Republic, confirmed the observation that thermal caves and other unusual caves in central Europe are more like eastern Australian caves than the 'normal' meteoric, fluvial caves of Slovenia, formed by the underground capture of surface water.

In the next few years cave profiles in dipping beds were taken into account and the giant keyholes at Jenolan were reinterpreted as being paragenetic in origin (Osborne 1999*a*). Osborne (1999*b*) presented the first new interpretation for the evolution of Jenolan Caves since Taylor (1923, 1958). This interpretation involved the overprinting of cavities produced by multiple phases of cave development, including thermal, paragenetic and fluvial phases. Jenolan Caves were no longer considered the product of a single process, nor do they have a single age. A whole series of new anomalies was recognized, including: (1) passages that end blindly in bedrock at both ends; in the past, these were interpreted as passages through which water once flowed; (2) caves where the chambers, passages and shafts decrease in size with depth; (3) downstream-narrowing maze caves; (4) maze-like caves in dipping strata; (5) poor connection between caves and surface hydrology; (6) the presence of cupolas, large dome-shaped cavities; (7) breakdown in small cavities; (8) etched cave walls and crystal linings.

These anomalies, together with the intersection of palaeokarst, which Ford (1995) noted was more likely in what he called *per ascensum* caves, suggested to Osborne (2001*a*) that rather than having a conventional fluvial origin, many eastern Australia caves were non-meteoric in origin. Ford (1995) used *per ascensum* to describe caves formed by rising water in contrast to *per descensum* caves formed by descending and then horizontally flowing surface water.

In many eastern Australian caves, elongate cavities, oriented along strike, blindly terminate in bedrock at both ends. In the 'fluvial' minds of eastern Australian workers (including Osborne 1993a), water at Bungonia Caves in the past must have flowed between blind elongate cavities at the same elevation to springs in the wall of Bungonia Gorge. It was difficult to get out of the mindset that an elongate horizontal cavity implied horizontal water flow. Osborne (2001b) introduced the term 'hall' for these cavities because they are blind (do not go anywhere), and so should not be called passages. He also described Hall and Narrows caves, a variety of network cave of probable thermal origin developed in dipping strata, common in eastern Australia (Fig. 3a, b).

Two other problems loomed. First, cupolas (Fig. 4) are the most striking features in many eastern Australian caves, but there was very little literature on them, and the word cupola was not in most karst glossaries. Osborne (2004b) discussed the origin, nomenclature and morphology of cupolas, and noted that they were associated with a range of smaller-scale speleogens. There were several explanations for the origin of cupolas, most involving upward-moving or eddying water, and cupolas are common features of thermal caves. Although it was possible to list 13 speleogens generally associated with cupolas and to describe a new form, a pseudonotch, further work showed that most of the accompanying speleogens had never been described and were unnamed.

The application of plane-table mapping using a laser rangefinder as alidade, combined with the development of section-measuring instruments based on laser rangefinders in the late 1990s, allowed the shape of cave walls to be mapped in detail and cross-sections to be measured. It was thus possible to determine the morphology of cupolas. Maps, sections, photographs and stereophotographs of cupolas were made at Jenolan and Wellington. These techniques were applied in Europe in 2005, with the first results from Slovenia now published (Osborne 2008).

Second, although breakdown was common, much occurred in small cavities and breakdown was active at high levels. This pattern did not fit with the beam failure model of White & White (1969, 2000), which that suggested breakdown should be most common in cavities with wide ceilings and should occur soon after caves entered the vadose zone (i.e. became dewatered and the cave ceiling lost hydraulic support). Osborne (2004a) reported that breakdown is an active process, often unrelated to the size or location of the cavity, and fallen blocks continue to disintegrate in the breakdown pile. Osborne recognized that the dominant mechanism for this type of breakdown is crystal wedging as a result of the weathering of pyrite and growth of gypsum, aragonite and hydromagnesite, and that most breakdown has no effect on the ground surface above the cave.

Osborne (2005) described another feature, partitions, which suggested that per ascensum speleogenesis is important in eastern Australian caves. Partitions are narrow, vertical insoluble rock masses or structures that act as aquicludes in karst rocks. Lannnigans Cave at Colong in NSW is a complex multistorey cave developed along strike in a narrow body of vertically bedded limestone. Mapping by Alan Pryke of Sydney University Speleological Society shows that Lannnigans Cave penetrates through a large number of dykes oriented perpendicular or oblique to the bedding. Osborne (2005) recognized that these dykes must have been intact partitions early in the history of the cave. The first cavities formed along cross-joints in the limestone between the dykes. The dykes were only breached later by vadose weathering. The current stream in the cave, flowing along strike, somehow flows through the dykes. Partitions are significant at Colong and Bungonia, and occur at Jenolan, Timor and Wombeyan. By recognizing partitions and portals (breaches in partitions or connections between genetically unrelated cavities), Osborne (2005) reinforced the idea that complex caves result from the integration of cavities with different origins and ages.

A detailed study of the morphology and geology of Cathedral Cave at Wellington documented new types of hypogene speleogens and revealed a complex morphostratigraphy indicating multiple phases of hypogene development (Osborne 2007b). Similar hypogene speleogens have now been recognized elsewhere in eastern Australia (Fig. 5).

Osborne et al. (2006) used K–Ar dating of illite to determine the age of clay sediments in Jenolan Caves. The Early Carboniferous age of unlithified clay deposits in the tourist sections of Jenolan Caves was unexpected, as palaeokarst-filled caves were thought to be Late Carboniferous or Early Permian in age, and the open and accessible caves at Jenolan were thought to be much younger, perhaps Late Cretaceous in age (Osborne 1999b). The K–Ar dates suggest that the caves containing the clays formed c. 340 Ma ago, making them older than the nearby Bathurst Granite (320–312 Ma) and close in age to the last folding event in the area, the Kanimblan Orogeny (350–330 Ma; Scheibner & Veevers 2000). These findings suggest that very ancient caves can survive without being destroyed by surface processes and can form parts of currently active complex caves. This conclusion raises the prospect of caves finally

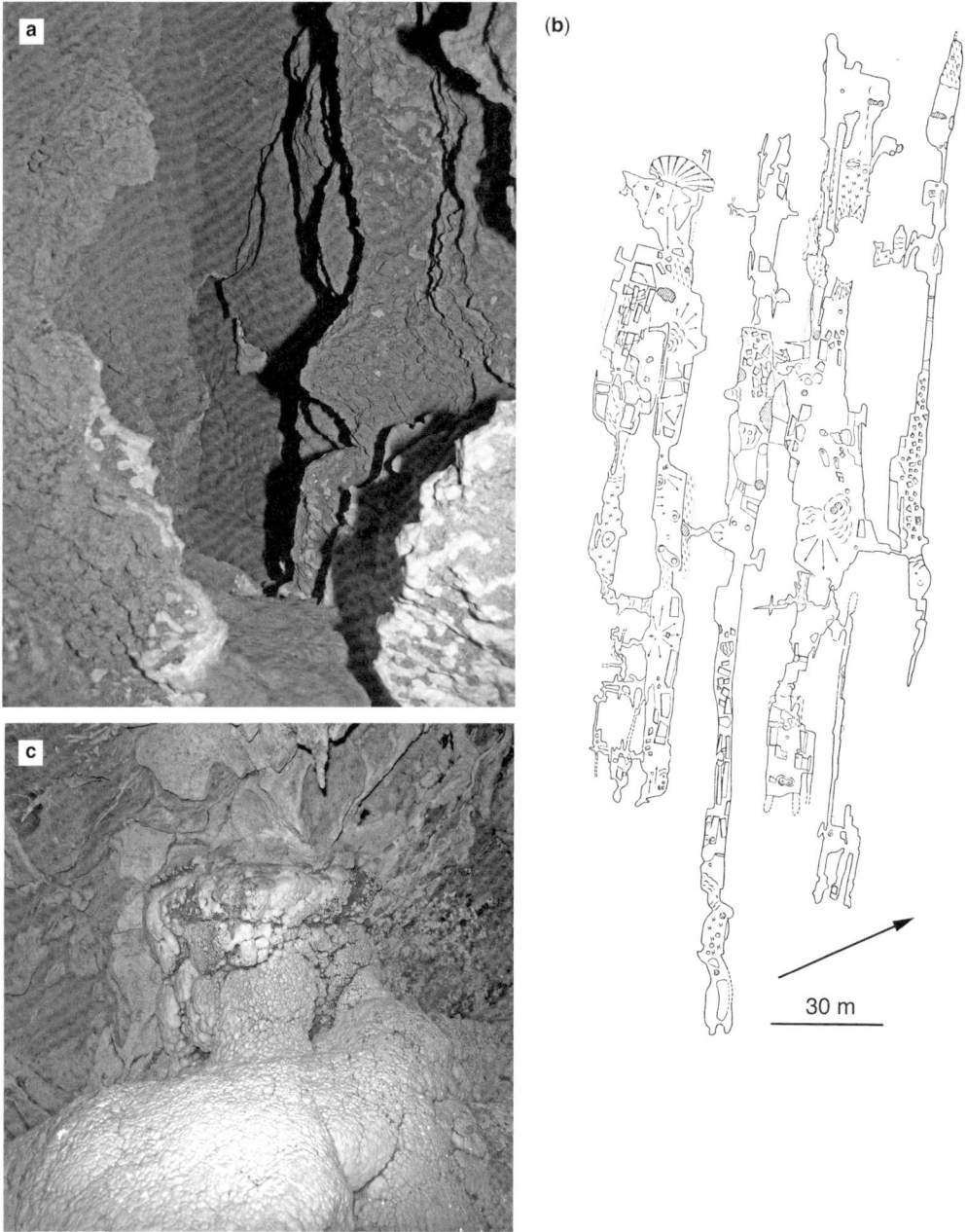

Fig. 3. (a) Blind hall, My Cave, Mole Creek Tasmania; (b) plan view of hall and narrows cave with blind terminations, Dip Cave, Wee Jasper, NSW, after Jennings (1963). (c) Clouds (mammillary speleothems) in Thermocline Cave, Wee Jasper, NSW, the only so far confirmed active hypogene cave in eastern Australia.

realizing their potential as repositories of geological records not preserved elsewhere. The ancient clays are similar in appearance to other cave sediments and their significance became apparent only after intensive mineralogical study and dating. It seems likely that there are more ancient sediments in the Tasmanic Caves and in caves elsewhere (Osborne 2007a).

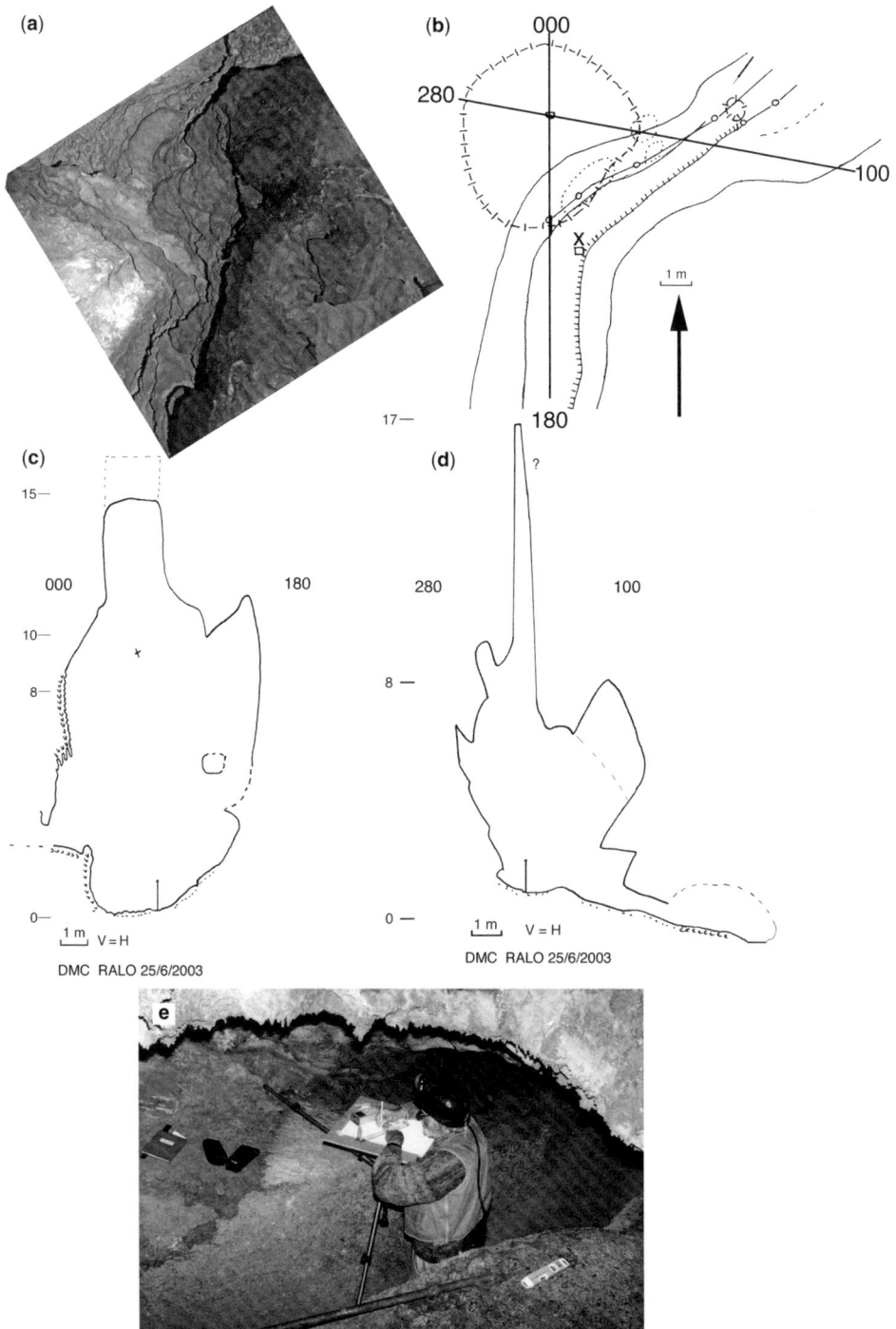

Fig. 4. (**a**) Looking vertically up into complex cupola, Pool of Cerberus Cave, Jenolan Caves, NSW, with same orientation as map; north is at top. (**b**) Plan view map showing wall contour and location of sections through cupola. (**c**) North–south vertical cross-section. (**d**) West–east vertical cross-section of cupola and path; ceiling hollow on eastern side of section is visible in mid-right of (a). (**e**) D. Colchester, plane-table mapping at station X on map shown in (b).

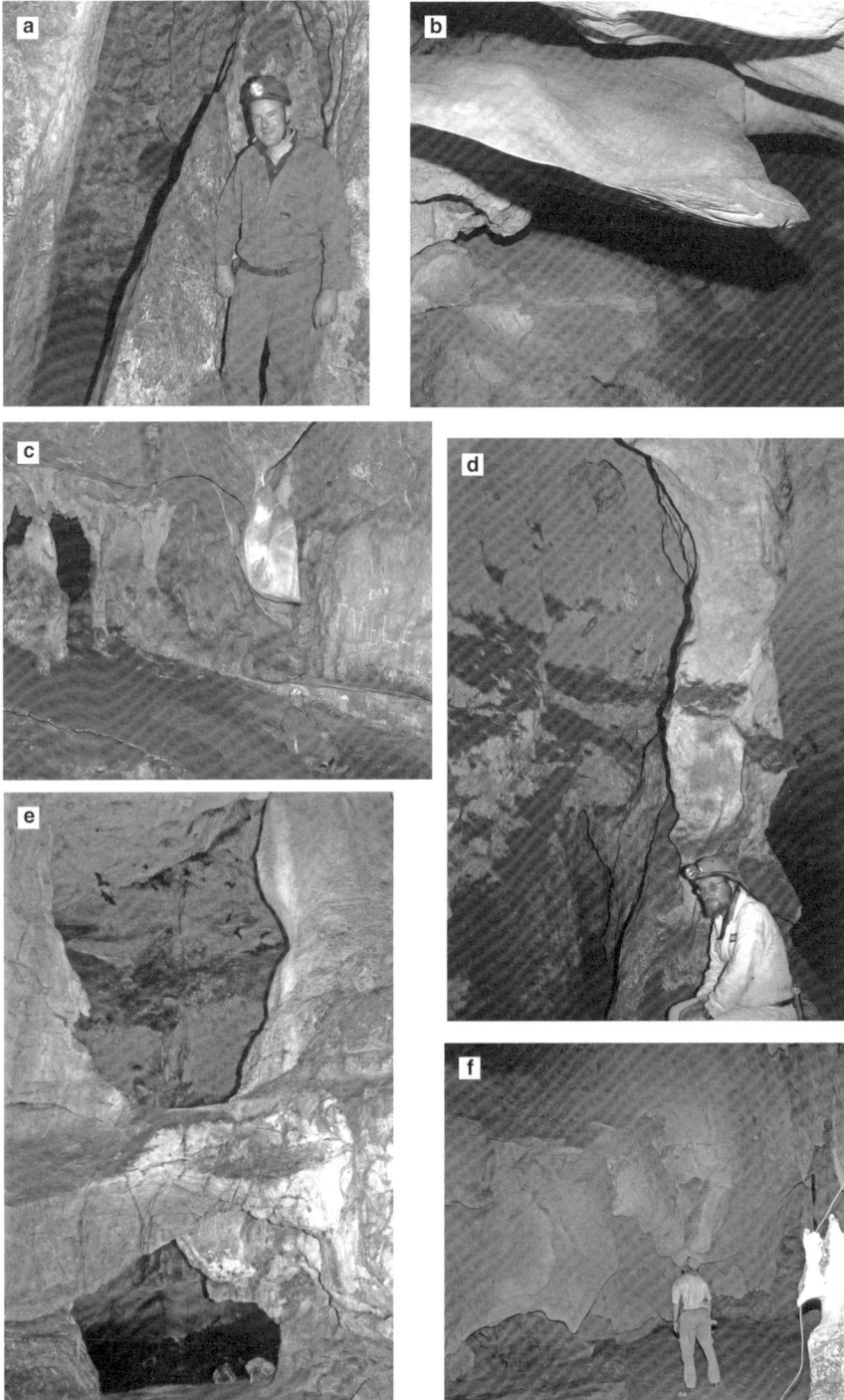

Fig. 5. (**a**) Vertical blade, Belfry Cave, Timor Caves, NSW. (**b**) Horizontal blade, Moparabah Cave west of Yessabah, NSW. (**c**) Pendant remnant left after integration, Signature Cave, Wee Jasper, NSW. (**d**) Projecting corner, Belfry Cave, Timor Caves, NSW. This bedrock projection is the remains of the wall that once separated cupolas to its left and right. (**e**) Looking through bridge to large cupola behind, Ashford Cave, NSW. (Note guiding joint in centre of frame.) (**f**) Chamber with blade-like pendants, Flogged Horse Cave, near Mt Etna, Queensland.

Changing international views 1999–2008

Over the last 10 years, two ideas have been changing international views on the origin and evolution of limestone caves. The rethinking of ideas about eastern Australian caves occurred against that background.

Idea 1: the hypogene revolution

Traditional views about speleogenesis, the process of cave formation, proposed that almost all caves formed by the dissolving action of meteoric water (rainwater) sinking into soluble rocks and excavating caves as the water moved vertically downwards and then horizontally through the rock emerging at a karst spring. Caves formed this way have been called meteoric, fluvial or *per descensum* caves.

Caves formed by groundwater moving upward through the rock mass, as a result of heat, density or pressure, are known by a number of terms including non-meteoric, hypogene, thermal, hydrothermal, artesian and *per ascensum*. Until the mid-1990s, these caves were seen as strange, rare or exceptional. Such caves had been reported from North Africa (Collignon 1983), Poland (Gradziński 1962; Bac-Moszaswill & Rudnicki 1978; Rudnicki 1978), Hungary (Muller & Sarvary 1977; Takács-Bolner & Kraus 1989), Russia (Dublyansky 1980) and the USA (Bakalowicz *et al.* 1987; Hill 1987), but Palmer (1991) estimated that only 10% of caves had non-meteoric origins, and Ford (1995) noted that 'this category of caves has received only limited attention in western literature, though rather more in eastern Europe'.

A turning point for hypogene caves in the western literature came in 1995: Ford (1995) reviewed the literature on hydrothermal caves and ore bodies in karst internationally, and Galdenzi & Menichetti (1995) described hypogene caves from central Italy. Ford (1995) noted that *per ascensum* caves were more likely to intersect palaeokarst than meteoric *per descensum* caves.

Interest in non-meteoric speleogenesis was further aroused by the publication of Klimchouk's (1996) paper on the giant gypsum caves of the Ukraine. Klimchouk drew attention to the mechanism of transverse speleogenesis, the morphology of hypogene caves and the use of the cave index to distinguish between hypogene and fluvial caves. The publication in 2000 of the monograph *Speleogenesis; Evolution of Karst Aquifers* (Klimchouk *et al.* 2000) with at least 16 out of 43 sections dealing with hypogene caves and speleogenesis, established hypogene speleogenesis as a topic of international interest (e.g. Audra *et al.* 2002, 2007).

In his monograph, *Hypogene Speleogenesis*, Klimchouk (2007) presented a contemporary picture of hypogene caves, morphology and speleogenetic processes. Following Ford (2006), Klimchouk presented a revised definition for hypogenic speleogenesis: 'the formation of caves by water that recharges the soluble formation from below, driven by hydrostatic pressure or other sources of energy, independent of recharge from the overlying or immediately adjacent surface' (Klimchouk 2007, p. 6). In Poland, Andrzej Tyc and his students from Sosnowiec have now taken up the pioneering work of Gradziński and Rudnicki and are re-examining the hypogene caves of the Krakow region.

Idea 2: the ancient cave revolution

From the late 1990s onwards, evidence from a range of dating techniques began to show that some caves accessible to humans, not just palaeokarst features filled with solid rock, are considerably older than the Quaternary. Polyak *et al.* (1998) dated alunite from caves in the Guadalupe Mountains of New Mexico, USA to 11.3 Ma by the ^{40}Ar/^{39}Ar method, Lundberg *et al.* (2001) dated calcite spar from a similar cave to 92 Ma by the U/Pb method. Similarly, and despite the fact that ages for active caves in the Alps extend back only hundreds of thousands of years (see Häuselmann 2002), Audra (2004) identified caves in the French Prealps that formed at 12 Ma, before the Alpine Orogeny.

Since 1997 a multinational group involving workers from the Czech Republic, Poland and Slovenia has been investigating the stratigraphy and chronology of cave sediments in central Europe. This work involves detailed lithostratigraphy and mineralogy coupled with palaeomagnetic, biostratigraphic and radiometric (U/Th) dating. Results from Slovenia show that fluvial sediments in large cave systems date back to 4.2 Ma, flowstone dates back to 3.4 Ma and clay from an unusual cave near the Slovenian–Italian border dates possibly from 34–35 Ma but more likely from 11–33 Ma. Because the sediments give a minimum age for the caves, these data suggest that Slovenian caves have Tertiary origins (Horáček *et al.* 2007; Zupan Hajna *et al.* 2008*a*, *b*).

A new synthesis

Until the 1980s, eastern Australian caves were seen as young and formed by a single phase of meteoric cave development, related to the evolution of the present surface landscape. During the 1980s and subsequently, it became clear that surface landscapes had considerable antiquity and the view developed that caves must be old, like the surface landscape. Although the presence of palaeokarst was recognized, caves were still regarded as single-phase, single-process, meteoric features related to the evolution of the present surface landscape.

A new understanding of the Tasmanic Caves is currently emerging, which differs considerably from previous conceptions. In this view, caves range in age from extremely old (hundreds of millions of years) to currently active. Most originate by internal, not surface, processes and have little relationship with the development of the present surface landscape. Complex caves are produced either by repeated phases of a single process or by a range of different processes acting at different times.

The origin of eastern Australian caves

An overall view of the Tasmanic Caves, but without the benefit of statistics, indicates that they fall into three groups based on their origin. A small group of caves owe their origin solely to meteoric fluvial processes. A large group of caves have an entirely hypogene origin. A third group, consisting of mostly large caves, are multiphase, multiprocess caves, formed by hypogene, fluvial and in some cases paragenetic processes.

Without doubt, however, most active speleogenesis in eastern Australia today is meteoric and fluvial. There are only a very few active, possibly active, or likely to have been recently active, hypogene caves in eastern Australia (Fig. 3c). There is little evidence of active paragenetic systems.

The solely hypogene caves and sections of caves are of four types: hall and narrows or maze caves; anticline caves; cupola-dominated caves; vertical caves that narrow with depth. Basal chamber caves, like Józef-hegyi Cave and Sátorköpuszta Cave in Hungary, have not yet been identified in eastern Australia.

In the complex, multiphase, multiprocess caves such as those at Colong, Jenolan and Wombeyan in NSW and Mole Creek in Tasmania, the active

and most recent speleogenesis is fluvial, producing cavities and speleogens distinct from those of the older hypogene and paragenetic sections of these caves. Although some complex caves do have ancient fluvial phases, for the Tasmanic Caves the present is usually not the key to the past, making them un-uniformitarian (or perhaps anti-uniformitarian) landforms.

Multiphase, multiprocess caves. Since the 1970s it has been recognized that karsts have complex histories with many phases of karstification (Avias 1972). These phases of karstification have generally been recognized by the exposure of palaeokarst features and deposits in natural surface landscapes and in mines and quarries (Bosák *et al.* 1989). Because of their geological and geomorphological history, the Tasmanic Caves preserve evidence of complex multiphase and multiprocess histories to a degree not found elsewhere. Particularly in the Lachlan–Thompson Fold Belt the combination of the following factors has contributed to the survival of ancient karst features: (1) last folding in the Carboniferous; (2) last significant glaciation in Permo-Carboniferous times; (3) last major uplift in the Cretaceous; (4) limited and slow geomorphological development in the Cenozoic; (5) limited meteoric cave development in the Cenozoic.

Complex multiphase caves in eastern Australia intersect two or more generations of palaeokarst and display morphological evidence for more than two distinct phases of speleogenesis (Table 1). The most complex of these cave systems, Jenolan Caves, contains the oldest dated open caves, formed over 340 Ma ago (Osborne *et al.* 2006). Main Cave at Timor contains the oldest dated lava flow palaeokarst deposit, dated at 73.5 Ma (Osborne 1986). Other possible locations of complex multiphase caves, requiring further investigation, include: Yarrangobilly, Wee Jasper and Cooleman Plain, NSW; Mt Etna, Chillagoe and Mitchell–Palmer, Queensland; Buchan, Victoria; Junee–Florentine and Ida Bay, Tasmania.

Table 1. *Multiphase–multiprocess caves mode of spelegenesis*

Number	Karst area	Hypogene	Meteoric	Paragenesis	Total
1	Wellington	6	3	0	9
2	Jenolan	1 (? 2)	4	1 +	6
3	Wombeyan	1	3	1 +	5
4	Colong	1	2	1	4
5	Borenore	1	2	1	4
6	Bungonia	1	2	?	3
7	Timor	1	2	1	3
8	Mole Creek	1 (? 2)	1	1	3

Particular characteristics of eastern Australian caves

Intersection of palaeokarst. An unusual feature of eastern Australian caves is the frequency with which they intersect palaeokarst deposits. Some of these deposits fill solution-enlarged joints and bedding planes but many are filled caves. Palaeokarst features and deposits have been rarely reported from caves elsewhere (Ford 1995; Osborne 2000, 2002). Caves at 18 eastern Australian karsts are known to intersect palaeokarst deposits and many intersect more than one generation of palaeokarst. The unusual frequency of caves intersecting palaeokarst in eastern Australia provides strong support to the morphological evidence for a hypogene origin of these caves.

Structural guidance of cave development. Among the Tasmanic Caves, hypogene and partly hypogene caves show a high degree of structural guidance. Osborne's (2001*b*) hall and narrows concept stresses the role of dipping beds in guiding cave development. Halls developed along dipping or vertical beds whereas narrows formed along vertical joints perpendicular to bedding. New observations suggest that joints, cleavage and fold axes, rather than bedding, guide most hypogene cavities and speleogens in the Tasmanic Caves (Fig. 6a, b). It is now clear that the main chamber of Cathedral Cave at Wellington has developed along a vertical joint and not a stratigraphic boundary as previously thought (Osborne 2007*b*). Downward-sloping development of Flying Fortress Cave and Hogans-Fossil Cave at Bungonia follows eastward-dipping joints, not the bedding that dips to the west.

Relationship between cave levels and age. The conventional view of cave levels, horizontal passages in caves, is that the higher levels of the caves are older than the lower levels, as a result of incision in the adjoining surface landscape (Droppa 1966). Consequently, palaeokarst intersections and old sediments should be more common in the higher levels, closer to the surface in the past than in the lower levels, younger and further from the surface in the past.

In the Tasmanic Caves, however, many palaeokarst bodies are intersected and ancient sediments occur in the lower levels of the caves. At Jenolan, a section of active stream cave is developed entirely within caymanite palaeokarst (Osborne 1993*b*), at Wombeyan, sandstone palaeokarst is exposed in the floor and wall of a seasonal flood channel (Fig. 6c), and at Colong caymanite is exposed in the low ceiling of an active stream channel (Fig. 6e). One of the Early Carboniferous clay deposits at Jenolan (Osborne *et al.* 2006) is located in the lowermost section of River Cave,

adjacent to a small perennial stream (Fig. 6d). These occurrences are difficult to explain if the cave levels are oldest at the top and youngest at the bottom as suggested by Taylor (1923, 1958). Surely, the palaeokarst intersections and the oldest sediments should be in the oldest caves.

There is no problem with the lower levels of the caves intersecting palaeokarst and containing old sediments if the caves are of hypogene origin. In hypogene caves, there is no direct relationship between cave elevation and age, as water rising through the rock will form cavities at a variety of levels at the same time (Osborne 2007*a*). This means that the upper levels of Bungonia Caves need not be the oldest sections of the caves. In short, the relative elevation of a cave or section of cave in the Tasmanic Caves is unreliable by itself as an indicator of age.

Processes modifying eastern Australian caves

Almost all the Tasmanic Caves are hypogene in origin, but the original forms of these caves have often been modified by a range of processes.

Meteoric invasion. In most of the complex caves in eastern Australia, original hypogene cavities have been invaded by surface streams. The underground paths of these streams are often complex, consisting partly of new stream passage and partly of intersected, and sometimes modified, pre-existing cavities. Complex stream paths and blockages make underground drainage inefficient, often leading to flow-shifting between surface and underground pathways and to whole or partial filling of the caves when the streams become blocked by sediment, debris or breakdown.

Paragenesis. Paragenesis has played an important role in karst areas where there has been significant sediment input from streams into the caves such as Colong, Jenolan, Mole Creek, Wee Jasper and Wombeyan (Fig. 7a). Where floodwater enters the caves from blind valleys such as Jenolan and Wombeyan there appear to have been several paragenetic events. During these events the caves became blocked, the blind valleys filled with sediment and the streams flowed over the top of the limestone barrier. Bridges, such as the Devils Coach House at Jenolan (Fig. 7b), owe their origin in part to such paragenesis.

Integration. Integration refers to the joining of originally isolated cavities by processes including later hypogene events, fluvial invasion, paragenesis, vadose solution and breakdown. Morphological evidence suggests that hypogene caves begin as a collection of unconnected adjacent cavities (Osborne

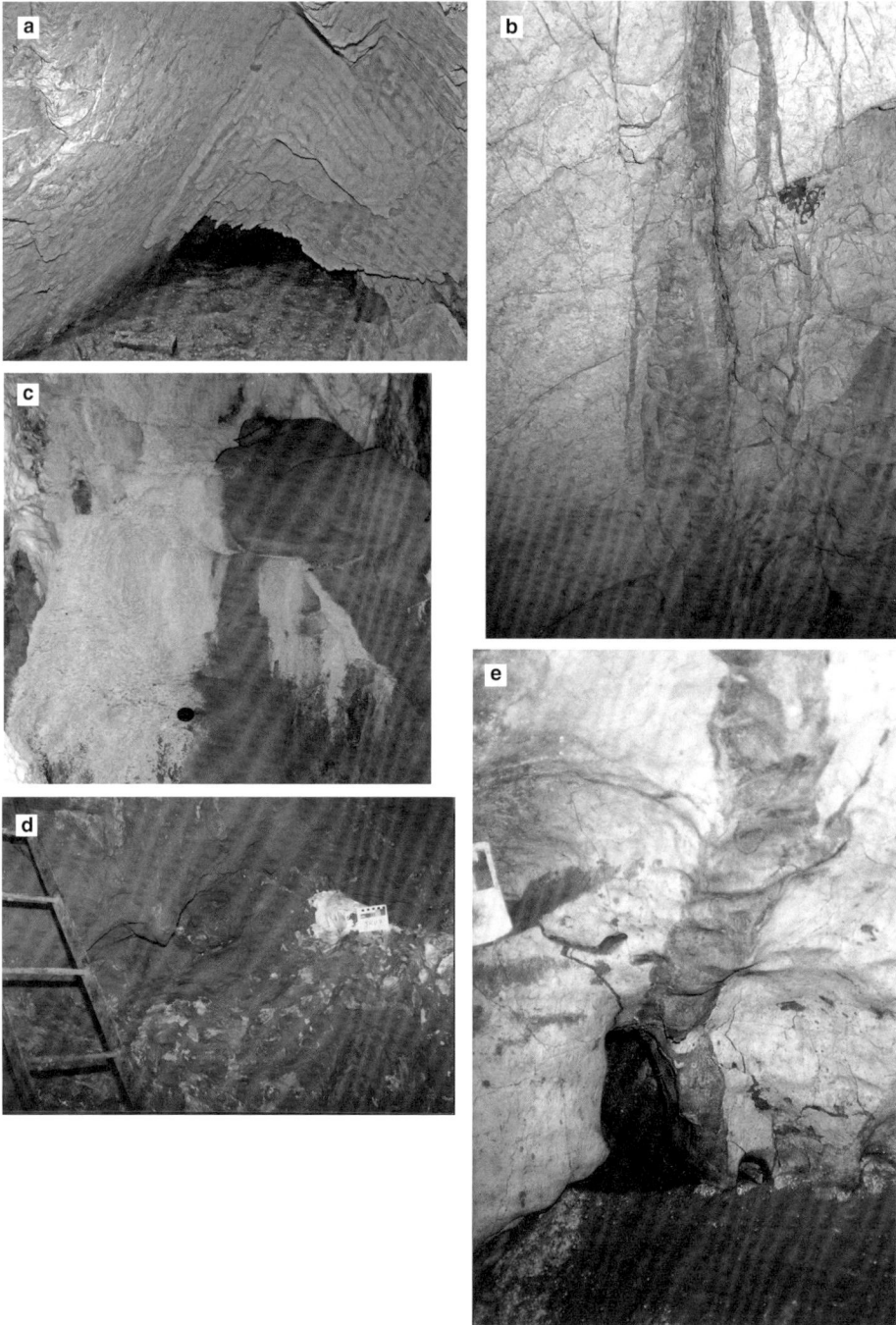

Fig. 6. (**a**) Anticline Cave at Wellington, NSW, developed along the axis of an anticline. (**b**) Ceiling of Rabbit Skull Cave, Mt Sebastopol, NSW, showing vein deposits in guiding joint. The guiding joint for the present cave has been used for fluid movement prior to the excavation of the cave. (**c**) Sandstone palaeokarst to right of lens cap, Creek Cave, Wombeyan Caves, NSW; lens cap is 50 mm. (**d**) Early Carboniferous clay deposit in River Lethe streamway, Jenolan Caves, NSW. Deposit is material to right of ladder with cleaned area and scale at top right. The base of this deposit is only a few hundred millimetres above the normal stream level. Black squares on scale are 10 mm wide. (**e**) Caymanite (dark grey) in ceiling of active stream passage, Lannigans Cave, Colong, NSW. Red square on scale is 20 mm wide.

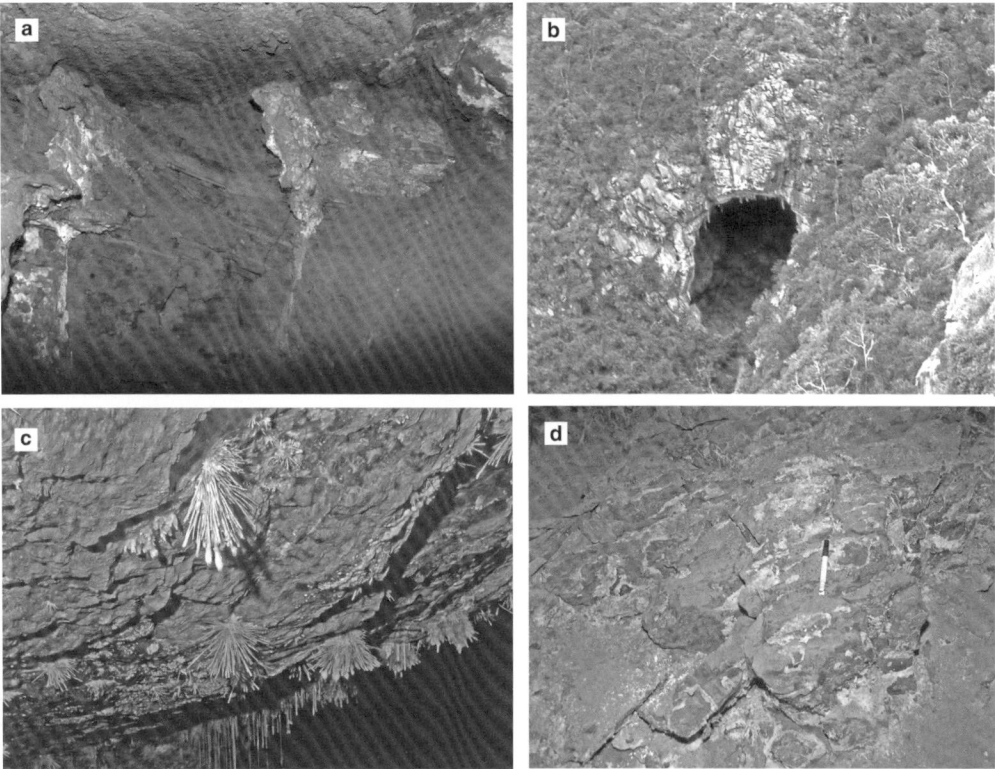

Fig. 7. (**a**) Sediment against flat paragenetic ceiling, Punchbowl Cave, Wee Jasper, NSW. (**b**) Northern entrance to Devils Coach House, Jenolan Caves, NSW. Paragenesis played an important role in the development of large bridges in eastern Australia. (**c**) Aragonite anthodites on bedding plane, Genghis Khan Cave, Mole Creek, Tasmania. The slab that has fallen off this plane forms the floor of this section of the cave. (**d**) Fallen block being broken up by the continuing growth of gypsum crystals, Marakoopa Cave, Mole Creek, Tasmania.

2007*b*, 2008). Many of the hypogene speleogens in Cathedral Cave such as projecting corners, blades, pseudonotches and curved juts are the remnants of dividing walls that separated originally isolated cavities (Osborne 2007*b*). These speleogens are widespread in eastern Australia, indicating the importance of integration in eastern Australian caves.

Renovation. The walls of active and relatively young hypogene caves are often coated with crystal linings and/or deeply etched by lithologically selective solution. These features are less common in many of the older hypogene caves of eastern Australia. Renovation refers to the loss of such crystal linings and to the smoothing of etched and selectively dissolved cave walls. The scarcity of old crystal linings is not surprising, as crystal coatings can be seen peeling off cave walls in relatively young thermal caves in Hungary.

In eastern Australia, etched and selectively dissolved cave walls are usually found only in recently active hypogene caves or in hypogene caves that have been isolated from surface processes. Elsewhere, processes such as solution by films of seepage water and condensation corrosion have renovated the etched walls, forming new smooth surfaces.

Filling and re-excavation. Because they lack an effective outlet, hypogene caves frequently fill with sediment when they become open to the land surface. Sediment fills can be removed directly by later meteoric excavation, indirectly by slumping or falling into lower-level passages or directly by a later phase of hypogene speleogenesis. These processes can be recognized in eastern Australian caves, and at some localities such as Wellington there have been repeated episodes of filling, excavation and refilling, resulting in sedimentary sequences with complex stratigraphy.

Breakdown. Breakdown is an important late-stage modification process in eastern Australian caves.

New work at Mole Creek is confirming the role of vadose weathering of pyrite and crystal growth in producing breakdown. Slab failure along bedding planes is occurring at Mole Creek on the limbs of anticlinal chambers in Marakoopa Cave and in gently dipping strata close to the nose of a plunging anticline in Genghis Khan Cave (Fig. 7c). There is a strong relationship at Mole Creek between breakdown and the growth of aragonite, gypsum and hydromagnesite. Fallen blocks continue to disintegrate as a result of crystal wedging (Fig. 7d).

The source of hypogene water for eastern Australian caves

Although a great deal of morphological evidence indicates that hypogene speleogenesis took place in the past, the small number of active and recently active hypogene caves in eastern Australia, as well as the lack of hydrological studies, severely limits what can be said about the source and chemistry of the rising water that excavated the caves.

Structural conditions in the Lachlan–Thompson and New England Fold Belts mostly rule out an artesian source for the rising water. Warm springs, active and old hypogene caves often occur in close proximity to large regional faults and/or cross faults intersecting limestone bodies. This arrangement suggests that faults are the most likely source for hypogene water excavating caves in eastern Australia.

The relationship of eastern Australian caves to surface geomorphology

One of the major problems facing cave science in eastern Australia has been linking cave development with the evolution of the surrounding surface landscape. The recognition that many caves are wholly or partly hypogene in origin and the discovery of Early Carboniferous clays in Jenolan Caves suggest that the caves, or parts of the caves, are much older that the present landscape and generally unrelated to its evolution. The Early Carboniferous sections of Jenolan Caves could have formed before the excavation of the ancient Carboniferous landscape represented by the unconformity surface at the base of the Sydney Basin.

The complex history of large cave systems such as Jenolan makes them difficult sites for investigating relationships between the caves and the present landscape. The best prospects for investigating cave–landscape relationships are probably the caves developed in the Early Permian limestone near Yessabah, NSW (Fig. 1). Caves in the Yessabah Limestone occur in a residual hill on the coastal plain and at a range of elevations along the

Great Escarpment. The bedrock here is younger than much of the palaeokarst in eastern Australia and the oldest open caves at Jenolan so there is no problem of confusing Tertiary gravel with Permian or Carboniferous gravel. The difference in tectonic history between the New England and the Lachlan–Thompson Fold Belts also assists. The final deformation of the New England Fold Belt was the Permo-Triassic Hunter–Bowen Orogeny, setting a maximum Triassic age for unconformable palaeokarst and caves.

National and international morphological comparisons

Simple morphological comparisons between the Tasmanic Caves and other Australian and European caves produce surprising results that may lead to further productive research. The best morphologically studied cave, Cathedral Cave at Wellington in NSW, is surprisingly similar in morphology to Cathedral Cave at Naracoorte in South Australia. Cathedral Cave at Wellington is developed in dense, deformed Devonian limestone, whereas the caves at Naracoorte are in undeformed, poorly indurated Oligocene–Miocene limestone, suggesting that rock age and degree of lithification are not important factors in determining cave morphology.

Some caves at Mole Creek in Tasmania are morphologically similar to caves to the west of Yessabah in New South Wales. The caves at Mole Creek are in dolomitized Ordovician limestone, those west of Yessabah are in Permian crinoidal–bryozoan limestone, and the areas have dramatically different climates. The common feature is that the caves are guided by the axes of small anticlines, developed in the limbs of larger folds.

Work in progress suggests morphological similarities between caves at Mt Etna, on the Tropic of Capricorn in Queensland in Devonian limestone, and caves in the residual limestone hills of the Kraków–Częstochowa Upland near Krakow, located at just over 50°N in Poland in Jurassic limestone. These comparisons suggest that rock age, rock type and climate have little to do with the morphology of hypogene caves, but that structural factors, particularly jointing, may be significant.

Conclusions

In the 30 years since 1979 there has been a complete rethinking about the nature, origin and evolution of eastern Australian caves. Before 1979, eastern Australian caves were considered simple, young, fluvial and short-lived features that developed as part of a young landscape. Now, caves are seen as complex, old, hypogene and long-lived features, mostly or

partly unrelated to the surrounding ancient landscape. The age and complexity of the Tasmanic Caves make them potential repositories for information at a geological time scale, but they are often poor sources for information about recent geomorphological history. Geomorphological information, traditionally sought from caves, is most likely to come from the caves in the youngest bedrock.

The age of the limestone, the stability of the Australian continent and the age of the surrounding landscape make the Tasmanic Caves valuable sites for a new type of cave science. This new science spans hundreds of millions of years and provides insight into times in the past when other information on the surface is scarce or has been eroded away. The Tasmanic Caves are rich and complex, with much still to be discovered at the well-trodden sites, and little is known about the less-studied localities.

It is not practical to acknowledge all those who have assisted with my research for more than 30 years. I should record my thanks to those who helped me get started, J. James, the late J. Jennings, L. Sutherland, B. Webby and my late parents. D. Branagan supervised my PhD and always thought that Jenolan Caves were Permian in age; it is strange now to think this was an underestimate. D. Lowe and P. Bosák assisted me in joining the international cave science community. Recent work has benefited from collaboration with R. Pogson and D. Colchester of the Australian Museum, and H. Zwingmann and P. Schmidt of the CSIRO. Over the last 10 years or so collaboration with and assistance from colleagues in central Europe has been critical. I would particularly like to thank T. Slabe and staff at the Karst Research Institute, Postojna; J. Hlaváč, P. Bella and staff at the Slovak Caves Administration, Liptovský-Mikuláš; P. Bosák, V. Cílek and K. Zak at the Czech Geological Institute, Prague; and colleagues at Krakow, Sosnoweic, Budapest and Vienna. P. Osborne proofread this paper and most of my work over the last 20 years. She has spent many hours assisting with fieldwork, logistics and politics. I must thank the unsung heroes of cave research: cavers of all types, cave guides, rangers, cave managers and cave owners, without whose assistance and dedication cave science would be impossible. This paper has greatly benefited from the suggestions of two anonymous referees and the attention of the editor.

References

ANDREWS, E. C. 1911. Geographical unity of eastern Australia in the late and post Tertiary times with applications to biological problems. *Journal and Proceedings of the Royal Society of New South Wales*, **44**, 420–480.

ANONYMOUS 1968. *Formation of Caves*. Geological Survey of NSW Department of Mines Information Brochure, **26**.

AUDRA, P. 2004. An overview of the current research carried out in the French Western Alps. *Acta carsologica*, **31**, 25–44.

AUDRA, P., BIGOT, J.-Y. & MOCOCHAIN, L. 2002. Hypogenic caves in Provence (France): specific features and sediments. *Acta carsologica*, **31**, 33–50.

AUDRA, P., HOBLEA, F., BIGOT, J.-Y. & NOBECOURT, J.-C. 2007. The role of condensation–corrosion in thermal speleogenesis: study of a hypogenic sulfidic cave in Aix-Les-Bains, France. *Acta carsologica*, **36**, 185–194.

AVIAS, J. 1972. Karst of France. *In*: HERAK, M. & SPRINGFIELD, V. T. (eds) *Karst, Important Karst Regions of the Northern Hemisphere*. Elsevier, Amsterdam, 129–188.

BAC-MOSZASWILL, M. & RUDNICKI, J. 1978. O mozliwosei hydrotermalnej genezy jaskini Dziura w. Tatrach. *Kras i Speleologia*, **2**, 84–91.

BAKALOWICZ, M. J., FORD, D. C., MILLER, T. E., PALMER, A. N. & PALMER, M. V. 1987. Thermal genesis of dissolution caves in the Black Hills, South Dakota. *Geological Society of America Bulletin*, **99**, 729–738.

BISHOP, P. 1985. Southeast Australian late Mesozoic and Cenozoic denudation rates: a test for late Tertiary increases in continental denudation. *Geology*, **13**, 497–482.

BISHOP, P. 1986. Horizontal stability of the Australian continental drainage divide in south central New South Wales during the Cainozoic. *Australian Journal of Earth Sciences*, **33**, 295–307.

BOSÁK, P., FORD, D. C., GLAZEK, J. & HORÁČEK, I. (eds) 1989. *Paleokarst: A Systematic and Regional Review*. Elsevier and Academia, Amsterdam and Prague.

COLDITZ, M. J. 1943. The physiography of the Wellington district, NSW. *Journal and Proceedings of the Royal Society of New South Wales*, **76**, 235–251.

COLLIGNON, B. 1983. Spélégenèse hydrothermal dan les Bibans (Atlas Tellien-Nord de l'Algérie). *Karstologia*, **2**, 45–54.

CONNOLLY, M. D. 1983. Reassessment of cave ages at Isaacs Creek. *Helictite*, **21**, 64–65.

CONNOLLY, M. D. & FRANCIS, G. 1979. Cave and landscape evolution at Isaacs Creek, New South Wales. *Helictite*, **17**, 5–24.

DAVID, T. W. E. 1897. *Report on the Wombeyan Caves, by T. W. E. David, Geological Surveyor*. Annual Report of the Department of Mines NSW 1896, 149–151.

DROPPA, A. 1966. The correlation of some horizontal caves with river terraces. *Studies in Speleology*, **1**, 186–192.

DUBLYANSKY, V. N. 1980. Hydrothermal karst in the alpine folded belt of the southern parts of U.S.S.R. *Kras i Speleologia*, **3**, 18–36.

ETHERIDGE, R., JR 1892. V. Notes made at the Kybean Caves, parish of Throsby, county of Beresford in October, 1890. *Records of the Geological Survey of New South Wales*, **3**, 21–24.

FABEL, D., HENRICKSEN, D., FINLAYSON, B. & WEBB, J. A. 1996. Nickpoint recession in karst terrains: an example from the Buchan Karst, southeastern Australia. *Earth Surface Processes and Landforms*, **21**, 453–466.

FIELD, M. S. 1999. *A Lexicon of Cave and Karst Terminology with Special Reference to Environmental Karst Hydrology*. US Environment Protection Agency, Washington, DC.

FORD, D. C. 1995. Paleokarst as a target for modern karstification. *Carbonates and Evaporites*, **10**, 138–147.

FORD, D. C. 2006. Karst geomorphology, caves and cave deposits: A review of North American contributions during the past half-century. *In*: HARMON, R. S. & WICKS, C. W. (eds) *Perspectives on Karst Geomorphology, Hydrology and Geochemistry*. Geological Society of America, Special Papers, **404**, 1–14.

FORD, T. D. 1995. Some thoughts on hydrothermal caves. *Cave and Karst Science*, **22**, 107–118.

FRANCIS, G. 1973. Evolution of the Wellington Caves landscape. *Helictite*, **11**, 79–91.

FRANCIS, G. & OSBORNE, R. A. L. 1983. Comment—Reassessment of cave ages at Isaacs Creek. *Helictite*, **21**, 66.

FRANK, R. M. 1969. The clastic sediments of Douglas Cave, Stuart Town, NSW. *Helictite*, **7**, 3–13.

FRANK, R. M. 1971. The clastic sediments of Wellington Caves, New South Wales. *Helictite*, **9**, 3–26.

FRANK, R. M. 1972*a*. *Sedimentological and morphological study of selected cave systems in eastern N.S.W., Australia*. PhD thesis, Australian National University, Canberra.

FRANK, R. M. 1972*b*. Sedimentary and morphological development of the Borenore Caves, New South Wales, I. *Helictite*, **10**, 75–91.

FRANK, R. M. 1973. Sedimentary and morphological development of the Borenore Caves, New South Wales, II. *Helictite*, **11**, 27–44.

FRANK, R. M. 1974. Sedimentary development of the Walli Caves, New South Wales. *Helictite*, **12**, 3–30.

FRANK, R. M. 1975. Late Quaternary climatic change: evidence from cave sediments in central eastern New South Wales. *Australian Geographical Studies*, **13**, 154–168.

GALDENZI, S. & MENICHETTI, M. 1995. Occurrence of hypogenic caves in a karst region: examples from central Italy. *Environmental Geology*, **26**, 29–47.

GILLIESON, D. S. 1981. Scanning electron microscope studies of cave sediments. *Helictite*, **19**, 22–27.

GILLIESON, D. S. 1985. Geomorphic development of limestone caves in the highlands of Papua New Guinea. *Zeitschrift für Geomorphologie*, **29**, 51–70.

GRADZIŃSKI, R. 1962. Origin and development of subterranean Karst in the southern part of the Krakow Upland. *Roczink Polskiego Towarzystwa Geologicznego*, **32**, 429–492.

GRIMES, K. G. 1978. The geology and geomorphology of the Texas Caves, southeastern Queensland. *Memoirs of the Queensland Museum*, **19**, 17–59.

HÄUSELMANN, P. 2002. Cave genesis and its relationship to surface processes: investigations in the Siebenhengste Region (BE, Switzerland). *Organ der Höhlenforschergemeinschaft Region Hohgant (HRH)*, **6**, 1–168.

HILL, C. A. 1987. Geology of Carlsbad Cavern and other caves in the Guadalupe Mountains, New Mexico and Texas. *New Mexico Bureau of Mines and Mineral Resources Bulletin*, **117**, 1–150.

HORÁČEK, I., MIHEVC, A., ZUPAN HAJNA, N., PRUNER, P. & BOSÁK, P. 2007. Fossil vertebrates and palaeomagnetism update of one of the earlier stages of cave evolution in the Classical Karst, Slovenia: Pliocene of Črnotiče II site and Račiške Pečina Cave. *Acta carsologica*, **36**, 453–468.

JAMES, J. M., FRANCIS, G. & JENNINGS, J. N. 1978. Bungonia Caves and Gorge; a new view of their geology and geomorphology. *Helictite*, **16**, 53–63.

JENNINGS, J. N. 1963. Geomorphology of Dip Cave Wee Jasper, New South Wales. *Helictite*, **1**, 43–58.

JENNINGS, J. N. 1965. Bungonia Caves and rejuvenation. *Helictite*, **3**, 79–84.

JENNINGS, J. N. 1966. Jiří V. Daneš and the Chillagoe Caves District. *Helictite*, **4**, 83–87.

JENNINGS, J. N. 1967. Karst in Australia. *In*: JENNINGS, J. N. & MABBUTT, J. A. (eds) *Landform Studies from Australia and New Guinea*. Australian University Press, Canberra, 256–292.

JENNINGS, J. N. 1977. Caves around Canberra. *In*: SPATE, A. P., BRUSH, J. & COGGAN, M. (eds) *Australia Speleological Federation. Proceedings of 11th Biennial Conference*. The Australian Speleological Federation, Broadway, NSW, 79–95.

JENNINGS, J. N. 1983. A map of karst areas in Australia. *Australian Geographical Studies*, **21**, 183–196.

JENNINGS, J. N., JAMES, J. M., COUNSELL, W. J. & WHAITE, T. M. 1972. Geomorphology of Bungonia Caves and Gorge. *Sydney Speleological Society Occasional Paper*, **4**, 113–143.

KLIMCHOUK, A. B. 1996. Gypsum karst of the western Ukraine. *International Journal of Speleology*, **25**, 263–278.

KLIMCHOUK, A. B. 2007. *Hypogene Speleogenesis: Hydrogeological and Morphogenic Perspective*. National Cave and Karst Research Institute Special Publication, **1**.

KLIMCHOUK, A. B., FORD, D. C., PALMER, A. N. & DREYBRODT, W. (eds) 2000. *Speleogenesis: Evolution of Karst Aquifers*. National Speleological Society, Huntsville, AL.

LASERON, C. F. 1954. *The Face of Australia*, 2nd edn. Angus and Robertson, Sydney.

LUNDBERG, J., FORD, D. C. & HILL, C. A. 2001. A preliminary U–Pb date on cave spar, Big A Canyon, Guadalupe Mountains, New Mexico, U.S.A. *Journal of Cave and Karst Studies*, **62**, 144–148.

MIDDLETON, G. J. 1991. Oliver Trickett, doyen of Australia's cave surveyors 1847–1934. *Sydney Speleological Society Occasional Paper*, **10**, 1–156.

MITCHELL, T. L. 1838. *Three Expeditions into the Interior of Eastern Australia*. T. and W. Boone, London.

MULLER, P. & SARVARY, I. 1977. Some aspects of developments in Hungarian speleology theories during the last ten years. *Karszt es Barlang*, Special Issue 1997, 53–59.

NURSE, B. 1972. Summary of the history of Bungonia Caves and Area. *Sydney Speleological Society Occasional Paper* **4**, 13–26.

OLLIER, C. D. 1966. Speleochronology. *Helictite*, **5**, 12–21.

OLLIER, C. D. 1975. Coral island geomorphology—the Trobriand Islands. *Zeitschrift für Geomorphologie*, **19**, 164–190.

OLLIER, C. D. 1978. Tectonics and geomorphology of the Eastern Highlands. *In*: DAVIES, J. L. & WILLIAMS, M. A. J. (eds) *Landform Evolution in Australia*, Australian National University Press, Canberra, 5–47.

OLLIER, C. D. 1982. The Great Escarpment of eastern Australia: tectonic and geomorphic significance.

Journal of the Geological Society of Australia, **29**, 13–22.

OLLIER, C. D. & HOLDSWORTH, D. K. 1969. Caves of Vakuta, Trobriand Islands, Papua. *Helictite*, **7**, 50–61.

OSBORNE, R. A. L. 1984. Multiple karstification in the Lachlan Fold Belt in New South Wales: reconnaissance evidence. *Journal and Proceedings of the Royal Society of New South Wales*, **107**, 15–34.

OSBORNE, R. A. L. 1986. Cave and landscape chronology at Timor Caves, New South Wales. *Journal and Proceedings of the Royal Society of New South Wales*, **119**, 55–76.

OSBORNE, R. A. L. 1991. Red earth and bones: the history of cave sediment studies in New South Wales, Australia. *Journal of Earth Sciences History*, **10**, 13–28.

OSBORNE, R. A. L. 1993a. A new history of cave development at Bungonia, N.S.W. *Australian Geographer*, **24**, 62–74.

OSBORNE, R. A. L. 1993b. Geological Note: cave formation by exhumation of Palaeozoic palaeokarst deposits at Jenolan Caves, New South Wales. *Australian Journal of Earth Sciences*, **40**, 591–593.

OSBORNE, R. A. L. 1996. Vadose weathering of sulfides and limestone cave development—Evidence from eastern Australia. *Helictite*, **34**, 5–15.

OSBORNE, R. A. L. 1999a. The inception horizon hypothesis in vertical to steeply-dipping limestone: applications in New South Wales, Australia. *Cave and Karst Science*, **26**, 5–12.

OSBORNE, R. A. L. 1999b. The origin of Jenolan Caves: elements of a new synthesis and framework chronology. *Proceedings of the Linnean Society of New South Wales*, **121**, 1–26.

OSBORNE, R. A. L. 2000. Paleokarst and its significance for speleogenesis. *In*: KLIMCHOUK, A. B., FORD, D. C., PALMER, A. N. & DREYBRODT, W. (eds) *Speleogenesis: Evolution of Karst Aquifers*. National Speleological Society, Huntsville, AL, 113–123.

OSBORNE, R. A. L. 2001a. Non-meteoric speleogenesis: evidence from eastern Australia. *In*: *Speleology in the Third Millenium: Sustainable Development of Karst Environments*. Proceedings of the 13th International Congress of Speleology, Sociedade Brasileria de Espeleologia, Barao Geraldo Campinas, Brazil, 37–101.

OSBORNE, R. A. L. 2001b. Halls and narrows: network caves in dipping limestone, examples from eastern Australia. *Cave and Karst Science*, **28**, 3–14.

OSBORNE, R. A. L. 2002. Paleokarst: cessation and rebirth? *In*: GABROVSEK, F. (ed.) *Evolution of Karst: from Prekarst to Cessation*. Založba ZRC, Ljubljana, 97–114.

OSBORNE, R. A. L. 2004a. Cave breakdown by vadose weathering. *International Journal of Speleology*, **31**, 37–53.

OSBORNE, R. A. L. 2004b. The troubles with cupolas. *Acta carsologica* **33**, 9–36.

OSBORNE, R. A. L. 2005. Partitions, compartments and portals: cave development in internally impounded karst masses. *International Journal of Speleology*, **34**, 71–78.

OSBORNE, R. A. L. 2007a. The world's oldest caves: how did they survive and what can they tell us? *Acta carsologica*, **36**, 133–142.

OSBORNE, R. A. L. 2007b. Cathedral Cave, Wellington Caves, New South Wales, Australia: a multiphase, non-fluvial cave. *Earth Surface Processes and Landforms*, **32**, 2075–2103.

OSBORNE, R. A. L. 2008. Detailed morphological studies in Netopirjev rov, Predjama Cave: a hypogene segment of a Slovenian cave. *Acta carsologica*, **37**, 65–82.

OSBORNE, R. A. L. & BRANAGAN, D. F. 1988. Karst landscapes in New South Wales. *Earth-Science Reviews*, **25** 467–480.

OSBORNE, R. A. L., ZWINGMANN, H., POGSON, R. E. & COLCHESTER, D. M. 2006. Carboniferous clay deposits from Jenolan Caves, New South Wales: implications for timing of speleogenesis and regional geology. *Australian Journal of Earth Sciences*, **53**, 377–405.

PALMER, A. N. 1991. Origin and morphology of limestone caves. *Geological Society of America Bulletin*, **103**, 1–21.

POLYAK, V. J., MCINTOSH, W. C., GÜVEN, N. & PROVENCIO, P. 1998. Age and origin of Carlsbad Cavern and related caves from $^{40}Ar/^{39}Ar$ of alunite. *Science*, **279**, 1919–1922.

PRATT, B. T. 1964. The origin of Bungonia Caves. *University of New South Wales Mining Geology Society Journal*, **2**, 44–51.

RENAULT, H. P. 1968. Contribution à l'étude des actions mécaniques et sédimentologiques dans la spéléogenèse. *Annales de Spéléologie*, **22**, 529–593.

RUDNICKI, J. 1978. Role of convection in shaping subterranean karst forms. *Kras i Speleologia*, **2**, 92–100.

SCHEIBNER, E. & VEEVERS, J. J. 2000. Tasman Fold Belt System. *In*: VEEVERS, J. J. (ed.) *Billion Year Earth History of Australia and Neighbours in Gondwanaland*. GEMOC Press, Macquarie University, Sydney, 154–234.

SCHMIDT, V. A., JENNINGS, J. N. & BAO, H. 1984. Dating of cave sediments at Wee Jasper, New South Wales, by magnetostratigraphy. *Australian Journal of Earth Sciences*, **31**, 361–370.

SUSSMILCH, C. A. 1923. Geological notes on the trip to the Jenolan Caves. *In*: *Pan-Pacific Science Congress, Australia 1923, Guide-book to the Excursion to the Blue Mountains, Jenolan Caves and Lithgow*. Alfred James Kent, Government Printer, Sydney, 12–17.

SUSSMILCH, C. A. & STONE, W. G. 1915. Geology of the Jenolan Caves district. *Journal and Proceedings of the Royal Society of New South Wales*, **49**, 332–348.

TAKÁCS-BOLNER, K. & KRAUS, S. 1989. The results of research into caves of thermal water origin. *Karszt es Barlang*, Special Issue 1989, 31–38.

TAYLOR, G. 1923. The Blue (Mountain) Plateau. *In*: *Pan-Pacific Science Congress, Australia 1923, Guide-book to the Excursion to the Blue Mountains, Jenolan Caves and Lithgow*. Alfred James Kent, Government Printer, Sydney, 17–29.

TAYLOR, G. 1958. *Sydneyside Scenery and How it Came About*. Angus and Robertson, Sydney.

WEBB, J. A., FINLAYSON, B. L., FABLE, D. & ELLAWAY, M. 1991. The geomorphology of the Buchan Karst—implications for the landscape history of the southeastern highlands of Australia. *In*: WILLIAMS, M. A. J., DE DECKKER, P. & KERSHAW, A. P. (eds) *The Cainozoic in Australia: A Re-appraisal of the*

Evidence. Geological Society of Australia Special Publications, **18**, 210–233.

WEBB, J. A., FABLE, D., FINLAYSON, B. L., ELLAWAY, M., SHU, L. & SPIERTZ, H.-P. 1992. Denudation chronology from cave and river terrace levels: the case of the Buchan karst, southeastern Australia. *Geological Magazine* **129**, 307–317.

WHITE, E. L. & WHITE, W. B. 1969. Processes of cavern breakdown. *National Speleological Society Bulletin*, **13**, 83–96.

WHITE, E. L. & WHITE, W. B. 2000. Breakdown morphology. *In*: KLIMCHOUK, A. B., FORD, D. C., PALMER, A. N. & DREYBRODT, W. (eds) *Speleogenesis: Evolution of Karst Aquifers*. National Speleological Society, Huntsville, AL, 427–429.

YOUNG, L. H. G. 1880. Report on the Wambian Caves, by Lamont H. G, Young, F.G.S., Geological Surveyor.

Annual Report of the Department of Mines NSW, **1879**, 227–229.

YOUNG, R. W. 1977. Landscape development in the Shoalhaven River catchment of southeastern New South Wales. *Zeitschrift für Geomorphologie*, **21**, 262–283.

YOUNG, R. W. 1982. The tempo of geomorphological change: evidence from southeastern Australia. *Journal of Geology*, **91**, 211–230.

ZUPAN HAJNA, N., MIHEVC, A., PRUNER, P. & BOSÁK, P. 2008a. *Palaeomagnetism and Magnetostratigraphy of Karst Sediments in Slovenia, Carsologica 8*. Založba ZRC, Ljubljana.

ZUPAN HAJNA, N., PRUNER, P., MIHEVC, A., SCHNABL, P. & BOSÁK, P. 2008b. Cave sediments from the Postojnska–Planinska Cave System (Slovenia): evidence of multiphase evolution in epiphreatic zone. *Acta carsologica*, **37**, 63–86.

Oxygen-isotope dating the Yilgarn regolith

ALLAN. R. CHIVAS[1]* & JULIUS. R. ATLHOPHENG[1,2]

[1]*GeoQuEST Research Centre, School of Earth and Environmental Sciences,
University of Wollongong, NSW 2522, Australia*

[2]*Present address: Department of Environmental Sciences, University of Botswana,
Private Bag 0022, Gabarone, Botswana*

*Corresponding author (e-mail: toschi@uow.edu.au)

Abstract: The broad-scale distribution of $\delta^{18}O$ values of kaolinite developed in weathering profiles in the Yilgarn Craton is interpreted as reflecting their age. As Australia progressively moved from a near-polar latitude in the Permian to lower latitude, with most translation during the past 60 Ma, the imprint of varying oxygen-isotope composition of meteoric water (rainwater and groundwater) has been preserved in weathering minerals such as clays and iron oxides. This correlation, namely $\delta^{18}O$ values of kaolinite v. palaeo-latitude (and therefore, age), is well understood for eastern Australia. We have applied the same approach to samples widely spaced across the entire Yilgarn Craton and find that kaolinite from the majority of partially dissected weathering profiles displays Neogene $\delta^{18}O$ ages. There are older profiles, some seemingly of pre-late Mesozoic age, and these are predominantly in the north and east of the craton. There is no evidence within the $\delta^{18}O$-derived age pattern for a northern older plateau and a younger southern plateau, at least in terms of their primary age of deep weathering as equated to planation. Instead, the difference between northern and southern areas is that the southern area is more dissected and displays more deeply stripped weathering profiles.

The Yilgarn Craton of Western Australia underlies an area of 657 000 km² and represents a topographically subdued planation surface etched into an Archaean basement of granite and greenstone. Deep and intense subarial weathering has produced a kaolinitic pallid zone and saprolite, particularly on granitic lithologies, over much of the craton. Much of the pallid zone is, in turn, capped by a duricrust of silcrete and/or ferricrete. In some localities an intervening or transitional mottled zone (kaolinite–iron oxide) occurs. Retreating erosional escarpments occur between tracts of more level land, from one to a few tens of metres high, capped by indurated and subindurated parts of the weathered mantle. These steps in the landscape are commonly termed breakaways (Fig. 1), in both western and central Australia (see Eggleton 2001). There is debate (e.g. Anand & Paine 2002) as to whether the breakaways divide an 'Old Plateau' from a lower 'New Plateau' (e.g. Jutson 1914, 1934), and whether their ages of weathering are broadly the same throughout the Yilgarn. More probably, the pattern of weathering ages as represented by kaolinite formation, within the faces of breakaways, might be expected to be more complex. For example, Ollier *et al.* (1988) and some others since, have argued for (potentially) repeated valley-fill, induration, then topographic inversion, in preserving duricrusted breakaways. Ollier & Galloway (1990)

argued that 'there is no reason to suppose that inversions of relief occurred once only; it may have been repeated many times. If this were so, individual mesas or breakaways cannot be readily correlated'.

Thus there is no consensus on the number of cycles or continuity of deep weathering (here broadly equated to kaolinite formation), duricrust formation, or whether ferricrete as a duricrust is temporally related to underlying parts of the weathering profile (e.g. Bourman 1993). Furthermore, the actual ages of deep weathering in the Yilgarn Craton are not reliably known.

The aim of this contribution is to address some of these difficulties. Accordingly, we have applied correlated $\delta^{18}O$ dating to 90 kaolinite samples distributed across the Yilgarn Craton from sites including breakaways, open-pit mine exposures, and road, railway and dam excavations.

Granitic geomorphology of the Yilgarn Craton

The subdued topography of the craton varies from 650 m elevation in the north to near sea level in the south. Breakaway erosional remnants dominate in the north, whereas exposed unweathered granitic sheets and rocks (bornhardts), although isolated, are more common in the south (Fig. 1). The latter give

From: BISHOP, P. & PILLANS, B. (eds) *Australian Landscapes*. Geological Society, London, Special Publications, **346**, 309–320. DOI: 10.1144/SP346.16 0305-8719/10/$15.00 © The Geological Society of London 2010.

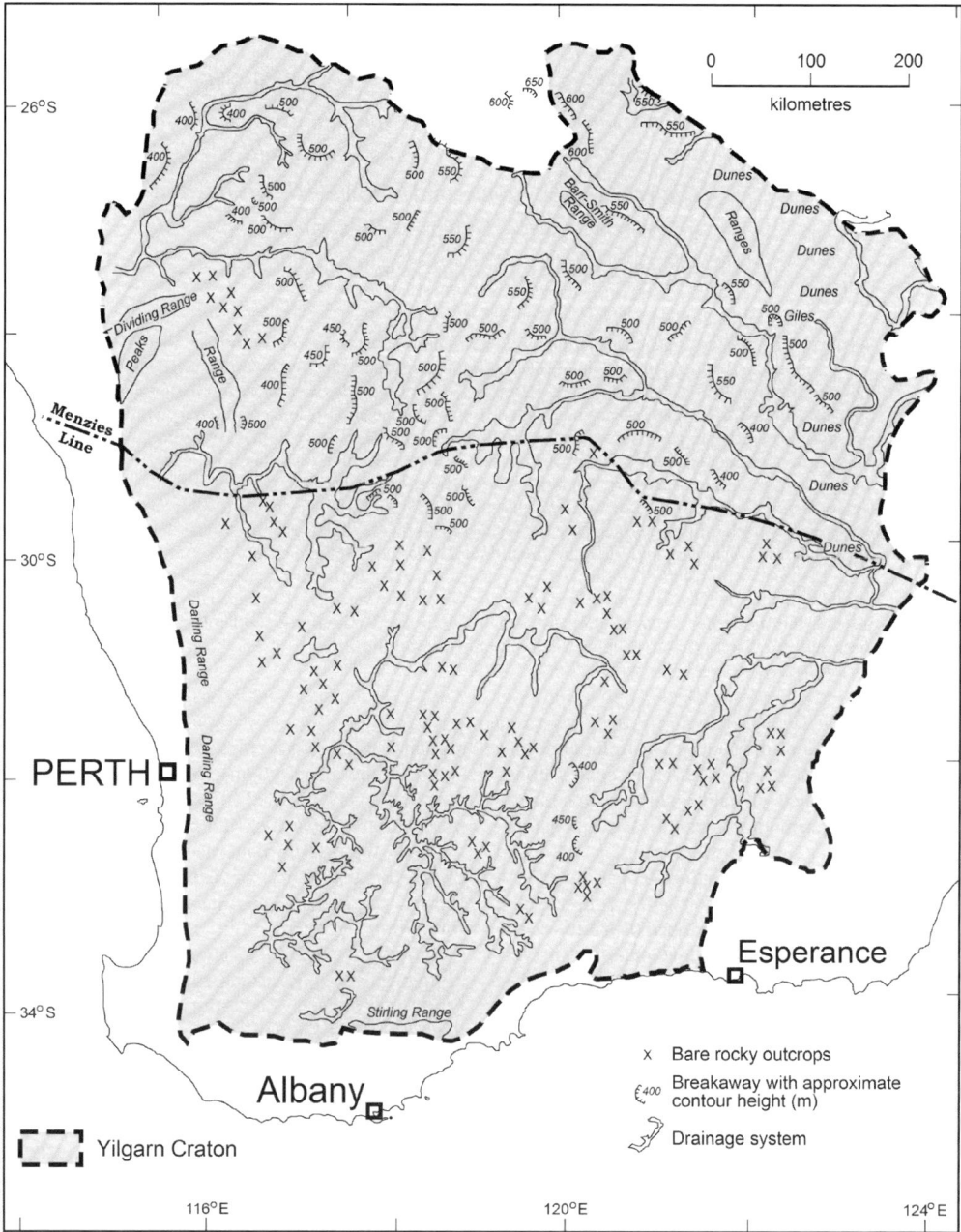

Fig. 1. Physiography of the Yilgarn Craton, Western Australia, showing the distribution of unweathered rocky outcrops (inselbergs and bornhardts) and breakaways. The broad valley drainage lines define a centrally located north–south drainage divide. Noteworthy features are the areas of extensive linear dune sands in the northeastern part, and significant areas of unweathered bedrock outcrop, labelled 'ranges'. The distribution of rocky areas (×) and breakaways is based on broad field reconnaissance supplemented by a summary of features marked on the 1:250 000-scale topographic maps (52 sheets). The position of the Menzies Line is from Butt *et al.* (1977).

rise to several town and locality names with the suffix 'rock' (e.g. Bonnie Rock, Bruce Rock, Merredin Rock, Wave Rock). The regional topography and degree of preservation of deep weathering profiles form the basis of the concept of a southward dipping or tilted 'Old Plateau'. The broad pattern can be seen as an eroding but partly preserved plateau in the north, and a stripped, etched plain in the south typified by granitic bornhardts, originally beneath an earlier land surface (Twidale & Bourne 1998), and surrounded by truncated weathering profiles, commonly without duricrust capping (see also Finkl 1973). In intensely weathered profiles, the upper granitic saprolite and lower pedolith (pallid zone material) are composed almost entirely of kaolinite and residual quartz. Minor amounts of iron oxides and residual zircon may be present. Kaolinite is the main weathering product used in this isotopic study and, accordingly, the palest coloured weathering zones overlying granitic parent materials are the commonest sample type, permitting simple mineral separation, and recovery of kaolinite by suspension in water. Some kaolinite samples are from greenstone parent materials, but the latter commonly weather to a more complicated clay mineral assemblage including smectite, and are less amenable to the extraction of pure kaolinite.

The younger elements of the landscape include Eocene channel-incision within existing broad valleys (de Broekert & Sandiford 2005). Increasing aridity since the Miocene led to infilling of the broad (now underfit) valleys with clastic sediment, formation of both pedogenic and groundwater (commonly within valley-fill) calcretes, and development of playa lakes and sand sheets. These features and their development have been comprehensively described by Anand & Paine (2002), but are not particularly relevant to the earlier history of weathering, and so are not pursued further here.

One prominent feature of the current surface of the Yilgarn Craton is the Menzies Line (Butt et al. 1977), an east–west environmental transitional boundary zone, through the town of Menzies (Fig. 1). North of this zone, soils are neutral to acid, comprising non-calcareous earths and sands with hardpans and calcreted drainages. To the south, neutral to alkaline soils are associated with saline and calcareous loams. There are vegetation differences: salmon and gimlet gums to the south; mulga shrublands to the north. Groundwaters tend to be saline and neutral to acidic in the south; whereas to the north, groundwaters are less saline and commonly neutral to alkaline. The Menzies Line is broadly coincident with a boundary within previously defined landform divisions (Mulcahy 1967; Bettenay et al. 1976). Even though the line corresponds to a zone of gradational changes in precipitation, temperature and evaporation (more severe to the north), its position is coincident with the change from remnant breakaway country to the north to the rocky outcrops and sand plains of the south (Fig. 1), and a component of bedrock control, or more correctly, degree of weathering and stripping, may be significant.

History of weathering of the Yilgarn Craton

The application of conventional stratigraphic principles to elucidate the ages of weathering on the Yilgarn Craton is limited. The bedrocks are almost exclusively Archaean greenstones and granites (commonly c. 2.65 Ga), except for a few early Proterozoic east–west dolerite dykes (the Jimberlana suite, 2.41 Ga, Fletcher et al. 1987) in the southern Yilgarn. The Craton has been commonly considered to have been subaerially exposed since some time in the Precambrian, perhaps for the past 1000 Ma (Beckmann 1983). Permian glacial deposits occur in the eastern parts (e.g. Eyles & de Broekert 2001), and there is a small Permian sedimentary coal-bearing basin (30 km × 10 km), the Collie Basin, which is located 170 km SE of Perth. Several palaeo-valleys in the south accumulated Eocene marine and estuarine sedimentation during a transgression at that time (e.g. Clarke 1994). Accordingly, the deep kaolinite weathering is little better described than post-Archaean, perhaps post-Proterozoic, or, if Permian glaciation stripped some of the contemporaneous regolith, part might be post-Permian. The very subdued current topography is also of little assistance as there is no widespread stepped multi-surface relief that might be indicative of progressive landscape development.

There can be little serious expectation that weathered material of pre-Palaeozoic age might be preserved intact; however, there is clearly scope in such a tectonically stable and slowly eroding terrain for the preservation of 'very old' weathering profiles, perhaps of Mesozoic and even Palaeozoic age.

Apatite fission-track data (Weber et al. 2005) from the northern Yilgarn Craton support the erosion of 2.5–3.1 km of material during Permian to mid-Jurassic time. However, the paucity of detrital Archaean zircon in the adjacent Perth Basin (Cawood & Nemchin 2000) deposited during the same interval suggests that the material was not of Archaean provenance. This finding suggests that the Yilgarn Craton may have been buried by reworked Proterozoic and Palaeozoic sediment in Early Permian time (Weber et al. 2005). Material of such age appears to be completely eroded from the Craton today. Perhaps the first planation of the Yilgarn Craton was a Proterozoic event!

Accordingly, geomorphological relationships rarely permit unambiguous or precise interpretation of landscape development, particularly for older weathering stages. A sound interpretation of the Kambalda region, south of Kalgoorlie, by Clarke (1994) posits weathering stages that are Permian–Jurassic, Jurassic–Eocene, and post-Eocene. The boundaries between these stages largely correspond, clearly, to markers provided by the Permian and Eocene deposition events noted above. Were there more preserved sedimentary episodes, one might subdivide the (continuous?) weathering into more stages.

Before reviewing the limited numerical data available for ages of weathering on the Yilgarn Craton, it is worth considering the composite nature of 'single' weathering profiles in a long-exposed terrain. Apart from the expected overprinting and repetition of weathering events at a given site, the relative probable ages of specific mineral phases amenable to dating is typically kaolinite $\geq$ iron oxides $>$ K bearing Mn oxides $>$ alunite. This series corresponds to initial deep weathering (kaolinite) in a humid environment with contemporaneous and/or later iron-oxide development through to the final onset of aridity as the groundwater table falls, leading to alunite $(KAl_3(SO_4)_2(OH)_6)$ precipitation as an essentially evaporative process. Accordingly, a given weathering profile may show a variety of mineral ages, spanning many tens to a few hundreds of million of years, depending on the phases analysed.

The oldest plausible weathering ages on the Yilgarn Craton are provided by palaeomagnetism of iron oxides. This technique relies upon comparing the palaeomagnetic poles from samples with the independently determined Australian apparent polar wander path (Schmidt & Embleton 1976; Idnurm & Senior 1978; Klootwijk 2009). Such data provide a range of ages, from Late Carboniferous, Jurassic and Late Cretaceous to Cenozoic (Anand & Paine 2002, p. 114; Pillans 2005, 2007). Within the Cenozoic, the rapid northward motion of Australia provides a sensitive palaeopole–age relationship, and iron oxide ages are commonly in two modes, 60 ± 10 Ma (early Palaeocene) and 10 ± 5 Ma (late Miocene). Several localities preserve multiple ages in the same profile (e.g. Bronzewing Mine, northeastern Yilgarn, 60 Ma and 10 Ma; Mt. Percy Pit at Kalgoorlie, Carboniferous, Jurassic and 60 Ma). Of all the samples measured, the most common palaeomagnetic age is Miocene, with the oldest ages being progressively less represented, and such older samples deriving from the northern and eastern portions of the Yilgarn Craton.

Cryptomelane–hollandite minerals (potassium-bearing manganese oxides) from weathering profiles in the Yilgarn Craton have K–Ar and $^{40}Ar/^{39}Ar$ ages of 36 Ma to 1 Ma (10 dates, Dammer et al. 1999) showing Mn remobilization and reprecipitation continuing virtually to the present day, especially in the more humid SW. Alunite K–Ar ages are generally younger, less than 8.5 Ma (14 dates, Bird et al. 1990; Dammer 1995), and $(U + Th)$–^4He determination from hematite fractions in pisolitic nodules within ferricrete from Moringup Hill (50 km NE of Perth) gives ages of about 10 Ma (Pidgeon et al. 2004).

Oxygen-isotopic dating of regolith minerals

The major control on the oxygen isotope composition ($^{18}O/^{16}O$, commonly expressed as $\delta^{18}O$ values) of meteoric waters, that is, waters in isotopic equilibrium with the atmosphere such as rainfall, snow, lake and river water and groundwater, is the mean annual temperature at the site of the rainfall's condensation. Thus, $\delta^{18}O_{water}$ variation, globally, is strongly latitude-dependent (Craig 1961; Dansgaard 1964), with topographic elevation and distance from coastlines being additional factors. Typical $\delta^{18}O$ values (a measure of the $^{18}O/^{16}O$ ratio) of meteoric waters vary from $-4‰$ (per mil; compared with V-SMOW, Standard Mean Ocean Water with a $\delta^{18}O$ value of $0‰$) near the Equator to $-10‰$ at $60°$ latitude to $-50‰$ near the South Pole. However, this relationship is not linear with latitude, as much of the $\delta^{18}O$ variation occurs at higher latitude with less 'sensitivity' nearer the Equator.

Kaolinite of weathering origin precipitates with $\delta^{18}O$ values that are about $26‰$ greater than their contemporary groundwater as demonstrated by the $\delta^{18}O$ values of modern and near-modern kaolinitic soils across the world (Savin & Epstein 1970). Importantly, the actual temperature of soil formation, or kaolinite precipitation, has a small effect on the $\Delta^{18}O_{kaolinite-water}$ fractionation factor (from $24‰$ at $25 °C$ to $28‰$ at $10 °C$; Sheppard & Gilg 1996). Accordingly, the $\delta^{18}O$ values of ancient kaolinites can be used to recover the $\delta^{18}O$ values of palaeo-groundwaters and, in the case of mobile continents, crude estimates of palaeolatitude. In the case of a continent whose palaeolatitude is independently known from palaeomagnetic results or plate-tectonic reconstructions based on sea-floor spreading ages, the $\delta^{18}O$ kaolinite values may be related to age of weathering, by correlation.

Not all continents will be amenable to investigation by this technique. Plate-tectonic reconstructions (e.g. Scotese 2004) show that North America and Eurasia may have been progressively separated by the opening of the Atlantic Ocean but barely changed their palaeolatitude during the last

200 Ma. The Brazilian shield, a key site for deep weathering, has remained near the Equator for the past 200 Ma, and its deep weathering profiles exhibit kaolinite with little $\delta^{18}O$ variation (Giral et al. 1993; Girard et al. 2000). India moved progressively from mid-Southern Hemisphere to mid-Northern Hemisphere during the past 250 Ma, with most translation being in the past 80 Ma. These low latitudes have a less sensitive $\delta^{18}O$–latitude gradient, although Indian sedimentary clays from Jurassic and Cretaceous strata can be distinguished isotopically (Dutta 1985).

Eastern Australia has translated northwards, almost progressively since the Permian, as it separated from Antarctica and other fragments of Gondwanaland, save minor stasis in the Jurassic. During the past 70 Ma, Australia moved across almost 40° of latitude, ensuring that its ancient weathering profiles were not removed by Quaternary glaciations. Accordingly, progressively younger kaolinite samples of weathering origin have successively higher $\delta^{18}O$ values, and indeed range from +4‰ to +23‰, an extraordinary range, not yet found elsewhere in the world.

In contrast to the Yilgarn Craton, eastern Australia has many younger sedimentary units and alkali–olivine basalt outpourings spanning the range from 70 Ma to 5 ka. Basalts commonly overlie weathering profiles, and in many cases, are, in turn, deeply weathered. From such stratigraphically dated weathering profiles, Bird & Chivas (1988a, 1989, 1993) and Chivas & Bird (1995) were able to calibrate kaolinite $\delta^{18}O$ values as a function of their age, and distinguish four broad groupings: (1) Early Permian clays, with $\delta^{18}O$ values less than +10‰; (2) Late Palaeozoic to pre-late Mesozoic clays, with $\delta^{18}O$ values between +10 and +15‰; (3) pre-mid-Cenozoic clays (late Cretaceous to early Neogene) with $\delta^{18}O$ values between +15 and +17.5‰; (4) post-mid-Cenozoic (Neogene) clays, with $\delta^{18}O$ values of +17.5 to +22.0‰.

This subdivision was applied to profiles of previously unknown stratigraphic age to demonstrate that much of the eastern and central Australian landscape developed during early and mid-Mesozoic time, and that it formed under cold conditions. The technique is also able to provide the original weathering ages for kaolinite that is subsequently transported and deposited in terrestrial sedimentary basins. Again, even in young basins, commonly Miocene, the kaolinite was typically sourced from much older weathered material, as the older landscape was eroded.

That the original $\delta^{18}O$ values of kaolinite in most profiles are retained is demonstrated by old kaolinite that retains its isotopic signature despite the continued presence of groundwater of progressively higher $\delta^{18}O$ value through time. Modern groundwater has $\delta^{18}O$ values well out of isotopic equilibrium with old kaolinite with which it has contact. The same is not true for the hydrogen isotopes within kaolinite (Bird & Chivas 1988b), which have partly and progressively exchanged hydrogen with later groundwaters.

Application of $\delta^{18}O$ dating to the Yilgarn Craton

Unlike eastern Australia, the Yilgarn Craton presents additional difficulties. As noted, there are few post-Archaean stratigraphic units and the ages of the majority of deeply weathered granitic profiles are unconstrained. An independent calibration of $\delta^{18}O_{kaolinite}$ to age is not available. In addition, the Yilgarn Craton does not have the same palaeogeographical history as eastern Australia. From about 80 Ma to 0 Ma, western and eastern Australia, although moving rapidly north, have the same azimuth, that is, the continent was not rotated compared with its present position (Scotese 2004). During this time, Australia was isolated, with both an Indian and a Pacific coastline. From 120 to 80 Ma, the continent was rotating, with the Yilgarn Craton at lower latitude than eastern Australia, and before 130 Ma, the Yilgarn Craton was the most northerly part of Australia (a 90° rotation compared with the current orientation). Before about 140 Ma ago and since c. 250 Ma the continent was incorporated in Gondwanaland, without a 'western' (then northern) coastline owing to its attachment to greater India.

Accordingly, ascribing weathering ages based on $\delta^{18}O_{kaolinite}$ from the Yilgarn has some certainty for the interval back to perhaps 80 Ma, using the calibration from eastern Australia. Before 80 Ma, one can expect kaolinite with lower $\delta^{18}O$ values, but it is unlikely that further age subdivision within the interval 80–250 Ma will be possible. Permian age kaolinites from the Yilgarn Craton can be expected to have very low $\delta^{18}O$ values, although no samples of this age, some of which can be identified stratigraphically, were analysed in the present study.

Despite the difficulties, the strength of the technique based around $\delta^{18}O_{kaolinite}$ values allows elucidation of ages from several parts of the time scale, and from a sample medium (deep pallid-zone weathering) not otherwise currently amenable to dating. Therefore, our approach in the Yilgarn has been to sample widely geographically, and prepare a reconnaissance $\delta^{18}O_{kaolinite}$ map, to allow consideration of whether the ubiquitous pallid zone is of the same general age throughout, and whether multiple ages of deep weathering can be recognized.

Results

Oxygen-isotope determinations (Table 1) of 92 kaolinite samples from across the Yilgarn are presented, including two earlier results from Bird & Chivas (1993). The samples are from breakaways, mine open-pits, road and railway cuttings, and farm dams. The latter correspond to square or rectangular bulldozer scrapings commonly 5–20 m wide, and 3–10 m deep, cut into kaolinite-rich saprolite that, upon compaction, forms an impermeable base capable of retaining surface water runoff.

Samples were crushed and timed water sedimentation was used to extract the <2 μm, or more commonly <20 μm, fraction. This fraction was identified by X-ray diffraction, and re-crushing and re-floating was commonly required to produce a pure kaolinite separate. Nevertheless, more that half of the collected samples could not be used, as a sufficiently pure kaolinite separate was not produced, with quartz and smectite being the major impurities. About 10–15 mg of each purified kaolinite was reacted with BrF_5 at 550–600 °C for 14 h. The recovered oxygen was converted to CO_2 over incandescent carbon and analysed for $^{18}O/^{16}O$ by dual-inlet isotope-ratio mass spectrometry (Finnigan MAT 251) at CSIRO, North Ryde, Sydney. The accuracy of the reported $\delta^{18}O$ value for each sample is $\pm 0.25‰$, based on replication (47 single analyses, 25 duplicates, 10 triplicates and seven in quadruplicate; Atlhopheng 2002). Precision is provided by standardization against NBS-28 quartz ($\delta^{18}O = +9.6‰$ v. V-SMOW, Standard Mean Ocean Water). The variation of $\delta^{18}O_{kaolinite}$ values from samples at similar depths within single profiles was investigated by eight paired samples (Table 1), each originally more than 1 m apart laterally. These sample pairs have differences in their $\delta^{18}O$ values that average about 1‰ and range from 0.2 to 2.6‰.

The spatial distribution of $\delta^{18}O_{kaolinite}$ values across the Yilgarn (Fig. 2) indicates that the majority of samples have $\delta^{18}O$ values in the range 17–22‰, indicative of a Neogene age of deep weathering, and that the whole of the Yilgarn Craton retains a Neogene weathering imprint. In several localities, $\delta^{18}O_{kaolinite}$ values are less than 17‰, and are considered to indicate older ages of weathering. These apparently older profiles are restricted to the northern areas of the Yilgarn and are largely in the northern central and northeastern segments.

Interpretations and conclusions

A systematic difference in weathering ages between the northern and southern portions of the Yilgarn Craton is not sustained. The kaolinite ages from the faces of breakaways in the north are the same, within the broad resolution of the technique, as in the more deeply eroded terrain of the south. The difference between the areas is reflected in their topography and relief, and is consistent with more deeply stripped profiles in the south. Thus some aspects of the concept of an older northern plateau may be broadly correct, although relative ages (merely a twofold division) refer to the stripping and erosion of the craton rather than the ages of formation of deep weathering that accompanied the formation of planation surfaces. It is plausible that the stripping is a continuous and progressive process, albeit with some differences in rates through time, and so specific ages of stripping continue to await investigation, possibly by cosmogenic dating techniques.

It had been proposed (Chivas 1983) that a detailed $\delta^{18}O_{kaolinite}$ map, if restricted to samples of the same age, would represent a climatic map. The sampling available in the present study is probably still short of that required to make an appropriate map, as the age control is only broad. Nevertheless, within the Neogene kaolinite grouping, several areal domains are recognized. For example, there are several samples around Paynes Find with $\delta^{18}O$ values around +20.4‰ and another group near Mount Magnet and Cue with $\delta^{18}O$ values around +18.3‰. It remains to be seen if these regional differences have climatic or age significance, or represent the variability of both.

Lower $\delta^{18}O_{kaolinite}$ values, interpreted to derive from remnants of older ages of weathering, are present in a few localities, mostly from mine exposures or drill core. Those samples with $\delta^{18}O$ values within the range +15 to +17.5‰ are plausibly of Late Cretaceous to Palaeogene age. Such localities include a drill hole 35 km north of Kalgoorlie (AW-63; +16.7‰), the HP Pit at Lawlers (WA-34; +16.2‰), the southern end of the Porphyry Pit at Mount Percy, Kalgoorlie (WA-82; +15.9‰), and the Bluff Point breakaway, 71 km north of Yeelirrie (AW-72; +15.5‰). Even lower $\delta^{18}O$ values are reported for three sample localities. A high-level breakaway from near the Scuddles Mine, south of Yalgoo (two samples, AW-16 and AW-17; +14.6‰ and +15.1‰), the Discovery Pit at Bronzewing (WA-66; +14.5‰) and from a drill hole, 22.5 km south of Meekatharra (AW-31; +13.2‰).

It is difficult to independently constrain the ages of these samples, but samples with $\delta^{18}O_{kaolinite}$ values less than +15‰ may be of pre-Cretaceous age. Of significance are palaeomagnetic ages on hematite from several of the same sites. Pillans (cited by Anand & Paine 2002) reported both

Table 1. *Oxygen-isotope analyses of kaolinite ($\delta^{18}O_{V-SMOW}$ values) from weathering profiles within the Yilgarn Craton*

Map number	Field number	Outcrop type	Locality	$\delta^{18}O$ of kaolinite (‰)
1	AW31	Drill core, 57 m depth	22.5 km S of Meekatharra	+13.4
2	AW29	Breakaway	52 km NW of Cue	+17.9
3	AW30b	Breakaway in greenstone	Glen to Beebyn road	+18.2
4	AW72	Breakaway	Bluff Point breakaway, 71 km NW of Yeelirrie	+15.4
5	WA55	Open pit	Bronzewing Central Pit, 36 m depth	+19.3
5	WA54	Open pit	Bronzewing Central Pit, 36 m depth	+19.4
5	WA52	Open pit	Bronzewing Central Pit, 45 m depth	+18.2
5	WA50	Open pit	Bronzewing Central Pit, 54 m depth	+18.1
5	WA47	Open pit	Bronzewing Central Pit, 60 m depth	+16.6
5	WA66	Open pit	Bronzewing Discovery Pit	+14.5
6	AW28	Breakaway	49 km from Cue on Lakeside Road	+17.9
7	AW26	Breakaway	7 km N of Mount Magnet	+18.5
8	AW32	Breakaway	42 km SE of Cue	+21.9
9	AW25	Breakaway	40 km E of Mount Magnet	+18.4
10	AW24	Breakaway	Sandstone–Youanmi road	+19.8
10	AW24a	Breakaway	Sandstone–Youanmi road	+18.2
11	WA11	Open pit	Lawlers North Pit	+18.9
11	WA13	Open pit	Lawlers North Pit	+18.9
11	WA33	Open pit	Lawlers HP Pit	+18.7
11	WA34	Open pit	Lawlers HP Pit	+16.2
11	WA35	Open pit	Lawlers HP Pit	+20.4
12	AW74/75	Breakaway	73 km N of Leonora (two samples)	+18.2, +19.2
13	AW70	Breakaway	Giles breakaway, NE of Laverton	+19.1
14	AW68	Breakaway	38 km NE of Laverton	+20.1
15	AW67	Open pit	Beasley Creek Mine, 10 km W of Laverton	+20.3
16	AW12	Creek exposure	Wooderarung gully, Mullewa	+21.3
17	AW13	Gas pipeline trench	11.5 km E of Pindar	+20.1
18	AW14	Gas pipeline trench	Wurarga	+21.4
19	AW15	Breakaway	15 km SE of Yalgoo	+19.0
20	AW16	Breakaway (upper)	Near Scuddles mine	+15.1
20	AW17	Breakaway (lower)	Same breakaway as AW16	+14.6
21	AW18	Breakaway	Burnabinmah	+20.5
22	AW10	Farm dam	15 km NE of Perenjori	+19.8
23	AW9	Farm dam	Perenjori	+19.3
24	AW20	Breakaway	Goodingnow, 13 km from Paynes Find	+20.4
25	AW21	Breakaway	Pindabunna breakaway	+20.3
26	AW22	Breakaway	Red Bluff breakaway	+20.3
27	AW23	Breakaway	Paynes Find–Sandstone road	+18.9
28	AW64	Breakaway in sedimentary rock	48 km S of Leonora	+17.6
28	AW64b	Breakaway in sedimentary rock	As above, more intense weathering	+19.3
29	WA5	Open pit	Murrin Murrin lateritic nickel mine	+17.9

(Continued)

Table 1. *Continued*

Map number	Field number	Outcrop type	Locality	$\delta^{18}O$ of kaolinite (‰)
30	AW33	Farm dam	N of Kelannie	+19.6
31	AW8	Road cutting	7 km N of Dalwallinu	+20.0
32	AW7	Road cutting	N of Koorda	+20.0
33	AW63	Drill core	35 km N of Kalgoorlie	+16.7
34	AW5	Railway cutting	Between Calingiri and Yerecoin	+20.4
35	AW4	Farm dam	Calingiri	+18.2
36	AW3b	Road cutting	Calingiri	+20.9
37	AW34	Breakaway	Bencubbin water supply dam	+19.8
38	AW82	Railway cutting (top)	15 km E of Merredin	+20.8
38	AW81	Railway cutting (middle)	From same profile as above	+19.7
38	AW80	Railway cutting (base)	From same profile as above	+20.1
39	AW79	Farm dam	E of Southern Cross	+19.0
40	AW78	Road cutting	21 km E of Southern Cross	+18.0
41	AW77	Breakaway	60 km E of Southern Cross	+20.1
42	AW76	Felsic dyke, road cutting	27 km W of Kalgoorlie	+18.7
43	BMR-98	Kaolinized granite	6 km N of Coolgardie (Bird & Chivas 1993)	+19.0
44	AW62	Breakaway in acid volcanic rock	16 km S of Kalgoorlie	+17.8
45	WA81	Open pit	Mount Percy (Mystery and Porphyry pits)	+17.1
45	WA82	Open pit	Mount Percy (Porphyry pit), Kalgoorlie	+15.9
46	AW1	Road cutting	Wooroloo	+21.9
47	AW37b	Breakaway	25 km W of Quairading	+18.8
48	AW36	Farm dam	43 km SW of Kellerberrin	+20.0
49	AW35	Farm dam	34 km SW of Kellerberrin	+19.7
50	AW38	Farm dam	10 km SW of Mount Kokeby	+19.3
51	AW39	Transported kaolinite	18.5 km SW of Mount Kokeby	+18.3
52	AW40	Breakaway	Brookton	+20.6
53	AW41	Farm dam	Jubuk, 20 km W of Corrigin	+19.7
54	AW42/42b	Kaolin open pit	SW Corrigin (two levels in pit)	+19.6, +18.8
55	AW43	Farm dam	E of Wickepin	+20.1
56	AW44	Farm dam	N of Ulleling	+19.2
57	AW45/46/ 47/48	Kaolin open pit	Ulleling: four samples, top to base of pit	+19.4, +19.2, +18.9, +18.3
58	AW49/49b	Farm dam	Wagin	+21.1, +18.5
59	AW55	Road cutting	N of Lake Grace	+19.8
60	AW56	Farm dam	Pingaring	+21.1
61	AW57	Farm dam	Pingaring	+19.3
62	AW58/58b	Open pit in basic rocks	Flying Fox nickel mine; two samples	+20.5, +19.9
63	AW59	Breakaway	130 km E of Hyden	+18.4
64	AW60	Drill core	16 km S of Norseman	+22.4
65	Ernies-3.88	Drill core or pit, 3.88 m deep	25 km E of Collie (Bird & Chivas 1993)	+19.2
66	AW50	Farm dam	11 km S of Katanning	+20.8
67	AW54	Farm dam	Pingrup	+19.0
68	AW51	Farm dam	10 km W of Tambellup	+20.8
69	AW52	Farm dam	SW of Gnowangerup	+19.8
70	AW53	Farm dam	6 km N of Ongerup	+19.1

Samples overlie granitic parent material, unless noted, and samples from open-pit gold and nickel mines are within greenstone.

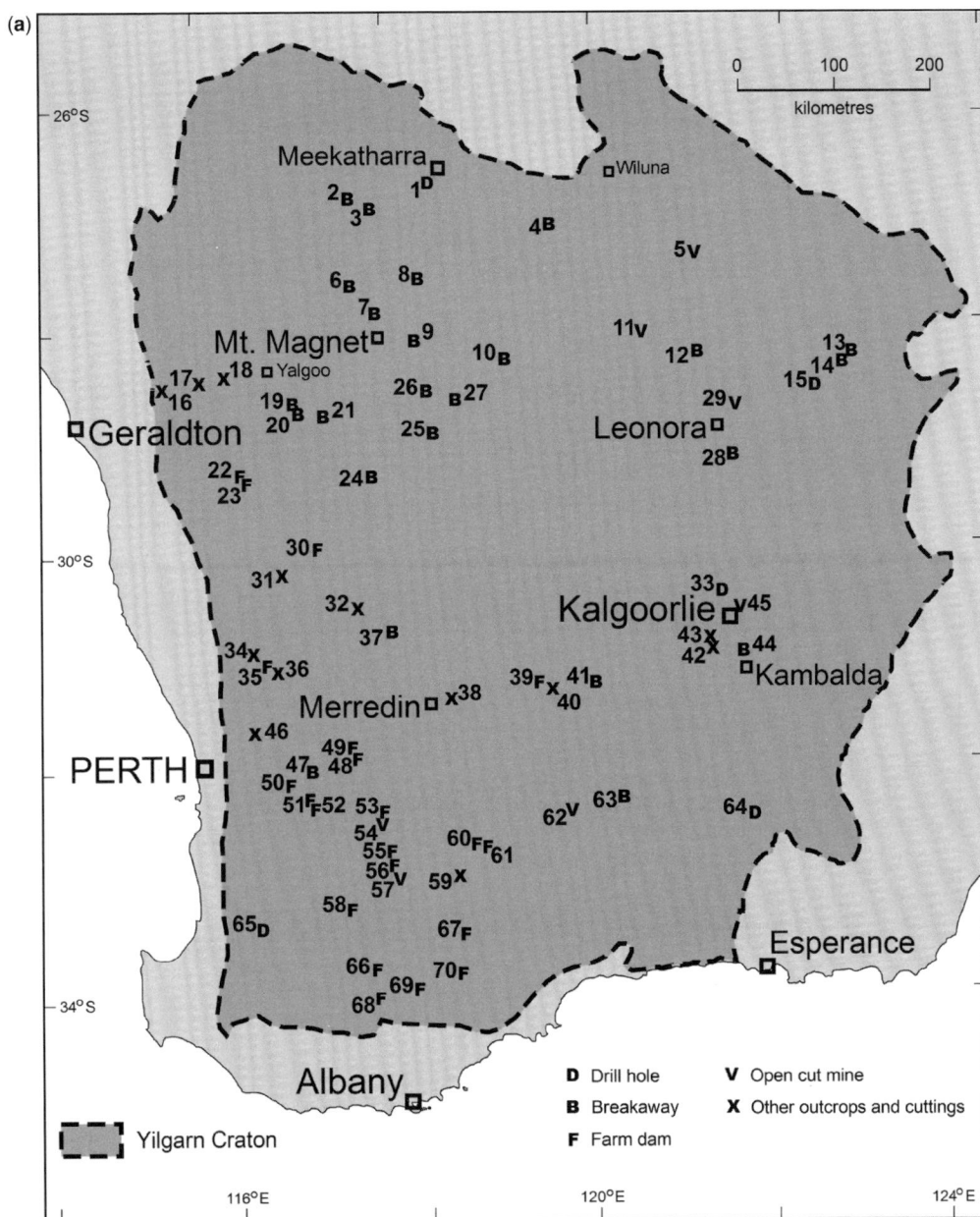

Fig. 2. (a) Map showing the locations of kaolinite samples from weathering profiles overlying the Yilgarn Craton. The samples derive from breakaways (B), mine open pits (V), exploration drill holes (D), road and railway cuttings (X), and farm dams (F). The map sample-numbers (from 1 to 70) correspond to those presented in the first column of Table 1.

Miocene and Palaeocene ages for weathering at Bronzewing; Palaeocene and Palaeozoic ages from Lawlers; and Palaeocene, Mesozoic and Palaeozoic ages for Mount Percy.

The older ages of weathering are restricted, at present, to the northern and eastern portions of the Yilgarn Craton, although these same areas also show evidence of Neogene weathering. Given

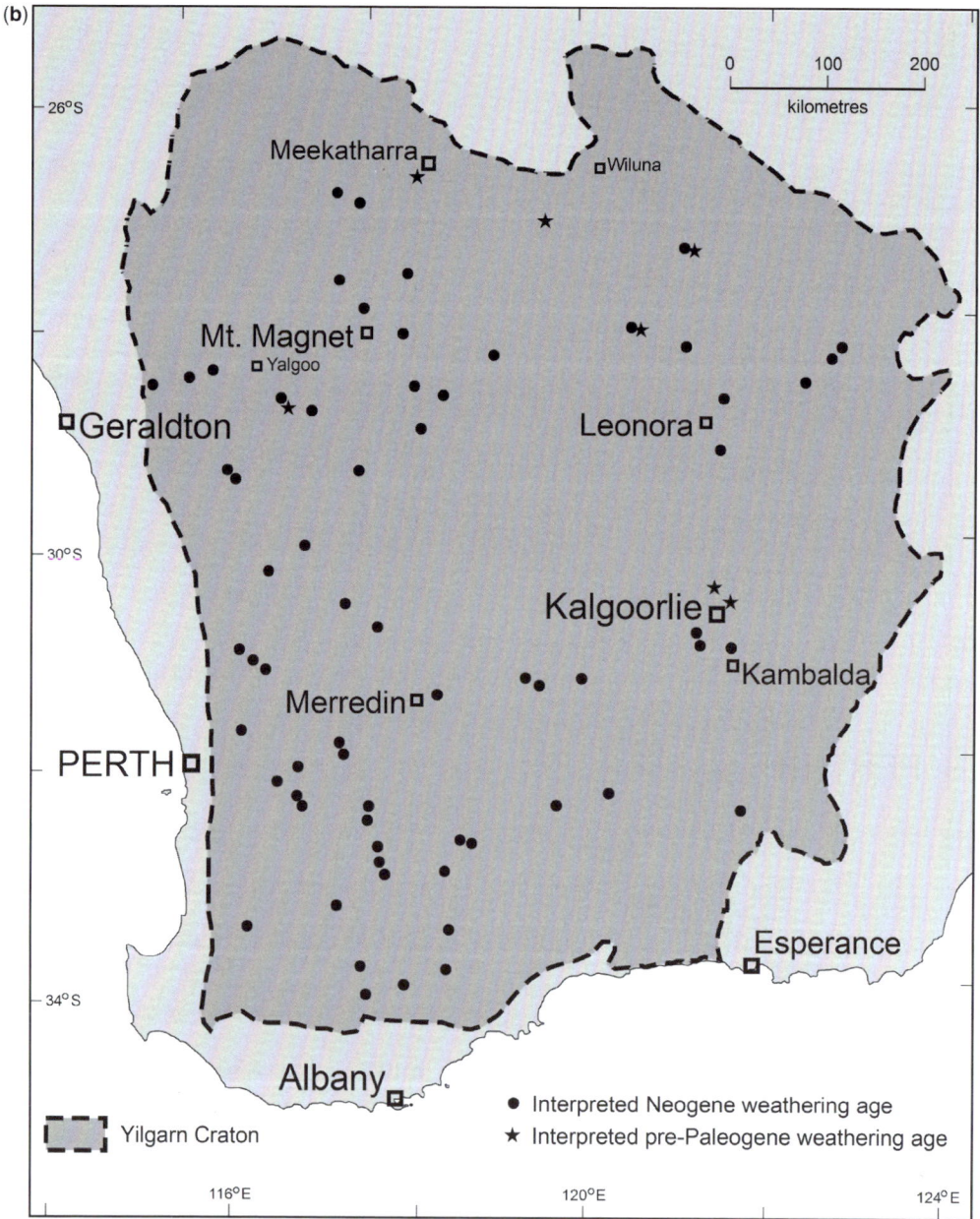

Fig. 2. (*Continued*) (**b**) Interpreted ages of kaolinitic weathering profiles in the Yilgarn Craton. Most profiles (dots) appear to be of Neogene age. Only seven profiles (stars), in the north and east of the craton, are interpreted to be of Palaeogene or older age.

that the apparently older remnant profiles are commonly recorded from a few high-level breakaways and mostly mine exposures and exploration drill holes, further older profiles might be sought in the subsurface rather than in outcrop. Further work might be directed to a more 3D approach rather

that relying on the 2D reconnaissance $\delta^{18}O$ map presented here.

The fieldwork and analyses were funded by the GeoQuEST Research Centre at the University of Wollongong and by the University of Botswana. For assistance in laboratory techniques and preparation we gratefully acknowledge B. G. Jones, D. Carrie, R. Miller, A. Andrew and B. McDonald.

References

ANAND, R. R. & PAINE, M. 2002. Regolith geology of the Yilgarn Craton, Western Australia: implications for exploration. *Australian Journal of Earth Sciences*, **49**, 3–162.

ATLHOPHENG, J. R. 2002. *Weathering Profiles—Their Development and Ages Using Oxygen Isotopes*. PhD thesis, University of Wollongong.

BECKMANN, G. G. 1983. Development of soil landscapes. In: *Soils—An Australian Perspective*. Division of Soils, CSIRO, Melbourne, 163–172.

BETTENAY, E., SMITH, R. E. & BUTT, C. R. M. 1976. Physical features of the Yilgarn Block. In: SMITH, R. E., BUTT, C. R. M. & BETTENAY, E. (eds) *Surficial Mineral Deposits and Exploration Geochemistry, Yilgarn Block, Western Australia, 25th Int. Geol. Congress (Sydney), Excursion Guide*, International Geological Congress, Sydney, **41c**, 5–11.

BIRD, M. I. & CHIVAS, A. R. 1988a. Oxygen isotope dating of the Australian regolith. *Nature*, **331**, 513–516 (Erratum, **332**, 568).

BIRD, M. I. & CHIVAS, A. R. 1988b. Stable-isotope evidence for low-temperature weathering and post-formational hydrogen-isotope exchange in Permian kaolinites. *Chemical Geology*, **72**, 249–265.

BIRD, M. I. & CHIVAS, A. R. 1989. Stable-isotope geochronology of the Australian regolith. *Geochimica et Cosmochimica Acta*, **53**, 3239–3256.

BIRD, M. I. & CHIVAS, A. R. 1993. Geomorphic and palaeoclimatic implications of an oxygen-isotope chronology for Australian deeply weathered profiles. *Australian Journal of Earth Sciences*, **40**, 345–358.

BIRD, M. I., CHIVAS, A. R. & McDOUGALL, I. 1990. An isotopic study of surficial alunite in Australia. 2. Potassium–argon geochronology. *Chemical Geology*, **80**, 133–145.

BOURMAN, R. 1993. Perennial problems in the study of laterite: a review. *Australian Journal of Earth Sciences*, **40**, 387–401.

BUTT, C. R. M., HOROWITZ, R. C. & MANN, A. W. 1977. *Uranium Occurrences in Calcretes and Associated Sediments in Western Australia*. CSIRO Division of Mineralogy, Perth, Report **FP16**.

CAWOOD, P. A. & NEMCHIN, A. A. 2000. Provenance record of a rift basin: U/Pb ages of detrital zircons from the Perth Basin, Western Australia. *Sedimentary Geology*, **134**, 209–234.

CHIVAS, A. R. 1983. The climatic conditions during regolith formation: Oxygen- and hydrogen-isotope evidence. In: WILFORD, G. E. (ed.) *Regolith in Australia: Genesis and Economic Significance*. Bureau of Mineral Resources, Australia, Record 1983/27, 42–47.

CHIVAS, A. R. & BIRD, M. I. 1995. Palaeoclimate from Gondwanaland clays. In: CHURCHMAN, G. J., FITZPATRICK, R. W. & EGGLETON, R. A. (eds) *Clays: Controlling the Environment. Proceedings 10th International Clay Conference, Adelaide, Australia, 1993*. CSIRO Publishing, Melbourne, 333–338.

CLARKE, J. D. A. 1994. Geomorphology of the Kambalda region, Western Australia. *Australian Journal of Earth Sciences*, **41**, 229–239.

CRAIG, H. 1961. Isotopic variations in meteoric waters. *Science*, **133**, 1702–1703.

DAMMER, D. 1995. *Geochronology of Chemical Weathering Processes in the Northern and Western Australian Regolith*. PhD thesis, Australian National University, Canberra.

DAMMER, D., McDOUGALL, I. & CHIVAS, A. R. 1999. Timing of weathering-induced alteration of manganese deposits in Western Australia: evidence from K/Ar and ^{40}Ar/^{39}Ar dating. *Economic Geology*, **94**, 87–108.

DANSGAARD, W. 1964. Stable isotopes in precipitation. *Tellus*, **16**, 436–468.

DE BROEKERT, P. & SANDIFORD, M. 2005. Buried inset-valleys in the eastern Yilgarn Craton, Western Australia: geomorphology, age, and allogenic control. *Journal of Geology*, **113**, 471–493.

DUTTA, P. K. 1985. In search of the origin of cement in siliciclastic sandstones: an isotopic approach. *Chemical Geology*, **52**, 337–348.

EGGLETON, R. A. (ed.) 2001. *The Regolith Glossary— Surficial Geology, Soils and Landscapes*. Cooperative Research Centre for Landscape Evolution and Mineral Exploration, Perth, WA.

EYLES, N. & DE BROEKERT, P. 2001. Glacial tunnel valleys in the Eastern Goldfields of Western Australia cut below the Late Paleozoic Pilbara ice sheet. *Palaeogeography, Palaeoclimatology, Palaeoecology*, **171**, 29–40.

FINKL, C. W. 1973. Stripped (etched) land surfaces in southern Western Australia. *Australian Geographical Studies*, **17**, 33–52.

FLETCHER, I. R., LIBBY, W. G. & ROSMAN, K. J. R. 1987. Sm–Nd dating of the 1411 Ma Jimberlana dyke, Yilgarn Block, Western Australia. *Australian Journal of Earth Sciences*, **34**, 523–525.

GIRAL, S., SAVIN, S. M., GIRARD, J.-P. & NAHON, D. B. 1993. The oxygen isotope geochemistry of kaolinites from lateritic profiles: implications for pedology and paleoclimatology. *Chemical Geology*, **107**, 237–240.

GIRARD, J.-P., FREYSSINET, P. & CHAZOT, G. 2000. Unravelling climatic changes from intraprofile variation in oxygen and hydrogen isotopic composition of goethite and kaolinite in laterites: an integrated study from Yaou, French Guiana. *Geochimica et Cosmochimica Acta*, **64**, 409–426.

IDNURM, M. & SENIOR, B. R. 1978. Palaeomagnetic ages of late Cretaceous and Tertiary weathered profiles in the Eromanga Basin, Queensland. *Palaeogeography, Palaeoclimatology, Palaeoecology*, **24**, 263–277.

JUTSON, J. T. 1914. An outline of the physiographical geology (physiography) of Western Australia. *Geological Survey of Western Australia, Bulletin*, **61**, 240.

JUTSON, J. T. 1934. The physiography (geomorphology) of Western Australia. *Geological Survey of Western Australia, Bulletin*, **95**, 366.

KLOOTWIJK, C. 2009. Sedimentary basins of eastern Australia: paleomagnetic constraints on geodynamic evolution in a global context. *Australian Journal of Earth Sciences*, **56**, 273–308.

MULCAHY, M. J. 1967. Landscapes, laterites and soils in south western Australia. *In*: JENNINGS, J. N. & MABBUTT, J. A. (eds) *Landform Studies from Australia and New Guinea*. Australian National University Press, Canberra, 211–230.

OLLIER, C. D. & GALLOWAY, R. W. 1990. The laterite profile, ferricrete and unconformity. *Catena*, **17**, 97–109.

OLLIER, C. D., CHAN, R. A., CRAIG, M. A. & GIBSON, D. L. 1988. Aspects of landscape history and regolith in the Kalgoorlie region, Western Australia. *BMR Journal of Australian Geology and Geophysics*, **10**, 309–321.

PIDGEON, R. T., BRANDER, T. & LIPPOLT, H. J. 2004. Late Miocene $(U + Th)-^4He$ ages of ferruginous nodules from lateritic duricrust, Darling Range, Western Australia. *Australian Journal of Earth Sciences*, **51**, 901–909.

PILLANS, B. 2005. Geochronology of the Australian regolith. *In*: ANAND, R. R. & DE BROEKERT, P. (eds) *Regolith Landscape Evolution Across Australia*. Co-operative Research Centre for Landscape Environments and Mineral Exploration (CRC LEME), Perth, WA, 41–52.

PILLANS, B. 2007. Pre-Quaternary landscape inheritance in Australia. *Journal of Quaternary Science*, **22**, 439–447.

SAVIN, S. M. & EPSTEIN, S. 1970. The oxygen and hydrogen isotope geochemistry of clay minerals. *Geochimica et Cosmochimica Acta*, **34**, 25–42.

SCHMIDT, P. W. & EMBLETON, B. J. J. 1976. Palaeomagnetic results from sediments of the Perth Basin, Western Australia, and their bearing on the timing of regional lateritisation. *Palaeogeography, Palaeoclimatology, Palaeoecology*, **19**, 257–273.

SCOTESE, C. R. 2004. A continental drift flipbook. *Journal of Geology*, **112**, 729–741.

SHEPPARD, S. M. F. & GILG, H. A. 1996. Stable isotope geochemistry of clay minerals. *Clay Minerals*, **31**, 1–24.

TWIDALE, C. R. & BOURNE, J. A. 1998. Origin and age of bornhardts, southwest Western Australia. *Australian Journal of Earth Sciences*, **45**, 903–914.

WEBER, U. D., KOHN, B. P., GLEADOW, A. J. W. & NELSON, D. R. 2005. Low temperature Phanerozoic history of the Northern Yilgarn Craton, Western Australia. *Tectonophysics*, **400**, 127–151.

Index

Note: Page numbers in *italic* denote figures. Page numbers in **bold** denote tables.